Comprehensive Volume and Capacity Measurements

Comprehensive Volume and Capacity Measurements

S.V. Gupta
Formerly, UNIDO Consultant in Metrology
Director, Weights and Measures, Govt. of India
Scientist, National Physical Laboratory, New Delhi

NEW AGE INTERNATIONAL (P) LIMITED, PUBLISHERS
(formerly Wiley Eastern Limited)
New Delhi • Bangalore • Chennai • Cochin • Guwahati • Hyderabad
Jalandhar • Kolkata • Lucknow • Mumbai • Ranchi
Visit us at www.newagepublishers.com

Published by New Age International (P) Ltd., Publishers
First Edition : 2006

ISBN : 81-224-1819-8

Rs. 195.00

C-05-12-507

Printed in India at Taj Press, Noida
Typesetter : Aakriti Graphics, New Delhi.

PUBLISHING FOR ONE WORLD
NEW AGE INTERNATIONAL (P) LIMITED, PUBLISHERS
(formerly Wiley Eastern Limited)
4835/24, Ansari Road, Daryaganj, New Delhi - 110002
Visit us at **www.newagepublishers.com**

Dedicated to my wife Mrs Prem Gupta

and

to my children

PREFACE

No teaching institute or University teaches measurement of basic parameters like volume. At school only preliminaries are dealt with about volume measurement. A new entrant, to a calibration laboratory dealing in calibration and testing of volumetric glassware does not find himself/ herself a comfortable starter. The reason is there is no book dedicated to such a subject. If someone refers to him the Dictionary of Applied Physics volume IV, which dates back to very early part of 20th century or to Notes on Applied Sciences of 1950 published by National Physical Laboratory, U.K. it certainly gives him an impression that he has come to a primitive field and has been trapped. Even a better experience scientist gives him direction to carry out his work in the prescribed manner without furnishing him the reasons to do so. Basic reason is no efforts have been made to consolidate the research work carried out during the last century and to make it assessable to a normal user.

No body talks about the solid artefacts which serve as primary standard of volume. Water is normally used as a medium for calibrating volumetric measures. But the corrections applicable are still based on old data of water density and temperature scale. Recent work on density measurement of water, taking in to account of isotopic composition of water and solubility for air, is not often used.

Efforts therefore have been made to use latest water and mercury density data to prepare correction tables. The coefficients of expansion, density of standard weights, air density and reference temperature are the variable, which comes in the equation in preparing the corrections tables. Therefore, a large variety of coefficients of expansions have been taken in preparing easy to use tables. The coefficient chosen are such that practically all material used in fabricating volumetric measures are covered. A separate table of corrections for unit difference in the coefficients of expansion has been constructed which will make it possible to find the corrections for any coefficients of expansion. There are two internationally accepted values for the density of mass standards. Similarly there are two reference temperatures to which the capacity of measures are referred to, so separate set of tables have been made for each possible combinations of density values of mass standard and reference temperatures. The corrections have been calculated to 4th decimal place instead of 3rd decimal place.

Solid based primary standards of volume have been discussed. Inter-comparison at international level and collating the results of measurements by various laboratories has been discussed. The principle of measurement has remained the same it is the technology which has changed. The establishment of solid base volume standards and their international inter-comparison has considerably improved reproducibility in volume measurements.

The chapter on the surface tension effect on the meniscus volume gives an insight story of physics and measurements. It is for the first time that analytical formula for meniscus volumes in tubes of different diameters has been worked out. It will go a long way in understanding the purpose of calibration and limitations associated with it. To measure volume of any liquid through a volumetric measure is meaningful if proper corrections are applied due to change in surface tension and density of the liquid. Meniscus volume and corrections applicable due to change in capillarity constant for tubes of diameter from 0.2 mm to 120 mm have been given in the form of tables at the end of the chapter 7. The subject matter is treated in a way, which can interest an undergraduate physics student.

The hierarchy in volume measurement and method of its realisation has been taken up. Design, fabrication and material requirements of standard capacity measures have been explained. Range of capacity of these measures is from a few cm^3 to several thousand dm^3. Secondary standard capacity measures in glass from 50 litres to 5 cm^3 have been discussed in respect of their design, calibration and use.

The methods of measurement and calibration of capacity of vertical, horizontal and spherical storage tanks, together with road tankers, vehicle tanks, ships and barges have been described for the first time in a consolidated way.

It is my pleasant duty to thank quite a number of people, who have encouraged me at each step to complete the book. I am grateful to Professor A.R. Verma, Dr. A.P. Mitra FRS, and Prof S.K. Joshi, all former Directors of National Physical Laboratory, New Delhi, who have been a constant source of encouragement to me during the preparation of the manuscript. I wish to thank Mrs Reeta Gupta and other colleagues at the National Physical Laboratory, New Delhi, who have been helpful in procuring material for the book. The work carried out at the National Physical Laboratory, New Delhi and mentioned in the book was teamwork, so every colleague of mine at that time, alive or dead, deserves my appreciation and thanks.

S. V. Gupta

CONTENTS

UNITS AND PRIMARY STANDARD OF VOLUME

1.1 INTRODUCTION

The accurate knowledge of volume of solids, liquids and gases is required in all walks of life including that of trade and commerce. In addition, the volume of a solid or liquid must be known to calculate its density. The frequency of the need of volume measurement is as much as that of measurement of mass. In this book, however, we will be restricting to measurement of volume of solids and liquids. Precise volumetric measurements are required in breweries, petroleum and dairy industry and in water management. More precise measurements are required in scientific research and chemical analysis. Liquids have to be contained in physical artefacts, which are called measures. So finding the capacity of these measures is also a part of volume measurement.

1.2 VOLUME AND CAPACITY

There are two terms, which are often used in volume measurements. One is capacity and the other is volume. Both terms represent the same quantity. The capacity is the property of a vessel or container and is characterised by how much liquid, it is able to hold or deliver. These vessels or containers are generally termed as volumetric measures. So capacity is the property of volumetric measures. While volume is the basic property of matter in relation to its occupation of space, so it applies to every material body.

1.3 REFERENCE TEMPERATURE

Both volume of a body and capacity of a volumetric measure depend upon temperature. Hence statement about the capacity of a volumetric measure or volume of a body should necessarily contain a statement of temperature. Saying only, the volume of a body is so many units of volume, does not carry much weight unless we specify temperature to which it is referring. Now if every body gives the results of a volume measurement at its temperature of measurement than it will be difficult to compare the results given by two persons for the same body but at different temperatures. To obviate this difficulty, one solution is that all measurements of volume are carried out at one temperature, which is again not possible. As in this case, all laboratories and work places, at which volume measurements are carried out, have to be maintained at the same temperature. So better viable solution is that measurements are carried out at different temperatures but all results are adjusted to a common agreed temperature. This agreed temperature is called as reference/standard temperature, which is kept same for a country or region. However reference temperature may be kept different for different commodities and regions of globe. Depending upon general climate of a country or region, it may be 27 °C, 20 °C or 15 °C. For all European countries including U.K. it is 20 °C for general purpose, and 15.5 °C for petroleum products. However, India due to its tropical climate, has adopted 27 °C for general purpose and 15.5 °C for petroleum industry. Other tropical countries have, similarly, adopted 27 °C for general purpose and 15.5 °C for petroleum industry.

1.3.1 Reference or Standard Temperature for Capacity Measurement

The capacity of a volumetric measure is defined by the volume of liquid, which it contains or delivers under specified conditions and at the standard temperature. The capacity of each measure, in India, is referred to 27 °C. However temperatures of 20 °C and 15 °C are also permitted for specific purposes.

1.3.2 Reference or Standard Temperature for Volume Measurement

The results of volume measurements of all solids generally refer to 27 °C, in India. However temperatures of 20 °C and 15 °C are also permitted for specific purposes.

1.4 UNIT OF VOLUME OR CAPACITY

In earlier days the unit of volume and capacity used to be different. The unit of volume was taken as the cube of the unit of length. The unit of capacity was defined as the space occupied by one kilogram of water at the temperature of its maximum density.

The Kilogram de Archives of 1799, the unit of mass was defined equal to the mass of water at its maximum density and occupying the space of one decimetre cube. But later on it was realised that there was some error in realising the decimetre cube. So in 1879 the unit of mass-the kilogram was de-linked with water and its volume. The kilogram was defined as the mass of the International Prototype Kilogram. The mass of the International Prototype Kilogram was itself made, as far as possible, equal to the mass of the Kilogram de Archives. The volume of one kilogram of water at its maximum density was found to be 1.000 028 dm^3. So in 1901, third General Conference for Weights and Measures (CGPM) decided a new unit of volume and named it as litre. The litre was defined as the volume occupied by one kilogram of water at its temperature of maximum density and at standard atmospheric pressure. The unit was termed as the unit of capacity. For finding the capacity of a measure, the unit litre was used and for volume, the unit decimetre cube continued to be used. The symbol l was assigned to the litre in 1948 by the 9th CGPM. However the controversy of having two units for essentially the same quantity remained and finally in 1964 the CGPM in its 11th conference abrogated the definition of the litre altogether but allowed the name litre to be used as another name of one decimetre cube. Keeping in view the fact that the letter l, the symbol of litre as adopted in 1948, may be confused with numeral one, the 16th CGPM, in 1979, sanctioned the use of the letter L also as symbol of litre.

So presently, in International System of Units (SI), the unit of volume as well as that of capacity is cubic metre with symbol m^3. The cubic metre is equal to the volume of a cube having an edge equal to one metre. But sub-multiples of cubic metre, like cubic decimetre (symbol dm^3), cubic centimetre (symbol cm^3) and cubic millimetre (symbol mm^3) may also be used. Litre (1), millilitre (ml) and micro-litre (μl) may be used as special names for dm^3, cm^3 and mm^3 respectively. L may also be used as symbol of litre.

1.5 PRIMARY STANDARD OF VOLUME

Volume of a solid is determined either by dimensional measurements or by hydrostatic weighing. Dimensional method gives the volume of the solid in base unit of length i.e. metre. Hydrostatic weighing method requires a medium of known density and gives the volume of the body in terms of mass and density of liquid displaced. The primary standard of volume, therefore, is a solid artefact of known geometry. Its volume is calculated from the measurements of its dimensions.

1.5.1 Solid Artefact as Primary Standard of Volume

Solids of known geometry are maintained as artefact standards of volume. Two simpler geometrical shapes are those of cube and sphere. Both these shapes are used for making solid artefacts as standard of volume.

1.5.1.1 Shape – Solid Artefacts of Spherical in Shape

The spherical shape is obtained by rolling mill process. Spheres of diameters around 85 mm have been made. Peak to peak difference between the diameters of the sphere, so far made, vary from 220 nm to 28 nm.

1.5.1.2 Shape – Solid Artefacts in the Shape of a Cube

The cubical shape is achieved by using the method of optical grinding, lapping and final polishing. The plainness of its faces is examined by using interference method or an auto-collimator.

1.5.2 Maintenance

Spherical shape is attainable and maintainable far more easily than the cubical shape. In cubical shape, the edges cannot be made perfect straight lines, or the corners as points. Further, there is always a danger of chipping of edges and corners causing change in volume if the artefact is in the shape of a cube.

1.5.3 Material

The material requirements for the two shapes are different. The material for cubical shape must be such that can be worked out using optical grinding, lapping techniques and is able to acquire high degree of polish. The material should not be brittle, otherwise edges will not be maintained but should have low coefficient of expansion. Quartz fulfils all the requirements. Other materials are silicon, low expansion glass and zerodur. For spherical shape steel is good except its rusting property. Silicon crystals are being used to determine the Avogadro's number so its physical constants like coefficient of expansion are well measured, hence Silicon is now preferred over any other materials. Avogadro's number is the number of molecules, atoms or entities in one gram molecule of substance.

1.5.4 Primary Volume Standards Maintained by National Laboratories

The shape, material, value of volume along with uncertainty of solid artefacts maintained as primary standard of density/volume are given in table 1.1

Table 1.1 Solid Artefacts as Primary Standards of Density/Volume

Country	*Laboratory*	*Shape*	*Material*	*Volume cm³*	*Uncertainty*
USA	NIST	Sphere	Steel	134.067 062	0.2 ppm
		Disc	Silicon	86.049 788	0.3 ppm
Australia	NML	Sphere	ULE glass	228.519 022	0.25 ppm
Japan	NRLM	Sphere	Quartz	319.996 801	0.36 ppm
Italy	IMGC	Sphere	Silicon	429.647 784	0.13 ppm
		Sphere	Zerodur	386.675 59	0.18 ppm
Germany	PTB	Cube	Zerodur	394/542 60	0.8 ppm
India	NPL	Sphere	Quartz	268.225 1	1.0 ppm

One such standard is shown below

Photo of a silicon sphere from NRLM, Japan

1.6 MEASUREMENT OF VOLUME OF SOLID ARTEFACTS

As seen above practically every national measurement laboratory maintains its volume/ density standard in the form of an artefact. Some determine its volume by dimensional method others derive the volume of their primary standard through hydrostatic weighing using water as density standard. In the latter case, the primary standard of mass is used as reference standard in hydrostatic weighing.

1.6.1 Dimensional Method

1.6.1.1 Sphere

Diameter of a solid artefact in the shape of sphere is measured by the use of Saunders type interferometer [1] with a parallel plate's etalon or by Spherical Fizeau's type interferometer [2].

For measurement of various diameters, a great circle is marked on the sphere. The diameter of this great circle is measured with the help of an interferometer. The circle is usually named as equator. N sets of equiangular points are chosen on this circle. Each set consists of two diametrically opposite points. M equiangular points divide each of the n great circles passing through these 2N points. The diameters of these M great circles are inter-compared to see the roundness of the sphere. Further details may be obtained from the book by the author [3].

1.6.1.2 Cube

Dimensions of a cube are determined by using commercially available interferometers and the errors due to roundness of edges and corners, out of plainness of faces are estimated and proper corrections are applied [4,5].

1.6.2 Volume of Solid Body by Hydrostatic Method

Hydrostatic method is based on the Archimedes Principle. The principle states that if a solid is immersed in a fluid, it loses its weight, and loss in weight is equal to the weight of the fluid displaced. If a solid body has a perfectly smooth surface and fluid wets the surface, then volume of the fluid displaced is equal to that of the body. If the density of the fluid is known then volume of fluid displaced i.e. volume of solid may be calculated by dividing the loss in mass of the solid by density of the fluid. Generally water is used as fluid for this purpose. The body is first weighed in air and then in water.

Let M_1, M_2 be respectively the apparent masses of the body when weighed against the weights of density D first in air and then in water. Let σ_1 and σ_2 be density of air at the time of two weighing while ρ be density of water at the temperature of measurement. Then

$$M_1(1-\sigma_1/D) = M - V\sigma_1 - \frac{\pi d T_1}{g}, \text{ and}$$

$$M_2(1-\sigma_2/D) = M - V\rho - \frac{\pi d T_2}{g}$$

Where T_1 and T_2 are values of surface tension of water at the time of two weighing and d is the diameter of the suspension wire and V is the volume of the body.

Subtracting the two equations we get

$$V(\rho - \sigma_1) = M_1(1-\sigma_1/D) - M_2(1-\sigma_2/D) + \pi d\, T_1/g - \pi d\, T_2/g, \text{ giving}$$

$$V = [M_1(1-\sigma_1/D) - M_2(1-\sigma_2/D) + (\pi d/g)\{T_1 - T_2\}]/(\rho - \sigma_1)$$

A good care is required to ensure that the length of the portion of wire submerged in water and surface tension of the liquid at its intersection remains unchanged in each of two weighing steps. The real problem comes in wetting the surface of the solid completely. If the solid is not wetted properly then the calculated value of volume of solid will be more than the actual. The problem may be greatly reduced by :

- Removing of air bubbles sticking to surface of the solid by mechanical means.
- Removing dissolved air by creating a partial vacuum through a water pump or any other vacuum pump.
- Boiling the water with solid inside it to remove air and then cooling after cutting off the air contact by suitable plugging the system containing water and the solid. This method is time consuming and it is difficult to ensure the temperature equilibrium inside the solid especially when it is made of ceramic like material.
- Thorough cleaning of the surface of the solid body.
- Having the solid with highly polished and smooth surface.

1.6.2.1 Effect of Surface Tension in Hydrostatic Weighing

Let the diameter of the wire from which the solid body is suspended be d mm, then an upward force equal to πdT will be acting on it at the air liquid intersection. So the loss in apparent mass of the body in water will be $\pi dT/g$.

For water, surface tension T = 72 mN/m, the error could be 23.08 mg for a wire of diameter 1 mm. However, the apparent mass of the body in water is determined by two weighing, namely (1) when the hanger alone is in water and (2) when body is placed in hanger. Apparent mass of the body will be the difference of two readings. There will be no error in apparent mass of the body in water if surface tension does not change during these two weighing. But surface tension of water changes drastically with contamination, so even with 10 percent change in surface tension, the error in volume measurement will be equal to the volume of water of mass 2.3 mg, which is roughly equivalent 2.3 mm^3. If the true volume of the body is 10 cm^3 then relative error will be 2.3 parts in 10000.

1.6.2.2 Effect of Different Immersion Length of the Suspended Wire

If the change in water level, in the two weighing, is 1 mm, then change in immersed volume of the wire of diameter 1 mm will be 0.7854 mm^3, which will amount to an error of 0.8 parts in 10000 in a body of true volume 10 cm^3. Normally much thinner wires of platinum are used for this purpose so error due to wire immersing at different length is further reduced.

The hydrostatic weighing method is quite often used for determining the purity of gold in ornaments. Let us assume a bangle of 15 g whose purity of gold is to be determined. If the bangle is of pure gold with density 17.31 gcm^{-3}, then its volume should be 15/17.31 = 0.86655 cm^3. An error of 0.000 8 cm^3 as calculated above will make the measured volume as 0.86575 cm^3 and giving the density of the bangle as 17.29 gcm^{-3}.

1.7 WATER AS A STANDARD

Water is being used as a liquid of known density from very long time. So measurement of its density has remained a concern to all metrologists. In the last decade of 19th century, Chappuis of BIPM, International Bureau of Weights & Measures, Paris and Thiesen of PTR Physikalisch Technische Reichsanstalt, Germany, measured the density of water at different temperatures. They expressed their results in terms of two totally different formulae. The two formulae give density of water at different temperatures which differed by 6 parts per million around 25 °C but by 9 parts per million at 40 °C. At that time, the idea of isotopic composition of water and its effect on the density was not clear. Hence isotopic composition of water was not taken in to account. Similarly air dissolves in water and lowers its density, but the extent to which dissolution of air affects the density of water was not known. With the development of new technology in measurement and the growing demand of accuracy in knowing the density of water, several national laboratories took up the job of measurement of water density with a precision better than one parts per million. Last 25 years of twentieth century were spent to measure the density of well defined and air free water. Each laboratory expressed its results in different forms. BIPM set up an international Committee for harmonising the results of various laboratories. Simultaneously the Author also took up the job of expressing the density of water at different temperatures using the recent results of measurement of water density by various laboratories. The author reported latest expression and values of density of water in the Second International Conference on Metrology in New Millennium and Global Trade, held at NPL, New Delhi, in February 2001 [6, 7]. Most recently the international Committee set by BIPM has also come to a conclusion and expressed density of water as a function of temperature [8]. But the values of water density obtained by the author and the Committee differ only by a few parts per ten million. The density table in terms of international temperature scale ITS 90 of SMOW has been given in table 1.1. Henceforth the table 1.1 should be used for gravimetric determination of capacity of all the capacity measures and volumetric glassware, when water is used as standard of known density.

1.7.1 SMOW

Standard Mean Ocean Water with acronym SMOW means pure water having different isotopes of water satisfying the following relations

$$R_D = (155.76 \pm 0.05) \times 10^{-6}$$

and

$$R_{18} = (2005.2 \pm 0.05) \times 10^{-6}$$

The international community has agreed to the aforesaid values after determining tne isotope abundance ratios of samples of water taken from different sources and locations in the sea. It may be mentioned that due to different isotopic composition of water, the density of water may differ only by a few parts in one million.

Pure water molecules are formed when one oxygen atom combines with two atoms of hydrogen. However oxygen as well as hydrogen is found to have different isotopes. Atoms of isotopes of an element have same number of electrons and protons but different number of neutrons in the nucleus. In other words, isotopes will have same chemical properties but different physical properties; especially the relative mass values of its atoms will be different. Atomic mass number is the ratio the mass of an atom to the mass of one hydrogen atom and is simply called as mass number. For example most of the atoms of oxygen have mass number 16 but there are some atoms having mass number 17 and 18. Similarly most of atoms of hydrogen have mass number 1 but there are some atoms with mass number 2. So in water we have most of the molecules having one atom of oxygen of mass number 16 and two hydrogen atoms of mass number 1. But there could be some molecules having one oxygen atom of mass number 17 or 18 combining with two hydrogen atoms of mass number 1. Similarly there will be some molecules of water having one oxygen atom of mass number 16 combined with two hydrogen atoms of mass number 2.

The abundance ratio is the ratio of the number of isotopic atoms of specific mass number, present in a given volume, to the number of atoms of the normal mass number. For example: oxygen has isotopes of mass number 18 and 17, while its normal mass number is 16. Then the abundance ratio denoted as R_{18} is the ratio of number of atoms of mass number 18 to those of mass number 16, present in a given volume. Similarly the abundance ratio of isotopes of water with oxygen of mass number 18 or hydrogen mass number 2 will respectively be

$$R_{18} = n(^{18}O)/n(^{16}O) \text{ and}$$
$$R_D = n(D)/n(H)$$

Density values given in table 1.1 are of air-free SMOW.

Corrections, if accuracy so demands, are applied for isotopic composition by the following relation

$$\rho - \rho_{(V\text{-}SMOW)} = 0.233\,\delta^{18}O + 0.0166\,\delta D$$

Similarly for the water having dissolved air, additional correction is applied to the density values given in the table 1.1 by the following relation:

$$\Delta\rho/\,\text{kgm}^{-3} = (-\,0.004612 + 0.000\,106t)\chi$$

Where

χ = degree of saturation.

t = temperature in °C.

ρ = density of sample water in kgm^{-3}.

R_D = ratio of number deuterium atoms to the number of hydrogen atoms.

R_{18} = ratio of oxygen atoms of mass number 18 to the number of oxygen atoms of mass number 16.

δ = deviation from unity of the ratio of abundance ratio of the sample to the abundance ratio of the SMOW.

For example $\delta^{18}O = [R_{18(sample)}/R_{18(SMOW)} - 1]$ and $\delta D = [R_{D(sample)}/R_{D\,(SMOW)} - 1]$

1.7.2 International Temperature Scale of 1990 (ITS90)

We know that elements and compounds change its phase (solid to liquid or liquid to gaseous state) at specified conditions only at a fixed temperature. International temperature scale is a set of such accurately determined temperatures at which phase transition takes place of certain pure elements and compounds water. The set covers the range of temperatures likely to be met in day to day life. We can measure thermodynamic temperature only through the thermometers whose equation of state can be written down explicitly without having to introduce unknown temperature dependent constants. These thermometers are called as primary standards which are only a few world-wide and also the reproducibility of measurements through such instrument are not quite satisfactory.

The use of such thermometers to high accuracy is difficult and time-consuming. However there exist secondary thermometers, such as the platinum resistance thermometer, whose reproducibility can be better by a factor of ten than that of any primary thermometer. So phase change temperatures are measured of several elements. The elements are such that these are available in the pure form. Such measurements are taken at national measurement laboratories world-wide. International Community then accepts a set of such temperatures. Such a set of temperatures is known as practical temperatures scale. In order to allow the maximum advantage to be taken of these secondary thermometers the General Conference of Weights and Measures (CGPM) has, in the course of time, adopted successive versions of an international temperature scale. The first of these was in 1927 as ITS 127. Subsequently depending upon new experiments carried out with better available technology, various temperature scale such as IPTS 48 in 1948 and IPTS68 in 11968 have been adopted. Finally in January, 1990, CGPM adopted a new set of temperatures, which is known as ITS 90.

Primary thermometers that have been used to provide accurate values of thermodynamic temperature include the constant-volume gas thermometer, the acoustic gas thermometer, the spectral and total radiation thermometers and the electronic noise thermometer.

1.8 INTERNATIONAL INTER-COMPARISON OF VOLUME STANDARDS

1.8.1 Principle

Like all other International inter-comparisons of standards of other quantities, standards of volume/ capacity are also inter-compared keeping a certain objective(s) in view. In these inter-comparisons, several national measurement laboratories participate. So participants list and identification of the pilot laboratory is the first thing to start such a project. The pilot laboratory takes upon it the responsibility of co-ordinating with other laboratories. Its job is to outline in clear-cut terms the following:

- The aims and objective(s) of the project.
- Preparation or procurement of the artefact.
- Method to be used in determination of the attribute of the artefact under investigation. In the present case it is volume of the artefact.
- Time schedule in consultation with the participating laboratories.
- Method of reporting the results with detailed analysis of uncertainty.
- Monitoring the progress of the measurements at different laboratories and the influence parameters like temperature.
- Quite often, the Pilot laboratory determines the attribute of the artefact before and after the determination of the attribute by each participating laboratory.
- Collating and correlating the results of determination by participating laboratories.

1.8.2 Participation

A preliminary meeting is held to prepare a list of likely participating laboratories and to assign the job of the pilot laboratory to one of the willing participating laboratories. The Pilot laboratory may contact the other laboratories whose participation is considered necessary. The laboratory will prepare the list of participating laboratories, address with communication facilities available at each laboratory and name of contact person in each laboratory.

1.8.3 Aims and Objectives of the Project

The aims and objective of the project may be any one, some or all the following points mentioned below:

1. To establish mutual recognition for the available measurement facilities with known and stated uncertainty of measurements.
2. To build up confidence in measurement capability for specific quantity (volume in this case) with the known uncertainty.
3. To ascertain and quantify the change in measured quantity due to specific influence parameter.
4. To ensure the user or user industry for the measurements carried out by the laboratory with specified uncertainty.
5. To ensure the maintenance of other standards for other quantities with the required uncertainty. For example calibration of standards of mass requires determination of its volume. So each laboratory requires the capability for measurement of volume of mass standard with the required uncertainty.

1.8.4 Preparation or Procurement of the Artefact

Before proceeding further, let us defines the word attribute as the property of the artefact, under investigation; for example, in the present case, volume of the artefact is measured. The artefact of stable volume and having a highly smooth and polished surface, whose volume can preferably be determined through dimensional method, is used as travelling standard; every participating laboratory assigns the value of the volume to the same artefact. For this purpose, a suitable artefact is prepared or procured by the pilot laboratory. The artefact should be such that the attribute under investigation., (volume in this case) does not change during its transport to different laboratories. Its carrying case along with its handling equipment should be properly designed and instruction for its use including cleaning etc. should be detailed out. Material of the travelling standard should be such that the attribute under investigation does not change with time, if it is not possible then a well-defined relation between the changes in the attribute with respect to time should be clearly stated and every participating laboratory should be requested to use the given relation only. Other parameters, which affect the value of the attribute, should be well documented and each laboratory should use the same document.

1.8.5 Method to be Used in Determination of the Parameter(s) of the Artefact

The method for determination of the required attribute should be clearly detailed out, unless the object is to study the compatibility of the different methods of measurements for the same attribute. Say in case of measurement of volume of a travelling standard, it should be specified as to which method is used, the dimensional or hydrostatic. Every measurement should be traceable to the national standards maintained in the country and it should be clearly specified in the report.

1.8.6 Time Schedule in Consultation with the Participating Laboratories

For the success of a project of this nature, a well-defined, optimum time schedule should be worked out in advance. Each laboratory should follow the time schedule and the Pilot laboratory should monitor it. One problem, which is commonly faced by the developing countries, is the custom clearance and handling of artefact at that stage. Each participating Laboratory should take special pains to sort out the custom clearance problem well in advance. The Pilot Laboratory should provide a set of clear instructions for handling the artefact especially by the custom people.

1.8.7 Method of Reporting the Results with Detailed Analysis of Uncertainty

A detailed procedure for calculating the uncertainty should be laid out. The influence parameters should be clearly defined and the associated uncertainty should be grouped in appropriate class (Type A or B) [9]. Each participating laboratory should be asked to report the uncertainty associated with the defined parameters, even if it is insignificant according to the participating laboratory. Uncertainty in base standards or national standards is to be stated and taken into account and should be grouped as Type B uncertainty.

1.8.8 Monitoring the Progress of the Measurements at Different Laboratories and the Influence Parameters Like Temperature

There are certain influence factors, which affect the value of the measured value of the parameter under investigation in a very complicated and unknown way. In this case the parameter should be monitored by each laboratory and reported to the pilot laboratory. Pilot laboratory should make arrangement for monitoring of such parameter during transport of the artefact.

1.8.9 Monitoring the Required Parameter(s) of the Artefact

In some cases, the Pilot Laboratory measures the parameter under investigation before and after a participating laboratory, so as to see for any change in the parameter and to assess any damage during transportation. For example, in case of mass standards, there may be a change in mass value of the travelling standard due to a scratch caused by rough handling.

1.8.10 Collating and Correlating the Results of Determination by Participating Laboratories

Finally all the results are statistically evaluated and assessed for their correctness within the stated uncertainty by the laboratory. Any bias component in a particular laboratory or an artefact is identified and accounted for. Great care should be taken that the sentiments of no laboratory are hurt. Adverse comments about a laboratory, if any, should be avoided.

1.8.11 Evaluation of Results from Participating Laboratories

Basic problem in collating the results of international inter-comparisons is the variation of results, though each laboratory may claim a reasonable uncertainty. If all the results reported are arranged in ascending order of their magnitudes, then results on either end may become susceptible and one starts wondering if those results should be considered or not in compiling the final value. One simple criterion is the Dixon's test, which may be used for ignoring or not ignoring the results on either end. As a policy one should not ignore or at least appear to ignore any result. It is, therefore, advisable to apply a method so that none of the result is ignored. Some laboratories have better equipment and manpower so will report the results with smaller uncertainty values, which are likely to be more reliable. One has to give some more respect to results obtained with smaller values of uncertainty. So do not ignore any results, but give more weight factor to results with smaller uncertainty, keeping in mind that outliers do not affect the result too much. Outliers can be identified by the Dixon outlier test as given below. For collating and analysing the results from different laboratories host of other statistical methods are available in the literature.

1.8.11.1 Outlier Dixon Test

Basic assumption of this test is that all reported results follow normal distribution. For application of the test, all observations are arranged in either ascending or descending order. If the lower value result is under suspicion, the results are arranged in descending order. The results are arranged in ascending order if the higher value result is to be tested for outlier. So that suspected result is the last i.e. n^{th} result is under scrutiny, n being the total number of results. Depending upon the value of n, the test parameter is taken as one of the following ratios:

$$(X_n - X_{n-1})/(X_n - X_1) \text{ for } 3 < n < 7$$

$$(X_n - X_{n-1})/(X_n - X_2) \text{ for } 8 < n < 10$$

$$(X_n - X_{n-2})/(X_n - X_2) \text{ for } 11 < n < 13$$

$$(X_n - X_{n-2})/(X_n - X_3) \text{ for } 14 < n < 24$$

For given n, the value of test parameter should not exceed the corresponding critical value given in the table 1.2.

If n^{th} - the last result happens to be an outlier then test is applied to the n-1st results. The process should continue till the test parameter is less than the critical values given in the table.

Table 1.2 Critical Values for Dixon Outlier Test

n	*Test parameter*	*Critical Value*
4 5 6 7	$(X_n - X_{n-1})/(X_n - X_1)$	0.765 0.620 0.560 0.507
8 9 10	$(X_n - X_{n-1})/(X_n - X_2)$	0.554 0.512 0.477
11 12 13	$(X_n - X_{n-2})/(X_n - X_2)$	0.576 0.546 0.521
14 15 16 17 18 19 20 21 22 23 24 25	$(X_n - X_{n-2})/(X_n - X_3)$	0.546 0.525 0.507 0.490 0.475 0.462 0.450 0.440 0.430 0.421 0.413 0.406

The result under test is X_n.

Generally speaking, to collate the results from participating laboratories, we may adopt any of the three methods as described below. The methods are:

- Arithmetic mean method,
- Median method, and
- Weighted mean method.

1.8.11.2 Arithmetic Mean Method

Simple mean or the arithmetic mean X_m is defined as

$$X_m = \Sigma X_i / n, \text{ where } i \text{ takes all values from 1 to } n \text{ and}$$

estimated standard deviation "*s*" of the single observation is given by

$$s = [\Sigma(X_i - X_m)^2/(n-1)]^{1/2}$$

While standard uncertainty of the mean $U(X_m)$ is given as

$$U(X_m) = [\Sigma(X_i - X_m)^2/\{n(n-1)\}]^{1/2}$$

Though taking arithmetic mean appears to be more reasonable in the first instance, but here extreme values of the results effect more than the ones, which are closer to mean values.

Standard deviation s and $U(X_m)$ is rather more sensitive to inclusion of reported extreme values. This point will be further clarified, when we discuss the results of the example later.

1.8.11.3 Median Method

In this method, all results are arranged in ascending order and the result, which comes exactly in midway is taken as median for the odd number of results. If the number of results is even, then the arithmetic mean of the two middle ones is taken as the median. In this method only one or two of the reported results are taken into consideration. The notations used are

$$X_{med} = \text{med}\{X_i\}$$

The uncertainty attributable, according to Muller [22], to median is based on the Median of the Absolute Deviations, which is abbreviated as MAD and defined as

$$\text{MAD} = \text{med}\ \{X_i - X_{med}\}$$

The standard uncertainty in this case is given by

$$U(X_{med}) = 1.9\ \text{MAD}/(n-1)^{1/2}$$

It may be noted that median is unaffected by outliers as long they exist, while arithmetic mean is greatly affected by an outlier. However median method does not distinguish between good and bad values. Equal importance is given to every result irrespective of uncertainty. Mean is affected equally by the result having very large uncertainty as by the one with very small uncertainty. To overcome this defect weighted mean method may be used.

1.8.11.4 Weighted Mean

Though it is natural that the results obtained with smaller uncertainty are more reliable than those with larger uncertainty, but no such distinction has been made while taking the arithmetic mean, which appears to be not fair. So to give due importance to the results obtained by smaller uncertainty, we may assign a weight equal to inverse of the square of the uncertainty to each result; i.e. a result X_i with uncertainty $U(X_i)$ will have the weight equal to $U^{-2}(X_i)$.

So weighted mean X_{wm}, is given by

$$X_{wm} = \{\Sigma\, U^{-2}(X_i).\, Xi\}/\{\,\Sigma U^{-2}(X_i)\}$$

While uncertainty of weighted means $U(X_{wm})$ is given by

$$U(X_{wm}) = \{\,\Sigma U^{-2}(X_i)\}^{-1/2}$$

1.8.11.5 Derivation of Standard Uncertainty in Case of Weighted Mean

Weighted uncertainty = weight factor w_i times uncertainty

Weighted variance = Square of weighted uncertainty

Mean variance of inter-comparison = Sum of weighted variances from all laboratories divided by the sum of the weight factors uncertainty is the square root of the variance

If U_i is uncertainty with weight factor U_i^{-2},

$$\text{so weighted uncertainty} = U_i \times U_i^{-2} = U_i^{-1}$$

$$\text{Weighted variance} = U_i^{-2},$$

$$\text{Total variance} = \Sigma U_i^{-2}$$

$$\text{Total uncertainty} = (\Sigma U_i^{-2})^{1/2},$$

$$\text{Mean uncertainty} = \text{uncertainty/sum of weight factors}$$

$$= (\Sigma U_i^{-2})^{1/2}/(\Sigma U_i^{-2})$$

$$= (\Sigma U_i^{-2})^{-1/2}.$$

1.8.11.6 Outlier Test for En

To look for the outlier if any, find *En*– the normalised deviation for each laboratory by the formula given below.

$$En = 0.5\,[\{X_i - X_{wm}\}/\{U^2(X_i) + U^2(X_{wm})\}^{1/2}]$$

A result having *En* value larger than 1.5 is excluded for the purpose of taking weighted mean. But as soon as a result is excluded, the value of $U(X_{wm})$ will change, so iterative process is applied, starting from the largest until all results contributing to the mean have $|En|$ values smaller than 1.5. Taking into account the individual uncertainties yields an objective criterion for "outliers" to be excluded. The limit value of $|En| = 1.5$ corresponds to a confidence level of 99.7% or to a limit of three times standard deviation.

The method assumes that the individual uncertainty has been estimated by following a common approach and taking same influence factors and sources of uncertainty in to account. So all parameters and influenced factors should be identified and classified either in Type A or in Type B should be sent along with other instructions. For estimating the uncertainty, every body should be told to follow the ISO Guide [9]. Otherwise a single wrong result with a wrongly underestimated (too small) standard uncertainty would strongly influence or even fully determine the weighted mean. On the other hand, a high quality measurement with overestimated (too large) standard uncertainty would only weakly contribute to the mean value so calculated.

1.9 EXAMPLE OF INTERNATIONAL INTER-COMPARISON OF VOLUME STANDARDS

Practically every national laboratory while calibrating their mass standards measures volume of the standard mass pieces by using hydrostatic method. The volume of the standard gives its true mass after applying the proper buoyancy correction. As the uncertainty available in comparison of two 1 kg mass pieces is as high as 1in 10^9, so the volume measurements should also be carried with a standard uncertainty of 1 in 10^6. It was, therefore, felt necessary to carry out round robin test between national laboratories for determination of volume of solid artefacts having volume corresponding to stainless steel weights of mass values between 2 kg and 500 g. So a project of inter-laboratory comparison of volume standards to access the volume measurement capability of various Laboratories was discussed in 7th Conference of Euromet Mass Contact Persons Meeting in 1995 at DFM, Lygby, Denmark. The project "Inter-laboratory comparison of measurement standards in field of density (Volume of solids) was proposed by Mr. J G Ulrich and was agreed to as the EUROMET Project No. 339. The final report on the project was published by EUROMET in August 2000, some portions of this project report [10] are discussed below.

1.9.1 Participation and Pilot Laboratory

The Laboratories of European countries, which took part in the inter-comparison [10] were:.

1. Swiss Federal Office of Metrology, (OFMET), Switzerland
2. Swedish National Testing and Research Institute (SP), Sweden
3. Physikalisch Technische Budesanstalt (PTB), Germany
4. Bundesamt fur Eich-und Vermessungswesen (BEV), Austria
5. Instituto di Metrologia "G Colonnetti" (IMGC), Italy
6. National Physical Laboratory (NPL), Great Britain
7. Service de Metrologia (SM), B Belgium

8. Centro Espanol de Metrologia (CEM), Spain
9. Laboratoire d'Essais (MNM-LNE), France,
10. National reference laboratory for Volume and Density (Force Institutet), (DK), Denmark
11. Orszagos Meresuugyi Hivatal (OMH), Hungary
12. Ulusel Metroloji Enstitusu (UME) Turkey.

Note: SM (Belgium) did performed the mass and volume measurements between March and April 1997, but due to restricted staff the test report was unfortunately not sent.

Swiss Federal Office of Metrology (OFMET) worked as a Pilot Laboratory, Dr Jeorges Ulrich was appointed as the contact person from the Laboratory, and Dr. Philippe Richard took over from him in January 1997.

1.9.2 Objective

The aim of the project was to determine the volume measurement capability of participating laboratories by inter-comparison of the measured volume of one or more transfer standards by hydrostatic weighing. In other words, basic aim was to access measurement capability of measuring the volume of solid objects and to access the efficacy of the method of hydrostatic weighing.

1.9.3 Artefacts

Three spheres were made of ceramic material composed mainly of 90 percent Si_3N_4 and 10 percent of MgO. The spheres were labelled according to the nominal diameters in millimetres, like CS 85, CS 75 and CS 55. The Ekasin 2000 was the trade name of the material used. The material had a cubical expansion of 4.8×10^{-6} K^{-1} between 18 ºC and 23 ºC with hardness of 1600 HV. The spheres were prepared by Messrs. SWIP, Saphirwerk, Erientstrasse 36, CH-2555 Brugg/Beil, Switzerland. Their nominal mass and volume were as follows:

Designation	CS 85	CS 75	CS 55
Mass	998.83 g	697.41g	277.14 g
Volume	315.50 cm^3	220.18 cm^3	87.165 cm^3

The spheres are shown below

Three spheres used in volume measurement
Courtesy OFMET, Switzerland

These spheres were named as transfer standard of volume as the volume values to these standards were assigned from primary standard of volume. In most of the cases, silicon spheres, whose diameters were measured using suitable interferometric techniques with laser and the volume calculated, in terms of base unit of length, were taken as primary standard while in other cases water was taken as reference standard. The spheres were transported in special wooden boxes. To avoid loss in mass and volume due to abrasion, the boxes were so made that there was no relative motion of sphere with respect of box. The boxes were packed in other boxes to avoid any mechanical and thermal shocks during transportation. As ceramic is bad conductor of heat and may take very long time to regain thermal uniformity, temperature of the each sphere was monitored with the help of data logger, during transportation and use in the laboratory. The temperature was separately plotted for each sphere and it was observed that temperature remained between 5 °C and 30 °C during all transportations except only one time from Italy to Switzerland the temperature went down beyond 5 °C.

1.9.3.1 Stability of the Artefact Standards

After each measurement carried out by a participating laboratory, volume of each sphere was measured at OFMET. The maximum deviation of all OFMET single monitoring measurements for each sphere was less than the uncertainty of the first measurement. A single crystal silicon sphere designated, as RAW08 was taken as reference standard. The volume of the reference standard was determined by IMGC against their standards, whose volume was measured by dimensional method. The difference in volume for each sphere was calculated between the volumes measured in

- Jan 99 and July 97
- July 97 and March 96
- Jan 99 and March 96

The change in volume for each sphere was determined between the end and middle of the period, at the middle and beginning and at the end and the beginning of the project.

The change in volume values observed is tabulated in the table below:

Table 1.3

Sphere volume at 20 °C	*ΔV in cm³*		
	$V_{Jan\ 99} - V_{Jul\ 97}$	$V_{Jul\ 97} - V_{Mar\ 96}$	$V_{Jan\ 99} - V_{Mar\ 96}$
CS 85 315.502 42 cm³	0.000 00	– 0.000 22	– 0.000 22
CS 75 220.178 27 cm³	– 0.000 05	0.000 1	0.000 05
CS 55 87.165 07 cm³	0.000 00	0.000 08	0.000 08

The figures in the table indicate that volume of the standards remained stable with in one part in one million i.e. 1 in 10^6.

Similarly the mass values of these standards were also monitored and the difference obtained was tabulated as given in table 1.4.

Table 1.4

Sphere	*Mass*	Δm *in mg*		
		$M_{Jan\,99} - M_{Jul\,97}$	$M_{Jul\,97} - M_{Mar\,96}$	$M_{Jan\,99} - M_{Mar\,96}$
CS 85	998.852 827 g	–0.130	0.062	–0.068
CS 75	697.413 510 g	–0.038	0.010	0.048
CS 55	277.139 191 g	0.026	0.022	0.004

Here the maximum difference in mass values corresponds to a relative difference of 0.13 in 10^6 (about 1 part in 10 million).

1.9.3.2 Visual Inspection

Each participating laboratory visually inspected the surface of each sphere.

Remarks were as follows:

Some scratches were observed before the first monitoring measurement at OFMET (May 1996) on the CS 85. At this time two heavy and three light scratches were observed on CS 85 sphere. Nothing more was reported until January 1997. NPL, UK reported six heavy and fifteen light scratches on CS 85. NPL also reported some three light scratches on CS 75. Two medium and eight light scratches were reported on CS 55 also. No other laboratory reported more defects than this very detailed report from NPL.

1.9.4 Method of Measurement

In the guidelines issued to the participating laboratories, it was clearly stated that volume of each transfer standard was to be calculated at 20 °C and at normal atmospheric pressure. No correction due to change in normal atmospheric pressure was to be applied. Temperature was to be measured on ITS 90. While calculating the volume at 20 °C, thermal coefficient of volume expansion supplied by Pilot laboratory was to be used. The guidelines contained data of standards, instructions for handling and transportation and a format for a unified reporting of the mass and volume measurement results. The guidelines also included forms for the estimation of uncertainty as well as the details of the hydrostatic method for determination of volume. At least 2 series of 10 weighing for each standard were to be carried out. The participants were requested to report for:

- The characteristics of the balance and suspension arrangements,
- If solid primary standard is used then its particulars and traceability,
- If not, source of water density table, along with the information about corrections applied for isotopic composition and dissolution of air and the formulae used,
- Mode for determination of apparent mass whether manual or automated,
- Visual examination in regard to scratches or any damage done during transport if any.

BEV of Austria used Nonane instead of water. Laboratory measured the density of Nonane using a sinker of known volume.

1.9.5 Time Schedule

Every laboratory followed the mutually agreed time schedule.

1.9.6 Equipment and Standard used by Participating Laboratories

1.9.6.1 *Laboratories Who Used Solid Standard as Reference*

OFMET [11]–used 1005 AT Mettler Toledo balance of capacity 1109 g and readability 0.01 mg. Suspension wire was 0.3 mm diameter platinum black coated stainless steel. Silicon sphere RAW 08 was used as reference. The volume of this sphere is traceable to the volume standard of Italy, while mass measurement was traceable to Swiss National standard of mass.

PTB [12]–used HK 1000 MC Mettler-Toledo balance of capacity 1001.12 g with readability of 0.001 mg. Suspension wire was of diameter 0.2 mm stainless steel uncoated wire. Volume and mass measurement were directly traceable to national standards of mass and length.

IMGC [13]–used mechanical two-knife edge balance constructed on a design of H315, capacity 1000 g and readability 0.001 mg. 0.125 mm stainless steel wire coated with platinum black was used for suspension purpose. Silicon spheres Si1 and Si2 were used as reference whose volume was measured directly in terms of base unit of length. The mass measurements were traceable to national standards of mass.

BEV–used two balances (1) MC1 Sartorius of capacity 1000 g and readability 1 mg and (2) AT 400 Mettler Toledo of 410 g capacity readability of 0.1 mg, 0.4 mm platinum uncoated wire was used for suspension. A glass sinker of known volume was used as reference and liquid Nonane instead of water was used as hydrostatic medium. Nonane has comparatively lower surface tension than water.

CEM–used AT 1005 Mettler Toledo balance of capacity 1109 g readability 0.01 mg. 0.5 mm stainless steel uncoated wire was used as suspension. Quartz- glass spheres CEM1 and CEM 2 were used as reference. Volume and mass measurements were respectively traceable to national standards of PTB and CEM.

FORCE–used LC 1200 S balance of capacity 1220 g and readability 1 mg and 0.2 mm stainless steel wire was used as suspension. Si_3N_4 ceramic sphere was used as reference. Volume and mass measurements were directly traceable to OFMET and PTB respectively.

1.9.6.2 *Laboratories Who Used Water as Reference*

SP–used a mass comparator PK200 of Mettler-Toledo of capacity of 2000 g with 1 mg readability. Suspension wire was of stainless steel of diameter 0.2 mm. Operation of 2 kg balance was manual. Deionised and degassed water was taken as density standard, Wagenbreth [14] density tables for ITS-90 was used; Correction due to hydrostatic pressure at different immersion depth was not applied. Conductivity of water was found to be 0.1 μS/cm.

NPL–used mass comparator H315 of Mettler-Toledo of capacity of 1000 g with readability of 0.1 mg; Platinum black plated wire was used for suspension. Operation of 1 kg balance was manual. Deionised and distilled water was taken as density standard, Patterson and Morris [15] density tables were used; Corrections due to hydrostatic pressure at different immersion depth and isotopic compositions were applied [21]. Conductivity of water was found to be between 1 to 2 μS/cm.

LNE–used mass comparator AT 1005 VC of Mettler-Toledo) of capacity of 1109 g with readability of 0.01 mg; Nylon wire was used as suspension wire. Mass comparator was manual. Bi-distilled water was taken as density standard, Masui [16] and Watanabe [17] density tables were used; Correction due to dissolution of air was applied using Bignell [18, 19]. The correction due to isotopic composition was applied taking Girard and Menache [20] formula. Correction due to hydrostatic pressure at different immersion depth was applied taking Kell's [21] relation.

OMH–used two mass comparators H315 of Mettler-Toledo of capacity of 1000 g with readability of 0.1 mg; and other Sartorius CS 500 of 500 g capacity with readability of 0.01mg. Suspension wire was of platinum–iridium of diameter 0.2 mm. Operation of 1 kg balance was manual but that of 500 g was automatic. Deionised and degassed water was taken as density standard and Wagenbreth [14] density tables were used. Correction due to hydrostatic pressure at different immersion depth was not applied. However the density of water was checked with two pyrex spheres.

UME–used a mass comparator H315 of Mettler- Toledo, having a capacity of 1000 g with readability of 0.1 mg; suspension wire was of platinum–iridium. No automation was used in measurement of mass repeatedly; Distilled water was taken as standard of known density, Kell [21] density tables were used; Correction due to hydrostatic pressure at different immersion depth was applied due to Kell [21].

1.9.7 Results of Measurement by Participating Laboratories

Each laboratory determined the mass and volume of each sphere. Reported volumes, of three spheres with associated uncertainties with date of examination, are tabulated below:

Table 1.5

S.No.	*Date*	*Laboratory*	*CS 85*		*CS 75*		*CS 55*	
			Volume cm^3	U_c mm^3	*Volume* cm^3	*Uc* mm^3	*Volume* cm^3	*Uc* mm^3
1.	Jan-Mar 1996	OFMET1	315.50242	0.23	220.17827	0.18	87.16507	0.13
2.	Apr-May 1996	SP	315.49955	2.84	220.17920	2.02	87.16523	0.67
3.	Jun 1996	PTB	315.50273	0.29	220.17807	0.21	87.16496	0.11
4.	Aug-Sep 1996	BEV	315.50815	0.68	220.18495	0.51	87.15880	0.19
5.	Oct-Nov 1996	IMGC	315.50272	0.17	220.17867	0.35	87.16556	0.13
6.	Jan-Feb 1997	NPL	315.5048	1.5	220.1778	1.2	87.1654	0.69
7.	May-Jun 1997	CEM1	—	—	—	—	—	—
8.	Oct-Nov 1997	LNE	315.50311	0.72	220.17989	0.56	87.16717	0.24
9.	Jan 1998	FORCE	315.50443	1.44	220.1804	0.92	87.1665	0.89
10.	Mar 1998	OMH	315.50417	1.02	220.17918	0.54	87.16604	0.30
11.	May-Jun 1998	UME	315.50575	0.76	220.1799	0.59	87.1673	0.37
12.	Oct 1998	CEM2	315.50275	0.5	220.1785	0.6	87.16545	0.7
13.	Dec-Jan 1999	OFMET2	315.50220	0.32	220.17832	0.26	87.16515	0.14
14.	OFMET Δ_{2-1}		− 0.22		+ 0.05		+ 0.08	

1.10 METHODS OF CALCULATING MOST LIKELY VALUE WITH EXAMPLE

1.10.1 Median and Arithmetic Mean of Volume of CS 85

Table 1.6

Data		*Median*		*Arithmetic Mean*	
S.No.	*Volume* X_i *cm*^{3}	$\lvert X_i - X_{med} \rvert$ *mm*^{3}	*Arrange* $\lvert X_i - X_{med} \rvert$ *mm*^{3}	$\lvert X_i - X_m \rvert$ *mm*^{3}	$(X_i - X_m)^2$ *mm*^{6}
1	315.49955	3.56	0.00	4.14	17.1396
2	315.50242	.69	0.36	1.27	1.6129
3	315.50272	.39	0.38	0.97	.9409
4	315.50273	.38	0.39	0.96	.9216
5	315.50275	.36	0.69	0.94	.9604
6	315.50311	0.00	1.06	0.58	.8817
7	315.50417	1.06	1.32	0.48	.2304
8	315.50443	1.32	1.69	0.74	.5476
9	315.5048	1.69	2.64	1.11	1.2321
10	315.50575	2.64	3.56	2.06	4.2436
11	315.50815	5.04	5.04	4.46	19.8916
Median	315.50311	MAD	1.06	Sum	48.6424

Median $X_{med} = 315.50311 \text{cm}^3$,

Uncertainty of Median $U_{med} = 1.9\text{MAD}/\sqrt{(n-1)} = 1.9 \times 1.06/3.1623 = 0.0637 \text{ mm}^3$

Arithmetic Mean $X_m = 315 + 55.4058/11 = 315.\ 50369 \text{ cm}^3$

S.D. from mean $= \sqrt{48.6424/10} = 2.2055 \text{ mm}^3$

Uncertainty of mean $U_m = 2.205 \text{ mm}^3$

1.10.2 Weighted Mean of Volume of CS 85

Table 1.7

S.No.	*Xi* *mm*^{3}	U_c *mm*^{3}	U^{-2} *mm*$^{-6}$	$(X_i - 315) \times U^{-2}$ 10^3 *mm*$^{-3}$
1	315.50242	0.23	18.903	9.4972
2	315.49955	2.84	0.1240	0.0619
3	315.50273	0.29	11.891	5.9780
4	315.50815	0.676	2.188	1.1110
5	315.50272	0.173	33.411	16.7963
6	315.50480	1.5	0.444	0.2241
7	315.50275	0.5	4.000	2.011
8	315.50311	0.72	1.929	0.9705
9	315.50417	1.02	0.961	0.4845
10	315.50575	0.757	1.745	0.882
11	315.50443	1.44	0.482	0.231
Sum	——	——	76.078	38.2609

$$X_{wm} = 315 + 38.9952/76.078 = 315.50292 \text{ cm}^3$$

$$U_{wm} = (76.078)^{-1/2} \text{ mm}^3 = 0.1146 \text{ mm}^3 = 0.115 \text{ mm}^3.$$

Similarly, from the data in table 1.5, we can calculate the mean, median and weighted means with associated uncertainties for the other two spheres. Summary of results is given below in the Table 1.8.

1.10.2.1 Mean, Median and Weighted Mean Values of the Three Spheres Volume

The values of mean, median and weighted mean of three spheres are given in Table 1.8.

Table 1.8

Sphere	*Mean*		*Median*		*Weighted mean*	
	Mean cm^3	U_m mm^3	*Median* cm^3	U_{med} Mm^3	*Weighted Mean* cm^3	U_{wm} mm^3
CS 85	315.503689	2.190	315.503110	0.637	315.50292	0.115
CS 75	220.179530	1.978	220.179180	0.433	220.178773	0.112
CS 55	87.165226	2.279	87.165450	0.294	87.164746	0.060

1.11 REALISATION OF VOLUME AND CAPACITY

So volume of a solid artefact is realised by the dimensional measurements directly in terms of base unit of length. From the volume of the solid artefact, density of water is obtained and water is used as a transfer standard. The capacity of the measure maintained at highest level is obtained by gravimetric method. Further volumetric measurements, standards (Capacity measures) maintained at lower levels are calibrated by volume transfer method. The water is normally used as medium for this purpose. Volume of liquids is measured by using calibrated capacity measures. Volume of solid bodies is either measured by dimensional methods or by hydrostatic weighing. Quite often, in industry, the volume of solid powder is also measured through the calibrated volumetric measures. The process of realisation (Hierarchy of volume measurment) is given in Figure 1.1.

1.11.1 International Inter-Comparison of Capacity Measures

Quite recently, Centro Nacional de Metrologia (CENAM), Mexico, Physikalisch Technische Bundesanstalt (PTB), Germany, Measurement Canada (MC), Canada and the National Institute of Standards and Technology (NIST), USA took part in an international inter-comparison of capacity measures. A report of the inter-comparison has been published in Metrologia [24]. Each of the aforesaid laboratories maintains the national primary standards facilities for the measurement of volume. A 50 dm^3 measure was circulated among each laboratory for measurement of its capacity by using gravimetric method and using water as density standard. The maximum departure between any two results was 0.0098%.

A worldwide program for measurement of capacity of three transfer standards of nominal values 50 ml, 100 ml and 20 litres is under way on the regional basis. The regions are Asia Pacific, Europe, North and South America. The program was started in 2002. Australia, Korea, Chinese Taipei, Japan and China are taking part in this endeavour under Asia Pacific Metrology Program APMP. Austria, Italy, South Africa, Poland, France, Switzerland, The Netherlands,

Hungary, Germany, Sweden, Turkey and Russia are taking part in measurement of capacity of the three transfer standards under European Co-operation in Measurement Standard EUROMET. Similarly Countries like Mexico, Brazil, USA and Canada are doing the same exercise under the Inter-American Metrology System SIM. General Conference on Weights and Measures CGPM has under taken the same project through its consultative committee on mass and related matters in which countries like Australia, Mexico and Sweden are cooperating on behalf of their respective regional organisations APMP, SIM and EUROMET. No results have been published of the said comparisons till the end of 2004.

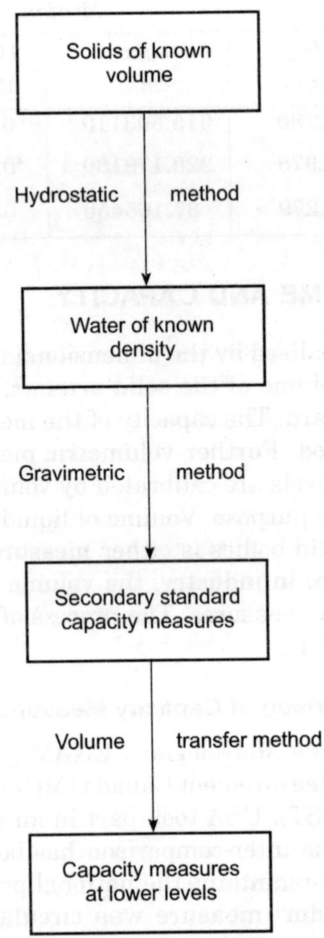

Figure 1.1 Hierarchy of volume measurement

Table 1.1 Density of Water (SMOW) on ITS-90

Temp		0.0	0.1	0.2	0.3	0.4	0.5	0.6	0.7	0.8	0.9
0	999	.8431	.8498	.8563	.8626	.8687	.8747	.8804	.8860	.8915	.8967
1	999	.9018	.9067	.9114	.9159	.9203	.9245	.9285	.9324	.9361	.9396
2	999	.9429	.9461	.9491	.9519	.9546	.9571	.9595	.9616	.9636	.9655
3	999	.9671	.9687	.9700	.9712	.9722	.9731	.9738	.9743	.9747	.9749
4	999	.9749	.9748	.9746	.9742	.9736	.9728	.9719	.9709	.9697	.9683
5	999	.9668	.9651	.9633	.9613	.9592	.9569	.9545	.9519	.9492	.9463
6	999	.9432	.9400	.9367	.9332	.9296	.9258	.9218	.9177	.9135	.9091
7	999	.9046	.8999	.8951	.8902	.8851	.8798	.8744	.8689	.8632	.8574
8	999	.8514	.8453	.8391	.8327	.8261	.8195	.8127	.8057	.7986	.7914
9	999	.7840	.7765	.7689	.7611	.7532	.7451	.7370	.7286	.7202	.7116
10	999	.7029	.6940	.6850	.6759	.6666	.6572	.6477	.6380	.6283	.6183
11	999	.6083	.5981	.5878	.5774	.5668	.5561	.5452	.5343	.5232	.5120
12	999	.5007	.4892	.4776	.4659	.4540	.4420	.4299	.4177	.4054	.3929
13	999	.3803	.3676	.3547	.3418	.3287	.3154	.3021	.2887	.2751	.2614
14	999	.2475	.2336	.2195	.2053	.1910	.1766	.1621	.1474	.1326	.1177
15	999	.1027	.0876	.0723	.0569	.0414	.0258	.0101	*.9943	.9783	.9623
16	998	.9461	.9298	.9133	.8968	.8802	.8634	.8465	.8296	.8125	.7952
17	998	.7779	.7605	.7429	.7253	.7075	.6896	.6716	.6535	.6353	.6170
18	998	.5985	.5800	.5613	.5425	.5237	.5047	.4856	.4664	.4471	.4276
19	998	.4081	.3885	.3687	.3489	.3289	.3089	.2887	.2684	.2480	.2275
20	998	.2069	.1863	.1654	.1445	.1235	.1024	.0812	.0599	.0384	.0169
21	997	.9953	.9735	.9517	.9297	.9077	.8855	.8633	.8409	.8185	.7959
22	997	.7733	.7505	.7276	.7047	.6816	.6585	.6352	.6118	.5884	.5648
23	997	.5412	.5174	.4936	.4696	.4455	.4214	.3971	.3728	.3483	.3238
24	997	.2992	.2744	.2496	.2247	.1996	.1745	.1493	.1240	.0986	.0731
25	997	.0475	.0218	*.9960	.9701	.9441	.9180	.8918	.8656	.8392	.8128
26	996	.7862	.7596	.7328	.7060	.6791	.6521	.6250	.5978	.5705	.5431
27	996	.5156	.4881	.4604	.4326	.4048	.3769	.3488	.3207	.2925	.2642
28	996	.2358	.2074	.1788	.1501	.1214	.0926	.0636	.0346	.0055	*.9763
29	995	.9470	.9177	.8882	.8587	.8290	.7993	.7695	.7396	.7096	.6795
30	995	.6494	.6191	.5888	.5583	.5278	.4972	.4666	.4358	.4049	.3740
31	995	.3430	.3118	.2806	.2494	.2180	.1865	.1550	.1234	.0917	.0599
32	995	.0280*	.9960	.9640	.9319	.8996	.8673	.8350	.8025	.7700	.7373
33	994	.7046	.6718	.6389	.6060	.5729	.5398	.5066	.4733	.4399	.4065
34	994	.3729	.3393	.3056	.2718	.2380	.2040	.1700	.1359	.1017	.0675
35	994	.0331*	.9987	.9642	.9296	.8949	.8602	.8254	.7905	.7555	.7204
36	993	.6853	.6501	.6148	.5794	.5439	.5084	.4728	.4371	.4013	.3655
37	993	.3296	.2936	.2575	.2213	.1851	.1488	.1124	.0760	.0394	.0028
38	992	.9661	.9294	.8925	.8556	.8186	.7815	.7444	.7072	.6699	.6325
39	992	.5951	.5576	.5200	.4823	.4446	.4067	.3688	.3309	.2928	.2547
40	992	.2166	.1783	.1400	.1016	.0631	.0245	*.9859	.9472	.9085	.8696
41	991	.8307									

***Note*:** Whenever an asterisk (*) appears, the integral value of density thereafter in the row will be one less than the integer given in second column.

Base density of V-SMOW is taken as 999.974 950 ± 0.000 84 kgm^{-3} at 3.983 035 °C.

REFERENCES

[1] Saunders J B, 1972, Ball and cylinder interferometer; J. Res. Natl. Stand. C 76 11-20.

[2] Nicolaus R A and Bonch G, 1997; A novel interferometer for dimensional measurements of a silicon sphere; IEEE Trans. Instrum. Meas. 46, 54-60.

[3] Gupta S V, 2002, Practical density measurements and hydrometery, Institute of Physics Publishing, Bristol and Philadelphia.

[4] Cook A H and Stone N W M, 1957, "Precise measurement of the density of mercury at 20 °C": I, absolute displacement method; Phil. Trans. R. Soc. A 250 279-323.

[5] Cook A H, 1961, Precise measurement of the density of mercury at 20 °C:-II Content method Phil. Trans. R. Soc. A 254 125-153.

[6] Gupta S V, 2001, Unified Method of expressing temperature dependence of water; Proceedings 3rd International Conference on Metrology in New millennium and Global trade (MMGT), Mapan- Journal of Metrology Society of India.

[7] Gupta S V, 2001, New water density table at ITS 90; Indian. J. Phys. 75B 427-432.

[8] Tanaka M et al; 2001 Recommend table for the density of water between 0 °C to 40 °C based on recent experimental report, Metrologia, 38 301-309.

[9] ISO Guide to the expression of uncertainty in measurement, 1993 ISO.

[10] Richard Philippe, 2000, Euro Project No. 339 Final Report on Inter-comparison of volume standards by hydrostatic weighing.

[11] Beer W and Ulrich "New volume comparator" OFMET Info, 1996 3, 7-10.

[12] Spieweck F, Kozdon A, Wagenbreth H, Toth H, Hoburg D "A computer Controlled Solid density measuring Apparatus, PTB Mitteillungen, 1990, 100 169-173.

[13] Mosca M, Birello G et al Calibration of a 1 kg automatic weighing system for density measurements" 13th Conference on Force and Mass Measurements, 1993, Helsinki, Finland.

[14] Wagen H, Blanke W "Die Dichte des wasser im international Einheiten sydtem und in der Internationalen Praktischen Temperatureskala von 1968, PTB Mitteillungen, 1971, 81, 412-415.

[15] Patterson J B and Morris E C, 1994 Measurement of absolute water density, 1 °C to 40 °C 1994, Metrologia, 31, 277-288.

[16] Masui R, Fujii K and Takenake M Determination of the absolute density of water at 16 °C and 0.101235 MPa, 1995/96, Metrologia, 35, 333-362.

[17] Watanabe H, Thermal dilatation of water between 4 °C and 44 °C, 1991, Metrologia, 28, 33-43.

[18] Bignell N, The effect of dissolved air on the density of water, 1983, Metrologia, 19, 57-59.

[19] Bignell N, The change in water density due to aeration in the range of 0 °C to 8 °C, 1986, Metrologia, 23, 207-211.

[20] Girard G and Menache M, Sur le calcul de la mass volumique de l'eau, 1972, C. R. Acad. Sc. Paris, 274 (Series B), 377-379.

[21] Kell G S Density, Thermal expansivity and compressibility of liquid water from 0 °C to 150 °C: corrections and tables for atmospheric pressure and saturation reviewed and expressed on 1968 temperature scale, 1975, J. Chem. Eng. Data, 20, 97-105.

[22] Muller J W, Possible advantage of a robust evaluation of comparisons BIPM –95/2, 1995, BIPM: Sevres.

[23] Peuto A et al. "Precision measurements of IMGC Zerodur spheres", IEEE Trans. Instrum 1984, 449.

[24] Maldonado J M, Arias R; Oelze H-H, Bean V E; Houser J F; Lachance C and Jacques C, international comparison of volume measuring standard at 50 L level at CENAM (Mexico), PTB (Germany), Measurement Canada and NIST (USA), 2002, Metrologia, 39, 91-95.

STANDARDS OF VOLUME/CAPACITY

2.1 REALISATION AND HIERARCHY OF STANDARDS

We have seen in the previous chapter that primary standard of volume is a solid whose volume has been determined by measurement of its dimensions in terms of the unit of length. From this primary standard, the density of well-characterised water has been obtained by hydrostatic weighing. The determination of mass of water delivered or contained in a measure gives the capacity of the measure by using mass density relationship of water. This method is known as Gravimetric method of determination of capacity of a measure. Thus the density of water is a link between mass of water delivered or contained in a measure and its capacity. Hence water acts as a transfer standard for capacity measurements.

Best standards of capacity, which can be maintained, are those whose capacity is determined by the gravimetric method. The capacity of the measures maintained at lower level is determined by volume transfer method. In this method a standard measure of same capacity as that of the measure under test is used and volume of water transferred from it to the measure under test gives the capacity of the measure under test. The water will be transferred from the standard measure to the measure under test if the measure under test is a content measure. In this case the standard must be a delivery measure. The reverse process is to be employed if the measure under test is a delivery measure. In that case standard measure has to be a content measure so that volume of water transferred from measure under test is delivered to the standard measure. This volume transfer method is also called as one to one comparison method. For measures of larger capacity, a standard measure, whose capacity is an exact sub-multiple of the measure under test, is taken. The water is transferred several times from standard measure to measure under test. This method is called multiple volume transfer method.

So we have the following modes of realisation of volume/ capacity and hierarchy of capacity and volume standards.

Primary level — Solid of known volume

Use of Hydrostatic weighing method gives

Water of known density, which is maintained at the transfer level

Use of Gravimetric method gives
Capacity of measures maintained at levels I and II,
By means of hydrostatic weighing in water of known density gives
Volume of solids of any shape
By means of hydrostatic weighing of solids of known volume gives
Density of liquids
One to one volume transfer gives
Working standard capacity measures
Multiple filling / volume transfer gives
Commercial capacity measures
Volume of all liquids is measured with the help of graduated capacity measures.

The hierarchy, realisation of volume and its measurements standards are represented in Figure 2.1. The arrow from one box to next downward box not only shows the standard at a lower level, but the language part indicates the technique used in realising it.

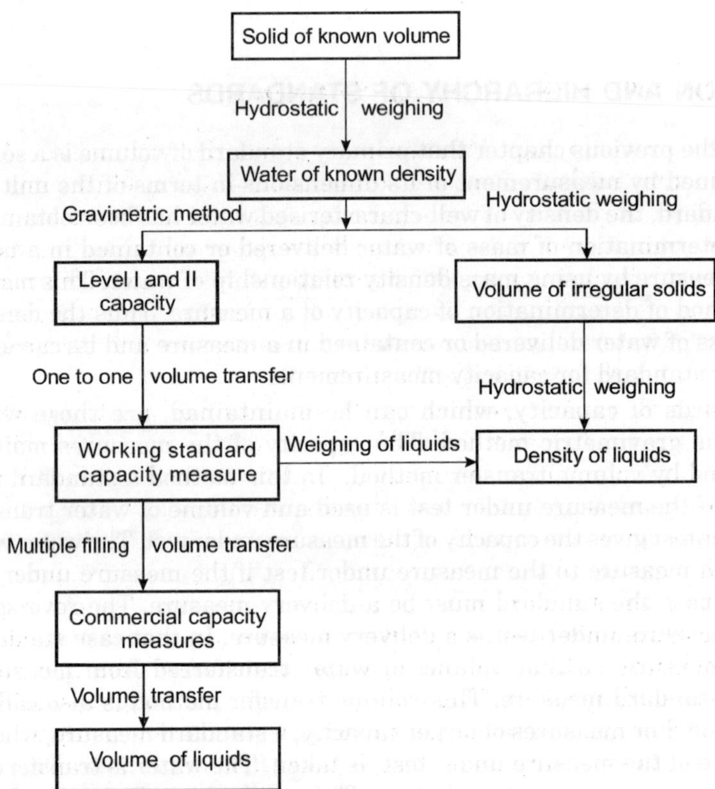

Figure 2.1 Hierarchy of volume standards, realisation of volume and volume measurement

Measurement of volume of liquids used in trade and commerce falls under the ambit of legal metrology. In legal metrology, every thing is documented and standards used for the purpose of measuring volumes of liquids are assigned appropriate names. Nomenclature of such standard measures may vary from country to country. For example, we in India call them as

- Secondary standard capacity measures
- Working standard capacity measures
- Test measures
- Commercial measures

An hierarchy of volumetric standards, nomenclature, range of capacity, maximum permissible error at one dm^3 level, method of realisation and period of verification as followed by the Indian Departments of Legal metrology [1] is depicted in Figure 2.2.

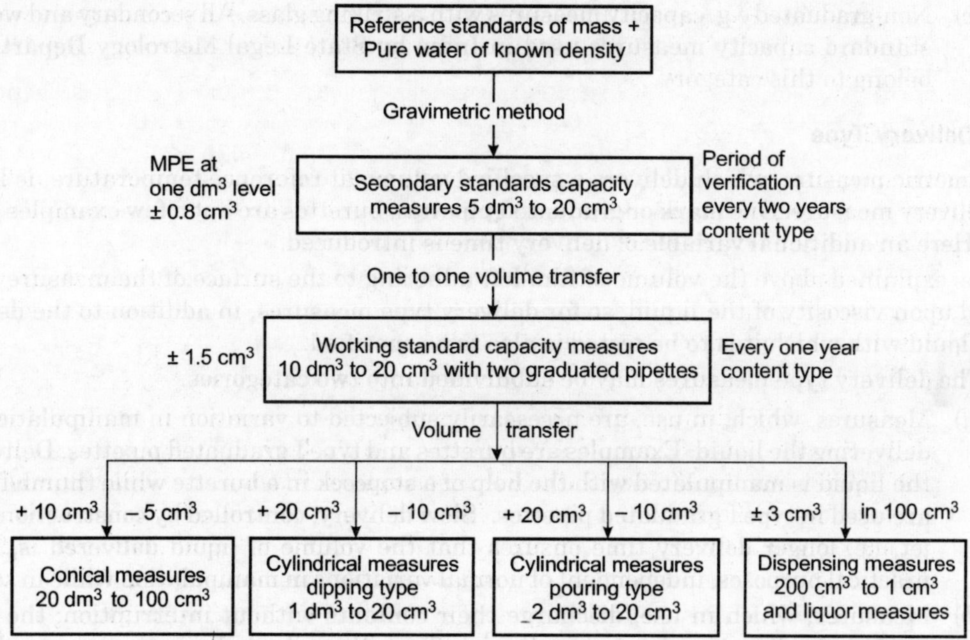

Figure 2.2 Volumetric measurement for legal metrology in India

2.2 CLASSIFICATION OF VOLUMETRIC MEASURES

When the content of a volumetric measure is transferred, then all liquid contained in it will not be transferred from it. Some liquid will be left out adhering to the inside surface of the measure. The volume of the liquid left will depend upon several factors such as viscosity of the liquid, surface roughness of the measure and the time taken in transferring the liquid. So a measure will contain more liquid than what it could transfer. Hence the capacity of measure is to be qualified by the word "Content" or "Delivery". Consequently any measure is designated as either a content type measure or a delivery type measure.

2.2.1 Content Type

A volumetric measure, which contains a specified volume at reference temperature, is known as a content measure.

A one-mark flask of say of denomination of 100 cm^3 will contain 100 cm^3 ± tolerance allowed when filled up to its mark. Similarly a measuring cylinder will contain a liquid equal to the value of graduation mark ± tolerance allowed at that mark in cm^3 at 27 °C. The word "tolerance" is quite often replaced by the expression "maximum permissible error".

A volumetric measure, which contains a volume equal to its nominal value at a reference temperature, is known as a content measure.

These may be further classified as

(*a*) One mark e.g. one mark flasks

(*b*) Graduated e.g. graduated cylinders

(*c*) Non-graduated e.g. capacity measures with a striking glass. All secondary and working standard capacity measures used in India by State Legal Metrology Departments belong to this category.

2.2.2 Delivery Type

A volumetric measure, which delivers a specified volume at reference temperature, is known as a delivery measure. One mark or graduated pipettes, burettes are but a few examples of this class. Here an additional variable of delivery time is introduced.

As explained above the volume of film left adhering to the surface of the measure would depend upon viscosity of the liquid, so for delivery type measures, in addition to the delivery time, liquid with which it is to be tested is also to be specified.

The delivery type measures may be subdivided into two categories.

(*a*) Measures, which, in use, are necessarily subjected to variation in manipulations for delivering the liquid. Examples are burettes and type I graduated pipettes. Delivery of the liquid is manipulated with the help of a stopcock in a burette while thumb/fingers are used in type I graduated pipettes. Slow delivery, controlled by construction of the jet, i.e. longer delivery time ensures that the volume of liquid delivered is, for all practical purposes, independent of normal variations in manipulation while in service.

(*b*) Measures, which in use, discharge their contents without interruption; the liquid surface finally comes to rest in the jet. Type II graduated pipette, one mark bulb pipette and one mark cylindrical pipette fall in this category. These measures have less delivery time but are allowed to drain into the receiving vessels for a specified time to secure consistent results.

2.3 PRINCIPLE OF MAINTENANCE OF HIERARCHY FOR CAPACITY MEASURES

The primary standard of volume is a solid of known geometry and its volume is calculated by dimensional measurements in terms of unit of length. Capacity measures, whose capacity is determined by finding the mass of water of known density contained or delivered, are maintained at various levels of hierarchy. Capacity of other measures is determined by volume transfer method, for which it is necessary that out of two measures to be compared one measure is of content type and the other is of delivery type. So measures maintained at successive levels are alternately content and delivery types or vice versa. Hence to determine if the measures maintained at first level should be of content type or the delivery type, we have to start from the commercial measures. Normally these measures are of delivery type, as a tradesman or a retailer will pour liquids in the vessel of a customer. So all commercial measures like milk measures, oil measures, or dispensing measures are delivery type. Commercial measures are verified against the standards maintained by the Inspectors (Agents) of Legal Metrology, which must be content type. These standards are normally termed as working standard measures. Now to verify these working standard measures by volume transfer method the measures used for the purpose must be of delivery type. Let us call them as secondary standard capacity measures. So, in India, we have commercial measures, working standard capacity measures and finally secondary standard capacity measures.

Reference standard of mass are used to determine the mass of water of known density contained in these measures, so these are rightly called Secondary Standard Measures.

In section 2.4 we will describe 25 l and 50 l automatic pipettes maintained at NPL and pipe provers. In section 2.5, secondary standard capacity measures, both single capacity and multiple capacities made of metals and glass, have been described. The working standard measures of all types are described in section 2.6.

2.4 FIRST LEVEL CAPACITY MEASURES

2.4.1 25 dm^3 Capacity Measure at NPL India

Mr Mohinder Nath, a colleague of the author at National Physical Laboratory, New Delhi, designed a 25 dm^3 automatic pipette. The pipette proved to be a very handy tool for calibration of large capacity measures using multiple volume transfer method. The pipette was fabricated in the NPL workshop. It is called pipette as it has a three-way stopcock for inlet and outlet of water and a position where the pipette is disconnected from outside. The pipette is called as automatic pipette, as no final setting of water on any graduated mark is required, as is normally required in one mark pipette, pipette is supposed to be full at the instant when water starts overflowing through its upper small bore tube 16.

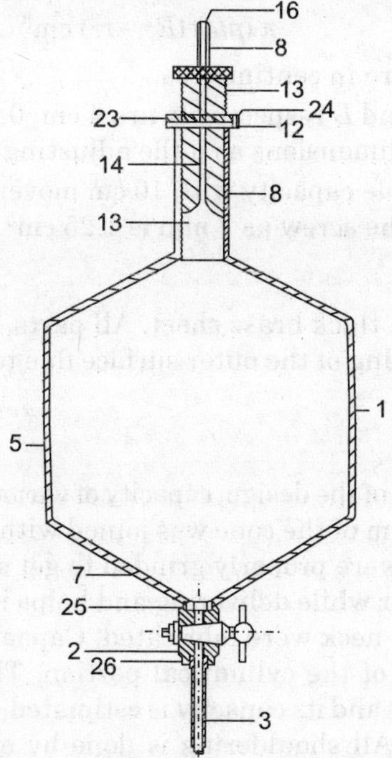

Figure 2.3 25 dm^3 automatic pipette (NPL, India)

2.4.1.1 Shape

The pipette consists of a cylinder surmounted by a frustum of a cone on either side Figure 2.3. The lower end terminates in to delivery tube and intake tube through a three-way stopcock. At the upper end there is a cylindrical neck having a novel system of adjusting the capacity of such a big measure within 1cm^3. The details of the adjusting device have also been shown in the Figure 2.3. Finally the neck terminates into a smaller bore tube of 10 mm in diameter.

The upper end of the tube is bevelled so that water drop formed due to surface tension is of the same shape and size at the top end of the tube. The part 3 is the delivery tube connected to the main body through three-way stopcock assembly 4. The numerals indicate the parts whose detailed drawings are to be made.

2.4.1.2 Adjusting Device

A threaded cylinder 14 with a through and through hole is screwed into the neck of the pipette. Top part of it is connected to the vertical tube 16. The pitch of the screw is 2 mm. There is a fixed flange 12 at the top of the neck and the moveable nut 23 on cylinder. The bore of the axial hole in the cylinder is same as that of the tube at its top. When the cylinder has reached the appropriate position, it can be locked with the neck through a quarter pin 24.

2.4.1.3 Capacity and Precision in Adjustment

The capacity is increased if the cylinder is screwed out and decreased if pushed in.

If the maximum travel of the threaded cylinder is L and it radius is R, if r is the radius of the hole in cylinder than the adjustment capacity of the pipette is given by

$$\pi L\,(R^2 - r^2)\ \text{cm}^3$$

If pitch of the screw is p cm and we can move the cylinder with a precision of 1/4th of revolution then precision in adjustment is give

$$\pi\,(p/4)\,(R^2 - r^2)\ \text{cm}^3$$

All linear dimensions are in centimetres.

Typical values of R, r and L respectively are 4 cm, 0.2 cm and 20 cm. Here it is assumed that initially we worked all dimensions as if the adjusting cylinder was in middle.

The amount of adjustable capacity with 10 cm movement is ± 482.5 cm^3 and precision in adjustment taking pitch of the screw as 1 mm is 1.25 cm^3.

2.4.1.4 Material

The pipette is made of 3 mm thick brass sheet. All parts, including adjusting cylinder, are of brass. To avoid the discolouring of the outer surface due to atmospheric oxygen, outer surface of brass sheet is tinned.

2.4.1.5 Fabrication

Apart from proper calculation of the design, capacity of various parts was continuously monitored. Starting from bottom, frustum of the cone was joined with the cylindrical portion by easy flow method. All shoulder joints were properly grinded to get smooth surface. Rough surface may hold varying amount of water while delivering and helps in forming air pockets. Similarly the top parts of the frustum and neck were fabricated. Capacity of each component was assessed before finalising the height of the cylindrical portion. The lower portion of the measure is shouldered to cylindrical part and its capacity is estimated. The upper cone and frustum portion are shouldered in the last. All shouldering is done by easy flow method. Final capacity is adjusted by proper positioning of the screwed cylinder in the neck. Adjustment of capacity may be carried out within one part in 10^4. Great care is taken to avoid any rough surface, tool marks or dents while working with sheet metal.

Considering the smaller size of the pipette, capacity only 25 dm^3, no ports have been provided to measure the inside water temperature. The temperature of water is measured at the entrance of the three-way stopcock.

2.4.1.6 *How to Use*

The reservoir of water and the pipette are kept in the same air-conditioned room. The reservoir may be 2 to 3 metres higher than the highest point of the pipette in use. The water should remain stored in the air-conditioned room for at least twelve hours prior to use. At the outlet of the reservoir and at the inlet of the pipette, two thermometers are used so that the temperature of water and change during the transit from reservoir to the pipette is monitored. If the change is not appreciable say is within 0.1°C then mean of the two gives the temperature of water. It should be ensured that temperature difference between the outlet of the reservoir and at inlet of the pipette is not more than 0.1°C. The temperature of water in the pipette should be noted with an over all uncertainty of not more than 0.1°C.

The pipette is filled under gravity from a water reservoir. Overflow of the water ensures that the pipette is full to its capacity. When the pipette is used as a standard delivery measure, it is placed well above the content measure to be tested.

Filling arrangement of the pipette is shown in Figure 2.4. Calibration procedure and other precaution to be taken will be dealt with in the Chapter on Calibration of Standard Measures. The pipette was patented after successful trials.

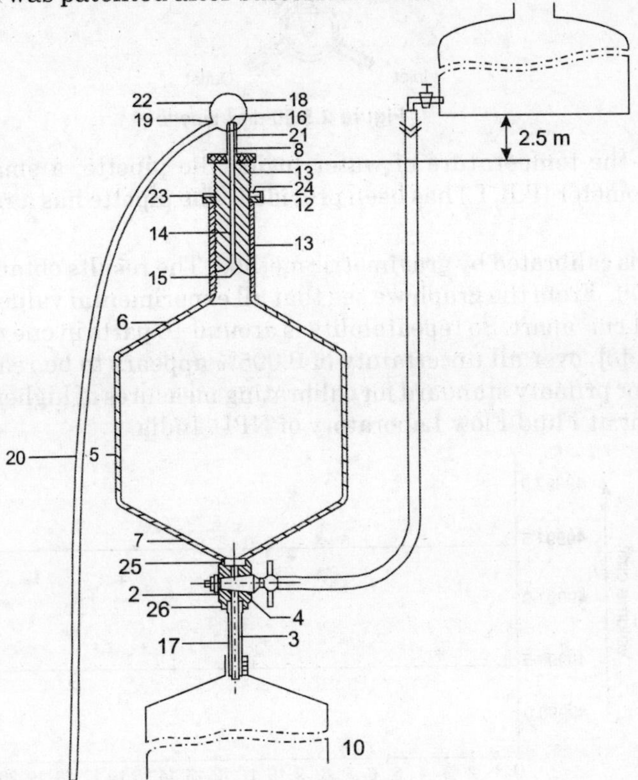

Figure 2.4 25 dm^3 pipette in use

2.4.2 50 dm^3 Capacity Measure

A measure similar to the one described above, of capacity 50 dm^3, was received from PTB Germany. It is shown in Figure 2.5. The measure is made from stainless steel sheet and similar in design as 25dm^3 pipette. Its neck is detachable from the body, which helps to check the inside cleanliness.

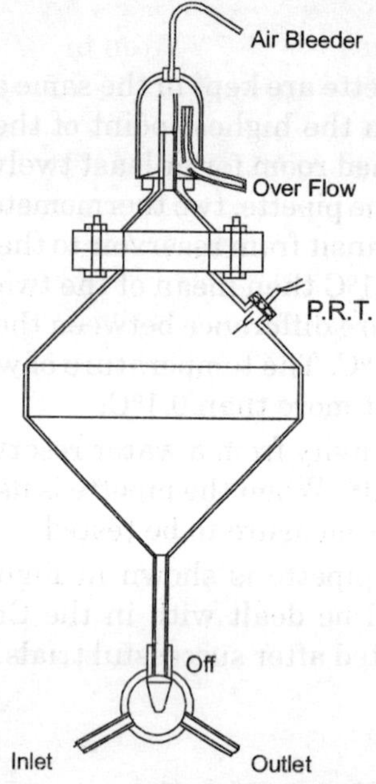

Figure 2.5 50-dm^3 pipette

To find out the temperature of water inside the pipette, a small port for a platinum resistance thermometer (P.R.T.) has been provided. The pipette has a repeatability of ten parts in a million.

The pipette is calibrated by gravimetric method. The results obtained are indicated in the graph of Figure 2.6. From the graph we see that all experimental values lie in between the two horizontal lines, 1 cm^3 apart. So repeatability is around 10 parts in one million. The pipette was first described in [3], over all uncertainty of 0.005% appears to be reasonable. The pipette is used as a master or primary standard for calibrating measures of higher capacity used in water flow measurement at Fluid Flow Laboratory of NPL, India.

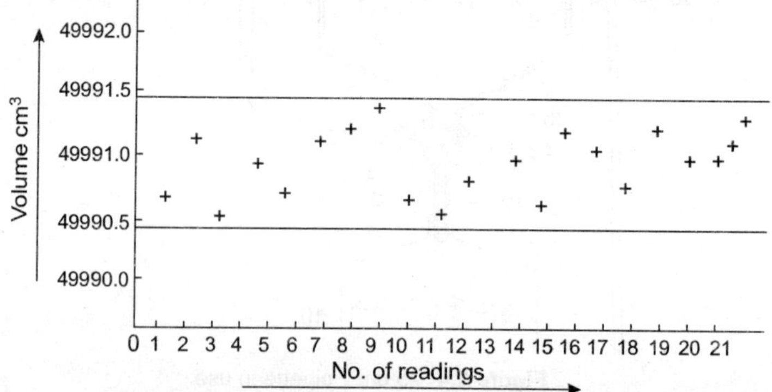

Figure 2.6 Uncertainty in capacity of 50 dm^3 pipette

2.4.3 Pipe Provers (Standard of Dynamic Volume Measurement)

We have discussed the standard measures for static volume measurement, suppose we wish to measure the rate of flow through a transport system than we need to measure time of transit for precisely known volume. Flow meter is such a device, which gives the ratio of volume of liquid passed and the time taken to do so. Quite often in oil fields, it is not feasible to stop the flow put the flow meter in series and measure the rate of flow. So flow meters are permanently installed in the pipeline itself. The flow meters are required to be calibrated without any disruption to flow. For on line measurement of liquid flow, pipe provers are used.

Consider a circular pipe, whose inside surface is smooth and a spherical ball can travel inside the pipe so that there is no slippage of liquid between it and walls of the pipe. This will amount to that the spherical ball will displace the volume of the liquid equal to the product of distance moved by the spherical ball and average cross-section of the pipe. If we measure the time taken by the sphere in moving between the two marks bounding the required volume, we can get flow rate. Such an arrangement is called a pipe prover. So to measure volume of fluid in motion per unit time, a pipe prover is used. Pipe prover is a reference standard for on-line calibration of the flow meters.

2.4.4 A Typical Pipe Prover

A pipe prover essentially [2] consists of the following components and is shown in Figure 2.7.

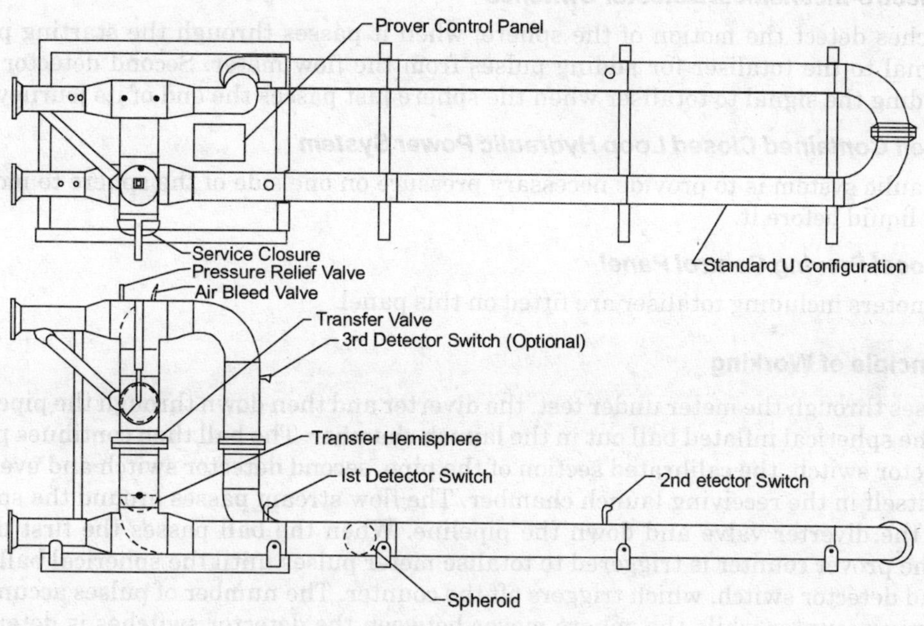

Figure 2.7 Typical pipe-prover

2.4.4.1 Prover Barrel (Volume Measuring Section)

It is a cylindrical pipe, whose inside surface is made smooth. It may be a straight pipe or in the form of 'U' to economise on space. To make the inside surface of the pipe smooth, it is sand blasted or coated with special friction reducing compounds. The coating not only improves the measurement accuracy but also extends the service life of the pipe. The capacity between the two detector switches (marked points) is measured with water with uncertainty better than 0.02%.

2.4.4.2 Transfer Valve and Actuator

The sphere comes to the transfer valve after travelling the distance between two actuators and rest there, till it is actuated again to complete its measurement run. Essentially it has two moving parts, namely the main valve and the transfer hemisphere. As soon as the main valve opens to allow the sphere to pass through, the hemisphere, which is placed in horizontal position to receive the sphere and to block the upward flow through the transfer assembly. Such a system is cost effective to build, to maintain and to operate. The main leak proof valve is closed before the sphere is launched. In this case we can have twice the velocity of conventional pipe prover, since under ideal conditions its sphere launching is a smooth, simple process. This way it will require less proving time.

2.4.4.3 Elastomeric Sphere

The sphere is made with neoprene or polyuerethane. The sphere is filled with glycol or glycol water mixture. Under sufficient pressure its free outside diameter is slightly larger than that of the pipe. It displaces the entire liquid on one side while travelling between the actuators. In fact, the system acts as a piston and cylinder. So the pipe prover is analogous to a positive displacement flow meter.

2.4.4.4 Electro-mechanical Detector Switches

The switches detect the motion of the sphere; when it passes through the starting point, it sends signal to the totaliser for adding pulses from the flow meter. Second detector switch stops sending the signal to totaliser when the sphere just passes the end of its journey.

2.4.4.5 Self Contained Closed Loop Hydraulic Power System

The hydraulic system is to provide necessary pressure on one side of the sphere to move and carry the liquid before it.

2.4.4.6 Local Proving Control Panel

Various meters including totaliser are fitted on this panel.

2.4.5 Principle of Working

Flow passes through the meter under test, the diverter and then down through the pipe prover moving the spherical inflated ball out in the launch chamber. The ball then continues past the first detector switch, the calibrated section of the pipe, second detector switch and eventually deposits itself in the receiving launch chamber. The flow stream passes around the spherical ball, out the diverter valve and down the pipeline. When the ball passes the first detector switch, the prover counter is triggered to totalise meter pulses until the spherical ball passes the second detector switch, which triggers off the counter. The number of pulses accumulated on the prover counter while the sphere moves between the detector switches is determined. The meter factor is the ratio of the calibrated volume to the number of pulses detected by the totaliser.

2.4.5.1 Bi-directional Pipe Prover

The proving cycle of the bi-directional pipe prover is one round trip of the sphere; equivalent to the sum of the pulses accumulated on the prover counter as the sphere travels in both direction between detector switches. The direction of travel of the spheroid is reversed by changing the direction of flow through the prover via a 4-way diverter valve.

2.4.6 Movement of Sphere During Proving Cycle

The Figures 2.8 (1) to 2.8 (5) depict the position of the sphere, transfer valve, transfer hemisphere during a proving cycle.

2.4.6.1 Idle State

Power is off, in this position the main valve actuator is fully extended and the sphere rests in the upper portion of the transfer valve. The transfer hemisphere is in position to receive the sphere.

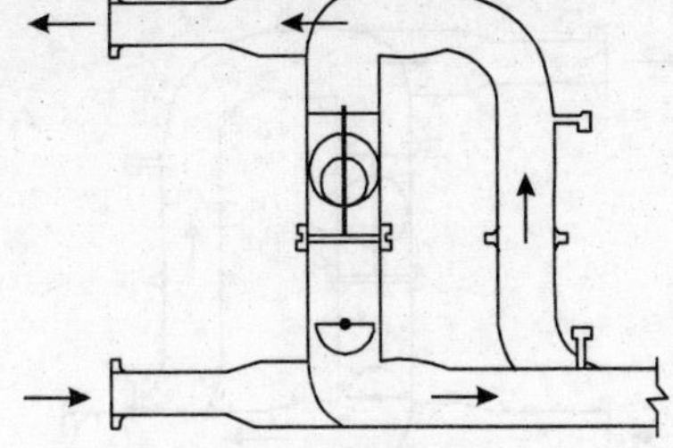

Figure 2.8 (1) Idle state

2.4.6.2 Starting the Unit

After selecting the flow meter to be proved and establishing valve alignment with the prover, the operator resets the prover counter and sets the LAUNCH/ TRANSFER switch to its transfer position. This sends power to unit and retracts the main valve.

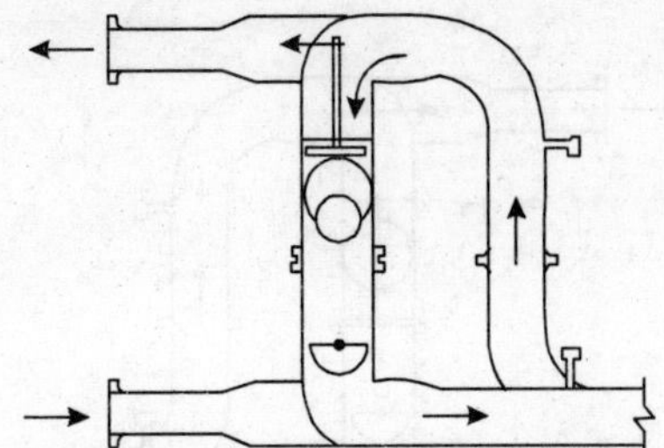

Figure 2.8 (2) Starting the unit

The LAUNCH/TRANSFER switch is set to LAUNCH position. After a slight delay, the actuator moves the main valve towards its seated position.

2.4.6.3 Launching of Sphere

(3A) When the main valve is completely seated and its double seals are compressed; more pressure is created between the seals than in the pipe line. A sensor detects this differential pressure and reacts by sending hydraulic power to hold the valve in its seated position.

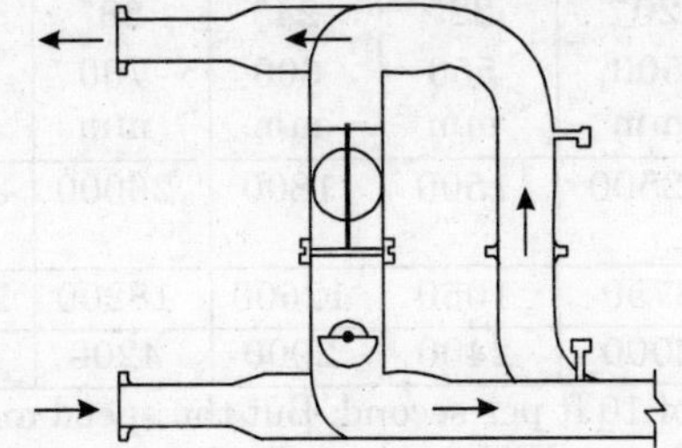

Figure 2.8 (3A) Launching of sphere

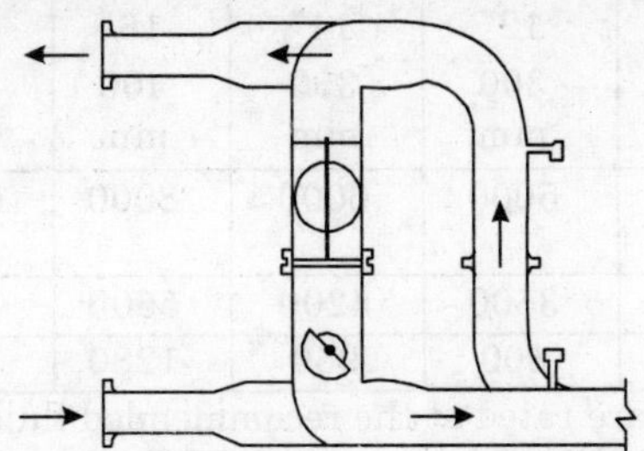

Figure 2.8 (3B) Turning of sphere

(3B) A hydraulic drive then rotates the hemisphere to launch the sphere. As soon as the sphere leaves clear to hemisphere, it returns to its receiving position and the hydraulic power is switched off.

2.4.6.4 Proving the Run

The sphere achieve flow velocity before it enters the measuring section and trips the first detector switch which in turn starts the proving counter. The counter impulses continue till the sphere trips the second detector switch at the end of measurement section.

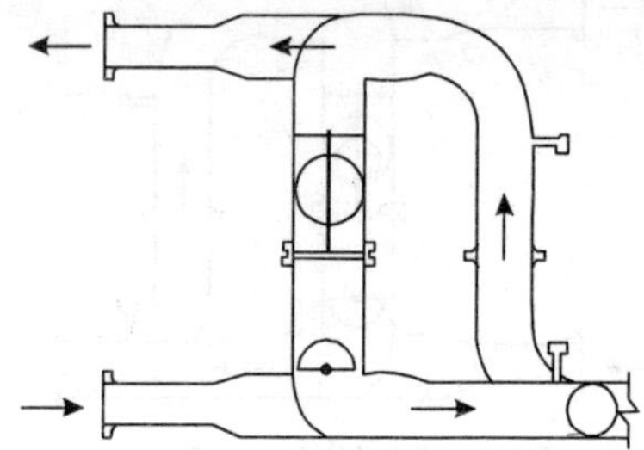

Figure 2.8 (4) Proving the run

2.4.6.5 Stopping the Sphere

As the sphere emerges from the measurement section, the adjacent pipe diameter increases; this increase slows down the sphere velocity and buffers it as it deflects into the upper portion of the transfer valve.

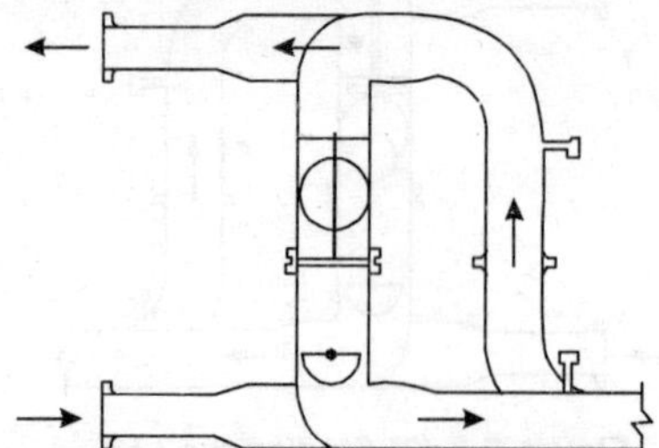

Figure 2.8 (5) Stopping the sphere

The sphere rests there till the operator starts the next proving cycle.

The pipe provers are available in variety of capacity, diameter and flow rate. Some typical examples are given below in Table 2.1.

Table 2.1 Particulars of Pipe Provers

Pipe diameter	12″ 300 mm	14″ 350 mm	16″ 400 mm	18″ 450 mm	20″ 500 mm	22″ 550 mm	24″ 600 mm	28″ 700 mm	30″ 750 mm
Flow rate bph	5000	6000	8000	10000	12500	1500	1800	26000	30000
GPM	3500	4200	5600	7000	8750	1050	12600	18200	21000
m^3/h	800	960	1280	1600	2000	2400	2900	4200	4800
Capacities are rated at the recommended fluid velocity of 10 ft per second. But the speed may be varied from 10 to 15 ft per second. BPH means Barrel per hour, GPM means gallons per minute and m^3/h means cubic metre per hour.									

2.5 SECONDARY STANDARDS CAPACITY MEASURES/LEVEL II STANDARDS

The departments of Legal Metrology in a country maintain level II standards of capacity that are calibrated by the National Metrology Laboratory of that country using the gravimetric method.

In general level II standard capacity measures are both of content type as in India and delivery type as in European countries like France, Germany etc. Content type measure may again be of two types namely single capacity measures as used in India or multiple capacity measures having a graduated scale attached to the neck or neck itself is graduated.

In India, we call these as secondary standard capacity measures. Volume is a derived unit and capacity of a measure is realised through weighing water of known density, using the standards of mass. Nomenclature used in mass measurement at legal metrology level for mass standards is reference, secondary and working standards, so capacity is realised through reference standards of mass, hence these are called one-step lower i.e. secondary standards. These are single capacity non graduated measures used for verifying working standard capacity measures and have the following capacities:

5 dm^3, 2 dm^3, 1dm^3, 500 cm^3, 200 cm^3, 100 cm^3, 50 cm^3, and 20 cm^3

2.5.1 Single Capacity Content Type Measures

2.5.1.1 Material

Normally good heat conducting materials like brass, bronze, copper or stainless steel are used for such purpose. Stainless steel, though, is not a good conductor but is used because of its chemical inertness and resistance to wear and tear. Surface of measures made of stainless steel remain clean for a longer period in comparison to those, which are made from copper, brass or bronze.

2.5.1.2 Shape and Design

Single capacity measures of content type are mostly made in the cylindrical form. The measures are cast or are made of thick sheets. If sheets are used for the measures, metal or wood strips are used to reinforce its vertical wall. This is done to avoid deformation and indentations. A secondary standard capacity measure is shown in Figure 2.9. Its capacity is defined by a cover plate of thick glass, having a through and through hole in its centre and is shown in figure 2.10. The glass plate is quite often called the striking glass. At the time of calibration or use, it is ensured that there is no air bubble in between the liquid and glass plate.

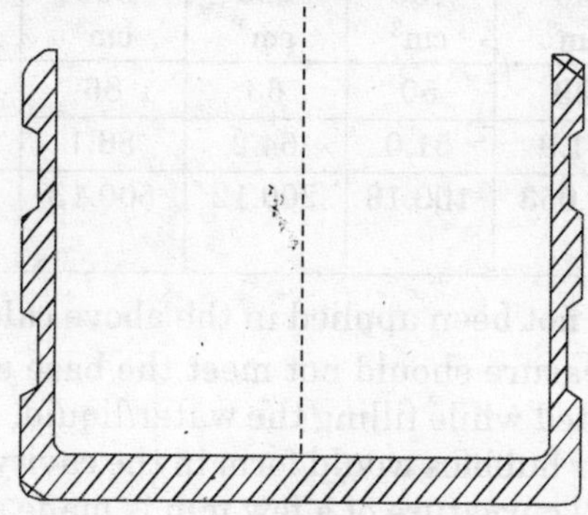

Figure 2.9 Single capacity content type cylindrical measure

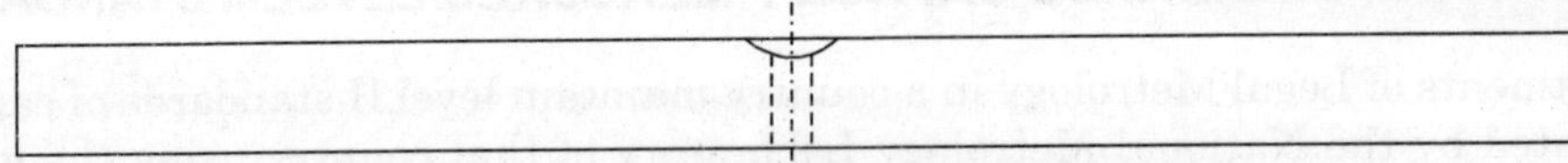

Figure 2.10 Striking glass for the capacity measure

2.5.1.3 Capacity Limit

Normally capacity of such measures is limited to 5 dm^3, otherwise the measure with its contents become too heavy to lift. However capacity of the measure may be as small as 10 cm^3.

Limitation in such measures comes not only from lifting point of view but also from the capacity of the balance required for calibration of such measures. For example a 5 dm^3 measure weighs as much as 10 kg so a balance of 20 kg capacity is required to calibrate such a measure.

2.5.1.4 Design

To keep the surface as small as possible, cylindrical measures are made in such a way that diameter and height of the cylinder are equal. Though wall of a measure is thick, but rim of its upper edge is made thin and well defined. This helps in defining the capacity of measure with better sensitivity. As the well-defined edge reduce the error due to surface tension of liquid. The edges at the rim should not be very sharp to avoid injury. Sharp edges break easily and vary the capacity. Broken edges create problem of seepages between glass plate and itself at the time of filling the measure. Similarly the reinforcing strips should also not have sharp edges otherwise water would remain attached at the junctions at the time of calibration.

The measure is so made that it drains readily and the liquid can be easily poured from it without any splashing or loosing any drop of it. The measure is provided with a cover plate so that it holds and delivers specified volume of water within very close limits and with finer repeatability. The measure holds a definite amount of water under the striking glass when the measure is held in the upright position. The fit is such that when the glass disc is held tightly and the measure is tipped, water does not come out from the measure unless the disc is slid off the opening.

Dimensions of measures from 5 dm^3 to 10 cm^3 are given in Table 2.2 on the assumption that diameter D and height H are almost equal.

Table 2.2 Dimensions of Single Capacity Cylindrical Measures

Capacity	10 cm^3	20 cm^3	50 cm^3	100 cm^3	200 cm^3	500 cm^3	1000 cm^3	2000 cm^3	5000 cm^3
D	23	29	39	50	63	86	108	136	185
H	24.1	30.3	41.9	51.0	64.2	86.1	109.2	137.7	186
Cal. Cap.	10.012	20.014	50.053	100.13	200.12	500.13	1000.3	2000.3	4999.9

Correction due to fillet has not been applied in the above calculations.

The vertical wall of the measure should not meet the base exactly at right angles [3, 4] otherwise a small crevice is created while filling the water/liquid. In small crevices so created, some irremovable and unseen air bubbles would form in the cavity at the base of the measure. To avoid such a situation a small curvature of a few mm is made see Figure 2.11. The volume of the fillet, which is to be subtracted from the calculated capacity, is derived below for a general case.

2.5.2 Volume of the Fillet

Taking radius of the cylindrical measure R, that of the quadrant of small circle r, axis of the measure as y-axis and the horizontal line in upper surface of the bottom of the measure as x-axis, the coordinates of the centre of the quadrant of the circle will be $(R - r, r)$ and equations of the circle as

$$\{x - (R - r)\}^2 + (y - r)^2 = r^2$$

V the volume of the fillet will be the volume of the solid generated by revolving the quadrant of the circle about axis of the measure Figure 2.11. An elementary strip of height y and width δx is revolved about the y-axis of the measure, generating a thin cylinder of radius x, thickness δx and height y, so V the volume of the fillet is given as

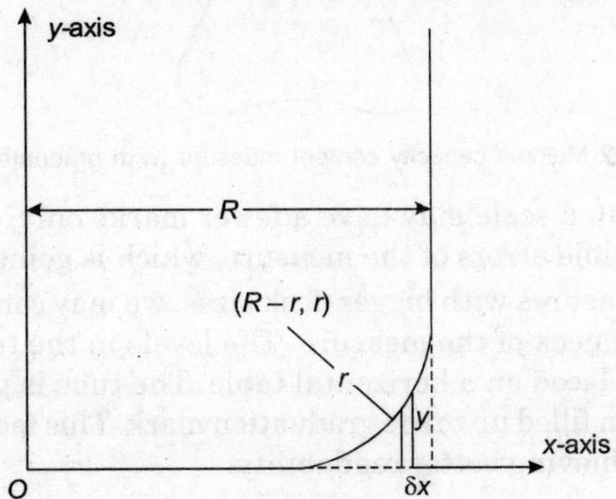

Figure 2.11 Enhanced vertical section of the measure with fillet

$V = 2\pi \int x\, y dx = 2\pi \int x[r - \{r^2 - \{x - (R - r)\}^2\}^{1/2}]dx$, limits of x are from $R - r$ to R

Put $x - (R - r) = r \sin \theta$, giving

$dx = r \cos \theta$ and limits for θ will be from 0 to $\pi/2$, so above integral becomes

$$V = 2\pi \int ((R - r) + r \sin\theta)(r - r\cos\theta)r \cos\theta \; d\theta$$

$$V = 2\pi r^2 \int [(R - r)(\cos\theta - \cos^2\theta) + r \sin\theta\, (\cos\theta - \cos^2\theta)]d\theta$$

$$= 2\pi r^2 \int [(R - r)(\cos\theta - (1 + \cos 2\theta)/2) + r \sin\theta\, (\cos\theta - \cos^2\theta)]d\theta \text{, giving}$$

$$= 2\pi r[(R - r)\{\sin\theta - (\theta + \sin 2\theta/2)/2\} - r\,(\cos^2\theta/2 - \cos^3\theta/3)]$$

Substituting the limits, we get

$$V = 2\pi r^2\, [(R - r)\{1 - \pi/4\} + r(1/2 - 1/3)]$$

$$V = 2\pi r^2[(R - r)(1 - \pi/4) + r/6]$$

2.5.3 Multiple Capacity Content Measures

These are made from metal sheet of stainless steel or galvanised iron. Main body of the measure is a frustum of cone having larger diameter at the base. The upper end of the frustum is joined with a cylindrical neck. Neck is made either of glass or of a metal sheet with a sealed glass window. The window glass is graduated with capacity markings. Normally the mark representing the nominal capacity is at the centre of the graduated scale. One of the designs is given in Figure 2.12.

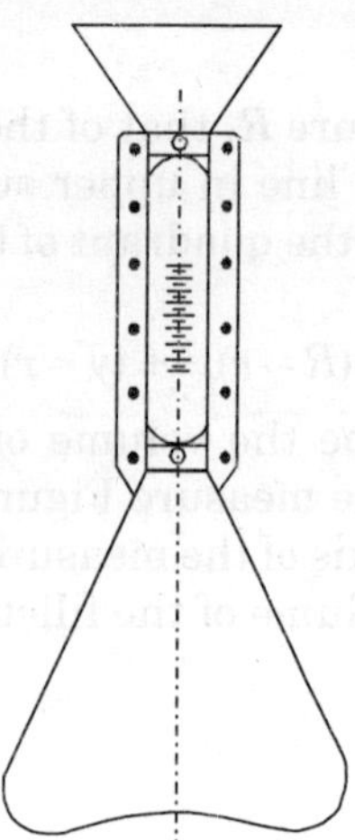

Figure 2.12 Multiple capacity content measure (with graduated neck)

Sometimes the graduated scale may have a fewer marks only, which may represent the limits of maximum permissible errors of the measure, which is going to be verified against it.

For larger capacity measures with bigger neck sizes, we may connect a vertical graduated glass tube in parallel to the neck of the measure. The levels in the tube and the measure will be equal if the measure is placed on a horizontal table. The tube is graduated in terms of the capacity of the measure when filled up to the graduation mark. This facilitates in better visibility of meniscus and helps in achieving better readability.

2.6 DELIVERY TYPE MEASURES

Capacity of a deliver type measure may be as high as 5000 dm^3 and as small as 5 cm^3, but normally a national metrology laboratory will maintain capacity measures from 50 dm^3 to 10 cm^3. The shapes depend upon the material and the way these are going to be used.

Delivery measures of capacity below 10 dm^3 and used in a laboratory are normally made of borosilicate glass. Soda glass is supposed to be inferior to borosilicate in respect of larger coefficient of expansion, inertness to various chemicals and in working out in the required shape.

A measure essentially consists of a suction and delivery tubes and its main body is in the form of a cylinder or sphere, this contains the major portion of volume of the measure. A stopcock is attached at the end of the delivery tube. It may be taken as a magnified version of a bulb pipette with a stopcock. A fixed mark or cut-off device for fixing the capacity is provided at the suction tube. Quite often an over-flow device is used for self-adjustment of water level. The measures may be of single capacity or of multiple capacities.

The measures used for the verification of other measures by volumetric method, may have two graduated marks at the deliver tube, which represent the positive and negative maximum permissible errors for the measure to be tested against it. Sometimes these marks may be on the suction tube. In that case, water level is to be adjusted up to the upper mark for verification of the maximum capacity and to lower mark for minimum capacity permitted for the measure under test.

All measures have circular symmetry. That is these are made by rotating a combination of plane curves including straight line about the axis of the measure. So to have smooth joints, the two curves should meet each other with a common tangent. The main body of the measures is cylindrical, which is obtained by rotating a straight line parallel to and at a distance equal to the desired radius of the measure from the axis of the measure, surmounted on either side with semi-spherical or conical ends. At the end of upper surface, a suction tube is axially attached, while at the bottom of the body, a delivery tube with a stopcock is attached axially. However in both these cases straight lines generating the circular tubes will not meet tangentially the part of the circle generating the spherical surface or the straight line generating the conical surface. So the suction and delivery tubes will not meet the surface of the body smoothly, which will hamper the drainage and flow of the liquid at these joints. To have better drainage along the surface of the body the vertical tubes are joined smoothly with the surface of the body by using a part of a small spherical surface. The method of joining it is to choose two plane curves meeting tangentially. The surface of revolution of this plane curve will have perfect smooth joints.

Two quadrants of the circles will meet each other with a common horizontal tangent if their centres are vertically above each other. Also in this case free ends of the two quadrants will have vertical tangents. Rotating the vertical lines from the ends of the two quadrants generates the vertical walls of the cylindrical body, delivery and suction tubes.

The vertical sections of such measures depicting the three types of measures are shown in Figures 2.13, 2.14, and 2.16.

2.6.1 Measures having Cylindrical Body with Semi-spherical Ends

A vertical section of a measure with semi-spherical shape on each side with a delivery and upper tube is shown in Figure 2.13. To minimise the surface area of the main body the cylindrical portion should have its diameter and height equal. Similarly semi-spherical portion on each

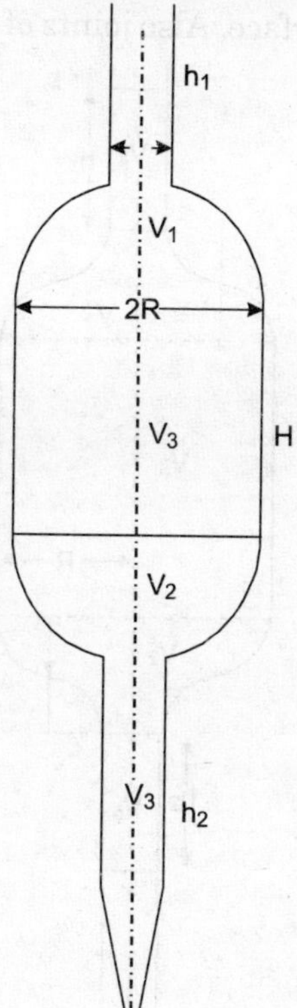

Figure 2.13 Cylindrical capacity measure with semi-spherical ends

side of the cylinder will have the radius equal to that of cylindrical portion of the body. Thus giving

$$V_1 = 2\pi R^3/3 = V_2$$

$$V_3 = \pi R^2 H$$

But H is taken as $2R$ to minimise the surface area.

Hence the total volume of the main body is given by

$$V_1 + V_2 + V_3 = 4\pi R^3/3 + 2\pi R^3 = 10\pi R^3/3$$

So we see the dimensions of glass measures would depend upon the radius of the available tubes of the cylindrical portion. Taking the main body is 90% by volume of the capacity of the measure. One can determine the value R and hence practically complete dimensions of the measure. Diameters and heights of measures of different capacity, main body having volume equal to 90% of its capacity, are given in Table 2.3.

Table 2.3 Dimensions of Capacity Measures with Semi-spherical Surface Figure 2.13

Capacity cm³	50	100	200	500	1000	2000	5000	10000
Diameter 2R	32.4	41	51.6	70.0	88.2	111.2	144.2	190.2
Height H	32.5	41	51.6	70.1	88.3	111.2	144.2	190.1

2.6.2 Measures having Cylindrical Body with no Discontinuity

It has been observed that aforesaid measures will have irregular drainage, especially from the top and bottom parts of spherical surface. Also joints of the suction and delivery tubes with the

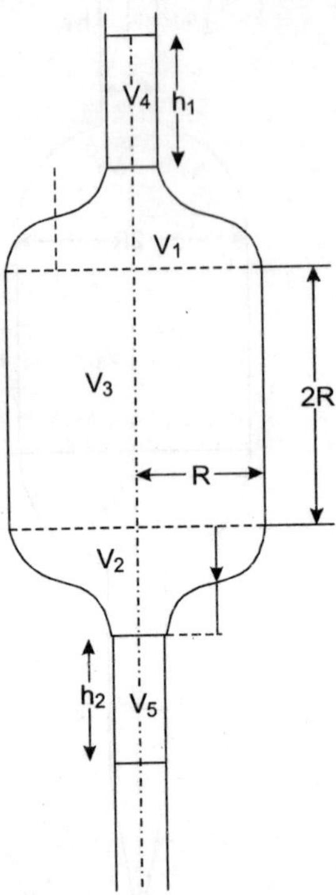

Figure 2.14 Cylindrical capacity measure with smooth surface joints

body are not smooth. To avoid these problems, the vertical section of the body will have two quadrants of the circle arranged in such a way that the two circles meet horizontally and tangents at the other two ends of the quadrants are vertical. The vertical lines extending from the two free ends of this combination of two quadrants will generate the vertical walls of the body and tubes. The surface of revolution made by such a section will naturally have a better drainage property. The vertical section of such a delivery measure is shown Figure 2.14

2.6.3 Volume of the Portion Bounded by Two Quadrants

The vertical section of the measure bounded by two horizontal lines passing through the extreme ends of the two quadrants is shown in Figure 2.15.

Take axis of the measure as y-axis and the horizontal line at the lower extreme end of the quadrant as x-axis, the coordinates of circles of radius r_1 and r_2 will respectively be $(R - r_1, 0)$ and $(R - r_1, r_1 + r_2)$ and corresponding equations of the two circles are

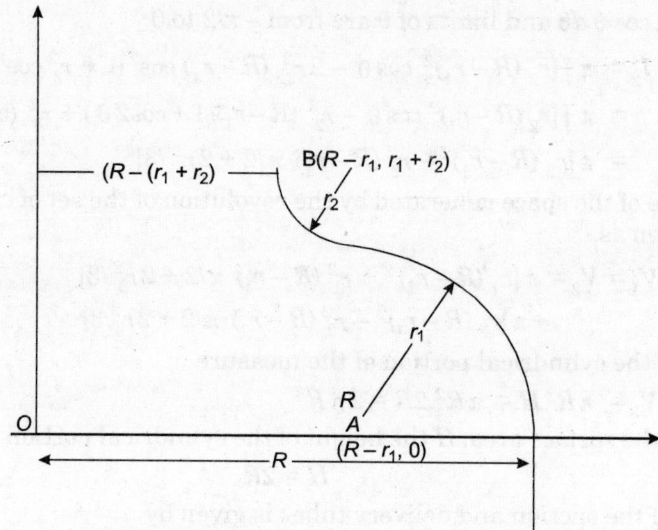

Figure 2.15 Vertical section of the measure at joints

$$\{(x - (R - r_1)\}^2 + y^2 = r_1^{\,2}, \text{ giving}$$

$$x = R - r_1 + (r_1^2 - y^2)^{\frac{1}{2}}$$

$$\{x - (R - r_1 - r_2)\}^2 + \{y - (r_1 + r_2)\}^2 = r_2^2, \text{ giving}$$

$$x = R - r_1 - r_2 - \{r_2^2 - (r_1 + r_2 - y)^2\}^{1/2}$$

Here R is the desired radius of the cylindrical portion of the measure.

The V_1 or V_2 volume of revolution by the area bounded by the tangents at extreme ends the y-axis is given as

$$V_1 = \int \pi x^2 dy = I_1 + I_2$$

Where $I_1 = \pi \int \{(R - r_1) + (r_1^2 - y^2)^{1/2}\}^2\, dy$. Limits for y in this integral are from 0 to r_1

$$I_2 = \pi \int [R - r_1 - r_2 \{r_2^2 - (r_1 + r_2 - y)^2\}^{1/2}]^2 dy$$

Limits for y in this integral are from r_1 to $r_1 + r_2$

For I_1,

Put $y = r_1 \sin\theta$, giving $dy = r_1 \cos\theta \, d\theta$, limits of θ will be 0 to $\pi/2$

So integral I_1 will become

$$I_1 = \pi \int \{(R - r_1) + r_1 \cos\theta\}^2 r_1 \cos\theta \, d\theta$$

$$I_1 = \pi \int [r_1 (R - r)^2 \cos\theta + 2r_1^2 (R - r_1)\cos^2\theta + r_3^2 \cos^3\theta] d\theta$$

$$I_1 = \pi \int [r_1 (R - r_1)^2 \cos\theta + r_1^2 (R - r_1)(1 + \cos 2\theta) + r_3^2 (\cos 3\theta + 3\cos\theta)/4] d\theta$$

$$I_1 = \pi [r_1 (R - r_1)^2 \sin\theta + r_1^2 (R - r_1)(\theta + \sin 2\theta/2) + r_3^2 (\sin 3\theta/3 + 3\sin\theta)/4]$$

Substituting 0 for lower limit of θ and $\pi/2$ for its upper limit, we get

$$I_1 = \pi [r_1 (R - r_1)^2 + r_1^2 (R - r_1)\, \pi/2 + 2r_3^2/3]$$

In integral I_2, we put $y - r_1 + r_2 = r_2 \sin\theta$

giving $dy = r_2 \cos\theta \, d\theta$ and limits of θ are from $-\pi/2$ to 0

$$I_2 = \pi \int [r_2 (R - r_1)^2 \cos\theta - 2r_2^2 (R - r_1)\cos^2\theta + r_3^2 \cos^3\theta] d\theta$$

$$= \pi \int [r_2 (R - r_1)^2 \cos\theta - r_2^2 (R - r_1)(1 + \cos 2\theta) + r_3^2 (\cos 3\theta + 3\cos\theta)/4] d\theta$$

$$= \pi [r_2 (R - r_1)^2 - r_2^2 (R - r_1)\, \pi/2 + 2r_3^2/3]$$

Hence volume of the space generated by the revolution of the set of curves about y-axis is V_1 or V_2 and is given as

$$V_1 = V_2 = \pi [r_1 (R - r_1)^2 + r_1^2 (R - r_1)\, \pi/2 + 2r_3^2/3] + \pi [r_2 (R - r_1)^2 - r_2^2 (R - r_1)\, \pi/2 + 2r_3^2\, 3]$$

Volume V_3 of the cylindrical portion of the measure

$$V_3 = \pi R^2 H = \pi R^2 . 2R = 2\pi R^3 \qquad ...(1)$$

To minimise the surface area, H the height of the cylindrical portion is taken equal to $2R$.

$$H = 2R$$

V_4 volumes of the suction and delivery tubes is given by

$$V_4 + V_5 = \pi r_3^2 h_1 + \pi r_3^2 h_2 \qquad ...(2)$$

Where h_1 and h_2 are lengths of the two tubes. The radius of the suction or delivery tubes r_3 is given as $R - r_1 - r_2$.

Total capacity of the measure will be $= V_1 + V_2 + V_3 + V_4 + V_5$ ...(3)

From equation (1) we calculate the value of R for the desired capacity of the cylindrical portion.The capacity of this portion may vary from 90% to 95% of the nominal capacity of the measure. Higher percentage is chosen for measures of larger capacity. Choose the glass tube having diameter as near to $2R$ as possible. Take R equal to the actual radius of the available tube and calculate the height of the cylindrical portion. Radii r_1 and r_2 are taken as some simple fractions of R. Appropriate values of r_1 and r_2 are substituted in equation (2) and new values of the length of suction and delivery tubes are calculated.

For example, let $r_1 = 7R/10$ and $r_2 = R/10$ giving r as

$$r_3 = R - R/10 - 7R/10 = R/5$$

So volume of the portion of the measure bounded by revolving quadrant of a circle of radius $7R/10$ is given by

$$I_1 = 0.5225746\, \pi R^3$$

And volume of the portion of the measure bounded by revolving upper quadrant of a circle of radius $R/10$ is given by integral I_2 giving us

$$I_2 = 0.00495\,546\;\pi R^3$$

So total volume V_1 is given by

$$V_1 = 0.5275292\,\pi R^3$$

So volume of the main body = $2V_1 + V_3 = 3.0550584\,\pi R^3$

Here V_3 is the volume of cylindrical portion of height $H = 2R$ so

$$V_3 = 2\pi R^3$$

Now assume that volume of the body of the measure is 92.5% of the nominal volume and height of the cylindrical portion is equal to $2R$ and calculate the value of R, say for 50 l measure R will come out to be 16.8998 cm, which when rounded off in mm will become 16.9, giving the main body volume $2V_1 + V_3 = 46326.6$ cm^3.

If the available tube is of diameter 34 cm, then volume of the main body will become 47153.8 cm^3.

If we wish to maintain volume of the main body as before, with $R = 17$, new value of H will be 33.

We have taken $r_3 = R/5 = 3.4$

If a tube of diameter 6.8 is available, then total lengths of suction and deliver tubes will be 3673.4/ π $(3.4)^2$ cm = 101.1 cm. This is rather too much.

So we assume that capacity of the body is 97.5 % of the nominal capacity. Giving us $3.0550584\pi R^3 = 48750$, which on simplification gives

$$R = 17.18969 \text{ cm}$$

If a tube of 17.2 cm diameter is available and we keep diameter of the suction and delivery tubes as 3.4, then lengths of the two tubes together will be

$$1250/\pi \times 4.0 \times 4.0 = 34.41$$

This appears to be reasonable.

We take radii of the two tubes respectively as 17.2 cm and 3.4 cm.

From here we can calculate the value of H as follows:

Volume of portion generated by revolving the two set of quadrants = 16866.0 cm^3

Cylindrical portion = 31 883.97544

Giving H = 34.3 cm

So dimensions are

$R = 17.2$, $r_3 = 3.4$ cm and height $H = 34.3$ cm

It may be noted that volume generated by the quadrant of the smaller radius is only 4.61 cm^3.

2.6.4 Measures having Cylindrical Body with Conical Ends

Another shape of the measure may be a cylindrical body surmounted by a frustum of the cone on each side. The slant side of the conical portion makes about 45° with horizontal. The vertical section of such a measure is shown in Figure 2.16. The joints of the suction and delivery tubes are made slightly rounded and smooth. Tangents at the ends of a quadrant of the circle are mutually perpendicular to each other. So for joining the horizontal and vertical parts of the section of a measure, the quadrant of circle is often used. Similarly a part of the circle may be used so that tangent at one end is vertical and at the other is in the direction of the slant height of the section of the cone. Part of the circle so that tangent at one end is vertical and the other makes an angle α with vertical is shown in Figure 2.16.

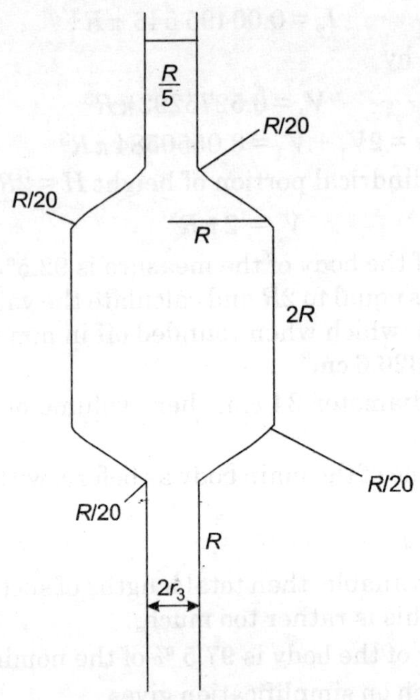

Figure 2.16 Cylindrical capacity measure with conical ends

Volume V_2 of the cylindrical portion is $\pi R^2 H$, if we take $H = 2R$, then V_2

$$V_2 = 2\pi R^3$$

V_1–Volume of the frustum of the cone, having a radius of the base equal R and that of top $= r_3$, as

$$V_1 = \pi(R^2 + r_3^2 + R\, r_3)\,(R - r_3)\tan \alpha/3$$

α is the semi-vertical angle of the cone.

Neglecting the contribution of change in volume due to rounding off the corners, the volume of the measure is $= 2V_1 + V_2 + V_3$.

Here V_3 is volume of the suction and delivery tubes.

Dimensions of the measures of this shape have been be worked out in a way similar to the one described in section 2.6.3 and are given in Table 2.4.

Table 2.4 Dimensions of Cylindrical Measures with Conical Ends

Capacity	50 dm^3	20 dm^3	10 dm^3	5 dm^3	2 dm^3	1 dm^3	0.5 dm^3	0.2 dm^3	0.1 dm^3	0.05 dm^3
R mm	172	126	100	80	59	46	38	28	22	17
r_3 mm	34	25	20	16	12	0.9	0.85	5	4.5	3.5
$h_1 + h_2$ mm	344	255	199	155	110	98	70	63.6	39.2	32.5
Volume of the body is 97.5 % of the total capacity, 2.5% of capacity is distributed in the suction and deliver tubes $r_1 = 7R/10$, $r_2 = R/10$.										

2.7 SECONDARY STANDARDS AUTOMATIC PIPETTES IN GLASS

2.7.1 Automatic Pipettes

Capacity measures in glass, which are commercially available, are manufactured on the basis of the principles discussed above. In addition to above discussed basic structure, a three-way stopcock and an overflow device to define the capacity of the measure are incorporated in the secondary standards automatic pipettes. Three-way stopcock is used for delivery and filling under gravity. The pipettes of nominal capacity 50 dm^3, 20 dm^3, 10 dm^3, 5 dm^3, 2 dm^3, 1 dm^3, 0.5 dm^3, 0.2 dm^3 and 0.1 dm^3 and 0.05 dm^3 are available [6]. These are made from a specific batch of borosilicate glass of known coefficient of linear expansion.

The pipette has three main parts:

(1) Main body is a cylindrical tube joined with the smaller tubes with no discontinuities.

(2) Delivery arrangement with a three-way stopcock and delivery jet and

(3) The upper tube with a device to collect over flown water.

The capacity of the pipette is the volume of water filling the delivery jet; body and outflow jet up to the brim, this condition is obtained by overflowing a small amount of water.

The collecting device is shown at the left hand top of Figure 2.17.

Dimensions of such pipettes may be worked out assuming the bulb-main body is cylinder joined with the smaller tubes with no discontinuity. For details section 2.6.3 may be referred to.

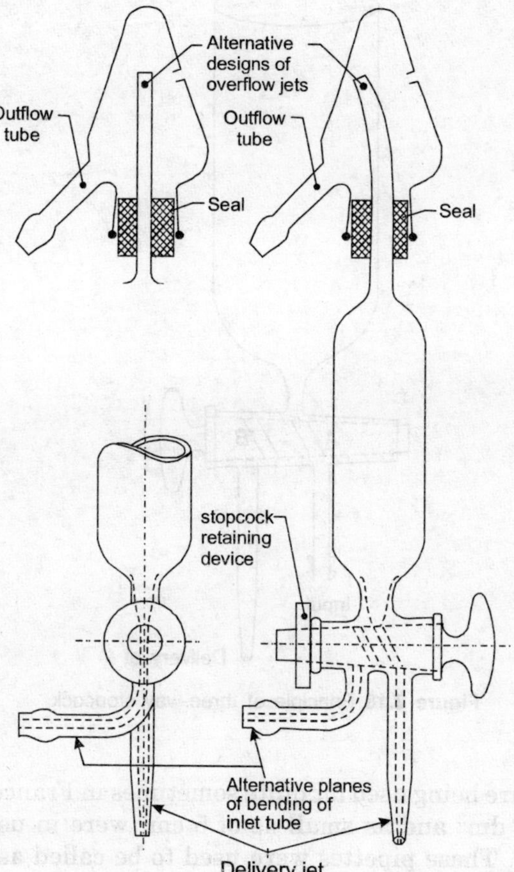

Figure 2.17 A typical automatic pipette

2.7.2 Three-way Stopcock

The pipette is to be filled and deliver water from below. That makes it necessary that two glass tubes are fixed to the lower side of the stopcock. One is called the input tube and other the delivery jet. The stopcock barrel has two parallel but slanting through and through holes say A and B. In present position of the stopcock in Figure 2.18, the input tube is connected with the body, so in this position, the pipette is filled with water from its reservoir under gravity. When the stopcock is turned through 180°, the present lower end of hole B connects to the body of the pipette and the upper end of hole B connects to the delivery jet, hence the pipette will deliver in this position. In all other positions none of the hole will be in position to connect any of the input or output tubes. So in those positions pipette is neither being filled nor delivering. The stopcock is normally kept horizontal i.e. 90° to the present position, when we wish to keep the pipette disconnected from delivery or input tubes. So the stopcock has three positions:

(1) body of the pipette is connected to the input tube.

(2) body of the pipette is connected to the delivery jet and finally

(3) not connected to either of the input tube or output jet.

This is why this is called as a three-way stopcock.

At the top of the pipette there is a tube, in an overflow pipette the tube extends so that capacity of the pipette is defined till the water overflows from it.

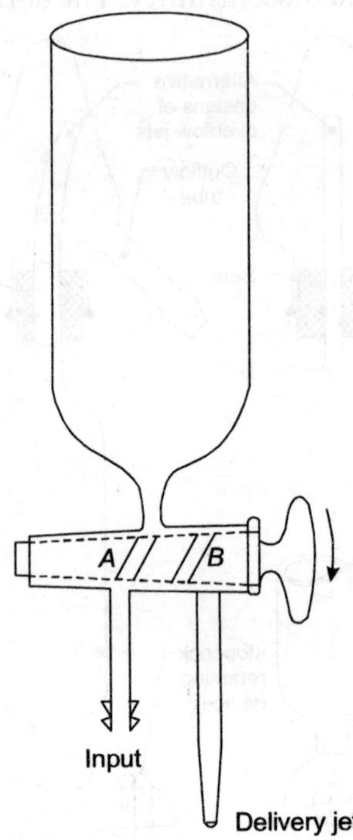

Figure 2.18 Principle of three-way stopcock

2.7.3 Old Pipettes

Glass automatic pipettes are being used for quite sometimes in France. In fact glass pipettes of capacities of as big as 50 dm^3 and as small as of 5 cm^3 were in use in different metrology laboratories [6] of France. These pipettes were used to be called as secondary standards of capacity. The unit of capacity used was litre, which was defined as the space occupied by 1 kg of water at the temperature of its maximum density. A set of such measures is mounted on a wooden board fixed to the wall. Rubber tubes are used to fill the measures and to take away waste water. Reservoir of water is kept about 2 metres higher than the highest end of the pipette. Rubber tubes are used to take water from reservoir to any of the pipettes in the set. A pinchcock is used to stop flow of water from the reservoir.

Some of the secondary standard measures used in France with dimension in mm are shown in Figure 2.19. Body of the similar pipettes is shown in Figure 2.20.

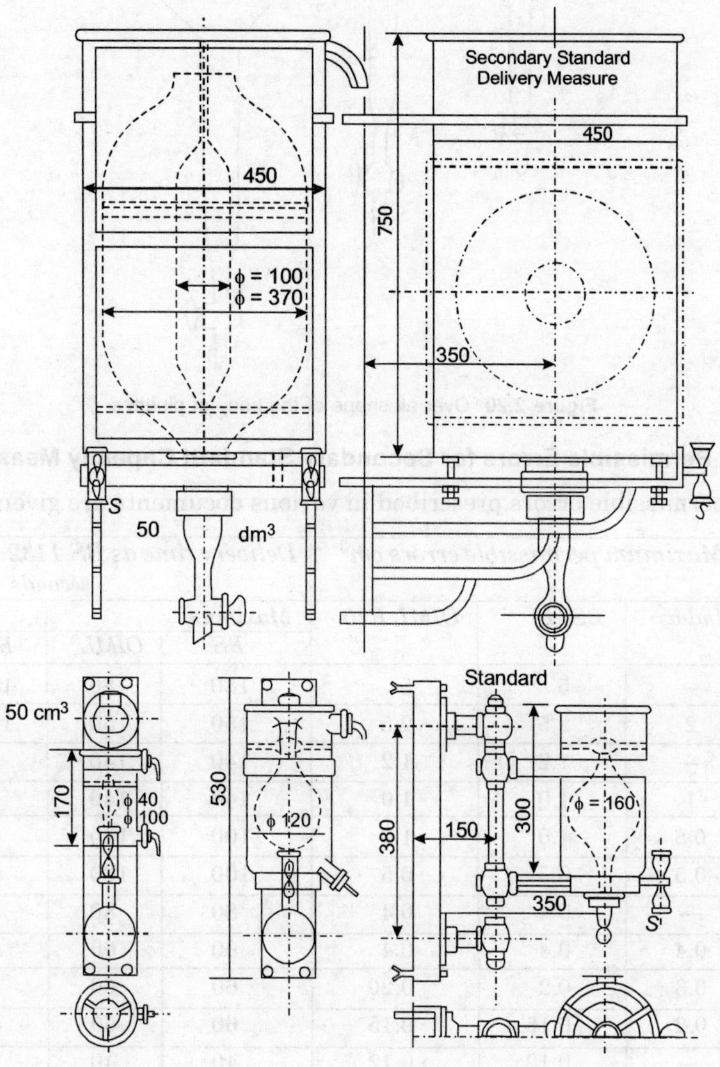

Figure 2.19 Dimensions of bulbs of 50 dm³ to 50 cm³ pipettes

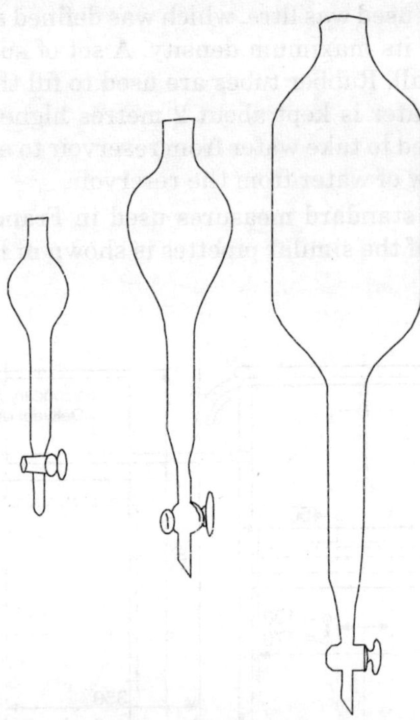

Figure 2.20 Over all shape of the body of pipettes

2.7.4 Maximum Permissible Errors for Secondary Standard Capacity Measure

The maximum permissible errors prescribed in various documents are given below.

Nominal Capacity dm^3	*Maximum permissible errors cm^3*			*Delivery time as BS 1132 / OIML R20 in seconds*			
	India	*BS113*	*OIML R20*	*Maximum*		*Minimum*	
				BS	*OIML*	*BS*	*OIML*
10	—	5	5	180	180	120	120
5	2	2.5	2.5	150	150	100	100
2.5	—	1.2	1.2	140	140		80
2	1	1.0	1.0	140	140	80	80
1	0.8	1.0	1.0	100	100	60	60
0.5	0.5	0.5	0.5	100	100	60	60
0.250	—	0.4	0.4	80	80	50	50
0.200	0.4	0.4	0.4	60	60	30	30
0.100	0.3	0.2	0.20	60	60	30	30
0.050	0.2	0.15	0.15	60	60	30	30
0.025	—	0.12	0.12	40	40	12	20
0.020	0.10	0.12	0.12	30	30	15	15
0.010	—	0.08	0.08	30	30	15	15
0.005	—	0.06	0.08	20	20	10	10

2.8 WORKING STANDARD AND COMMERCIAL CAPACITY MEASURES

2.8.1 Working Standard Capacity Measures used in India

In India, the working standard capacity measures of the state Departments of Legal Metrology are just simple cylinders with a striking glass, almost similar to Secondary Standard Measures as shown in Figures 2.9 and 2.10.

These are made of thick sheets of copper reinforced with wooden rings. Capacity of these measures is from 10 dm^3 to 20 cm^3. A set of graduated pipettes is also provided. The measures are used to verify all types of commercial measures, by volume transfer method. Working standard measures are verified every year against secondary standard measures.

In addition, there are some check measures, of capacity from 5 dm^3 to 1000 dm^3. The shape of the measure depends upon its capacity. For 5 dm^3 to 20 dm^3 capacity, these are conical type. The shape is similar to those of commercial conical measures. Beyond 20 dm^3, these are delivery measures of different shapes.

2.8.2 Commercial Measures

In general, commercial measures are designed keeping in view its end use. For example measures used for trading petroleum liquids are conical in shape. One such measure is given in Figure 2.21.

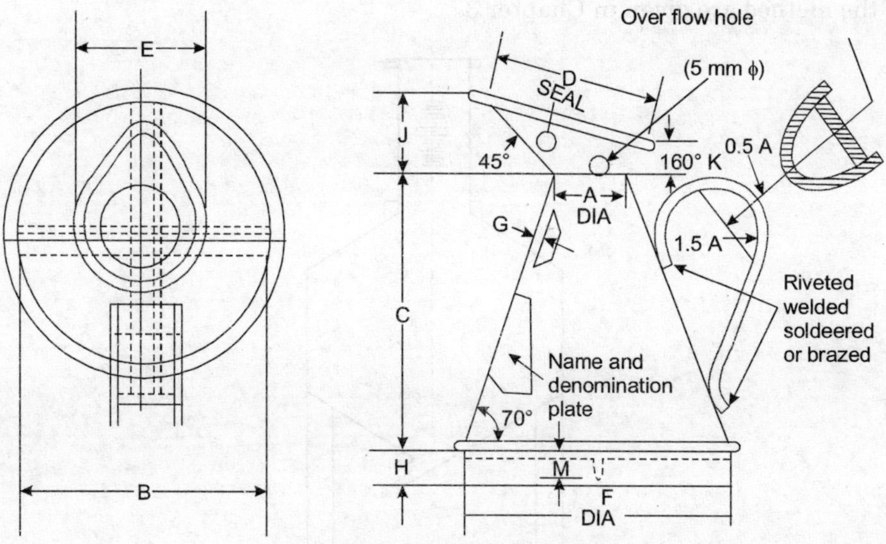

Figure 2.21 Commercial conical measure

The measures, which are used by dipping in to the liquids, are cylindrical in shape with long handles. One such measure is shown in Figure 2.22. Their capacity ranges from 1 dm^3 to 50 cm^3.

In some measures, the liquid is poured in or taken from the wide mouth vessels by swiping and then delivered by pouring. Such measures are also cylindrical but with smaller handles. Capacity of these measures is also 1 dm^3 to 50 cm^3. One of them is shown in Figure 2.23.

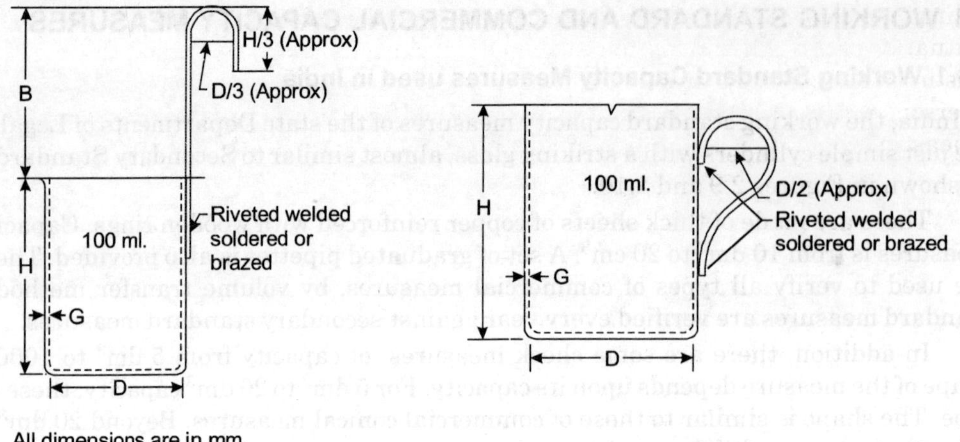

Figure 2.22 Dipping type

Figure 2.23 Pouring type

2.9 CALIBRATION OF STANDARD MEASURES

2.9.1 Secondary Standard Capacity Measures

These measures are calibrated by gravimetric method using distilled water as medium. The details of the method are given in Chapter 3.

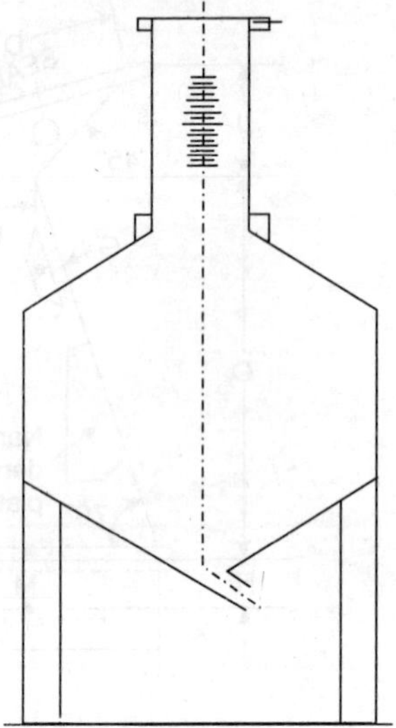

Figure 2.24 Secondary standard capacity measures

2.9.2 Working Standard Measures

These measures are verified against secondary standard capacity measures, by volume transfer method. Details of the method and applicable corrections are given in Chapter 4.

Sometimes, the capacity of secondary standard measure is much smaller than the measure under test so a multiple volume transfer method is used. In this case, it is very important to eliminate the un-forced errors in calibration of the secondary standard measure. A similar situation occurs when verifying a commercial capacity measure against the working standard. As in this process, a small error is multiplied linearly; hence proper training is vital for the persons engaged in verification of working standard measures against secondary standard measures.

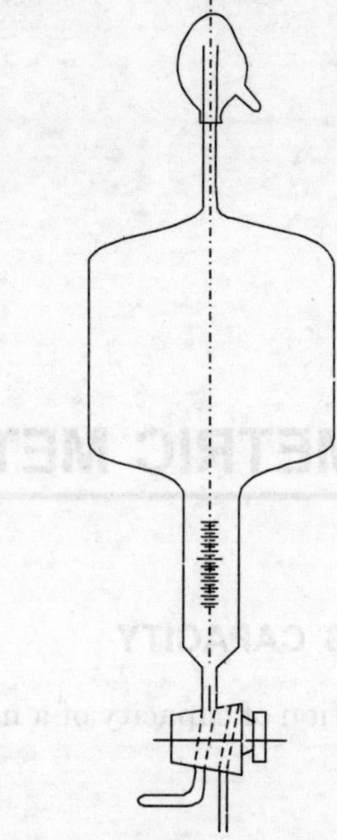

Figure 2.25 Working standard measures

REFERENCES

[1] Gupta S V 2003, A Treatise on Standards of Weights and Measures, pp 159,166, 626 and 627, Commercial Law Publishers, New Delhi.

[2] Pamphlet, 1985, Smith Meter Incorporation, Pennsylvania.

[3] Raj Singh et al. Study of a 50 Litre Automatic Overflow Pipette. The primary standard for volumetric vessels MAPAN- The Journal of Metrology Society of India, 6, 1991, 41-54.

[4] Cook A H and Stone N W M, 1957 Precise measurement of the density of mercury at 20°C: I, absolute displacement method Phil. Trans. R. Soc. A 250 279-323.

[5] Cook A H, 1961; Precise measurement of the density of mercury at 20°C: II Content method; Phil. Trans. R. Soc. A 254 125-153.

[6] Renovation des Etalons de Capacite, Chapter 24 to 26, Des Bureaux de verification avant la revision (in French).

[7] BS 1132:1987 British Standard Specifications for Automatic Pipettes.

GRAVIMETRIC METHOD

3.1 METHODS OF DETERMINING CAPACITY

There are two methods for determination of capacity of a measure, namely:

(*i*) Gravimetric Method, and

(*ii*) Volumetric Method.

3.2 PRINCIPLE OF GRAVIMETRIC METHOD

For precise determination of capacity of volumetric measures, the gravimetric method is used. In this method capacity is determined by weighing the volume of distilled water, which the measure contains or delivers, at the temperature of measurement and then a correction is applied to apparent mass of water to convert the result in the capacity of the measure at the reference temperature. In case of very small measures, mercury is used in place of water to achieve the desired precision.

To calculate the correction to be added to the observed mass of water the following parameters are taken in to account:

Density of water at different temperatures at atmospheric pressure,

Coefficient of volume expansion of the material of the measure,

Density of the material of mass standards used,

Density of air at the temperature and pressure of measurement, and

Reference temperature.

3.3 DETERMINATION OF CAPACITY OF MEASURES MAINTAINED AT LEVEL I OR II

Standard capacity measures maintained at levels II or I are calibrated by using gravimetric method. As reference standard weights are used for calibrating these, so these may be called as secondary standards rather than the reference standard capacity measures. As mentioned earlier, in gravimetric method, mass of water, required to fill completely or to a predetermined graduation line of the measure, for content measures, or mass of water delivered from the delivery measure, is determined. To change the apparent mass of water so obtained to the actual capacity of the measure reference temperature there are two methods. First method is to find out a factor, which is to be multiplied to the mass of water to give the capacity of the measure at reference temperature. Another method is to find out a correction to be added to mass of water to give the capacity of the measure at reference temperature.

To determine capacity of the measure at its reference temperature, normally additive correction is used when water is taken as medium but when mercury is taken as medium, a multiplying factor is used.

The formulae for the correction to be added or factor to be multiplied are being derived in section 3.4.

In addition of temperature, density of water depends upon the purity of water. Therefore, distilled water is used for calibrating the measures.

3.3.1 Determination of the Capacity of a Delivery Measure

The pure distilled water is filled against gravity as shown in the Figure 2.4 of Chapter 2 to a level well above the graduation mark. The rise of water level in the measure is minutely observed. The meniscus formed by water should rise uniformly without any kink at any place. Uniform rise in meniscus ensures the cleanliness of the measure. The filling rate should be such that time required to fill the measure is almost equal to the delivery time. A cleaned vessel is taken. Its capacity should be greater than the expected volume of water to be delivered by the measure. For determination of mass of water delivered, method of substitution weighing should be followed.

In case of a two-pan balance, standard weights equivalent to the mass of water to be delivered by the measure along with the empty cleaned vessel are placed on the same pan. One gram of standard weight for every one cm^3 of the nominal capacity of the measure under test is placed on the pan with the empty vessel. Placing similar weights on the opposite pan counterpoises the balance. Three turning points– two at the extreme left and one of the extreme right are taken and recorded, let the rest point of the balance be R_s and corrected apparent mass values of the weights placed be M_s. The water level is adjusted to the predetermined graduation mark of the measure and then delivered into the vessel. Care is taken that water jet falls on the wall of a slightly inclined vessel to avoid splashing and dissolution of air. Time equal to the drainage time as provided in the relevant specification is allowed and the last drop of water is taken by touching the tip of the measure with the wall of glass vessel. It is then placed on the pan of the balance. To restore the balance some standard weights will have to be removed. The standard weights removed should be such that equilibrium positions in the two weighing are almost equal. This way error in the balance scale or error due to sensitivity figure of the balance is very much reduced. If the corrected value of the apparent mass of weights left in the pan be M_u and equilibrium point be R_u then mass of water m delivered is given by

$$m = M_s - M_u + (R_s - R_u)S$$

S is the sensitivity figure of the balance. For the purpose of calculating rest points, the extreme left of scale of the two pan balance has been taken as zero.

In case of single pan balance, put the clean empty vessel on the pan and find equilibrium point by adjusting the knob-weights. Let the equilibrium point be I_s. Corrected value of the knob weights is M_s. Deliver the water in to the vessel as described above and put it on the pan. Now some knob-weights are to be lifted to bring back the equilibrium, let it be I_u. If M_u be the mass value of the knob-weights the apparent mass m of water delivered, for a two pan balance, is then given by

$m = M_u - M_s + (I_u - I_s)$, where I_u and I_s are indications of the balance in terms of the same unit of mass as used for mass standards.

The temperature of water was taken when the measure was filled, so it should be ensured that the temperature of water does not change while adjusting its level up to the desired graduation mark and collecting it in the vessel. The correction obtained from the intersection of the appropriate row and column of the relevant Table 3.1 to 3.24 is multiplied by the nominal capacity of the measure and then added to the mass m of water to obtain the capacity of the measure at the graduation-mark at the reference temperature.

When mercury is used instead of water then mass m of mercury is multiplied by the proper factor obtained from the Tables 3.31 to 3.46.

The corrections, in the aforesaid tables, are given for unit capacity of the measure.

3.3.2 Determination of the Capacity of a Content Measure

A factor, which affects the repeatability of determining the capacity of a measure, is the cleanliness and condition of the surface of the measure. It may be emphasized here that both outer and inner surface of the measure will matter in the repeatable capacity determination. Outer surface if not clean, the mass of the thin film of water remaining in contact, will be varying , due to change in humidity and also outer surface may catch up some dust or any other foreign material particles during the weighing process.

In India, the secondary standards are of content type and each has its own striking glass. So the striking glass provided with the measure should also be properly cleaned on both sides.

3.3.2.1 Determination of Apparent Mass of Water

Step by step method given below though is with specific reference to secondary standard capacity measures used in India, is applicable to calibrate any content measure of this type. Mass of water required to completely fill the capacity measure is determined as follows:

Step 1 : The measure under test with its striking glass and the measure having similar outer surface together with, if possible, are taken. The other measure is not required if a single pan balance is used.

Step 2 : On the right hand pan of the balance, place the measure under test with its striking glass and the standard weights at the rate of 1 g per cm^3. While on the left pan the similar measure and sufficient weights are placed so that the pointer of the beam balance swings within the scale almost equally to the midpoint of the scale. In case of single pan balance also reference standard weights at the rate of 1 g per cm^3 should be placed with secondary standard measure. This way, mass values of the built-in weights will not be required, mass of water will be obtained in terms of reference weights. In case of smaller measures like any volumetric glassware, built-in weights may be used provided these are of OIML F1 class or better. MPE in integeral gram weight should not be more that one part in 10^5.

Step 3 : Take at least three turning points– two at the extreme left and one of the extreme right. Record the scale readings and mass of standard weights. Let the rest point be (R_s) and mass of standard weights be M_s.

Step 4 : Take out the measure and fill it with triple distilled water. The water was kept in the same room over night so that it acquires the room temperature. Take a cleaned glass rod, the water is filled in such a way that it moves along the glass rod. No splashing or entrapping of air should take place.

Step 5 : Remove all air bubbles sticking to the walls and bottom of the measure with the clean glass rod.

Step 6 : Measure the temperature of water with a thermometer graduated in steps of 0.1 °C, let it be T_1.

Step 7 : Continue filling water up to the brim so that the water forms a slight convex surface.

Step 8 : Slide horizontally the striking glass, supplied with the measure to remove excess of water. Ensure that there is no air bubble between the surface of water and striking glass. Presence of air bubbles indicates that more water is needed. So add water in the spherical cavity of the striking glass and press it, air will come out and water will go in. If the method of putting water in the cavity and the striking glass does not remove all the bubbles. Remove the striking glass, fill more water and repeat the process.

Step 9 : Clean the measure from all sides with an ash–less filter paper. Special attention is to be paid to the bottom and the sides of the rings provided to strengthen the measure. Top of the striking glass is also properly cleaned. Ensure that there are no traces of water on any side especially on bottom and on the striking glass. The handling of the measure should be minimal. As handling changes the temperature of the measure and air bubble will appear, prolonged handling may also change water temperature then excess water will come out.

Step 10 : Put the measure in the right hand pan, remove the necessary weights, so that pointer swings within the scale. This way, water has been substituted by the standard weights. Take observations and calculate the rest point. Let it be R_u and mass of weights remained in the pan is M_u.

Then apparent mass of water m is given as

$$m = M_s - M_u + (R_s - R_u).S$$

Here S is the sensitivity figure of the two pan balance for that load.

In case of single pan balance, if the secondary standard measures are being calibrated, then reference standard weights at the rate of 1 g per cm^3 should be placed with measure.

This way, mass of water will be obtained in terms of reference weights and mass values of the built-in weights will not be required. In case of other volumetric measures, built-in weights may be used provided these are of OIML F1 class or better or MPE in integral gram weight is not to be more than one part in 10^5. Mass of water will be give as

$$m = M_s - M_u + (I_u - I_s), \text{ if reference weights are used.}$$

$$m = M_u - M_s + (I_u - I_s), \text{ if dial weights are used.}$$

Step 11 : Take out the measure, remove the striking glass by sliding and take the temperature of water. Let it be T_2.

Step 12 : Take the mean of T_1 and T_2 say it is T.

Step 13 : From the knowledge of T–the mean temperature of water, the correction, at the intersection of proper column and row from the table corresponding to reference temperature, density of standard weights and coefficient of expansion of the material of the capacity measure, is taken. The correction value so obtained is multiplied by the capacity of the measure and is added to apparent mass of water obtained. The resultant sum divided by 1000 will then give the capacity of the measure at the reference temperature. The corrections in all tables pertain to unit volume, which may be one m^3, dm^3 or cm^3 then corresponding corrections are in kg, g and mg respectively and apparent mass of water should correspondingly be calculated in kg, g or mg giving us

$V_{27} = (m + c)/1000$, where c is the correction arrived at by the aforesaid method.

If necessary, additional correction due to change in air density is also applied from the appropriate table from tables 3.25 to 3.26.

3.4 CORRECTIONS TO BE APPLIED

3.4.1 Temperature Correction

Let apparent mass of water as weighed against standard weights of density D be m. Let ρ be density of water and V_t be the volume of delivered/contained water at temperature t°C, then $m(1-\sigma/D) = V_t\,(\rho-\sigma)$, here σ is the density of air at the temperature and pressure of measurement.

If V_{ts} is the capacity of the measure at reference temperature t_s°C and α is the coefficient of cubical expansion of the material of the measure, then

$$V_t = V_{ts}\,\{1 + \alpha\,(t - t_s)\}$$

Substituting V_t in the above equation, we get

$$m(1 - \sigma/D) = V_{ts}\{1 + \alpha\,(t - t_s)\}\,(\rho - \sigma) \qquad ...(1)$$

V_{ts} can be calculated from equation (1), if the values of α , ρ , σ and D are known.

However to use this equation for each measurement is rather not practical. Let us consider a quantity c in kilograms such that when it is added to m – the mass of water in kilograms, than the sum is equal to 1000 times of the capacity of the measure V_{ts} at reference temperature t_s. V_{ts} is in cubic metre. The explanation is as follows:

Had the density of water been exactly 1000 kg/m^3, then V_{ts} in cubic metres should have been numerically equal to the mass of water in kilograms divided by 1000. So there is a quantity c in kilograms, which when added to m–the actual mass of water in kilograms to give a number equal to 1000 times of V_{ts}.

Giving $\quad m + c = 1000\,V_{ts}$. ...(2)

This explanation is necessary to justify the equation dimensionally. In this case both arms of the equation are in terms of unit of mass. Substituting the value of m from the above, we get

$$c = 1000\,V_{ts} - V_{ts}\,\{1 + \alpha\,(t - t_s)\}\,(\rho - \sigma)\}/(1 - \sigma/D)$$

$$= 1000\,V_{ts}[1 - \{1 + \alpha\,(t - t_s)\}\,\{(\rho - \sigma)/1000\}\}/(1 - \sigma/D)] \text{ in kilogram} \qquad ...(3)$$

Here we see that value of correction c is directly proportional to capacity of the measure but is a function of coefficient of volume expansion of material of the measures, density of mass standards used, reference temperature and of course on density of water at the temperature of measurement.

We discussed in Chapter 1 that volume/capacity measurements are carried out at two reference temperatures namely 20°C and 27°C. So we should calculate the correction c values for 27°C and 20°C separately.

Similarly, two values of density are taken for density of materials used for standard weights. Larger number of countries uses standard weights of stainless steel having density of 8000 kgm^{-3}. But still there are some developing countries, which use brass or similar materials for their standard weights and take 8400 kgm^{-3} for density D. Therefore it is necessary to prepare correction tables for two types of standard weights. In addition there are quite a few materials used in construction of these measures and coefficient of volume expansion of each material is different. For example glass used for volumetric measures has four different coefficients of expansion. Besides there are metallic measures. Hence correction Tables 3.1 to 3.24 have been constructed for all combinations of two values each of density of standard weights and reference temperature for different values of coefficients of volume expansion (ALPHA). The values of ALPHA, reference temperature and density of mass standards used have been indicated on the top of each of the tables from 3.1 to 3.24. Every correction table is for one unit volume, which may be m^3, dm^3 or cm^3.

The correction found at the intersection of the temperature added shall be in kg, g or mg according to mass of water expressed in kg, g or mg. From equation (2), we see that the sum of mass of water plus the applicable correction divided by 1000 will respectively give capacity in m^3, dm^3 or cm^3 at reference temperature indicated at the top of the table.

That is, the correction will be in grams and added to the apparent mass of water in grams weighed against the density of the standard weights as indicated on the top of the tables, the sum divided by 1000 will give capacity in dm^3. If correction is taken in kilogram and added to the mass of water in kg, the sum divided by 1000 will give capacity of the measure in metre cube (m^3), similarly if correction is taken in milligram and added to mass of water measured in milligram, then sum divided by 1000 will give capacity in centimetre cube.

Values of c have been calculated for temperatures from 5°C to 41°C in steps of 0.1°C for various values of α=ALPHA and taking latest values of density of water [1]. The values of coefficients of expansion are taken to cover the most widely used materials for construction of the capacity measures. The tables are suitable for most of the materials used in manufacturing of capacity measures. The materials covered include different types of glass, admiralty bronze and galvanised iron sheet.

Coefficient of expansion varies from 30×10^{-6}/°C to 25×10^{-6}/°C for soda glass, 15×0.10^{-6}/°C for neutral glass and 10×10^{-6}/°C for borosilicate glass.

Coefficient of expansion for admiralty bronze is 54×10^{-6}/°C, and is 33×10^{-6}/°C for galvanised iron sheet mostly used for larger capacity measures. Aluminium sheet or carboen steel also has a similar value of coefficient of expansion. Stainless steel has volume expansion close to 52×10^{-6}/°C. Coefficients of expansion of aluminium bronze, cupro-nickel alloy and red brass varies from 49×10^{-6}/°C to 61×10^{-6}/°C [5].

Equation (1) can be rewritten as

$$V_{ts} = m.K, \text{ where } K \text{ is given as}$$

$$K = (1 - \sigma/D)/\{1 + \alpha(t - t_s)\}(\rho - \sigma) \qquad \text{...(4)}$$

By calculating the values of K for different combinations of different parameters and multiplying it to the mass of water delivered/contained will give the volume or the capacity of the measure at the reference temperature. If density of water (medium used) and air is expressed in SI units viz. kgm^{-3} the V_{ts} in m^3 will be equal to m.K/1000.

3.4.2 Correction Due to Variation of Air Density

In driving the equation (3), σ the air density has been taken as constant. So for calculation of c, the density of air at t_s °C and 101 305 Pa, is taken. It is also assumed that air contains 0.004 percent of carbon dioxide.

However, in actual practice σ varies with temperature and pressure. To account for it, let c' be the additional correction due to change in air density. Then c' will be the difference between the two corrections, one calculated for density of air at temperature and pressure of measurements and the other for density of air at standard temperature and pressure, so we get c' as

$$c' = 1000.V_{ts}[1-\{1+\alpha(t-t_s)\}\{(\rho-\sigma)/1000\}\}/(1-\sigma/D)]$$
$$-1000.V_{ts}[1-\{1+\alpha(t-t_s)\}\{(\rho-\sigma_s)/1000\}\}/(1-\sigma/D)]$$
$$= V_{ts}\{1+\alpha(t-t_s)\}[(\rho-\sigma_s)/(1-\sigma_s/D)-(\rho-\sigma)/1-\sigma/D)]$$
$$c' = V_{ts}\{1+\alpha(t-t_s)\}D(D-\rho)/(D-\sigma)(D-\sigma_s)](\sigma-\sigma_s)$$
$$c' = V_{ts}\{1+\alpha(t-t_s)\}](1-\rho/D)(\sigma-\sigma_s)/\{(1-\sigma/D)(1-\sigma_s/D)$$

As $(\sigma-\sigma_s)$ is small and also keeping in view that $\alpha(t-t_s)$ and σ/D or σ_s/D each is very much smaller than unity, each of the terms $\{1+\alpha(t-t_s)\}$, $(1-\sigma/D)$ and $(1-\sigma_s/D)$ may be taken as unity, giving us

$$c' = V_{ts}\{(1-\rho/D)(\sigma-\sigma_s) \qquad ...(5)$$

The unit of c' will also be that of mass so the correction c' will be in kg, g or mg to be added to mass of water plus the correction c taken in kg, g or mg to give respectively the capacity of the measure in m^3, dm^3 or cm^3.

It may be reminded that the following relationship should be used to get capacity/volume

$$V_{ts} = (m+c+c')/1000 \qquad ...(6)$$

The values of σ and σ_s, for different values of temperature and pressure are calculated by using equations of air density given by BIPM [2].

Here the value of c' depends upon capacity of the measure, density of standard weights used and temperature and pressure of air but not on the coefficient of volume expansion of the material so. So correction tables, for two different values of density of standard weights used, have been prepared and are given as Tables 3.25 to 3.26.

3.4.3 Correction Due to a Unit Difference in Coefficients of Expansion

As given in section 4.1, coefficients of volume expansion of materials used for fabrication of capacity measures varies in the range of 61 × 10^{-6}/°C to 33 × 10^{-6}/°C, so to cover all materials, the following relation is derived.

Let α_1, α_2 be coefficients of expansion of two materials of two capacity measures, then corresponding corrections at the same temperature and pressure with same standard weights will be

$$c_1 = 1000.V_{ts}[1-\{1+\alpha_1(t-t_s)\}\{(\rho-\sigma)/1000\}\}/(1-\sigma/D)] \text{ and}$$
$$c_2 = 1000.V_{ts}[1-\{1+\alpha_2(t-t_s)\}\{(\rho-\sigma)/1000\}\}/(1-\sigma/D)] \text{ giving us}$$
$$c_1-c_2 = 1000.V_{ts}[(\alpha_2-\alpha_1)(t-t_s)\}(\rho-\sigma)/1000.(1-\sigma/D)] = 1000.c_{coef}$$
$$c_2 = c_1-1000.c_{coef} \qquad ...(7)$$

Where c_{coef} is given by

$$c_{coef} = V_{ts}[(\alpha_2-\alpha_1)(t-t_s)\}\{(\rho-\sigma)/1000\}\}/(1-\sigma/D)] \qquad ...(8)$$

It may be noted that the units of mass of c_{coef} and c will be the same. The values of c_{coef} have been calculated for unit capacity from temperatures 5 °C to 41 °C, for unit difference in coefficients of expansion. However, there are two reference temperatures and two values of density of standard weights, hence there are 4 combinations; hence values of c_{coef} are given in Tables 3.27 to 3.30 for unit value of V_{ts}.

To illustrate the use of the equations (6) and (7), an example is give below.

Nominal capacity of measure is 2 dm³; the value of the coefficient of expansion be 48 × 10^{-6}/ °C. Mean temperature of water filled is 24.3 °C, whose apparent mass is 1992.234 g. However the tables are available for a equal to 54.10^{-6}/°C.

Taking

$$\alpha_1 = 54 \times 10^{-6}/°\text{C}$$
$$\alpha_2 = 48 \times 10^{-6}/°\text{C}$$
$$\alpha_2 - \alpha_1 = -6 \times 10^{-6}/°\text{C}$$
$$V_{ts} = 2 \text{ dm}^3$$

But c_{coef} from the table 3.27 for $D = 8400$ kg/m³, $t_s = 27$°C and $t = 24.3$ °C is – 2.6897, hence

$1000.c_{coef} = 1000.2.(-6 \times 10^{-6})(-2.6897)\text{g} = 0.032196$ g

Correction for α_1 (table 3.1) = 2 × 3.9502 = 7.9004

Hence correction for α_2 = 7.9004 – 0.0322 = 7.8682 g

Capacity of the measure at 27 °C = (1992.334 +7.8682)/1000 = 2.0001022 dm³

3.5 USE OF MERCURY IN GRAVIMETRIC METHOD

When capacity of the measure is very small, mass of water delivered or contained in it will be comparatively small. Finding mass value of small mass entails more fractional error. So to reduce error in weighing we use mercury instead of water, increasing the mass of liquid delivered or contained in it by about 13.5 times. Moreover mercury being a bright opaque liquid is easy to see so that setting the mercury meniscus in very small bore tube of micro-pipettes will also be easier in comparison of setting water meniscus. Mercury is available in pure state and its density is also well known at different temperatures. Sometimes water leaves tiny water droplets, which are not easy to detect thereby increasing uncertainty in measurement. On the other hand mercury does not wet the glass and its droplets are easily seen and thus can be removed. In this case also, mercury delivered or contained in the measure from pre-defined graduation mark is weighed in air and its apparent mass is determined, then the mass of mercury so obtained is multiplied by a factor to give the capacity of the measure at reference temperature. Here you may notice that instead of finding correction to be added to the mass of water it is the correction factor, which we calculate and multiply to the apparent mass of mercury.

3.5.1 Temperature Correction

Let apparent mass of mercury as weighed against standard weights of density D be m. Let ρ be density of mercury and V_t be the volume of mercury delivered/ contained at temperature t°C and atmospheric pressure, then

$$m(1-\sigma/D) = V_t(\rho - \sigma)$$

If V_{ts} is the capacity of the measure up to the graduation mark at standards reference temperature t_s°C and α is the coefficient of cubical expansion of the material of the measure, then

$$V_t = V_{ts}\{1 + \alpha(t - t_s)\}$$

Substituting V_t in the above equation, we get

$$m(1 - \sigma/D) = V_{ts}\{1 + \alpha(t - t_s)\}(\rho - \sigma) \quad ...(9)$$

V_{ts} can be calculated from (7), if the values of α, ρ, σ and D are known.

K is a factor such that when multiplied to mass of mercury gives V_{ts} capacity at reference temperature.

$$K.m = V_{ts}$$

Substituting the V_{ts} from (7), we get

$$K.m = m(1 - \sigma/D)/[\{1 + \alpha(t - t_s)\}(\rho - \sigma)] \text{ giving}$$

$$K = (1 - \sigma/D)/[\{1 + \alpha(t - t_s)\}(\rho - \sigma)] \quad ...(10)$$

From (8) one can see that K has units of the inverse of the density i.e. K may be in terms of m^3/kg, or dm^3/g or cm^3/mg. Therefore if K is multiplied to the mass of mercury, contained or delivered by a measure, in kg it will give us its capacity in m^3, similarly if mass of mercury is taken in g or in mg the product will respectively give capacity of the measure in dm^3or cm^3.

One may notice that equation (8) is identical to the expression of K derived for water (4). However from equation (8), the value of K is very small value say of the order of 10^{-5}. So for the sake of brevity in writing, the values of $10^3 K$ have been calculated and tabulated in tables 3.31 to 3.46. So K.m/1000 will give us V_{ts} in $m^3/dm^3/cm^3$ according to the mass of mercury is taken in kg/g/mg respectively.

Density of standard weights, air and mercury may be taken in any consistent system of units.

Here we see that K depends upon

- Reference temperature.
- Air density at the temperature and pressure of measurement.
- Density of standard weights used.
- Density of mercury at the temperature and pressure of measurement.
- Coefficient of volume expansion of the material of the measure under test.

The factor K has been calculated for all combinations of the following parameters

Reference temperatures 20 °C and 27 °C

Density of standard weights viz 8400 kgm^{-3} and 8000 kg/m^{-3}

For density of mercury at different temperatures but at constant pressure and

Coefficient of volume expansion of the material of the measure under test

The values of 10^3 K factors are given in Tables 3.31 to 3.46.

As K factor has density of mercury in the denominator and is very large, so variation of air density with respect of temperature and pressure is neglected and values of air density corresponding to the reference temperature is taken.

3.6 DESCRIPTION OF TABLES

We have constructed correction tables for all combinations of reference temperatures, density of standard weights used and for different values of ALPHA the coefficients of cubical expansion of various materials used in constructing the capacity measures.

3.6.1 Correction Tables using Water as Medium

The Tables 3.1 to 3.24 are based on the density of water given by the author [1], nominal density of weights as recommended by OIML [3] and coefficients of expansion of glass as reported in ISO [4] and handbook [5].

1. All corrections are in grams and are to be added to the apparent mass of water delivered /contained in the measure when expressed in grams and for a capacity of 1 dm^3.

   When the unit of mass for corrections and mass of water is taken in milligram, then the unit of volume will be cm^3 and if unit of mass is taken in kilogram then unit of volume will be m^3.
2. Reference temperature 27 °C and density of standard weights 8400 kg/m^3.

   Coefficients of expansion are: 54×10^{-6}/ °C, 33×10^{-6}/°C, 30×10^{-6}/ °C, 25×10^{-6}/ °C, 15×10^{-6}/ °C and 10×10^{-6}/°C Tables 3.1 to 3.6.
3. Reference temperature 27 °C and density of standard weights 8000 kg/m^3.

   Coefficients of expansion taken are: 54×10^{-6}/ °C, 33 $\times 10^{-6}$/ °C, 30×10^{-6}/°C, 25×10^{-6}/ °C, 15×10^{-6}/°C and 10×10^{-6}/°C Tables 3.7 to 3.12.
4. Reference temperature 20 °C and density of standard weights 8400 kg/m^3.

   Coefficients of expansion taken are: 54×10^{-6}/ °C, 33×10^{-6}/°C, 30×10^{-6}/°C, 25×10^{-6}/ °C, 15×10^{-6}/°C and 10×10^{-6}/°C. Tables 3.13 to 3.18.
5. Reference temperature 20 °C and density of standard weights 8000 kg/m^3.

   Coefficients of expansion taken are: 54×10^{-6}/ °C, 33×10^{-6}/ °C, 30×10^{-6}/°C, 25×10^{-6}/°C, 15×10^{-6}/°C and 10×10^{-6}/°C. Tables 3.19 to 3.24.
6. In calculating the above corrections, density of air has been taken as constant, which is not quite true, so additional correction due to variation in air density with temperature and pressure have also been given. Corrections due to variation of air density have been given for the following:

   Density of mass standard used, 8400 kg m^{-3} Table 3.25.

   Density of mass standard used, 8000 kg m^{-3} Table 3.26.
7. Keeping in view the fact that a large variety of materials being used to fabricate the capacity measures, the values of c_{coef} unit difference in coefficients and unit capacity of the measure have been tabulated from equation (8) for the following cases:

   Reference temperature 27°C, density of standard weights, 8400 kg/m^3 Table 3.27.

   Reference temperature 27°C, density of standard weights, 8000 kg/m^3 Table 3.28.

   Reference temperature 20°C, density of standard weights, 8400 kg/m^3 Table 3.29.

   Reference temperature 20°C, density of standard weights, 8000 kg/m^3 Table 3.30.

It may be noted that 54×10^{-6}/°C is the coefficient of expansion of admiralty bronze, the material used in India for Secondary Standard Capacity Measures.

3.6.2 Correction Tables using Mercury as Medium

Reference temperature 20°C and density of standard weights 8000 kg/m^3.

Coefficients of expansion taken are: 10×10^{-6}/°C, 15×10^{-6}/°C, 25×10^{-6}/ °C and 30×10^{-6}/°C Tables 3.31 to 3.34.

Reference temperature 20°C and density of standard weights 8400 kg/m^3.

Coefficients of expansion taken are: 10×10^{-6}/°C, 15×10^{-6}/°C, 25×10^{-6}/°C and 30 ×

10^{-6}/°C. Tables 3.35 to 3.38.

Reference temperature 27°C and density of standard weights 8400 kg/m³.

Coefficients of expansion taken are: 10×10^{-6}/°C, 15×10^{-6}/ °C, 25×10^{-6}/°C and 30×10^{-6}/ °C. Tables 3.39 to 3.42.

Reference temperature 27 °C and density of standard weights 8000 kg/m³.

Coefficients of expansion taken are: 10×10^{-6}/°C, 15×10^{-6}/°C, 25×10^{-6}/°C and 30×10^{-6}/ °C. Tables 3.43 to 3.46.

After reducing each measurement carried in 20th century to a common temperature scale of ITS–90, the mean value of density of mercury has been taken as 13545.848 kg/m³ at 20 °C. Beattie's formula [6] as revised by Sommer and Proziemski [7] has been used to give the density–temperature relationship of mercury. The final mercury density–temperature relation is same as given in [1]. Mercury density table is given as 3.47.

3.7 RECORDING AND CALCULATIONS OF CAPACITY

3.7.1 Example

Let us consider a calibration of capacity measures of 1 dm³ and 50 cm³ of admiralty bronze for which alpha is 54×10^{-6} °C. Reference temperature for the measure is 27 °C and density of standard weights used is 8400 kg/m³.

Using two pan balance

Calibration of Secondary Standard Capacity Measure

Particulars of the measure alpha = 54.10^{-6}/°C, Capacity 1 dm³ and 50 cm³

Observer:

Date | Time of start | Time of finish

Air temperature

Pressure

Balance | Capacity | Sensitivity figure 1mg/div

Nominal capacity	*Temp T1*	*Weights in RHP*	*Scale readings*	*Mean*	*Rest point*	*Mass of water m*	*Temp T2*	*Mean Temp*	*c*
1 dm³		1000.3	4.3 4.5	4.4	11.45				
			18.5	18.5					
	30.5	5.6	2.5 2.7	2.6	9.6	994.7018	30.5	30.5	5.3446
			16.6	16.6					
50 cm³		55.6	3.5 3.7	3.7	10.15				
			16.6	16.6					
	30.5	5.85	2.7 3.9	2.8	8.8	49.7514	30.5	30.5	0.2672
			14.8	14.8					

The capacity in $dm^3 = (m + c)/1000 = (994.7018 + 5.3446)/1000 = 1.0000\,046\ dm^3$.

Correction due to air density variation

From table 3.25

– 0.02210 at 30 °C

– 0.01828 at 30 °C giving

– 0.02019 at 30.5 °C for 1 dm^3 measure

– 0.0010 at 30.5 for 50 cm^3 measure, which may be neglected for all practical purposes.

The capacity in dm^3 after this correction = $(m + c + c') = (994.7018 + 5.3446 – 0.02019)/1000$

$= 1.000\,026\ dm^3$

With Single Pan Balance

Nominal capacity	*Initial t emp °C*	*Weights on pan g*	*Balance indication mg*	*Mass of water g*	*Temp 0°C*	*Mean temp °C*	*Correction from table 3.1 in g*	*Corrected volume cm^3*
1 dm^3	—	1000.3	77.5	—				
1 dm^3	30.5	5.6	84.8	994.7073	30.5	30.5	5.3446	1000.052
50 cm^3		55.6	53.5	—	—	—	—	—
50 cm^3	29.6	5.85	65.7	49.7622	29.6	29.6	0.2560	50.0182

For correction due to change in density of air, we find the following entries from table (3. 25).

Temperature	*Corrections*	*Difference*	*for 1 dm^3 measure*
29	– 0.01447 g		
		0.00381	for 0.4 0.001524 giving –0.01828 + 0.001524 = –0.01676 g at 29.6 °C so for 50 cm^3 net correction is – 0.000 8 g, which is negligible in comparison of 0.05 cm^3 MPE for the measure
30	– 0.01828 g		
		0.00382	for 0.5 °C – 0.00191 giving net correction for 1 dm^3 = –0.02019
31	– 0.02210 g		

Thumb rule for calculating and applying corrections due to variation in air density.

To decide if the correction due to air density variation is necessary to apply, we should consider the MPE– maximum permissible error, if the correction is less than one-tenth of the MPE then we may not apply it, especially while in the field.

CORRECTION TABLES WHEN WATER IS USED (TABLE 3.1 TO 3.24)

Corrections are in kg/g/mg and are to be added to the mass of water measured in kg/g/mg as the unit of capacity is in $m^3/dm^3/cm^3$ respectively

Table 3.1 ALPHA = 54×10^{-6}/°C, DEN = 8400 kg/m^3, REFERENCE TEMP = 27 °C

Temp	0.0	0.1	0.2	0.3	0.4	0.5	0.6	0.7	0.8	0.9
5	2.2040	2.2002	2.1967	2.1933	2.1900	2.1869	2.1839	2.1811	2.1784	2.1759
6	2.1736	2.1714	2.1693	2.1674	2.1656	2.1640	2.1626	2.1612	2.1601	2.1591
7	2.1582	2.1574	2.1569	2.1564	2.1561	2.1560	2.1560	2.1561	2.1564	2.1568
8	2.1574	2.1581	2.1589	2.1599	2.1610	2.1623	2.1637	2.1652	2.1669	2.1687
9	2.1707	2.1728	2.1750	2.1774	2.1799	2.1826	2.1854	2.1883	2.1913	2.1945
10	2.1978	2.2013	2.2049	2.2086	2.2125	2.2165	2.2206	2.2248	2.2292	2.2337
11	2.2384	2.2432	2.2481	2.2531	2.2583	2.2636	2.2690	2.2746	2.2803	2.2861
12	2.2920	2.2981	2.3043	2.3106	2.3170	2.3236	2.3303	2.3371	2.3441	2.3512
13	2.3584	2.3657	2.3731	2.3807	2.3884	2.3962	2.4041	2.4122	2.4204	2.4287
14	2.4371	2.4457	2.4543	2.4631	2.4720	2.4810	2.4902	2.4995	2.5088	2.5183
15	2.5280	2.5377	2.5476	2.5575	2.5676	2.5778	2.5882	2.5986	2.6092	2.6198
16	2.6306	2.6415	2.6525	2.6637	2.6749	2.6863	2.6978	2.7094	2.7211	2.7329
17	2.7448	2.7569	2.7690	2.7813	2.7937	2.8061	2.8187	2.8315	2.8443	2.8572
18	2.8703	2.8834	2.8967	2.9101	2.9235	2.9371	2.9508	2.9647	2.9786	2.9926
19	3.0067	3.0210	3.0353	3.0498	3.0644	3.0790	3.0938	3.1087	3.1237	3.1388
20	3.1540	3.1693	3.1847	3.2002	3.2159	3.2316	3.2474	3.2634	3.2794	3.2955
21	3.3118	3.3281	3.3446	3.3612	3.3778	3.3946	3.4114	3.4284	3.4455	3.4626
22	3.4799	3.4973	3.5148	3.5323	3.5500	3.5678	3.5857	3.6036	3.6217	3.6399
23	3.6582	3.6765	3.6950	3.7136	3.7323	3.7510	3.7699	3.7889	3.8079	3.8271
24	3.8463	3.8657	3.8852	3.9047	3.9244	3.9441	3.9639	3.9839	4.0039	4.0240
25	4.0443	4.0646	4.0850	4.1055	4.1261	4.1468	4.1676	4.1885	4.2095	4.2306
26	4.2517	4.2730	4.2943	4.3158	4.3373	4.3590	4.3807	4.4025	4.4245	4.4465
27	4.4686	4.4908	4.5131	4.5354	4.5579	4.5805	4.6031	4.6259	4.6487	4.6716
28	4.6946	4.7177	4.7409	4.7642	4.7876	4.8111	4.8346	4.8583	4.8820	4.9058
29	4.9298	4.9538	4.9779	5.0020	5.0263	5.0507	5.0751	5.0997	5.1243	5.1490
30	5.1738	5.1987	5.2236	5.2487	5.2739	5.2991	5.3244	5.3498	5.3753	5.4009
31	5.4266	5.4523	5.4782	5.5041	5.5301	5.5562	5.5824	5.6086	5.6350	5.6614
32	5.6879	5.7145	5.7412	5.7680	5.7949	5.8218	5.8488	5.8760	5.9031	5.9304
33	5.9578	5.9852	6.0128	6.0404	6.0681	6.0958	6.1237	6.1516	6.1797	6.2078
34	6.2360	6.2642	6.2926	6.3210	6.3495	6.3781	6.4068	6.4356	6.4644	6.4933
35	6.5223	6.5514	6.5806	6.6098	6.6391	6.6685	6.6980	6.7276	6.7572	6.7870
36	6.8168	6.8466	6.8766	6.9066	6.9368	6.9670	6.9972	7.0276	7.0580	7.0885
37	7.1191	7.1498	7.1805	7.2113	7.2422	7.2732	7.3043	7.3354	7.3666	7.3979
38	7.4293	7.4607	7.4922	7.5238	7.5555	7.5872	7.6191	7.6509	7.6829	7.7150
39	7.7471	7.7793	7.8116	7.8439	7.8763	7.9088	7.9414	7.9741	8.0068	8.0396
40	8.0725	8.1054	8.1384	8.1715	8.2047	8.2379	8.2712	8.3046	8.3381	8.3716
41	8.4052	8.4389	8.4726	8.5065	8.5404	8.5743	8.6084	8.6425	8.6767	8.7109

Corrections are in kg/g/mg and are to be added to the mass of water measured in kg/g/mg as the unit of capacity is in $m^3/dm^3/cm^3$ respectively

Table 3.2 ALPHA = 33 × 10^{-6}/°C, DEN = 8400 kg/m^3, REFERENCE TEMP = 27 °C

Temp	0.0	0.1	0.2	0.3	0.4	0.5	0.6	0.7	0.8	0.9
5	1.7424	1.7408	1.7393	1.7380	1.7369	1.7358	1.7350	1.7343	1.7337	1.7333
6	1.7330	1.7329	1.7330	1.7332	1.7335	1.7340	1.7346	1.7354	1.7363	1.7374
7	1.7386	1.7400	1.7415	1.7432	1.7450	1.7469	1.7490	1.7512	1.7536	1.7561
8	1.7588	1.7616	1.7646	1.7676	1.7709	1.7742	1.7777	1.7814	1.7852	1.7891
9	1.7932	1.7974	1.8017	1.8062	1.8108	1.8155	1.8204	1.8254	1.8306	1.8359
10	1.8413	1.8469	1.8526	1.8584	1.8643	1.8704	1.8767	1.8830	1.8895	1.8961
11	1.9029	1.9097	1.9168	1.9239	1.9312	1.9386	1.9461	1.9537	1.9615	1.9694
12	1.9775	1.9857	1.9939	2.0024	2.0109	2.0196	2.0284	2.0373	2.0464	2.0555
13	2.0648	2.0743	2.0838	2.0935	2.1033	2.1132	2.1232	2.1334	2.1437	2.1541
14	2.1646	2.1752	2.1860	2.1969	2.2079	2.2190	2.2303	2.2416	2.2531	2.2647
15	2.2764	2.2883	2.3002	2.3123	2.3245	2.3368	2.3492	2.3618	2.3744	2.3872
16	2.4001	2.4131	2.4262	2.4394	2.4528	2.4663	2.4798	2.4935	2.5073	2.5212
17	2.5353	2.5494	2.5637	2.5780	2.5925	2.6071	2.6218	2.6366	2.6515	2.6666
18	2.6817	2.6970	2.7123	2.7278	2.7434	2.7591	2.7749	2.7908	2.8068	2.8229
19	2.8392	2.8555	2.8720	2.8885	2.9052	2.9220	2.9388	2.9558	2.9729	2.9901
20	3.0074	3.0248	3.0423	3.0599	3.0777	3.0955	3.1134	3.1314	3.1496	3.1678
21	3.1862	3.2046	3.2232	3.2418	3.2606	3.2794	3.2984	3.3175	3.3366	3.3559
22	3.3753	3.3947	3.4143	3.4340	3.4537	3.4736	3.4936	3.5136	3.5338	3.5541
23	3.5745	3.5949	3.6155	3.6362	3.6569	3.6778	3.6988	3.7198	3.7410	3.7622
24	3.7836	3.8050	3.8266	3.8482	3.8700	3.8918	3.9137	3.9358	3.9579	3.9801
25	4.0024	4.0248	4.0473	4.0699	4.0926	4.1154	4.1383	4.1613	4.1844	4.2076
26	4.2308	4.2542	4.2776	4.3012	4.3248	4.3485	4.3724	4.3963	4.4203	4.4444
27	4.4686	4.4929	4.5172	4.5417	4.5663	4.5909	4.6157	4.6405	4.6654	4.6904
28	4.7155	4.7407	4.7660	4.7914	4.8169	4.8424	4.8681	4.8938	4.9196	4.9455
29	4.9716	4.9976	5.0238	5.0501	5.0765	5.1029	5.1294	5.1561	5.1828	5.2096
30	5.2365	5.2634	5.2905	5.3176	5.3449	5.3722	5.3996	5.4271	5.4547	5.4823
31	5.5101	5.5379	5.5659	5.5939	5.6220	5.6501	5.6784	5.7068	5.7352	5.7637
32	5.7923	5.8210	5.8498	5.8786	5.9076	5.9366	5.9657	5.9949	6.0242	6.0536
33	6.0830	6.1125	6.1421	6.1718	6.2016	6.2315	6.2614	6.2914	6.3215	6.3517
34	6.3820	6.4123	6.4428	6.4733	6.5039	6.5346	6.5653	6.5962	6.6271	6.6581
35	6.6892	6.7203	6.7516	6.7829	6.8143	6.8458	6.8773	6.9090	6.9407	6.9725
36	7.0044	7.0363	7.0684	7.1005	7.1327	7.1650	7.1973	7.2297	7.2623	7.2948
37	7.3275	7.3603	7.3931	7.4260	7.4589	7.4920	7.5251	7.5583	7.5916	7.6250
38	7.6584	7.6919	7.7255	7.7592	7.7929	7.8268	7.8606	7.8946	7.9287	7.9628
39	7.9970	8.0313	8.0656	8.1000	8.1345	8.1691	8.2037	8.2385	8.2732	8.3081
40	8.3431	8.3781	8.4132	8.4483	8.4836	8.5189	8.5543	8.5897	8.6252	8.6608
41	8.6965	8.7323	8.7681	8.8040	8.8399	8.8760	8.9121	8.9483	8.9845	9.0208

Corrections are in kg/g/mg and are to be added to the mass of water measured in kg/g/mg as the unit of capacity is in $m^3/dm^3/cm^3$ respectively

Table 3.3 ALPHA = 30×10^{-6}/°C, DEN = 8400 kg/m^3, REFERENCE TEMP = 27 °C

Temp	0.0	0.1	0.2	0.3	0.4	0.5	0.6	0.7	0.8	0.9
5	1.6765	1.6752	1.6740	1.6730	1.6721	1.6714	1.6708	1.6704	1.6702	1.6701
6	1.6701	1.6703	1.6706	1.6711	1.6718	1.6725	1.6735	1.6746	1.6758	1.6772
7	1.6787	1.6804	1.6822	1.6841	1.6862	1.6885	1.6909	1.6934	1.6961	1.6989
8	1.7019	1.7050	1.7082	1.7116	1.7151	1.7188	1.7226	1.7266	1.7306	1.7349
9	1.7392	1.7437	1.7484	1.7531	1.7580	1.7631	1.7683	1.7736	1.7791	1.7846
10	1.7904	1.7962	1.8022	1.8083	1.8146	1.8210	1.8275	1.8342	1.8410	1.8479
11	1.8549	1.8621	1.8694	1.8769	1.8844	1.8921	1.9000	1.9079	1.9160	1.9242
12	1.9326	1.9410	1.9496	1.9583	1.9672	1.9762	1.9853	1.9945	2.0038	2.0133
13	2.0229	2.0326	2.0425	2.0524	2.0625	2.0727	2.0831	2.0935	2.1041	2.1148
14	2.1257	2.1366	2.1477	2.1588	2.1702	2.1816	2.1931	2.2048	2.2166	2.2285
15	2.2405	2.2526	2.2649	2.2773	2.2898	2.3024	2.3151	2.3279	2.3409	2.3540
16	2.3672	2.3805	2.3939	2.4074	2.4211	2.4348	2.4487	2.4627	2.4768	2.4910
17	2.5053	2.5198	2.5343	2.5490	2.5638	2.5787	2.5937	2.6088	2.6240	2.6393
18	2.6548	2.6703	2.6860	2.7018	2.7177	2.7336	2.7497	2.7660	2.7823	2.7987
19	2.8152	2.8319	2.8486	2.8655	2.8824	2.8995	2.9167	2.9340	2.9514	2.9689
20	2.9865	3.0042	3.0220	3.0399	3.0579	3.0760	3.0943	3.1126	3.1310	3.1496
21	3.1682	3.1870	3.2058	3.2248	3.2438	3.2630	3.2822	3.3016	3.3211	3.3406
22	3.3603	3.3801	3.3999	3.4199	3.4400	3.4601	3.4804	3.5008	3.5213	3.5418
23	3.5625	3.5833	3.6041	3.6251	3.6462	3.6673	3.6886	3.7099	3.7314	3.7530
24	3.7746	3.7964	3.8182	3.8401	3.8622	3.8843	3.9066	3.9289	3.9513	3.9738
25	3.9964	4.0192	4.0420	4.0649	4.0879	4.1110	4.1341	4.1574	4.1808	4.2043
26	4.2278	4.2515	4.2752	4.2991	4.3230	4.3470	4.3712	4.3954	4.4197	4.4441
27	4.4686	4.4932	4.5178	4.5426	4.5675	4.5924	4.6175	4.6426	4.6678	4.6931
28	4.7185	4.7440	4.7696	4.7953	4.8210	4.8469	4.8728	4.8989	4.9250	4.9512
29	4.9775	5.0039	5.0304	5.0570	5.0836	5.1104	5.1372	5.1641	5.1911	5.2182
30	5.2454	5.2727	5.3000	5.3275	5.3550	5.3826	5.4103	5.4381	5.4660	5.4940
31	5.5220	5.5502	5.5784	5.6067	5.6351	5.6636	5.6921	5.7208	5.7495	5.7783
32	5.8072	5.8362	5.8653	5.8944	5.9237	5.9530	5.9824	6.0119	6.0415	6.0711
33	6.1009	6.1307	6.1606	6.1906	6.2207	6.2508	6.2811	6.3114	6.3418	6.3723
34	6.4029	6.4335	6.4642	6.4950	6.5259	6.5569	6.5880	6.6191	6.6503	6.6816
35	6.7130	6.7445	6.7760	6.8076	6.8393	6.8711	6.9030	6.9349	6.9669	6.9990
36	7.0312	7.0634	7.0958	7.1282	7.1607	7.1933	7.2259	7.2586	7.2914	7.3243
37	7.3573	7.3903	7.4234	7.4566	7.4899	7.5233	7.5567	7.5902	7.6238	7.6574
38	7.6912	7.7250	7.7589	7.7928	7.8269	7.8610	7.8952	7.9294	7.9638	7.9982
39	8.0327	8.0672	8.1019	8.1366	8.1714	8.2063	8.2412	8.2762	8.3113	8.3465
40	8.3817	8.4170	8.4524	8.4879	8.5234	8.5590	8.5947	8.6304	8.6663	8.7022
41	8.7381	8.7742	8.8103	8.8465	8.8827	8.9191	8.9555	8.9919	9.0285	9.0651

Corrections are in kg/g/mg and are to be added to the mass of water measured in kg/g/mg as the unit of capacity is in $m^3/dm^3/cm^3$ respectively

Table 3.4 ALPHA = 25 × 10^{-6}/°C, DEN = 8400 kg/m^3, REFERENCE TEMP = 27°C

Temp	0.0	0.1	0.2	0.3	0.4	0.5	0.6	0.7	0.8	0.9
5	1.5666	1.5658	1.5651	1.5646	1.5642	1.5640	1.5639	1.5640	1.5643	1.5647
6	1.5652	1.5659	1.5667	1.5677	1.5689	1.5702	1.5716	1.5732	1.5749	1.5768
7	1.5788	1.5810	1.5833	1.5857	1.5883	1.5911	1.5940	1.5970	1.6002	1.6035
8	1.6070	1.6106	1.6143	1.6182	1.6222	1.6264	1.6307	1.6352	1.6397	1.6445
9	1.6493	1.6543	1.6595	1.6647	1.6702	1.6757	1.6814	1.6872	1.6932	1.6993
10	1.7055	1.7118	1.7183	1.7250	1.7317	1.7386	1.7456	1.7528	1.7601	1.7675
11	1.7750	1.7827	1.7905	1.7985	1.8065	1.8147	1.8231	1.8315	1.8401	1.8488
12	1.8577	1.8666	1.8757	1.8849	1.8943	1.9038	1.9134	1.9231	1.9329	1.9429
13	1.9530	1.9632	1.9736	1.9840	1.9946	2.0054	2.0162	2.0272	2.0382	2.0494
14	2.0608	2.0722	2.0838	2.0955	2.1073	2.1192	2.1312	2.1434	2.1557	2.1681
15	2.1806	2.1933	2.2060	2.2189	2.2319	2.2450	2.2582	2.2715	2.2850	2.2986
16	2.3123	2.3261	2.3400	2.3540	2.3682	2.3824	2.3968	2.4113	2.4259	2.4406
17	2.4554	2.4704	2.4854	2.5006	2.5159	2.5313	2.5468	2.5624	2.5781	2.5939
18	2.6099	2.6259	2.6421	2.6584	2.6748	2.6913	2.7079	2.7246	2.7414	2.7583
19	2.7753	2.7925	2.8097	2.8271	2.8445	2.8621	2.8798	2.8976	2.9155	2.9335
20	2.9516	2.9698	2.9881	3.0065	3.0250	3.0436	3.0624	3.0812	3.1001	3.1192
21	3.1383	3.1576	3.1769	3.1964	3.2159	3.2356	3.2553	3.2752	3.2952	3.3152
22	3.3354	3.3556	3.3760	3.3965	3.4171	3.4377	3.4585	3.4794	3.5003	3.5214
23	3.5426	3.5638	3.5852	3.6067	3.6282	3.6499	3.6716	3.6935	3.7155	3.7375
24	3.7597	3.7819	3.8043	3.8267	3.8492	3.8719	3.8946	3.9174	3.9404	3.9634
25	3.9865	4.0097	4.0330	4.0564	4.0799	4.1035	4.1272	4.1509	4.1748	4.1988
26	4.2228	4.2470	4.2712	4.2956	4.3200	4.3446	4.3692	4.3939	4.4187	4.4436
27	4.4686	4.4937	4.5188	4.5441	4.5695	4.5949	4.6204	4.6461	4.6718	4.6976
28	4.7235	4.7495	4.7756	4.8018	4.8280	4.8544	4.8808	4.9073	4.9340	4.9607
29	4.9875	5.0144	5.0413	5.0684	5.0956	5.1228	5.1501	5.1775	5.2050	5.2326
30	5.2603	5.2881	5.3159	5.3439	5.3719	5.4000	5.4282	5.4565	5.4849	5.5134
31	5.5419	5.5705	5.5993	5.6281	5.6570	5.6859	5.7150	5.7441	5.7734	5.8027
32	5.8321	5.8616	5.8911	5.9208	5.9505	5.9803	6.0102	6.0402	6.0703	6.1005
33	6.1307	6.1610	6.1914	6.2219	6.2525	6.2831	6.3139	6.3447	6.3756	6.4066
34	6.4376	6.4688	6.5000	6.5313	6.5627	6.5942	6.6257	6.6573	6.6891	6.7208
35	6.7527	6.7847	6.8167	6.8488	6.8810	6.9133	6.9456	6.9781	7.0106	7.0432
36	7.0759	7.1086	7.1414	7.1743	7.2073	7.2404	7.2735	7.3068	7.3401	7.3734
37	7.4069	7.4404	7.4740	7.5077	7.5415	7.5753	7.6093	7.6433	7.6773	7.7115
38	7.7457	7.7800	7.8144	7.8489	7.8834	7.9180	7.9527	7.9874	8.0223	8.0572
39	8.0922	8.1272	8.1624	8.1976	8.2329	8.2682	8.3037	8.3392	8.3748	8.4104
40	8.4461	8.4819	8.5178	8.5538	8.5898	8.6259	8.6621	8.6983	8.7346	8.7710
41	8.8075	8.8440	8.8806	8.9173	8.9541	8.9909	9.0278	9.0648	9.1018	9.1389

Corrections are in kg/g/mg and are to be added to the mass of water measured in kg/g/mg as the unit of capacity is in m³/dm³/cm³ respectively

Table 3.5 ALPHA = 15 × 10^{-6}/°C, DEN = 8400 kg/m^3, REFERENCE TEMP = 27 °C

Temp	0.0	0.1	0.2	0.3	0.4	0.5	0.6	0.7	0.8	0.9
5	1.3468	1.3470	1.3473	1.3478	1.3484	1.3492	1.3502	1.3513	1.3525	1.3539
6	1.3554	1.3571	1.3590	1.3609	1.3631	1.3654	1.3678	1.3704	1.3731	1.3760
7	1.3790	1.3822	1.3855	1.3890	1.3926	1.3963	1.4002	1.4042	1.4084	1.4127
8	1.4172	1.4218	1.4265	1.4314	1.4365	1.4416	1.4469	1.4524	1.4580	1.4637
9	1.4695	1.4755	1.4817	1.4880	1.4944	1.5009	1.5076	1.5144	1.5214	1.5285
10	1.5357	1.5431	1.5505	1.5582	1.5659	1.5738	1.5818	1.5900	1.5983	1.6067
11	1.6153	1.6239	1.6327	1.6417	1.6508	1.6600	1.6693	1.6787	1.6883	1.6980
12	1.7079	1.7179	1.7279	1.7382	1.7485	1.7590	1.7696	1.7803	1.7912	1.8021
13	1.8132	1.8245	1.8358	1.8473	1.8589	1.8706	1.8824	1.8944	1.9065	1.9187
14	1.9310	1.9434	1.9560	1.9687	1.9815	1.9944	2.0075	2.0206	2.0339	2.0473
15	2.0608	2.0745	2.0882	2.1021	2.1161	2.1302	2.1444	2.1588	2.1732	2.1878
16	2.2025	2.2173	2.2322	2.2472	2.2624	2.2777	2.2930	2.3085	2.3241	2.3398
17	2.3557	2.3716	2.3877	2.4038	2.4201	2.4365	2.4530	2.4696	2.4863	2.5032
18	2.5201	2.5372	2.5543	2.5716	2.5890	2.6065	2.6241	2.6418	2.6596	2.6775
19	2.6955	2.7137	2.7319	2.7503	2.7687	2.7873	2.8060	2.8248	2.8437	2.8627
20	2.8817	2.9010	2.9203	2.9397	2.9592	2.9788	2.9985	3.0184	3.0383	3.0583
21	3.0785	3.0987	3.1191	3.1395	3.1601	3.1807	3.2015	3.2224	3.2433	3.2644
22	3.2855	3.3068	3.3282	3.3496	3.3712	3.3929	3.4146	3.4365	3.4585	3.4805
23	3.5027	3.5250	3.5473	3.5698	3.5924	3.6150	3.6378	3.6606	3.6836	3.7066
24	3.7298	3.7530	3.7764	3.7998	3.8233	3.8470	3.8707	3.8945	3.9184	3.9425
25	3.9666	3.9908	4.0151	4.0395	4.0640	4.0885	4.1132	4.1380	4.1629	4.1878
26	4.2129	4.2380	4.2633	4.2886	4.3141	4.3396	4.3652	4.3909	4.4167	4.4426
27	4.4686	4.4947	4.5208	4.5471	4.5734	4.5999	4.6264	4.6530	4.6798	4.7066
28	4.7335	4.7604	4.7875	4.8147	4.8419	4.8693	4.8967	4.9243	4.9519	4.9796
29	5.0074	5.0353	5.0632	5.0913	5.1194	5.1477	5.1760	5.2044	5.2329	5.2615
30	5.2902	5.3189	5.3478	5.3767	5.4057	5.4348	5.4640	5.4933	5.5227	5.5521
31	5.5817	5.6113	5.6410	5.6708	5.7007	5.7307	5.7607	5.7909	5.8211	5.8514
32	5.8818	5.9123	5.9428	5.9735	6.0042	6.0350	6.0659	6.0969	6.1279	6.1591
33	6.1903	6.2216	6.2530	6.2845	6.3161	6.3477	6.3794	6.4112	6.4431	6.4751
34	6.5072	6.5393	6.5715	6.6038	6.6362	6.6686	6.7012	6.7338	6.7665	6.7993
35	6.8322	6.8651	6.8981	6.9312	6.9644	6.9977	7.0310	7.0645	7.0980	7.1315
36	7.1652	7.1989	7.2328	7.2667	7.3006	7.3347	7.3688	7.4030	7.4373	7.4717
37	7.5061	7.5407	7.5753	7.6099	7.6447	7.6795	7.7144	7.7494	7.7845	7.8196
38	7.8548	7.8901	7.9255	7.9609	7.9965	8.0321	8.0677	8.1035	8.1393	8.1752
39	8.2112	8.2472	8.2833	8.3195	8.3558	8.3922	8.4286	8.4651	8.5016	8.5383
40	8.5750	8.6118	8.6487	8.6856	8.7226	8.7597	8.7969	8.8341	8.8714	8.9088
41	8.9462	8.9837	9.0213	9.0590	9.0967	9.1345	9.1724	9.2104	9.2484	9.2865

Corrections are in kg/g/mg and are to be added to the mass of water measured in kg/g/mg as the unit of capacity is in m³/dm³/cm³ respectively

Table 3.6 ALPHA = 10 × 10^{-6}/°C, DEN = 8400 kg/m^3, REFERENCE TEMP = 27 °C

Temp	0.0	0.1	0.2	0.3	0.4	0.5	0.6	0.7	0.8	0.9
5	1.2370	1.2376	1.2385	1.2394	1.2406	1.2418	1.2433	1.2449	1.2466	1.2485
6	1.2505	1.2527	1.2551	1.2576	1.2602	1.2630	1.2659	1.2690	1.2722	1.2756
7	1.2791	1.2828	1.2866	1.2906	1.2947	1.2989	1.3033	1.3078	1.3125	1.3173
8	1.3223	1.3274	1.3326	1.3380	1.3436	1.3492	1.3550	1.3610	1.3671	1.3733
9	1.3797	1.3862	1.3928	1.3996	1.4065	1.4135	1.4207	1.4280	1.4355	1.4431
10	1.4508	1.4587	1.4667	1.4748	1.4830	1.4914	1.5000	1.5086	1.5174	1.5263
11	1.5354	1.5445	1.5539	1.5633	1.5729	1.5826	1.5924	1.6024	1.6124	1.6227
12	1.6330	1.6435	1.6541	1.6648	1.6756	1.6866	1.6977	1.7089	1.7203	1.7317
13	1.7433	1.7551	1.7669	1.7789	1.7910	1.8032	1.8155	1.8280	1.8406	1.8533
14	1.8661	1.8790	1.8921	1.9053	1.9186	1.9320	1.9456	1.9592	1.9730	1.9869
15	2.0010	2.0151	2.0293	2.0437	2.0582	2.0728	2.0875	2.1024	2.1173	2.1324
16	2.1476	2.1629	2.1783	2.1939	2.2095	2.2253	2.2411	2.2571	2.2732	2.2895
17	2.3058	2.3222	2.3388	2.3554	2.3722	2.3891	2.4061	2.4232	2.4404	2.4578
18	2.4752	2.4928	2.5104	2.5282	2.5461	2.5641	2.5822	2.6004	2.6187	2.6371
19	2.6556	2.6743	2.6930	2.7119	2.7308	2.7499	2.7691	2.7884	2.8078	2.8272
20	2.8468	2.8665	2.8864	2.9063	2.9263	2.9464	2.9666	2.9870	3.0074	3.0279
21	3.0486	3.0693	3.0902	3.1111	3.1322	3.1533	3.1746	3.1959	3.2174	3.2390
22	3.2606	3.2824	3.3043	3.3262	3.3483	3.3704	3.3927	3.4151	3.4375	3.4601
23	3.4828	3.5055	3.5284	3.5514	3.5744	3.5976	3.6208	3.6442	3.6676	3.6912
24	3.7148	3.7386	3.7624	3.7864	3.8104	3.8345	3.8587	3.8831	3.9075	3.9320
25	3.9566	3.9813	4.0061	4.0310	4.0560	4.0811	4.1063	4.1315	4.1569	4.1824
26	4.2079	4.2336	4.2593	4.2851	4.3111	4.3371	4.3632	4.3894	4.4157	4.4421
27	4.4686	4.4952	4.5218	4.5486	4.5754	4.6024	4.6294	4.6565	4.6837	4.7110
28	4.7384	4.7659	4.7935	4.8212	4.8489	4.8768	4.9047	4.9327	4.9608	4.9890
29	5.0173	5.0457	5.0742	5.1027	5.1314	5.1601	5.1889	5.2178	5.2468	5.2759
30	5.3051	5.3343	5.3637	5.3931	5.4226	5.4522	5.4819	5.5117	5.5416	5.5715
31	5.6016	5.6317	5.6619	5.6922	5.7226	5.7530	5.7836	5.8142	5.8449	5.8757
32	5.9066	5.9376	5.9687	5.9998	6.0310	6.0623	6.0937	6.1252	6.1568	6.1884
33	6.2201	6.2519	6.2838	6.3158	6.3479	6.3800	6.4122	6.4445	6.4769	6.5094
34	6.5419	6.5746	6.6073	6.6401	6.6729	6.7059	6.7389	6.7720	6.8052	6.8385
35	6.8719	6.9053	6.9388	6.9724	7.0061	7.0399	7.0737	7.1076	7.1416	7.1757
36	7.2099	7.2441	7.2784	7.3128	7.3473	7.3818	7.4165	7.4512	7.4859	7.5208
37	7.5557	7.5908	7.6259	7.6610	7.6963	7.7316	7.7670	7.8025	7.8380	7.8737
38	7.9094	7.9452	7.9810	8.0170	8.0530	8.0891	8.1252	8.1615	8.1978	8.2342
39	8.2707	8.3072	8.3438	8.3805	8.4173	8.4541	8.4910	8.5280	8.5651	8.6022
40	8.6394	8.6767	8.7141	8.7515	8.7890	8.8266	8.8642	8.9020	8.9398	8.9776
41	9.0156	9.0536	9.0917	9.1298	9.1681	9.2064	9.2447	9.2832	9.3217	9.3603

Corrections are in kg/g/mg and are to be added to the mass of water measured in kg/g/mg as the unit of capacity is in m³/dm³/cm³ respectively

Table 3.7 ALPHA = 54 × 10^{-6}/°C, DEN = 8000 kg/m^3, REFERENCE TEMP = 27 °C

Temp	0.0	0.1	0.2	0.3	0.4	0.5	0.6	0.7	0.8	0.9
5	2.1973	2.1936	2.1900	2.1866	2.1834	2.1802	2.1773	2.1745	2.1718	2.1693
6	2.1669	2.1647	2.1627	2.1608	2.1590	2.1574	2.1559	2.1546	2.1534	2.1524
7	2.1515	2.1508	2.1502	2.1498	2.1495	2.1493	2.1493	2.1495	2.1497	2.1502
8	2.1507	2.1514	2.1523	2.1533	2.1544	2.1556	2.1571	2.1586	2.1603	2.1621
9	2.1641	2.1662	2.1684	2.1708	2.1733	2.1759	2.1787	2.1816	2.1847	2.1879
10	2.1912	2.1947	2.1983	2.2020	2.2058	2.2098	2.2140	2.2182	2.2226	2.2271
11	2.2318	2.2365	2.2415	2.2465	2.2517	2.2570	2.2624	2.2679	2.2736	2.2794
12	2.2854	2.2914	2.2976	2.3040	2.3104	2.3170	2.3237	2.3305	2.3375	2.3445
13	2.3517	2.3590	2.3665	2.3741	2.3818	2.3896	2.3975	2.4056	2.4137	2.4221
14	2.4305	2.4390	2.4477	2.4565	2.4654	2.4744	2.4836	2.4928	2.5022	2.5117
15	2.5213	2.5311	2.5409	2.5509	2.5610	2.5712	2.5815	2.5920	2.6025	2.6132
16	2.6240	2.6349	2.6459	2.6570	2.6683	2.6797	2.6911	2.7027	2.7144	2.7262
17	2.7382	2.7502	2.7624	2.7746	2.7870	2.7995	2.8121	2.8248	2.8377	2.8506
18	2.8636	2.8768	2.8901	2.9034	2.9169	2.9305	2.9442	2.9580	2.9719	2.9860
19	3.0001	3.0144	3.0287	3.0432	3.0577	3.0724	3.0872	3.1021	3.1171	3.1322
20	3.1474	3.1627	3.1781	3.1936	3.2092	3.2250	3.2408	3.2567	3.2728	3.2889
21	3.3052	3.3215	3.3380	3.3545	3.3712	3.3879	3.4048	3.4218	3.4388	3.4560
22	3.4733	3.4907	3.5081	3.5257	3.5434	3.5612	3.5790	3.5970	3.6151	3.6333
23	3.6515	3.6699	3.6884	3.7070	3.7256	3.7444	3.7633	3.7822	3.8013	3.8205
24	3.8397	3.8591	3.8785	3.8981	3.9177	3.9375	3.9573	3.9772	3.9973	4.0174
25	4.0376	4.0580	4.0784	4.0989	4.1195	4.1402	4.1610	4.1819	4.2029	4.2239
26	4.2451	4.2664	4.2877	4.3092	4.3307	4.3524	4.3741	4.3959	4.4178	4.4399
27	4.4620	4.4842	4.5064	4.5288	4.5513	4.5738	4.5965	4.6192	4.6421	4.6650
28	4.6880	4.7111	4.7343	4.7576	4.7810	4.8045	4.8280	4.8517	4.8754	4.8992
29	4.9231	4.9472	4.9712	4.9954	5.0197	5.0441	5.0685	5.0930	5.1177	5.1424
30	5.1672	5.1921	5.2170	5.2421	5.2672	5.2925	5.3178	5.3432	5.3687	5.3943
31	5.4199	5.4457	5.4715	5.4975	5.5235	5.5496	5.5758	5.6020	5.6284	5.6548
32	5.6813	5.7079	5.7346	5.7614	5.7883	5.8152	5.8422	5.8693	5.8965	5.9238
33	5.9512	5.9786	6.0062	6.0338	6.0615	6.0892	6.1171	6.1450	6.1731	6.2012
34	6.2294	6.2576	6.2860	6.3144	6.3429	6.3715	6.4002	6.4290	6.4578	6.4867
35	6.5157	6.5448	6.5740	6.6032	6.6325	6.6619	6.6914	6.7210	6.7506	6.7804
36	6.8102	6.8400	6.8700	6.9000	6.9302	6.9604	6.9906	7.0210	7.0514	7.0819
37	7.1125	7.1432	7.1739	7.2047	7.2356	7.2666	7.2977	7.3288	7.3600	7.3913
38	7.4227	7.4541	7.4856	7.5172	7.5489	7.5806	7.6125	7.6444	7.6763	7.7084
39	7.7405	7.7727	7.8050	7.8373	7.8697	7.9022	7.9348	7.9675	8.0002	8.0330
40	8.0659	8.0988	8.1318	8.1649	8.1981	8.2313	8.2646	8.2980	8.3315	8.3650
41	8.3986	8.4323	8.4660	8.4999	8.5338	8.5677	8.6018	8.6359	8.6701	8.7043

REFERENCE TEMP = 27 ALPHA = .000054

Corrections are in kg/g/mg and are to be added to the mass of water measured in kg/g/mg as the unit of capacity is in m³/dm³/cm³ respectively

Table 3.8 ALPHA = 33×10^{-6}/°C, DEN = 8000 kg/m³, REFERENCE TEMP = 27 °C

Temp	0.0	0.1	0.2	0.3	0.4	0.5	0.6	0.7	0.8	0.9
5	1.7358	1.7342	1.7327	1.7314	1.7302	1.7292	1.7283	1.7276	1.7271	1.7267
6	1.7264	1.7263	1.7263	1.7265	1.7269	1.7273	1.7280	1.7288	1.7297	1.7308
7	1.7320	1.7334	1.7349	1.7365	1.7383	1.7403	1.7424	1.7446	1.7470	1.7495
8	1.7522	1.7550	1.7579	1.7610	1.7642	1.7676	1.7711	1.7748	1.7785	1.7825
9	1.7865	1.7907	1.7951	1.7995	1.8041	1.8089	1.8138	1.8188	1.8240	1.8292
10	1.8347	1.8402	1.8459	1.8517	1.8577	1.8638	1.8700	1.8764	1.8829	1.8895
11	1.8962	1.9031	1.9101	1.9173	1.9245	1.9319	1.9395	1.9471	1.9549	1.9628
12	1.9708	1.9790	1.9873	1.9957	2.0043	2.0130	2.0218	2.0307	2.0397	2.0489
13	2.0582	2.0676	2.0772	2.0868	2.0966	2.1065	2.1166	2.1267	2.1370	2.1474
14	2.1579	2.1686	2.1794	2.1902	2.2013	2.2124	2.2236	2.2350	2.2465	2.2581
15	2.2698	2.2816	2.2936	2.3057	2.3179	2.3302	2.3426	2.3551	2.3678	2.3806
16	2.3935	2.4065	2.4196	2.4328	2.4462	2.4596	2.4732	2.4869	2.5007	2.5146
17	2.5286	2.5428	2.5570	2.5714	2.5859	2.6005	2.6152	2.6300	2.6449	2.6599
18	2.6751	2.6903	2.7057	2.7212	2.7368	2.7525	2.7683	2.7842	2.8002	2.8163
19	2.8325	2.8489	2.8653	2.8819	2.8986	2.9153	2.9322	2.9492	2.9663	2.9835
20	3.0008	3.0182	3.0357	3.0533	3.0710	3.0888	3.1068	3.1248	3.1430	3.1612
21	3.1795	3.1980	3.2165	3.2352	3.2539	3.2728	3.2918	3.3108	3.3300	3.3493
22	3.3686	3.3881	3.4077	3.4273	3.4471	3.4670	3.4869	3.5070	3.5272	3.5475
23	3.5678	3.5883	3.6089	3.6295	3.6503	3.6712	3.6921	3.7132	3.7343	3.7556
24	3.7770	3.7984	3.8199	3.8416	3.8633	3.8852	3.9071	3.9291	3.9513	3.9735
25	3.9958	4.0182	4.0407	4.0633	4.0860	4.1088	4.1317	4.1547	4.1778	4.2009
26	4.2242	4.2476	4.2710	4.2945	4.3182	4.3419	4.3657	4.3897	4.4137	4.4378
27	4.4620	4.4862	4.5106	4.5351	4.5596	4.5843	4.6090	4.6339	4.6588	4.6838
28	4.7089	4.7341	4.7594	4.7848	4.8103	4.8358	4.8615	4.8872	4.9130	4.9389
29	4.9649	4.9910	5.0172	5.0435	5.0698	5.0963	5.1228	5.1494	5.1762	5.2030
30	5.2298	5.2568	5.2839	5.3110	5.3383	5.3656	5.3930	5.4205	5.4481	5.4757
31	5.5035	5.5313	5.5592	5.5873	5.6153	5.6435	5.6718	5.7001	5.7286	5.7571
32	5.7857	5.8144	5.8432	5.8720	5.9010	5.9300	5.9591	5.9883	6.0176	6.0469
33	6.0764	6.1059	6.1355	6.1652	6.1950	6.2249	6.2548	6.2848	6.3149	6.3451
34	6.3754	6.4057	6.4362	6.4667	6.4973	6.5280	6.5587	6.5896	6.6205	6.6515
35	6.6826	6.7137	6.7450	6.7763	6.8077	6.8392	6.8707	6.9024	6.9341	6.9659
36	6.9978	7.0297	7.0618	7.0939	7.1261	7.1584	7.1907	7.2231	7.2557	7.2882
37	7.3209	7.3537	7.3865	7.4194	7.4523	7.4854	7.5185	7.5517	7.5850	7.6184
38	7.6518	7.6853	7.7189	7.7526	7.7863	7.8202	7.8540	7.8880	7.9221	7.9562
39	7.9904	8.0247	8.0590	8.0934	8.1279	8.1625	8.1971	8.2319	8.2667	8.3015
40	8.3365	8.3715	8.4066	8.4417	8.4770	8.5123	8.5477	8.5831	8.6187	8.6543
41	8.6899	8.7257	8.7615	8.7974	8.8334	8.8694	8.9055	8.9417	8.9779	9.0142

Corrections are in kg/g/mg and are to be added to the mass of water measured in kg/g/mg as the unit of capacity is in m³/dm³/cm³ respectively

Table 3.9 ALPHA = 30 × 10^{-6}/°C, DEN = 8000 kg/m³, REFERENCE TEMP = 27 °C

Temp	0.0	0.1	0.2	0.3	0.4	0.5	0.6	0.7	0.8	0.9
5	1.6699	1.6685	1.6674	1.6663	1.6655	1.6648	1.6642	1.6638	1.6635	1.6634
6	1.6635	1.6637	1.6640	1.6645	1.6651	1.6659	1.6668	1.6679	1.6692	1.6705
7	1.6721	1.6737	1.6755	1.6775	1.6796	1.6818	1.6842	1.6868	1.6895	1.6923
8	1.6952	1.6983	1.7016	1.7050	1.7085	1.7122	1.7160	1.7199	1.7240	1.7282
9	1.7326	1.7371	1.7417	1.7465	1.7514	1.7565	1.7616	1.7670	1.7724	1.7780
10	1.7837	1.7896	1.7956	1.8017	1.8080	1.8144	1.8209	1.8275	1.8343	1.8412
11	1.8483	1.8555	1.8628	1.8702	1.8778	1.8855	1.8933	1.9013	1.9094	1.9176
12	1.9259	1.9344	1.9430	1.9517	1.9605	1.9695	1.9786	1.9878	1.9972	2.0067
13	2.0163	2.0260	2.0358	2.0458	2.0559	2.0661	2.0764	2.0869	2.0975	2.1082
14	2.1190	2.1300	2.1410	2.1522	2.1635	2.1749	2.1865	2.1982	2.2099	2.2218
15	2.2339	2.2460	2.2583	2.2706	2.2831	2.2957	2.3085	2.3213	2.3343	2.3473
16	2.3605	2.3738	2.3872	2.4008	2.4144	2.4282	2.4421	2.4561	2.4702	2.4844
17	2.4987	2.5131	2.5277	2.5424	2.5571	2.5720	2.5870	2.6021	2.6174	2.6327
18	2.6482	2.6637	2.6794	2.6951	2.7110	2.7270	2.7431	2.7593	2.7756	2.7921
19	2.8086	2.8252	2.8420	2.8589	2.8758	2.8929	2.9101	2.9273	2.9447	2.9622
20	2.9798	2.9975	3.0153	3.0333	3.0513	3.0694	3.0876	3.1060	3.1244	3.1429
21	3.1616	3.1803	3.1992	3.2181	3.2372	3.2564	3.2756	3.2950	3.3144	3.3340
22	3.3537	3.3734	3.3933	3.4133	3.4334	3.4535	3.4738	3.4942	3.5146	3.5352
23	3.5559	3.5766	3.5975	3.6185	3.6395	3.6607	3.6820	3.7033	3.7248	3.7463
24	3.7680	3.7897	3.8116	3.8335	3.8556	3.8777	3.8999	3.9223	3.9447	3.9672
25	3.9898	4.0125	4.0353	4.0582	4.0812	4.1043	4.1275	4.1508	4.1742	4.1976
26	4.2212	4.2449	4.2686	4.2925	4.3164	4.3404	4.3645	4.3888	4.4131	4.4375
27	4.4620	4.4865	4.5112	4.5360	4.5608	4.5858	4.6108	4.6360	4.6612	4.6865
28	4.7119	4.7374	4.7630	4.7887	4.8144	4.8403	4.8662	4.8923	4.9184	4.9446
29	4.9709	4.9973	5.0238	5.0503	5.0770	5.1037	5.1306	5.1575	5.1845	5.2116
30	5.2388	5.2661	5.2934	5.3209	5.3484	5.3760	5.4037	5.4315	5.4594	5.4874
31	5.5154	5.5435	5.5718	5.6001	5.6285	5.6570	5.6855	5.7142	5.7429	5.7717
32	5.8006	5.8296	5.8587	5.8878	5.9171	5.9464	5.9758	6.0053	6.0349	6.0645
33	6.0943	6.1241	6.1540	6.1840	6.2141	6.2442	6.2745	6.3048	6.3352	6.3657
34	6.3962	6.4269	6.4576	6.4884	6.5193	6.5503	6.5814	6.6125	6.6437	6.6750
35	6.7064	6.7379	6.7694	6.8010	6.8327	6.8645	6.8963	6.9283	6.9603	6.9924
36	7.0246	7.0568	7.0892	7.1216	7.1541	7.1866	7.2193	7.2520	7.2848	7.3177
37	7.3507	7.3837	7.4168	7.4500	7.4833	7.5167	7.5501	7.5836	7.6172	7.6508
38	7.6846	7.7184	7.7523	7.7862	7.8203	7.8544	7.8886	7.9228	7.9572	7.9916
39	8.0261	8.0606	8.0953	8.1300	8.1648	8.1997	8.2346	8.2696	8.3047	8.3399
40	8.3751	8.4104	8.4458	8.4813	8.5168	8.5524	8.5881	8.6238	8.6597	8.6956
41	8.7315	8.7676	8.8037	8.8399	8.8761	8.9125	8.9489	8.9854	9.0219	9.0585

Corrections are in kg/g/mg and are to be added to the mass of water measured in kg/g/mg as the unit of capacity is in $m^3/dm^3/cm^3$ respectively

Table 3.10 ALPHA = $25 \times 10^{-6}/^{\circ}C$, DEN = 8000 kg/m^3, REFERENCE TEMP = 27 °C

Temp	0.0	0.1	0.2	0.3	0.4	0.5	0.6	0.7	0.8	0.9
5	1.6052	1.6043	1.6037	1.6031	1.6028	1.6026	1.6025	1.6026	1.6028	1.6032
6	1.6038	1.6044	1.6053	1.6063	1.6074	1.6087	1.6101	1.6117	1.6134	1.6153
7	1.6173	1.6195	1.6218	1.6243	1.6269	1.6296	1.6325	1.6356	1.6387	1.6421
8	1.6455	1.6491	1.6529	1.6568	1.6608	1.6650	1.6693	1.6737	1.6783	1.6830
9	1.6879	1.6929	1.6980	1.7033	1.7087	1.7143	1.7199	1.7258	1.7317	1.7378
10	1.7440	1.7504	1.7569	1.7635	1.7703	1.7772	1.7842	1.7913	1.7986	1.8060
11	1.8136	1.8213	1.8291	1.8370	1.8451	1.8533	1.8616	1.8701	1.8787	1.8874
12	1.8962	1.9052	1.9143	1.9235	1.9329	1.9423	1.9519	1.9616	1.9715	1.9815
13	1.9916	2.0018	2.0121	2.0226	2.0332	2.0439	2.0548	2.0657	2.0768	2.0880
14	2.0993	2.1108	2.1223	2.1340	2.1458	2.1578	2.1698	2.1820	2.1943	2.2067
15	2.2192	2.2318	2.2446	2.2574	2.2704	2.2835	2.2968	2.3101	2.3236	2.3371
16	2.3508	2.3646	2.3786	2.3926	2.4067	2.4210	2.4354	2.4499	2.4645	2.4792
17	2.4940	2.5090	2.5240	2.5392	2.5545	2.5698	2.5854	2.6010	2.6167	2.6325
18	2.6485	2.6645	2.6807	2.6970	2.7133	2.7298	2.7464	2.7631	2.7800	2.7969
19	2.8139	2.8311	2.8483	2.8657	2.8831	2.9007	2.9184	2.9362	2.9540	2.9720
20	2.9901	3.0083	3.0267	3.0451	3.0636	3.0822	3.1009	3.1198	3.1387	3.1577
21	3.1769	3.1961	3.2155	3.2349	3.2545	3.2742	3.2939	3.3138	3.3337	3.3538
22	3.3740	3.3942	3.4146	3.4351	3.4556	3.4763	3.4971	3.5180	3.5389	3.5600
23	3.5812	3.6024	3.6238	3.6453	3.6668	3.6885	3.7102	3.7321	3.7541	3.7761
24	3.7983	3.8205	3.8429	3.8653	3.8878	3.9105	3.9332	3.9560	3.9790	4.0020
25	4.0251	4.0483	4.0716	4.0950	4.1185	4.1421	4.1658	4.1896	4.2134	4.2374
26	4.2615	4.2856	4.3099	4.3342	4.3586	4.3832	4.4078	4.4325	4.4573	4.4822
27	4.5072	4.5323	4.5574	4.5827	4.6081	4.6335	4.6591	4.6847	4.7104	4.7362
28	4.7621	4.7881	4.8142	4.8404	4.8666	4.8930	4.9194	4.9460	4.9726	4.9993
29	5.0261	5.0530	5.0800	5.1070	5.1342	5.1614	5.1887	5.2162	5.2437	5.2713
30	5.2989	5.3267	5.3546	5.3825	5.4105	5.4387	5.4669	5.4952	5.5235	5.5520
31	5.5805	5.6092	5.6379	5.6667	5.6956	5.7246	5.7536	5.7828	5.8120	5.8413
32	5.8707	5.9002	5.9298	5.9594	5.9892	6.0190	6.0489	6.0789	6.1089	6.1391
33	6.1693	6.1997	6.2301	6.2606	6.2911	6.3218	6.3525	6.3833	6.4142	6.4452
34	6.4763	6.5074	6.5386	6.5699	6.6013	6.6328	6.6644	6.6960	6.7277	6.7595
35	6.7914	6.8233	6.8554	6.8875	6.9197	6.9519	6.9843	7.0167	7.0492	7.0818
36	7.1145	7.1473	7.1801	7.2130	7.2460	7.2791	7.3122	7.3454	7.3787	7.4121
37	7.4456	7.4791	7.5127	7.5464	7.5802	7.6140	7.6479	7.6819	7.7160	7.7502
38	7.7844	7.8187	7.8531	7.8875	7.9221	7.9567	7.9914	8.0261	8.0609	8.0959
39	8.1308	8.1659	8.2010	8.2363	8.2715	8.3069	8.3423	8.3778	8.4134	8.4491
40	8.4848	8.5206	8.5565	8.5925	8.6285	8.6646	8.7008	8.7370	8.7733	8.8097
41	8.8462	8.8827	8.9193	8.9560	8.9928	9.0296	9.0665	9.1034	9.1405	9.1776

Corrections are in kg/g/mg and are to be added to the mass of water measured in kg/g/mg as the unit of capacity is in m³/dm³/cm³ respectively

Table 3.11 ALPHA = 15 × 10^{-6}/°C, DEN = 8000 kg/m³, REFERENCE TEMP = 27 °C

Temp	0.0	0.1	0.2	0.3	0.4	0.5	0.6	0.7	0.8	0.9
5	1.3402	1.3404	1.3407	1.3412	1.3418	1.3426	1.3435	1.3446	1.3459	1.3472
6	1.3488	1.3505	1.3523	1.3543	1.3564	1.3587	1.3612	1.3637	1.3665	1.3694
7	1.3724	1.3755	1.3789	1.3823	1.3859	1.3897	1.3936	1.3976	1.4018	1.4061
8	1.4106	1.4152	1.4199	1.4248	1.4298	1.4350	1.4403	1.4457	1.4513	1.4570
9	1.4629	1.4689	1.4750	1.4813	1.4877	1.4943	1.5010	1.5078	1.5147	1.5218
10	1.5291	1.5364	1.5439	1.5515	1.5593	1.5672	1.5752	1.5834	1.5917	1.6001
11	1.6086	1.6173	1.6261	1.6351	1.6441	1.6533	1.6626	1.6721	1.6817	1.6914
12	1.7012	1.7112	1.7213	1.7315	1.7419	1.7524	1.7630	1.7737	1.7845	1.7955
13	1.8066	1.8178	1.8292	1.8406	1.8522	1.8639	1.8758	1.8877	1.8998	1.9120
14	1.9244	1.9368	1.9494	1.9621	1.9749	1.9878	2.0008	2.0140	2.0273	2.0407
15	2.0542	2.0678	2.0816	2.0955	2.1095	2.1236	2.1378	2.1521	2.1666	2.1812
16	2.1959	2.2107	2.2256	2.2406	2.2558	2.2710	2.2864	2.3019	2.3175	2.3332
17	2.3490	2.3650	2.3810	2.3972	2.4135	2.4299	2.4464	2.4630	2.4797	2.4965
18	2.5135	2.5305	2.5477	2.5650	2.5823	2.5998	2.6174	2.6351	2.6530	2.6709
19	2.6889	2.7071	2.7253	2.7437	2.7621	2.7807	2.7994	2.8181	2.8370	2.8560
20	2.8751	2.8943	2.9136	2.9330	2.9526	2.9722	2.9919	3.0117	3.0317	3.0517
21	3.0719	3.0921	3.1125	3.1329	3.1535	3.1741	3.1949	3.2157	3.2367	3.2578
22	3.2789	3.3002	3.3215	3.3430	3.3646	3.3862	3.4080	3.4299	3.4518	3.4739
23	3.4961	3.5183	3.5407	3.5632	3.5857	3.6084	3.6311	3.6540	3.6770	3.7000
24	3.7232	3.7464	3.7697	3.7932	3.8167	3.8403	3.8641	3.8879	3.9118	3.9358
25	3.9599	3.9841	4.0085	4.0328	4.0573	4.0819	4.1066	4.1314	4.1562	4.1812
26	4.2063	4.2314	4.2567	4.2820	4.3074	4.3330	4.3586	4.3843	4.4101	4.4360
27	4.4620	4.4880	4.5142	4.5405	4.5668	4.5933	4.6198	4.6464	4.6731	4.6999
28	4.7268	4.7538	4.7809	4.8081	4.8353	4.8627	4.8901	4.9176	4.9453	4.9730
29	5.0008	5.0286	5.0566	5.0847	5.1128	5.1411	5.1694	5.1978	5.2263	5.2549
30	5.2836	5.3123	5.3412	5.3701	5.3991	5.4282	5.4574	5.4867	5.5161	5.5455
31	5.5751	5.6047	5.6344	5.6642	5.6941	5.7241	5.7541	5.7843	5.8145	5.8448
32	5.8752	5.9057	5.9362	5.9669	5.9976	6.0284	6.0593	6.0903	6.1213	6.1525
33	6.1837	6.2150	6.2464	6.2779	6.3095	6.3411	6.3728	6.4046	6.4365	6.4685
34	6.5006	6.5327	6.5649	6.5972	6.6296	6.6620	6.6946	6.7272	6.7599	6.7927
35	6.8256	6.8585	6.8915	6.9246	6.9578	6.9911	7.0244	7.0579	7.0914	7.1249
36	7.1586	7.1923	7.2262	7.2601	7.2940	7.3281	7.3622	7.3964	7.4307	7.4651
37	7.4995	7.5341	7.5687	7.6033	7.6381	7.6729	7.7078	7.7428	7.7779	7.8130
38	7.8482	7.8835	7.9189	7.9543	7.9899	8.0255	8.0611	8.0969	8.1327	8.1686
39	8.2046	8.2406	8.2767	8.3129	8.3492	8.3856	8.4220	8.4585	8.4951	8.5317
40	8.5684	8.6052	8.6421	8.6790	8.7160	8.7531	8.7903	8.8275	8.8648	8.9022
41	8.9396	8.9771	9.0147	9.0524	9.0901	9.1279	9.1658	9.2038	9.2418	9.2799

Corrections are in kg/g/mg and are to be added to the mass of water measured in kg/g/mg as the unit of capacity is in m³/dm³/cm³ respectively

Table 3.12 ALPHA = 10 × 10^{-6}/°C, DEN = 8000 kg/m³, REFERENCE TEMP = 27 °C

Temp	0.0	0.1	0.2	0.3	0.4	0.5	0.6	0.7	0.8	0.9
5	1.2303	1.2310	1.2318	1.2328	1.2339	1.2352	1.2366	1.2382	1.2400	1.2419
6	1.2439	1.2461	1.2484	1.2509	1.2536	1.2563	1.2593	1.2624	1.2656	1.2690
7	1.2725	1.2762	1.2800	1.2839	1.2880	1.2923	1.2967	1.3012	1.3059	1.3107
8	1.3157	1.3208	1.3260	1.3314	1.3369	1.3426	1.3484	1.3543	1.3604	1.3667
9	1.3730	1.3795	1.3862	1.3929	1.3998	1.4069	1.4141	1.4214	1.4288	1.4364
10	1.4442	1.4520	1.4600	1.4681	1.4764	1.4848	1.4933	1.5020	1.5108	1.5197
11	1.5287	1.5379	1.5472	1.5567	1.5662	1.5759	1.5858	1.5957	1.6058	1.6160
12	1.6264	1.6368	1.6474	1.6581	1.6690	1.6800	1.6911	1.7023	1.7136	1.7251
13	1.7367	1.7484	1.7603	1.7722	1.7843	1.7966	1.8089	1.8214	1.8339	1.8466
14	1.8595	1.8724	1.8855	1.8987	1.9120	1.9254	1.9389	1.9526	1.9664	1.9803
15	1.9943	2.0085	2.0227	2.0371	2.0516	2.0662	2.0809	2.0958	2.1107	2.1258
16	2.1410	2.1563	2.1717	2.1872	2.2029	2.2186	2.2345	2.2505	2.2666	2.2828
17	2.2991	2.3156	2.3321	2.3488	2.3656	2.3825	2.3995	2.4166	2.4338	2.4511
18	2.4686	2.4861	2.5038	2.5216	2.5394	2.5574	2.5755	2.5937	2.6121	2.6305
19	2.6490	2.6677	2.6864	2.7053	2.7242	2.7433	2.7625	2.7817	2.8011	2.8206
20	2.8402	2.8599	2.8797	2.8996	2.9197	2.9398	2.9600	2.9803	3.0008	3.0213
21	3.0419	3.0627	3.0835	3.1045	3.1255	3.1467	3.1680	3.1893	3.2108	3.2323
22	3.2540	3.2758	3.2976	3.3196	3.3417	3.3638	3.3861	3.4085	3.4309	3.4535
23	3.4761	3.4989	3.5218	3.5447	3.5678	3.5910	3.6142	3.6376	3.6610	3.6846
24	3.7082	3.7320	3.7558	3.7797	3.8038	3.8279	3.8521	3.8764	3.9009	3.9254
25	3.9500	3.9747	3.9995	4.0244	4.0494	4.0745	4.0996	4.1249	4.1503	4.1757
26	4.2013	4.2269	4.2527	4.2785	4.3044	4.3305	4.3566	4.3828	4.4091	4.4355
27	4.4620	4.4885	4.5152	4.5420	4.5688	4.5957	4.6228	4.6499	4.6771	4.7044
28	4.7318	4.7593	4.7869	4.8145	4.8423	4.8701	4.8981	4.9261	4.9542	4.9824
29	5.0107	5.0391	5.0676	5.0961	5.1248	5.1535	5.1823	5.2112	5.2402	5.2693
30	5.2985	5.3277	5.3571	5.3865	5.4160	5.4456	5.4753	5.5051	5.5350	5.5649
31	5.5950	5.6251	5.6553	5.6856	5.7160	5.7464	5.7770	5.8076	5.8383	5.8691
32	5.9000	5.9310	5.9621	5.9932	6.0244	6.0557	6.0871	6.1186	6.1502	6.1818
33	6.2135	6.2453	6.2772	6.3092	6.3413	6.3734	6.4056	6.4379	6.4703	6.5028
34	6.5353	6.5680	6.6007	6.6335	6.6663	6.6993	6.7323	6.7654	6.7986	6.8319
35	6.8653	6.8987	6.9322	6.9658	6.9995	7.0333	7.0671	7.1010	7.1350	7.1691
36	7.2033	7.2375	7.2718	7.3062	7.3407	7.3752	7.4099	7.4446	7.4793	7.5142
37	7.5491	7.5842	7.6193	7.6544	7.6897	7.7250	7.7604	7.7959	7.8315	7.8671
38	7.9028	7.9386	7.9744	8.0104	8.0464	8.0825	8.1187	8.1549	8.1912	8.2276
39	8.2641	8.3006	8.3372	8.3739	8.4107	8.4475	8.4844	8.5214	8.5585	8.5956
40	8.6328	8.6701	8.7075	8.7449	8.7824	8.8200	8.8577	8.8954	8.9332	8.9710
41	9.0090	9.0470	9.0851	9.1232	9.1615	9.1998	9.2381	9.2766	9.3151	9.3537

Corrections are in kg/g/mg and are to be added to the mass of water measured in kg/g/mg as the unit of capacity is in m³/dm³/cm³ respectively

Table 3.13 ALPHA = 54 × 10^{-6}/°C, DEN = 8400 kg/m³, REFERENCE TEMP = 20 °C

Temp	0.0	0.1	0.2	0.3	0.4	0.5	0.6	0.7	0.8	0.9
5	1.8996	1.8958	1.8923	1.8889	1.8856	1.8825	1.8795	1.8767	1.8740	1.8715
6	1.8692	1.8670	1.8649	1.8630	1.8613	1.8596	1.8582	1.8569	1.8557	1.8547
7	1.8538	1.8531	1.8525	1.8521	1.8518	1.8516	1.8516	1.8517	1.8520	1.8524
8	1.8530	1.8537	1.8546	1.8556	1.8567	1.8580	1.8594	1.8609	1.8626	1.8644
9	1.8664	1.8685	1.8707	1.8731	1.8756	1.8783	1.8811	1.8840	1.8870	1.8902
10	1.8936	1.8970	1.9006	1.9043	1.9082	1.9122	1.9163	1.9206	1.9250	1.9295
11	1.9342	1.9389	1.9439	1.9489	1.9541	1.9594	1.9648	1.9704	1.9761	1.9819
12	1.9878	1.9939	2.0001	2.0064	2.0129	2.0194	2.0262	2.0330	2.0399	2.0470
13	2.0542	2.0615	2.0690	2.0766	2.0843	2.0921	2.1000	2.1081	2.1163	2.1246
14	2.1330	2.1416	2.1502	2.1590	2.1679	2.1770	2.1861	2.1954	2.2048	2.2143
15	2.2239	2.2337	2.2435	2.2535	2.2636	2.2738	2.2842	2.2946	2.3052	2.3159
16	2.3267	2.3376	2.3486	2.3597	2.3710	2.3824	2.3938	2.4054	2.4172	2.4290
17	2.4409	2.4530	2.4651	2.4774	2.4898	2.5023	2.5149	2.5276	2.5404	2.5534
18	2.5664	2.5796	2.5929	2.6063	2.6198	2.6334	2.6471	2.6609	2.6748	2.6889
19	2.7030	2.7172	2.7316	2.7461	2.7606	2.7753	2.7901	2.8050	2.8200	2.8351
20	2.8503	2.8656	2.8811	2.8966	2.9122	2.9280	2.9438	2.9598	2.9758	2.9920
21	3.0082	3.0246	3.0410	3.0576	3.0743	3.0910	3.1079	3.1249	3.1420	3.1591
22	3.1764	3.1938	3.2113	3.2289	3.2466	3.2643	3.2822	3.3002	3.3183	3.3365
23	3.3548	3.3732	3.3916	3.4102	3.4289	3.4477	3.4666	3.4855	3.5046	3.5238
24	3.5430	3.5624	3.5819	3.6014	3.6211	3.6409	3.6607	3.6806	3.7007	3.7208
25	3.7411	3.7614	3.7818	3.8023	3.8230	3.8437	3.8645	3.8854	3.9064	3.9275
26	3.9486	3.9699	3.9913	4.0127	4.0343	4.0560	4.0777	4.0995	4.1215	4.1435
27	4.1656	4.1878	4.2101	4.2325	4.2550	4.2775	4.3002	4.3230	4.3458	4.3687
28	4.3918	4.4149	4.4381	4.4614	4.4848	4.5083	4.5318	4.5555	4.5792	4.6031
29	4.6270	4.6510	4.6751	4.6993	4.7236	4.7480	4.7724	4.7970	4.8216	4.8463
30	4.8712	4.8961	4.9210	4.9461	4.9713	4.9965	5.0219	5.0473	5.0728	5.0984
31	5.1241	5.1498	5.1757	5.2016	5.2276	5.2537	5.2799	5.3062	5.3326	5.3590
32	5.3856	5.4122	5.4389	5.4657	5.4925	5.5195	5.5465	5.5737	5.6009	5.6282
33	5.6555	5.6830	5.7105	5.7382	5.7659	5.7937	5.8215	5.8495	5.8775	5.9056
34	5.9338	5.9621	5.9905	6.0189	6.0475	6.0761	6.1048	6.1335	6.1624	6.1913
35	6.2203	6.2494	6.2786	6.3079	6.3372	6.3666	6.3961	6.4257	6.4554	6.4851
36	6.5149	6.5448	6.5748	6.6048	6.6350	6.6652	6.6955	6.7258	6.7563	6.7868
37	6.8174	6.8481	6.8789	6.9097	6.9406	6.9716	7.0027	7.0338	7.0650	7.0963
38	7.1277	7.1592	7.1907	7.2223	7.2540	7.2857	7.3176	7.3495	7.3815	7.4135
39	7.4457	7.4779	7.5102	7.5426	7.5750	7.6075	7.6401	7.6728	7.7055	7.7383
40	7.7712	7.8042	7.8372	7.8703	7.9035	7.9367	7.9701	8.0035	8.0369	8.0705
41	8.1041	8.1378	8.1716	8.2054	8.2393	8.2733	8.3074	8.3415	8.3757	8.4100

Corrections are in kg/g/mg and are to be added to the mass of water measured in kg/g/mg as the unit of capacity is in $m^3/dm^3/cm^3$ respectively

Table 3.14 ALPHA = 33×10^{-6}/°C, DEN = 8400 kg/m^3, REFERENCE TEMP = 20 °C

Temp	0.0	0.1	0.2	0.3	0.4	0.5	0.6	0.7	0.8	0.9
5	1.5849	1.5833	1.5818	1.5805	1.5793	1.5783	1.5774	1.5767	1.5762	1.5758
6	1.5755	1.5754	1.5754	1.5756	1.5760	1.5765	1.5771	1.5779	1.5788	1.5799
7	1.5811	1.5825	1.5840	1.5857	1.5875	1.5894	1.5915	1.5937	1.5961	1.5986
8	1.6013	1.6041	1.6071	1.6102	1.6134	1.6167	1.6203	1.6239	1.6277	1.6316
9	1.6357	1.6399	1.6442	1.6487	1.6533	1.6581	1.6630	1.6680	1.6731	1.6784
10	1.6838	1.6894	1.6951	1.7009	1.7069	1.7130	1.7192	1.7256	1.7321	1.7387
11	1.7454	1.7523	1.7593	1.7665	1.7737	1.7811	1.7887	1.7963	1.8041	1.8120
12	1.8201	1.8283	1.8366	1.8450	1.8535	1.8622	1.8710	1.8799	1.8890	1.8982
13	1.9075	1.9169	1.9264	1.9361	1.9459	1.9558	1.9659	1.9760	1.9863	1.9967
14	2.0072	2.0179	2.0287	2.0396	2.0506	2.0617	2.0729	2.0843	2.0958	2.1074
15	2.1191	2.1310	2.1429	2.1550	2.1672	2.1795	2.1920	2.2045	2.2172	2.2299
16	2.2428	2.2558	2.2690	2.2822	2.2956	2.3090	2.3226	2.3363	2.3501	2.3640
17	2.3781	2.3922	2.4065	2.4208	2.4353	2.4499	2.4646	2.4794	2.4944	2.5094
18	2.5245	2.5398	2.5552	2.5707	2.5862	2.6019	2.6177	2.6337	2.6497	2.6658
19	2.6821	2.6984	2.7149	2.7314	2.7481	2.7649	2.7817	2.7987	2.8158	2.8330
20	2.8503	2.8677	2.8853	2.9029	2.9206	2.9384	2.9564	2.9744	2.9926	3.0108
21	3.0291	3.0476	3.0662	3.0848	3.1036	3.1224	3.1414	3.1605	3.1796	3.1989
22	3.2183	3.2378	3.2573	3.2770	3.2968	3.3167	3.3366	3.3567	3.3769	3.3972
23	3.4176	3.4380	3.4586	3.4793	3.5000	3.5209	3.5419	3.5630	3.5841	3.6054
24	3.6267	3.6482	3.6697	3.6914	3.7131	3.7350	3.7569	3.7790	3.8011	3.8233
25	3.8456	3.8681	3.8906	3.9132	3.9359	3.9587	3.9816	4.0046	4.0277	4.0508
26	4.0741	4.0975	4.1209	4.1445	4.1681	4.1919	4.2157	4.2396	4.2636	4.2877
27	4.3119	4.3362	4.3606	4.3851	4.4096	4.4343	4.4591	4.4839	4.5088	4.5339
28	4.5590	4.5842	4.6095	4.6348	4.6603	4.6859	4.7115	4.7373	4.7631	4.7890
29	4.8150	4.8411	4.8673	4.8936	4.9200	4.9464	4.9730	4.9996	5.0263	5.0531
30	5.0800	5.1070	5.1341	5.1612	5.1885	5.2158	5.2432	5.2707	5.2983	5.3260
31	5.3537	5.3816	5.4095	5.4375	5.4656	5.4938	5.5221	5.5505	5.5789	5.6074
32	5.6360	5.6647	5.6935	5.7224	5.7513	5.7804	5.8095	5.8387	5.8680	5.8974
33	5.9268	5.9563	5.9860	6.0157	6.0454	6.0753	6.1053	6.1353	6.1654	6.1956
34	6.2259	6.2562	6.2867	6.3172	6.3478	6.3785	6.4093	6.4401	6.4710	6.5020
35	6.5331	6.5643	6.5956	6.6269	6.6583	6.6898	6.7214	6.7530	6.7847	6.8166
36	6.8484	6.8804	6.9125	6.9446	6.9768	7.0091	7.0414	7.0739	7.1064	7.1390
37	7.1717	7.2044	7.2372	7.2701	7.3031	7.3362	7.3693	7.4025	7.4358	7.4692
38	7.5027	7.5362	7.5698	7.6035	7.6372	7.6710	7.7049	7.7389	7.7730	7.8071
39	7.8413	7.8756	7.9099	7.9444	7.9789	8.0135	8.0481	8.0828	8.1177	8.1525
40	8.1875	8.2225	8.2576	8.2928	8.3280	8.3633	8.3987	8.4342	8.4697	8.5054
41	8.5410	8.5768	8.6126	8.6485	8.6845	8.7205	8.7567	8.7929	8.8291	8.8654

Corrections are in kg/g/mg and are to be added to the mass of water measured in kg/g/mg as the unit of capacity is in m³/dm³/cm³ respectively

Table 3.15 ALPHA = 30 × 10^{-6}/°C, DEN = 8400 kg/m³, REFERENCE TEMP = 20 °C

Temp	0.0	0.1	0.2	0.3	0.4	0.5	0.6	0.7	0.8	0.9
5	1.5400	1.5386	1.5375	1.5364	1.5356	1.5349	1.5343	1.5339	1.5336	1.5335
6	1.5336	1.5337	1.5341	1.5346	1.5352	1.5360	1.5369	1.5380	1.5393	1.5406
7	1.5422	1.5438	1.5456	1.5476	1.5497	1.5520	1.5544	1.5569	1.5596	1.5624
8	1.5654	1.5685	1.5717	1.5751	1.5786	1.5823	1.5861	1.5900	1.5941	1.5984
9	1.6027	1.6072	1.6119	1.6166	1.6216	1.6266	1.6318	1.6371	1.6426	1.6482
10	1.6539	1.6597	1.6657	1.6719	1.6781	1.6845	1.6911	1.6977	1.7045	1.7114
11	1.7185	1.7257	1.7330	1.7404	1.7480	1.7557	1.7635	1.7715	1.7796	1.7878
12	1.7961	1.8046	1.8132	1.8219	1.8308	1.8397	1.8488	1.8581	1.8674	1.8769
13	1.8865	1.8962	1.9061	1.9160	1.9261	1.9364	1.9467	1.9572	1.9677	1.9785
14	1.9893	2.0002	2.0113	2.0225	2.0338	2.0452	2.0568	2.0684	2.0802	2.0921
15	2.1042	2.1163	2.1286	2.1409	2.1534	2.1661	2.1788	2.1916	2.2046	2.2177
16	2.2309	2.2442	2.2576	2.2711	2.2848	2.2985	2.3124	2.3264	2.3405	2.3547
17	2.3691	2.3835	2.3981	2.4128	2.4275	2.4424	2.4574	2.4726	2.4878	2.5031
18	2.5186	2.5341	2.5498	2.5656	2.5815	2.5975	2.6136	2.6298	2.6461	2.6625
19	2.6791	2.6957	2.7125	2.7293	2.7463	2.7634	2.7805	2.7978	2.8152	2.8327
20	2.8503	2.8680	2.8859	2.9038	2.9218	2.9399	2.9582	2.9765	2.9949	3.0135
21	3.0321	3.0509	3.0697	3.0887	3.1078	3.1269	3.1462	3.1656	3.1850	3.2046
22	3.2243	3.2440	3.2639	3.2839	3.3040	3.3241	3.3444	3.3648	3.3853	3.4058
23	3.4265	3.4473	3.4682	3.4891	3.5102	3.5314	3.5526	3.5740	3.5955	3.6170
24	3.6387	3.6604	3.6823	3.7042	3.7263	3.7484	3.7707	3.7930	3.8154	3.8380
25	3.8606	3.8833	3.9061	3.9290	3.9520	3.9751	3.9983	4.0216	4.0450	4.0685
26	4.0920	4.1157	4.1394	4.1633	4.1872	4.2113	4.2354	4.2596	4.2839	4.3083
27	4.3328	4.3574	4.3821	4.4069	4.4317	4.4567	4.4818	4.5069	4.5321	4.5574
28	4.5829	4.6084	4.6339	4.6596	4.6854	4.7113	4.7372	4.7633	4.7894	4.8156
29	4.8419	4.8683	4.8948	4.9214	4.9480	4.9748	5.0016	5.0286	5.0556	5.0827
30	5.1099	5.1371	5.1645	5.1920	5.2195	5.2471	5.2748	5.3026	5.3305	5.3585
31	5.3865	5.4147	5.4429	5.4712	5.4996	5.5281	5.5567	5.5854	5.6141	5.6429
32	5.6718	5.7008	5.7299	5.7591	5.7883	5.8176	5.8471	5.8766	5.9061	5.9358
33	5.9656	5.9954	6.0253	6.0553	6.0854	6.1155	6.1458	6.1761	6.2065	6.2370
34	6.2676	6.2983	6.3290	6.3598	6.3907	6.4217	6.4528	6.4839	6.5151	6.5464
35	6.5778	6.6093	6.6408	6.6725	6.7042	6.7360	6.7678	6.7998	6.8318	6.8639
36	6.8961	6.9283	6.9607	6.9931	7.0256	7.0582	7.0908	7.1236	7.1564	7.1893
37	7.2223	7.2553	7.2884	7.3216	7.3549	7.3883	7.4217	7.4552	7.4888	7.5225
38	7.5562	7.5900	7.6239	7.6579	7.6920	7.7261	7.7603	7.7945	7.8289	7.8633
39	7.8978	7.9324	7.9671	8.0018	8.0366	8.0715	8.1064	8.1414	8.1765	8.2117
40	8.2469	8.2823	8.3177	8.3531	8.3887	8.4243	8.4600	8.4957	8.5316	8.5675
41	8.6035	8.6395	8.6756	8.7118	8.7481	8.7844	8.8208	8.8573	8.8939	8.9305

Corrections are in kg/g/mg and are to be added to the mass of water measured in kg/g/mg as the unit of capacity is in m³/dm³/cm³ respectively

Table 3.16 ALPHA = 25 × 10^{-6}/°C, DEN = 8400 kg/m^3, REFERENCE TEMP = 20°C

Temp	0.0	0.1	0.2	0.3	0.4	0.5	0.6	0.7	0.8	0.9
5	1.4650	1.4642	1.4635	1.4630	1.4627	1.4624	1.4624	1.4625	1.4627	1.4631
6	1.4636	1.4643	1.4652	1.4662	1.4673	1.4686	1.4700	1.4716	1.4733	1.4752
7	1.4772	1.4794	1.4817	1.4842	1.4868	1.4895	1.4924	1.4955	1.4986	1.5020
8	1.5054	1.5090	1.5128	1.5167	1.5207	1.5249	1.5292	1.5336	1.5382	1.5429
9	1.5478	1.5528	1.5579	1.5632	1.5686	1.5742	1.5799	1.5857	1.5916	1.5977
10	1.6040	1.6103	1.6168	1.6234	1.6302	1.6371	1.6441	1.6513	1.6586	1.6660
11	1.6735	1.6812	1.6890	1.6970	1.7050	1.7132	1.7216	1.7300	1.7386	1.7473
12	1.7562	1.7652	1.7742	1.7835	1.7928	1.8023	1.8119	1.8216	1.8315	1.8415
13	1.8516	1.8618	1.8721	1.8826	1.8932	1.9039	1.9148	1.9257	1.9368	1.9480
14	1.9593	1.9708	1.9824	1.9940	2.0059	2.0178	2.0298	2.0420	2.0543	2.0667
15	2.0792	2.0919	2.1046	2.1175	2.1305	2.1436	2.1568	2.1702	2.1836	2.1972
16	2.2109	2.2247	2.2386	2.2527	2.2668	2.2811	2.2955	2.3100	2.3246	2.3393
17	2.3541	2.3691	2.3841	2.3993	2.4146	2.4300	2.4455	2.4611	2.4768	2.4926
18	2.5086	2.5246	2.5408	2.5571	2.5735	2.5900	2.6066	2.6233	2.6401	2.6570
19	2.6741	2.6912	2.7085	2.7258	2.7433	2.7609	2.7785	2.7963	2.8142	2.8322
20	2.8503	2.8685	2.8869	2.9053	2.9238	2.9424	2.9612	2.9800	2.9989	3.0180
21	3.0371	3.0564	3.0757	3.0952	3.1147	3.1344	3.1542	3.1740	3.1940	3.2141
22	3.2342	3.2545	3.2749	3.2954	3.3159	3.3366	3.3574	3.3783	3.3992	3.4203
23	3.4415	3.4627	3.4841	3.5056	3.5271	3.5488	3.5706	3.5924	3.6144	3.6365
24	3.6586	3.6809	3.7032	3.7257	3.7482	3.7708	3.7936	3.8164	3.8393	3.8624
25	3.8855	3.9087	3.9320	3.9554	3.9789	4.0025	4.0262	4.0500	4.0739	4.0978
26	4.1219	4.1461	4.1703	4.1947	4.2191	4.2436	4.2683	4.2930	4.3178	4.3427
27	4.3677	4.3928	4.4179	4.4432	4.4686	4.4940	4.5196	4.5452	4.5709	4.5968
28	4.6227	4.6487	4.6747	4.7009	4.7272	4.7536	4.7800	4.8065	4.8332	4.8599
29	4.8867	4.9136	4.9406	4.9676	4.9948	5.0220	5.0494	5.0768	5.1043	5.1319
30	5.1596	5.1874	5.2152	5.2432	5.2712	5.2993	5.3275	5.3558	5.3842	5.4127
31	5.4412	5.4699	5.4986	5.5274	5.5563	5.5853	5.6144	5.6435	5.6727	5.7021
32	5.7315	5.7610	5.7905	5.8202	5.8499	5.8798	5.9097	5.9397	5.9697	5.9999
33	6.0301	6.0605	6.0909	6.1214	6.1520	6.1826	6.2133	6.2442	6.2751	6.3061
34	6.3371	6.3683	6.3995	6.4308	6.4622	6.4937	6.5253	6.5569	6.5886	6.6204
35	6.6523	6.6843	6.7163	6.7484	6.7806	6.8129	6.8453	6.8777	6.9102	6.9428
36	6.9755	7.0083	7.0411	7.0740	7.1070	7.1401	7.1732	7.2064	7.2398	7.2731
37	7.3066	7.3401	7.3738	7.4075	7.4412	7.4751	7.5090	7.5430	7.5771	7.6113
38	7.6455	7.6798	7.7142	7.7487	7.7832	7.8178	7.8525	7.8873	7.9221	7.9570
39	7.9920	8.0271	8.0622	8.0975	8.1327	8.1681	8.2036	8.2391	8.2747	8.3103
40	8.3461	8.3819	8.4178	8.4537	8.4898	8.5259	8.5620	8.5983	8.6346	8.6710
41	8.7075	8.7440	8.7806	8.8173	8.8541	8.8909	8.9278	8.9648	9.0018	9.0390

Corrections are in kg/g/mg and are to be added to the mass of water measured in kg/g/mg as the unit of capacity is in m³/dm³/cm³ respectively

Table 3.17 ALPHA = 15.10^{-6}/°C, DEN = 8400 kg/m³, REFERENCE TEMP = 20°C

Temp	0.0	0.1	0.2	0.3	0.4	0.5	0.6	0.7	0.8	0.9
5	1.3152	1.3154	1.3157	1.3162	1.3168	1.3176	1.3185	1.3196	1.3209	1.3223
6	1.3238	1.3255	1.3273	1.3293	1.3315	1.3337	1.3362	1.3388	1.3415	1.3444
7	1.3474	1.3506	1.3539	1.3573	1.3609	1.3647	1.3686	1.3726	1.3768	1.3811
8	1.3856	1.3902	1.3949	1.3998	1.4048	1.4100	1.4153	1.4208	1.4263	1.4321
9	1.4379	1.4439	1.4501	1.4564	1.4628	1.4693	1.4760	1.4828	1.4898	1.4969
10	1.5041	1.5115	1.5189	1.5266	1.5343	1.5422	1.5502	1.5584	1.5667	1.5751
11	1.5837	1.5923	1.6012	1.6101	1.6192	1.6284	1.6377	1.6472	1.6568	1.6665
12	1.6763	1.6863	1.6964	1.7066	1.7169	1.7274	1.7380	1.7487	1.7596	1.7706
13	1.7817	1.7929	1.8042	1.8157	1.8273	1.8390	1.8509	1.8628	1.8749	1.8871
14	1.8994	1.9119	1.9245	1.9371	1.9500	1.9629	1.9759	1.9891	2.0024	2.0158
15	2.0293	2.0430	2.0567	2.0706	2.0846	2.0987	2.1129	2.1273	2.1417	2.1563
16	2.1710	2.1858	2.2007	2.2157	2.2309	2.2462	2.2615	2.2770	2.2926	2.3084
17	2.3242	2.3401	2.3562	2.3723	2.3886	2.4050	2.4215	2.4381	2.4549	2.4717
18	2.4886	2.5057	2.5229	2.5401	2.5575	2.5750	2.5926	2.6103	2.6281	2.6461
19	2.6641	2.6822	2.7005	2.7188	2.7373	2.7559	2.7746	2.7933	2.8122	2.8312
20	2.8503	2.8695	2.8888	2.9083	2.9278	2.9474	2.9671	2.9870	3.0069	3.0269
21	3.0471	3.0673	3.0877	3.1081	3.1287	3.1494	3.1701	3.1910	3.2119	3.2330
22	3.2542	3.2754	3.2968	3.3183	3.3398	3.3615	3.3833	3.4052	3.4271	3.4492
23	3.4714	3.4936	3.5160	3.5385	3.5610	3.5837	3.6064	3.6293	3.6523	3.6753
24	3.6985	3.7217	3.7451	3.7685	3.7920	3.8157	3.8394	3.8632	3.8872	3.9112
25	3.9353	3.9595	3.9838	4.0082	4.0327	4.0573	4.0820	4.1067	4.1316	4.1566
26	4.1816	4.2068	4.2320	4.2574	4.2828	4.3083	4.3340	4.3597	4.3855	4.4114
27	4.4374	4.4634	4.4896	4.5159	4.5422	4.5687	4.5952	4.6218	4.6486	4.6754
28	4.7023	4.7293	4.7563	4.7835	4.8108	4.8381	4.8656	4.8931	4.9207	4.9484
29	4.9762	5.0041	5.0321	5.0601	5.0883	5.1165	5.1449	5.1733	5.2018	5.2304
30	5.2590	5.2878	5.3167	5.3456	5.3746	5.4037	5.4329	5.4622	5.4916	5.5211
31	5.5506	5.5802	5.6100	5.6398	5.6696	5.6996	5.7297	5.7598	5.7900	5.8203
32	5.8507	5.8812	5.9118	5.9424	5.9732	6.0040	6.0349	6.0659	6.0969	6.1281
33	6.1593	6.1906	6.2220	6.2535	6.2851	6.3167	6.3485	6.3803	6.4122	6.4441
34	6.4762	6.5083	6.5406	6.5729	6.6052	6.6377	6.6703	6.7029	6.7356	6.7684
35	6.8012	6.8342	6.8672	6.9003	6.9335	6.9668	7.0001	7.0336	7.0671	7.1007
36	7.1343	7.1681	7.2019	7.2358	7.2698	7.3038	7.3380	7.3722	7.4065	7.4408
37	7.4753	7.5098	7.5444	7.5791	7.6139	7.6487	7.6836	7.7186	7.7537	7.7888
38	7.8240	7.8593	7.8947	7.9302	7.9657	8.0013	8.0370	8.0727	8.1085	8.1444
39	8.1804	8.2165	8.2526	8.2888	8.3251	8.3614	8.3979	8.4344	8.4709	8.5076
40	8.5443	8.5811	8.6180	8.6549	8.6919	8.7290	8.7662	8.8034	8.8407	8.8781
41	8.9155	8.9531	8.9907	9.0283	9.0661	9.1039	9.1418	9.1797	9.2178	9.2559

Corrections are in kg/g/mg and are to be added to the mass of water measured in kg/g/mg as the unit of capacity is in m³/dm³/cm³ respectively

Table 3.18 ALPHA = 10 × 10^{-6}/°C, DEN = 8400 kg/m³, REFERENCE TEMP = 20 °C

Temp	0.0	0.1	0.2	0.3	0.4	0.5	0.6	0.7	0.8	0.9
5	1.2403	1.2410	1.2418	1.2428	1.2439	1.2452	1.2466	1.2482	1.2499	1.2518
6	1.2539	1.2561	1.2584	1.2609	1.2635	1.2663	1.2693	1.2723	1.2756	1.2789
7	1.2825	1.2861	1.2899	1.2939	1.2980	1.3023	1.3066	1.3112	1.3159	1.3207
8	1.3256	1.3307	1.3360	1.3414	1.3469	1.3526	1.3584	1.3643	1.3704	1.3766
9	1.3830	1.3895	1.3961	1.4029	1.4098	1.4169	1.4241	1.4314	1.4388	1.4464
10	1.4542	1.4620	1.4700	1.4781	1.4864	1.4948	1.5033	1.5120	1.5208	1.5297
11	1.5387	1.5479	1.5572	1.5667	1.5762	1.5859	1.5958	1.6057	1.6158	1.6260
12	1.6364	1.6468	1.6574	1.6682	1.6790	1.6900	1.7011	1.7123	1.7237	1.7351
13	1.7467	1.7585	1.7703	1.7823	1.7944	1.8066	1.8189	1.8314	1.8440	1.8567
14	1.8695	1.8824	1.8955	1.9087	1.9220	1.9354	1.9490	1.9627	1.9764	1.9903
15	2.0044	2.0185	2.0328	2.0471	2.0616	2.0762	2.0910	2.1058	2.1208	2.1358
16	2.1510	2.1663	2.1818	2.1973	2.2129	2.2287	2.2446	2.2606	2.2767	2.2929
17	2.3092	2.3257	2.3422	2.3589	2.3757	2.3925	2.4095	2.4267	2.4439	2.4612
18	2.4787	2.4962	2.5139	2.5317	2.5495	2.5675	2.5856	2.6038	2.6222	2.6406
19	2.6591	2.6778	2.6965	2.7154	2.7343	2.7534	2.7726	2.7918	2.8112	2.8307
20	2.8503	2.8700	2.8898	2.9098	2.9298	2.9499	2.9701	2.9905	3.0109	3.0314
21	3.0521	3.0728	3.0937	3.1146	3.1357	3.1568	3.1781	3.1995	3.2209	3.2425
22	3.2641	3.2859	3.3078	3.3297	3.3518	3.3740	3.3962	3.4186	3.4411	3.4636
23	3.4863	3.5091	3.5319	3.5549	3.5780	3.6011	3.6244	3.6477	3.6712	3.6947
24	3.7184	3.7421	3.7660	3.7899	3.8140	3.8381	3.8623	3.8866	3.9111	3.9356
25	3.9602	3.9849	4.0097	4.0346	4.0596	4.0847	4.1098	4.1351	4.1605	4.1860
26	4.2115	4.2372	4.2629	4.2887	4.3147	4.3407	4.3668	4.3930	4.4193	4.4457
27	4.4722	4.4988	4.5254	4.5522	4.5791	4.6060	4.6330	4.6602	4.6874	4.7147
28	4.7421	4.7696	4.7971	4.8248	4.8526	4.8804	4.9084	4.9364	4.9645	4.9927
29	5.0210	5.0494	5.0778	5.1064	5.1351	5.1638	5.1926	5.2215	5.2505	5.2796
30	5.3088	5.3380	5.3674	5.3968	5.4263	5.4560	5.4856	5.5154	5.5453	5.5753
31	5.6053	5.6354	5.6656	5.6959	5.7263	5.7568	5.7873	5.8180	5.8487	5.8795
32	5.9104	5.9414	5.9724	6.0036	6.0348	6.0661	6.0975	6.1290	6.1605	6.1922
33	6.2239	6.2557	6.2876	6.3196	6.3516	6.3838	6.4160	6.4483	6.4807	6.5132
34	6.5457	6.5784	6.6111	6.6439	6.6768	6.7097	6.7427	6.7759	6.8091	6.8424
35	6.8757	6.9092	6.9427	6.9763	7.0100	7.0437	7.0776	7.1115	7.1455	7.1796
36	7.2137	7.2480	7.2823	7.3167	7.3511	7.3857	7.4203	7.4550	7.4898	7.5247
37	7.5596	7.5947	7.6298	7.6649	7.7002	7.7355	7.7709	7.8064	7.8420	7.8776
38	7.9133	7.9491	7.9850	8.0209	8.0569	8.0930	8.1292	8.1654	8.2017	8.2381
39	8.2746	8.3112	8.3478	8.3845	8.4212	8.4581	8.4950	8.5320	8.5691	8.6062
40	8.6434	8.6807	8.7181	8.7555	8.7930	8.8306	8.8682	8.9060	8.9438	8.9816
41	9.0196	9.0576	9.0957	9.1338	9.1721	9.2104	9.2488	9.2872	9.3257	9.3643

Corrections are in kg/g/mg and are to be added to the mass of water measured in kg/g/mg as the unit of capacity is in m³/dm³/cm³ respectively

Table 3.19 ALPHA = 54 × 10^{-6}/°C, DEN = 8000 kg/m³, REFERENCE TEMP = 20 °C

Temp	0.0	0.1	0.2	0.3	0.4	0.5	0.6	0.7	0.8	0.9
5	1.8924	1.8887	1.8851	1.8817	1.8785	1.8753	1.8724	1.8696	1.8669	1.8644
6	1.8621	1.8598	1.8578	1.8559	1.8541	1.8525	1.8511	1.8497	1.8486	1.8475
7	1.8467	1.8459	1.8454	1.8449	1.8446	1.8445	1.8445	1.8446	1.8449	1.8453
8	1.8459	1.8466	1.8474	1.8484	1.8496	1.8508	1.8522	1.8538	1.8555	1.8573
9	1.8593	1.8614	1.8636	1.8660	1.8685	1.8711	1.8739	1.8769	1.8799	1.8831
10	1.8864	1.8899	1.8935	1.8972	1.9011	1.9051	1.9092	1.9135	1.9179	1.9224
11	1.9270	1.9318	1.9367	1.9418	1.9469	1.9522	1.9577	1.9632	1.9689	1.9747
12	1.9807	1.9868	1.9930	1.9993	2.0057	2.0123	2.0190	2.0259	2.0328	2.0399
13	2.0471	2.0544	2.0619	2.0694	2.0771	2.0850	2.0929	2.1010	2.1092	2.1175
14	2.1259	2.1344	2.1431	2.1519	2.1608	2.1699	2.1790	2.1883	2.1977	2.2072
15	2.2168	2.2265	2.2364	2.2464	2.2565	2.2667	2.2770	2.2875	2.2981	2.3087
16	2.3195	2.3304	2.3415	2.3526	2.3639	2.3752	2.3867	2.3983	2.4100	2.4218
17	2.4338	2.4458	2.4580	2.4703	2.4827	2.4952	2.5078	2.5205	2.5333	2.5463
18	2.5593	2.5725	2.5858	2.5991	2.6126	2.6262	2.6399	2.6538	2.6677	2.6817
19	2.6959	2.7101	2.7245	2.7389	2.7535	2.7682	2.7830	2.7979	2.8129	2.8280
20	2.8432	2.8585	2.8739	2.8895	2.9051	2.9208	2.9367	2.9526	2.9687	2.9848
21	3.0011	3.0174	3.0339	3.0505	3.0671	3.0839	3.1008	3.1178	3.1348	3.1520
22	3.1693	3.1867	3.2042	3.2218	3.2394	3.2572	3.2751	3.2931	3.3112	3.3294
23	3.3477	3.3660	3.3845	3.4031	3.4218	3.4406	3.4594	3.4784	3.4975	3.5167
24	3.5359	3.5553	3.5748	3.5943	3.6140	3.6337	3.6536	3.6735	3.6936	3.7137
25	3.7339	3.7543	3.7747	3.7952	3.8158	3.8366	3.8574	3.8783	3.8993	3.9203
26	3.9415	3.9628	3.9842	4.0056	4.0272	4.0488	4.0706	4.0924	4.1143	4.1364
27	4.1585	4.1807	4.2030	4.2254	4.2479	4.2704	4.2931	4.3159	4.3387	4.3616
28	4.3847	4.4078	4.4310	4.4543	4.4777	4.5012	4.5247	4.5484	4.5721	4.5960
29	4.6199	4.6439	4.6680	4.6922	4.7165	4.7409	4.7653	4.7899	4.8145	4.8392
30	4.8640	4.8889	4.9139	4.9390	4.9642	4.9894	5.0147	5.0402	5.0657	5.0913
31	5.1169	5.1427	5.1686	5.1945	5.2205	5.2466	5.2728	5.2991	5.3255	5.3519
32	5.3785	5.4051	5.4318	5.4586	5.4854	5.5124	5.5394	5.5666	5.5938	5.6211
33	5.6484	5.6759	5.7034	5.7311	5.7588	5.7866	5.8144	5.8424	5.8704	5.8985
34	5.9267	5.9550	5.9834	6.0118	6.0404	6.0690	6.0977	6.1264	6.1553	6.1842
35	6.2132	6.2423	6.2715	6.3008	6.3301	6.3595	6.3890	6.4186	6.4483	6.4780
36	6.5078	6.5377	6.5677	6.5977	6.6279	6.6581	6.6884	6.7187	6.7492	6.7797
37	6.8103	6.8410	6.8718	6.9026	6.9335	6.9645	6.9956	7.0267	7.0579	7.0892
38	7.1206	7.1521	7.1836	7.2152	7.2469	7.2787	7.3105	7.3424	7.3744	7.4065
39	7.4386	7.4708	7.5031	7.5355	7.5679	7.6004	7.6330	7.6657	7.6984	7.7312
40	7.7641	7.7971	7.8301	7.8632	7.8964	7.9296	7.9630	7.9964	8.0299	8.0634
41	8.0970	8.1307	8.1645	8.1983	8.2322	8.2662	8.3003	8.3344	8.3686	8.4029

Corrections are in kg/g/mg and are to be added to the mass of water measured in kg/g/mg as the unit of capacity is in $m^3/dm^3/cm^3$ respectively

Table 3.20 ALPHA = 33×10^{-6}/°C, DEN = 8000 kg/m^3, REFERENCE TEMP = 20 °C

Temp	0.0	0.1	0.2	0.3	0.4	0.5	0.6	0.7	0.8	0.9
5	1.5778	1.5762	1.5747	1.5734	1.5722	1.5712	1.5703	1.5696	1.5690	1.5686
6	1.5684	1.5683	1.5683	1.5685	1.5688	1.5693	1.5700	1.5708	1.5717	1.5728
7	1.5740	1.5754	1.5769	1.5785	1.5803	1.5823	1.5844	1.5866	1.5890	1.5915
8	1.5942	1.5970	1.5999	1.6030	1.6062	1.6096	1.6131	1.6168	1.6206	1.6245
9	1.6286	1.6328	1.6371	1.6416	1.6462	1.6509	1.6558	1.6608	1.6660	1.6713
10	1.6767	1.6823	1.6880	1.6938	1.6998	1.7059	1.7121	1.7184	1.7249	1.7315
11	1.7383	1.7452	1.7522	1.7593	1.7666	1.7740	1.7815	1.7892	1.7970	1.8049
12	1.8130	1.8211	1.8294	1.8378	1.8464	1.8551	1.8639	1.8728	1.8819	1.8910
13	1.9003	1.9098	1.9193	1.9290	1.9388	1.9487	1.9587	1.9689	1.9792	1.9896
14	2.0001	2.0108	2.0215	2.0324	2.0434	2.0546	2.0658	2.0772	2.0887	2.1003
15	2.1120	2.1239	2.1358	2.1479	2.1601	2.1724	2.1848	2.1974	2.2100	2.2228
16	2.2357	2.2487	2.2618	2.2751	2.2884	2.3019	2.3155	2.3292	2.3430	2.3569
17	2.3709	2.3851	2.3993	2.4137	2.4282	2.4428	2.4575	2.4723	2.4872	2.5023
18	2.5174	2.5327	2.5481	2.5635	2.5791	2.5948	2.6106	2.6265	2.6426	2.6587
19	2.6749	2.6913	2.7077	2.7243	2.7410	2.7577	2.7746	2.7916	2.8087	2.8259
20	2.8432	2.8606	2.8781	2.8958	2.9135	2.9313	2.9492	2.9673	2.9854	3.0037
21	3.0220	3.0405	3.0590	3.0777	3.0965	3.1153	3.1343	3.1534	3.1725	3.1918
22	3.2112	3.2306	3.2502	3.2699	3.2897	3.3096	3.3295	3.3496	3.3698	3.3901
23	3.4104	3.4309	3.4515	3.4722	3.4929	3.5138	3.5348	3.5558	3.5770	3.5983
24	3.6196	3.6411	3.6626	3.6843	3.7060	3.7279	3.7498	3.7718	3.7940	3.8162
25	3.8385	3.8609	3.8835	3.9061	3.9288	3.9516	3.9745	3.9975	4.0205	4.0437
26	4.0670	4.0904	4.1138	4.1374	4.1610	4.1847	4.2086	4.2325	4.2565	4.2806
27	4.3048	4.3291	4.3535	4.3780	4.4025	4.4272	4.4519	4.4768	4.5017	4.5267
28	4.5519	4.5771	4.6024	4.6277	4.6532	4.6788	4.7044	4.7302	4.7560	4.7819
29	4.8079	4.8340	4.8602	4.8865	4.9129	4.9393	4.9659	4.9925	5.0192	5.0460
30	5.0729	5.0999	5.1270	5.1541	5.1814	5.2087	5.2361	5.2636	5.2912	5.3189
31	5.3466	5.3745	5.4024	5.4304	5.4585	5.4867	5.5150	5.5434	5.5718	5.6003
32	5.6289	5.6576	5.6864	5.7153	5.7442	5.7733	5.8024	5.8316	5.8609	5.8903
33	5.9197	5.9492	5.9789	6.0086	6.0383	6.0682	6.0982	6.1282	6.1583	6.1885
34	6.2188	6.2491	6.2796	6.3101	6.3407	6.3714	6.4022	6.4330	6.4639	6.4949
35	6.5260	6.5572	6.5885	6.6198	6.6512	6.6827	6.7143	6.7459	6.7776	6.8095
36	6.8413	6.8733	6.9054	6.9375	6.9697	7.0020	7.0343	7.0668	7.0993	7.1319
37	7.1646	7.1973	7.2301	7.2631	7.2960	7.3291	7.3622	7.3955	7.4287	7.4621
38	7.4956	7.5291	7.5627	7.5964	7.6301	7.6639	7.6978	7.7318	7.7659	7.8000
39	7.8342	7.8685	7.9029	7.9373	7.9718	8.0064	8.0410	8.0758	8.1106	8.1454
40	8.1804	8.2154	8.2505	8.2857	8.3209	8.3563	8.3917	8.4271	8.4627	8.4983
41	8.5340	8.5697	8.6055	8.6414	8.6774	8.7135	8.7496	8.7858	8.8220	8.8584

Corrections are in kg/g/mg and are to be added to the mass of water measured in kg/g/mg as the unit of capacity is in m³/dm³/cm³ respectively

Table 3.21 ALPHA = 30 × 10⁻⁶/°C, DEN = 8000 kg/m³, REFERENCE TEMP = 20°C

Temp	0.0	0.1	0.2	0.3	0.4	0.5	0.6	0.7	0.8	0.9
5	1.5328	1.5315	1.5303	1.5293	1.5284	1.5277	1.5272	1.5268	1.5265	1.5264
6	1.5264	1.5266	1.5270	1.5274	1.5281	1.5289	1.5298	1.5309	1.5321	1.5335
7	1.5350	1.5367	1.5385	1.5405	1.5426	1.5448	1.5472	1.5498	1.5524	1.5553
8	1.5582	1.5613	1.5646	1.5680	1.5715	1.5752	1.5790	1.5829	1.5870	1.5912
9	1.5956	1.6001	1.6047	1.6095	1.6144	1.6195	1.6247	1.6300	1.6354	1.6410
10	1.6468	1.6526	1.6586	1.6647	1.6710	1.6774	1.6839	1.6906	1.6974	1.7043
11	1.7113	1.7185	1.7258	1.7333	1.7409	1.7486	1.7564	1.7643	1.7724	1.7806
12	1.7890	1.7975	1.8061	1.8148	1.8236	1.8326	1.8417	1.8509	1.8603	1.8698
13	1.8794	1.8891	1.8989	1.9089	1.9190	1.9292	1.9396	1.9500	1.9606	1.9713
14	1.9822	1.9931	2.0042	2.0154	2.0267	2.0381	2.0496	2.0613	2.0731	2.0850
15	2.0970	2.1092	2.1214	2.1338	2.1463	2.1589	2.1717	2.1845	2.1975	2.2105
16	2.2237	2.2370	2.2505	2.2640	2.2776	2.2914	2.3053	2.3193	2.3334	2.3476
17	2.3620	2.3764	2.3910	2.4056	2.4204	2.4353	2.4503	2.4654	2.4807	2.4960
18	2.5114	2.5270	2.5427	2.5584	2.5743	2.5903	2.6064	2.6226	2.6390	2.6554
19	2.6719	2.6886	2.7053	2.7222	2.7392	2.7562	2.7734	2.7907	2.8081	2.8256
20	2.8432	2.8609	2.8787	2.8967	2.9147	2.9328	2.9510	2.9694	2.9878	3.0064
21	3.0250	3.0438	3.0626	3.0816	3.1006	3.1198	3.1391	3.1584	3.1779	3.1975
22	3.2172	3.2369	3.2568	3.2768	3.2969	3.3170	3.3373	3.3577	3.3782	3.3987
23	3.4194	3.4402	3.4611	3.4820	3.5031	3.5243	3.5455	3.5669	3.5884	3.6099
24	3.6316	3.6533	3.6752	3.6971	3.7192	3.7413	3.7636	3.7859	3.8083	3.8308
25	3.8535	3.8762	3.8990	3.9219	3.9449	3.9680	3.9912	4.0145	4.0379	4.0613
26	4.0849	4.1086	4.1323	4.1562	4.1801	4.2042	4.2283	4.2525	4.2768	4.3012
27	4.3257	4.3503	4.3750	4.3998	4.4246	4.4496	4.4746	4.4998	4.5250	4.5503
28	4.5757	4.6012	4.6268	4.6525	4.6783	4.7042	4.7301	4.7561	4.7823	4.8085
29	4.8348	4.8612	4.8877	4.9143	4.9409	4.9677	4.9945	5.0214	5.0485	5.0756
30	5.1028	5.1300	5.1574	5.1848	5.2124	5.2400	5.2677	5.2955	5.3234	5.3514
31	5.3794	5.4076	5.4358	5.4641	5.4925	5.5210	5.5496	5.5782	5.6070	5.6358
32	5.6647	5.6937	5.7228	5.7520	5.7812	5.8105	5.8400	5.8695	5.8990	5.9287
33	5.9585	5.9883	6.0182	6.0482	6.0783	6.1084	6.1387	6.1690	6.1994	6.2299
34	6.2605	6.2912	6.3219	6.3527	6.3836	6.4146	6.4457	6.4768	6.5080	6.5393
35	6.5707	6.6022	6.6337	6.6654	6.6971	6.7289	6.7607	6.7927	6.8247	6.8568
36	6.8890	6.9213	6.9536	6.9860	7.0185	7.0511	7.0838	7.1165	7.1493	7.1822
37	7.2152	7.2482	7.2813	7.3145	7.3478	7.3812	7.4146	7.4481	7.4817	7.5154
38	7.5491	7.5829	7.6168	7.6508	7.6849	7.7190	7.7532	7.7875	7.8218	7.8562
39	7.8907	7.9253	7.9600	7.9947	8.0295	8.0644	8.0993	8.1343	8.1694	8.2046
40	8.2399	8.2752	8.3106	8.3460	8.3816	8.4172	8.4529	8.4887	8.5245	8.5604
41	8.5964	8.6324	8.6685	8.7047	8.7410	8.7774	8.8138	8.8503	8.8868	8.9234

Corrections are in kg/g/mg and are to be added to the mass of water measured in kg/g/mg as the unit of capacity is in m³/dm³/cm³ respectively

Table 3.22 ALPHA = 25 × 10⁻⁶/°C, DEN = 8000 kg/m³, REFERENCE TEMP = 20 °C

Temp	0.0	0.1	0.2	0.3	0.4	0.5	0.6	0.7	0.8	0.9
5	1.4579	1.4571	1.4564	1.4559	1.4555	1.4553	1.4552	1.4553	1.4556	1.4560
6	1.4565	1.4572	1.4580	1.4590	1.4602	1.4615	1.4629	1.4645	1.4662	1.4681
7	1.4701	1.4723	1.4746	1.4770	1.4797	1.4824	1.4853	1.4883	1.4915	1.4948
8	1.4983	1.5019	1.5056	1.5095	1.5136	1.5177	1.5220	1.5265	1.5311	1.5358
9	1.5407	1.5457	1.5508	1.5561	1.5615	1.5670	1.5727	1.5785	1.5845	1.5906
10	1.5968	1.6032	1.6097	1.6163	1.6231	1.6300	1.6370	1.6441	1.6514	1.6589
11	1.6664	1.6741	1.6819	1.6898	1.6979	1.7061	1.7144	1.7229	1.7315	1.7402
12	1.7491	1.7580	1.7671	1.7763	1.7857	1.7952	1.8048	1.8145	1.8243	1.8343
13	1.8444	1.8547	1.8650	1.8755	1.8861	1.8968	1.9076	1.9186	1.9297	1.9409
14	1.9522	1.9637	1.9752	1.9869	1.9987	2.0107	2.0227	2.0349	2.0472	2.0596
15	2.0721	2.0847	2.0975	2.1104	2.1234	2.1365	2.1497	2.1630	2.1765	2.1901
16	2.2038	2.2176	2.2315	2.2455	2.2597	2.2740	2.2883	2.3028	2.3174	2.3322
17	2.3470	2.3619	2.3770	2.3922	2.4074	2.4228	2.4383	2.4540	2.4697	2.4855
18	2.5015	2.5175	2.5337	2.5500	2.5663	2.5828	2.5994	2.6162	2.6330	2.6499
19	2.6669	2.6841	2.7013	2.7187	2.7362	2.7537	2.7714	2.7892	2.8071	2.8251
20	2.8432	2.8614	2.8797	2.8981	2.9167	2.9353	2.9540	2.9729	2.9918	3.0109
21	3.0300	3.0493	3.0686	3.0881	3.1076	3.1273	3.1470	3.1669	3.1869	3.2069
22	3.2271	3.2474	3.2678	3.2882	3.3088	3.3295	3.3503	3.3711	3.3921	3.4132
23	3.4344	3.4556	3.4770	3.4985	3.5200	3.5417	3.5635	3.5853	3.6073	3.6293
24	3.6515	3.6738	3.6961	3.7185	3.7411	3.7637	3.7865	3.8093	3.8322	3.8552
25	3.8784	3.9016	3.9249	3.9483	3.9718	3.9954	4.0191	4.0429	4.0667	4.0907
26	4.1148	4.1389	4.1632	4.1875	4.2120	4.2365	4.2611	4.2859	4.3107	4.3356
27	4.3606	4.3857	4.4108	4.4361	4.4615	4.4869	4.5125	4.5381	4.5638	4.5896
28	4.6155	4.6415	4.6676	4.6938	4.7201	4.7464	4.7729	4.7994	4.8261	4.8528
29	4.8796	4.9065	4.9334	4.9605	4.9877	5.0149	5.0423	5.0697	5.0972	5.1248
30	5.1525	5.1803	5.2081	5.2361	5.2641	5.2922	5.3204	5.3487	5.3771	5.4056
31	5.4341	5.4628	5.4915	5.5203	5.5492	5.5782	5.6072	5.6364	5.6656	5.6950
32	5.7244	5.7539	5.7834	5.8131	5.8428	5.8727	5.9026	5.9326	5.9626	5.9928
33	6.0230	6.0534	6.0838	6.1143	6.1449	6.1755	6.2062	6.2371	6.2680	6.2990
34	6.3300	6.3612	6.3924	6.4237	6.4551	6.4866	6.5182	6.5498	6.5815	6.6133
35	6.6452	6.6772	6.7092	6.7413	6.7735	6.8058	6.8382	6.8706	6.9031	6.9357
36	6.9684	7.0012	7.0340	7.0669	7.0999	7.1330	7.1661	7.1994	7.2327	7.2660
37	7.2995	7.3331	7.3667	7.4004	7.4341	7.4680	7.5019	7.5359	7.5700	7.6042
38	7.6384	7.6727	7.7071	7.7416	7.7761	7.8107	7.8454	7.8802	7.9150	7.9499
39	7.9849	8.0200	8.0551	8.0904	8.1257	8.1610	8.1965	8.2320	8.2676	8.3032
40	8.3390	8.3748	8.4107	8.4466	8.4827	8.5188	8.5550	8.5912	8.6275	8.6639
41	8.7004	8.7369	8.7736	8.8103	8.8470	8.8838	8.9207	8.9577	8.9948	9.0319

Corrections are in kg/g/mg and are to be added to the mass of water measured in kg/g/mg as the unit of capacity is in $m^3/dm^3/cm^3$ respectively

Table 3.23 ALPHA = 15×10^{-6}/°C, DEN = 8000 kg/m^3, REFERENCE TEMP = 20 °C

Temp	0.0	0.1	0.2	0.3	0.4	0.5	0.6	0.7	0.8	0.9
5	1.3081	1.3082	1.3086	1.3090	1.3097	1.3105	1.3114	1.3125	1.3137	1.3151
6	1.3167	1.3183	1.3202	1.3222	1.3243	1.3266	1.3290	1.3316	1.3344	1.3372
7	1.3403	1.3434	1.3467	1.3502	1.3538	1.3575	1.3614	1.3655	1.3697	1.3740
8	1.3784	1.3830	1.3878	1.3927	1.3977	1.4029	1.4082	1.4136	1.4192	1.4249
9	1.4308	1.4368	1.4429	1.4492	1.4556	1.4622	1.4689	1.4757	1.4826	1.4897
10	1.4970	1.5043	1.5118	1.5194	1.5272	1.5351	1.5431	1.5513	1.5596	1.5680
11	1.5765	1.5852	1.5940	1.6030	1.6120	1.6212	1.6306	1.6400	1.6496	1.6593
12	1.6692	1.6791	1.6892	1.6995	1.7098	1.7203	1.7309	1.7416	1.7525	1.7634
13	1.7745	1.7858	1.7971	1.8086	1.8202	1.8319	1.8437	1.8557	1.8678	1.8800
14	1.8923	1.9048	1.9173	1.9300	1.9428	1.9558	1.9688	1.9820	1.9953	2.0087
15	2.0222	2.0358	2.0496	2.0635	2.0775	2.0916	2.1058	2.1201	2.1346	2.1492
16	2.1639	2.1787	2.1936	2.2086	2.2238	2.2390	2.2544	2.2699	2.2855	2.3012
17	2.3171	2.3330	2.3491	2.3652	2.3815	2.3979	2.4144	2.4310	2.4477	2.4646
18	2.4815	2.4986	2.5157	2.5330	2.5504	2.5679	2.5855	2.6032	2.6210	2.6389
19	2.6570	2.6751	2.6934	2.7117	2.7302	2.7488	2.7674	2.7862	2.8051	2.8241
20	2.8432	2.8624	2.8817	2.9011	2.9207	2.9403	2.9600	2.9798	2.9998	3.0198
21	3.0400	3.0602	3.0806	3.1010	3.1216	3.1422	3.1630	3.1839	3.2048	3.2259
22	3.2471	3.2683	3.2897	3.3112	3.3327	3.3544	3.3762	3.3980	3.4200	3.4421
23	3.4642	3.4865	3.5089	3.5313	3.5539	3.5766	3.5993	3.6222	3.6451	3.6682
24	3.6913	3.7146	3.7379	3.7614	3.7849	3.8086	3.8323	3.8561	3.8800	3.9041
25	3.9282	3.9524	3.9767	4.0011	4.0256	4.0502	4.0749	4.0996	4.1245	4.1495
26	4.1745	4.1997	4.2249	4.2503	4.2757	4.3012	4.3268	4.3526	4.3784	4.4043
27	4.4302	4.4563	4.4825	4.5088	4.5351	4.5616	4.5881	4.6147	4.6415	4.6683
28	4.6952	4.7222	4.7492	4.7764	4.8037	4.8310	4.8585	4.8860	4.9136	4.9413
29	4.9691	4.9970	5.0250	5.0530	5.0812	5.1094	5.1378	5.1662	5.1947	5.2233
30	5.2519	5.2807	5.3096	5.3385	5.3675	5.3966	5.4258	5.4551	5.4845	5.5140
31	5.5435	5.5731	5.6028	5.6326	5.6625	5.6925	5.7226	5.7527	5.7829	5.8132
32	5.8436	5.8741	5.9047	5.9353	5.9661	5.9969	6.0278	6.0588	6.0898	6.1210
33	6.1522	6.1835	6.2149	6.2464	6.2780	6.3096	6.3414	6.3732	6.4051	6.4370
34	6.4691	6.5012	6.5335	6.5658	6.5981	6.6306	6.6632	6.6958	6.7285	6.7613
35	6.7941	6.8271	6.8601	6.8932	6.9264	6.9597	6.9930	7.0265	7.0600	7.0936
36	7.1272	7.1610	7.1948	7.2287	7.2627	7.2967	7.3309	7.3651	7.3994	7.4338
37	7.4682	7.5027	7.5373	7.5720	7.6068	7.6416	7.6765	7.7115	7.7466	7.7817
38	7.8169	7.8522	7.8876	7.9231	7.9586	7.9942	8.0299	8.0656	8.1014	8.1374
39	8.1733	8.2094	8.2455	8.2817	8.3180	8.3543	8.3908	8.4273	8.4638	8.5005
40	8.5372	8.5740	8.6109	8.6478	8.6848	8.7219	8.7591	8.7963	8.8336	8.8710
41	8.9085	8.9460	8.9836	9.0213	9.0590	9.0968	9.1347	9.1727	9.2107	9.2488

Corrections are in kg/g/mg and are to be added to the mass of water measured in kg/g/mg as the unit of capacity is in $m^3/dm^3/cm^3$ respectively

Table 3.24 ALPHA = 10 × 10^{-6}/°C, DEN = 8000 kg/m^3, REFERENCE TEMP = 20 °C

Temp	0.0	0.1	0.2	0.3	0.4	0.5	0.6	0.7	0.8	0.9
5	1.2332	1.2338	1.2346	1.2356	1.2368	1.2380	1.2395	1.2411	1.2428	1.2447
6	1.2467	1.2489	1.2513	1.2538	1.2564	1.2592	1.2621	1.2652	1.2684	1.2718
7	1.2753	1.2790	1.2828	1.2868	1.2909	1.2951	1.2995	1.3040	1.3087	1.3135
8	1.3185	1.3236	1.3289	1.3342	1.3398	1.3454	1.3513	1.3572	1.3633	1.3695
9	1.3759	1.3824	1.3890	1.3958	1.4027	1.4097	1.4169	1.4243	1.4317	1.4393
10	1.4470	1.4549	1.4629	1.4710	1.4793	1.4877	1.4962	1.5048	1.5136	1.5225
11	1.5316	1.5408	1.5501	1.5595	1.5691	1.5788	1.5886	1.5986	1.6087	1.6189
12	1.6292	1.6397	1.6503	1.6610	1.6719	1.6828	1.6939	1.7052	1.7165	1.7280
13	1.7396	1.7513	1.7632	1.7751	1.7872	1.7995	1.8118	1.8243	1.8368	1.8495
14	1.8624	1.8753	1.8884	1.9016	1.9149	1.9283	1.9419	1.9555	1.9693	1.9832
15	1.9972	2.0114	2.0256	2.0400	2.0545	2.0691	2.0838	2.0987	2.1136	2.1287
16	2.1439	2.1592	2.1746	2.1902	2.2058	2.2216	2.2374	2.2534	2.2695	2.2858
17	2.3021	2.3185	2.3351	2.3518	2.3685	2.3854	2.4024	2.4195	2.4368	2.4541
18	2.4715	2.4891	2.5068	2.5245	2.5424	2.5604	2.5785	2.5967	2.6150	2.6335
19	2.6520	2.6706	2.6894	2.7082	2.7272	2.7463	2.7654	2.7847	2.8041	2.8236
20	2.8432	2.8629	2.8827	2.9026	2.9227	2.9428	2.9630	2.9833	3.0038	3.0243
21	3.0450	3.0657	3.0866	3.1075	3.1286	3.1497	3.1710	3.1923	3.2138	3.2354
22	3.2570	3.2788	3.3007	3.3226	3.3447	3.3669	3.3891	3.4115	3.4340	3.4565
23	3.4792	3.5020	3.5248	3.5478	3.5708	3.5940	3.6173	3.6406	3.6641	3.6876
24	3.7113	3.7350	3.7589	3.7828	3.8068	3.8310	3.8552	3.8795	3.9039	3.9285
25	3.9531	3.9778	4.0026	4.0275	4.0525	4.0776	4.1027	4.1280	4.1534	4.1788
26	4.2044	4.2301	4.2558	4.2816	4.3076	4.3336	4.3597	4.3859	4.4122	4.4386
27	4.4651	4.4917	4.5183	4.5451	4.5719	4.5989	4.6259	4.6531	4.6803	4.7076
28	4.7350	4.7625	4.7900	4.8177	4.8455	4.8733	4.9012	4.9293	4.9574	4.9856
29	5.0139	5.0423	5.0707	5.0993	5.1279	5.1567	5.1855	5.2144	5.2434	5.2725
30	5.3017	5.3309	5.3603	5.3897	5.4192	5.4489	5.4785	5.5083	5.5382	5.5681
31	5.5982	5.6283	5.6585	5.6888	5.7192	5.7497	5.7802	5.8109	5.8416	5.8724
32	5.9033	5.9343	5.9653	5.9965	6.0277	6.0590	6.0904	6.1219	6.1534	6.1851
33	6.2168	6.2486	6.2805	6.3125	6.3445	6.3767	6.4089	6.4412	6.4736	6.5061
34	6.5386	6.5713	6.6040	6.6368	6.6697	6.7026	6.7357	6.7688	6.8020	6.8353
35	6.8686	6.9021	6.9356	6.9692	7.0029	7.0366	7.0705	7.1044	7.1384	7.1725
36	7.2066	7.2409	7.2752	7.3096	7.3441	7.3786	7.4132	7.4480	7.4827	7.5176
37	7.5525	7.5876	7.6227	7.6578	7.6931	7.7284	7.7638	7.7993	7.8349	7.8705
38	7.9062	7.9420	7.9779	8.0138	8.0498	8.0859	8.1221	8.1583	8.1947	8.2311
39	8.2675	8.3041	8.3407	8.3774	8.4142	8.4510	8.4879	8.5249	8.5620	8.5991
40	8.6363	8.6736	8.7110	8.7484	8.7859	8.8235	8.8612	8.8989	8.9367	8.9746
41	9.0125	9.0505	9.0886	9.1268	9.1650	9.2033	9.2417	9.2801	9.3186	9.3572

CORRECTIONS DUE TO VARIATION IN AIR DENSITY TABLES 3.25 TO 3.26

Additional corrections are in kg/g/mg and are to be added to the mass of water measured in kg/g/mg as the unit of capacity is in $m^3/dm^3/cm^3$ respectively.

Table 3.25 Additional Correction in Grams to be Applied to Measure of 1 dm^3 for Variation in Air Density D = 8400 kg/m^3

	Pressure in mm of Mercury/Pascals												
T	730	735	740	745	750	755	760	765	770	775	780	785	790
°*C*	97.3	98.0	98.7	99.3	100	100.7	101.3	102.0	102.7	103.3	104.0	104.7	105.3
5	0.04308	0.05043	0.05779	0.06514	0.07249	0.07985	0.08720	0.09455	0.10191	0.10926	0.11661	0.12397	0.13132
6	0.03910	0.04643	0.05376	0.06108	0.06841	0.07574	0.08306	0.09039	0.09772	0.10505	0.11238	0.11970	0.12703
7	0.03514	0.04244	0.04975	0.05705	0.06435	0.07165	0.07895	0.08625	0.09355	0.10086	0.10816	0.11546	0.12276
8	0.03120	0.03848	0.04576	0.05303	0.06031	0.06758	0.07486	0.08213	0.08941	0.09669	0.10396	0.11124	0.11852
9	0.02728	0.03453	0.04178	0.04904	0.05629	0.06354	0.07079	0.07804	0.08529	0.09254	0.09979	0.10704	0.11429
10	0.02338	0.03061	0.03783	0.04506	0.05228	0.05951	0.06673	0.07396	0.08118	0.08841	0.09563	0.10286	0.11008
11	0.01950	0.02670	0.03390	0.04110	0.04830	0.05550	0.06270	0.06990	0.07710	0.08430	0.09150	0.09870	0.10590
12	0.01563	0.02280	0.02998	0.03715	0.04433	0.05150	0.05868	0.06585	0.07303	0.08020	0.08738	0.09456	0.10173
13	0.01178	0.01893	0.02608	0.03323	0.04038	0.04753	0.05468	0.06183	0.06898	0.07613	0.08328	0.09043	0.09758
14	0.00794	0.01507	0.02219	0.02932	0.03644	0.04357	0.05069	0.05782	0.06495	0.07207	0.07920	0.08632	0.09345
15	0.00412	0.01122	0.01832	0.02542	0.03252	0.03962	0.04672	0.05383	0.06093	0.06803	0.07513	0.08223	0.08933
16	0.00031	0.00738	0.01446	0.02154	0.02862	0.03569	0.04277	0.04985	0.05693	0.06400	0.07108	0.07816	0.08523
17	–.00349	0.00356	0.01062	0.01767	0.02472	0.03178	0.03883	0.04588	0.05294	0.05999	0.06704	0.07410	0.08115
18	–.00727	–.00024	0.00678	0.01381	0.02084	0.02787	0.03490	0.04193	0.04896	0.05599	0.06302	0.07005	0.07708
19	–.01105	–.00404	0.00296	0.00997	0.01698	0.02398	0.03099	0.03799	0.04500	0.05200	0.05901	0.06602	0.07302
20	–.01481	–.00783	–.00085	0.00614	0.01312	0.02010	0.02708	0.03406	0.04105	0.04803	0.05501	0.06199	0.06898
21	–.01856	–.01161	–.00465	0.00231	0.00927	0.01623	0.02319	0.03015	0.03711	0.04407	0.05103	0.05798	0.06494
22	–.02231	–.01537	–.00844	–.00150	0.00543	0.01237	0.01931	0.02624	0.03318	0.04011	0.04705	0.05399	0.06092
23	–.02605	–.01913	–.01222	–.00531	0.00160	0.00852	0.01543	0.02234	0.02926	0.03617	0.04308	0.05000	0.05691
24	–.02978	–.02289	–.01600	–.00911	–.00222	0.00467	0.01156	0.01845	0.02534	0.03223	0.03912	0.04601	0.05290
25	–.03350	–.02663	–.01977	–.01290	–.00603	0.00084	0.00770	0.01457	0.02144	0.02831	0.03517	0.04204	0.04891
26	–.03722	–.03038	–.02353	–.01669	–.00984	–.00300	0.00385	0.01069	0.01754	0.02438	0.03123	0.03807	0.04492
27	–.04094	–.03411	–.02729	–.02047	–.01365	–.00682	–.00000	0.00682	0.01365	0.02047	0.02729	0.03411	0.04094
28	–.04465	–.03785	–.03105	–.02425	–.01745	–.01065	–.00385	0.00296	0.00976	0.01656	0.02336	0.03016	0.03696
29	–.04836	–.04158	–.03480	–.02802	–.02124	–.01447	–.00769	–.00091	0.00587	0.01265	0.01943	0.02621	0.03299
30	–.05207	–.04531	–.03855	–.03180	–.02504	–.01828	–.01153	–.00477	0.00199	0.00874	0.01550	0.02226	0.02902
31	–.05577	–.04904	–.04230	–.03557	–.02883	–.02210	–.01536	–.00863	–.00189	0.00484	0.01158	0.01831	0.02505
32	–.05948	–.05277	–.04606	–.03934	–.03263	–.02592	–.01920	–.01249	–.00578	0.00094	0.00765	0.01437	0.02108
33	–.06319	–.05650	–.04981	–.04312	–.03643	–.02973	–.02304	–.01635	–.00966	–.00297	0.00373	0.01042	0.01711
34	–.06691	–.06024	–.05357	–.04690	–.04022	–.03355	–.02688	–.02021	–.01354	–.00687	–.00020	0.00647	0.01314
35	–.07063	–.06398	–.05733	–.05068	–.04403	–.03738	–.03073	–.02408	–.01743	–.01078	–.00413	0.00252	0.00917
36	–.07435	–.06772	–.06109	–.05446	–.04783	–.04121	–.03458	–.02795	–.02132	–.01469	–.00806	–.00143	0.00520
37	–.07808	–.07147	–.06486	–.05825	–.05165	–.04504	–.03843	–.03182	–.02521	–.01861	–.01200	–.00539	0.00122
38	–.08182	–.07523	–.06864	–.06205	–.05547	–.04888	–.04229	–.03570	–.02912	–.02253	–.01594	–.00935	–.00277
39	–.08556	–.07899	–.07243	–.06586	–.05929	–.05273	–.04616	–.03959	–.03303	–.02646	–.01989	–.01332	–.00676
40	–.08932	–.08277	–.07622	–.06968	–.06313	–.05658	–.05004	–.04349	–.03694	–.03040	–.02385	–.01730	–.01076
41	–.09308	–.08656	–.08003	–.07350	–.06698	–.06045	–.05393	–.04740	–.04087	–.03435	–.02782	–.02129	–.01477

Temperature corrections are in kg/g/mg and are to be added to the mass of water measured in kg/g/mg as the unit of capacity is in m^3/dm^3/cm^3 respectively

Table 3.26 Additional Correction in Grams to be applied to Measure of 1 dm^3 for Variation in Air Density D = 8000 kg m^3

	Pressure in mm of Mercury/Pascals												
T	730	735	740	745	750	755	760	765	770	775	780	785	790
°C	97.3	98.0	98.7	99.3	100	100.7	101.3	102.0	102.7	103.3	104.0	104.7	105.3
5	0.04279	0.05009	0.05739	0.06470	0.07200	0.07931	0.08661	0.09391	0.10122	0.10852	0.11583	0.12313	0.13043
6	0.03884	0.04611	0.05339	0.06067	0.06795	0.07523	0.08250	0.08978	0.09706	0.10434	0.11162	0.11889	0.12617
7	0.03491	0.04216	0.04941	0.05666	0.06391	0.07117	0.07842	0.08567	0.09292	0.10018	0.10743	0.11468	0.12193
8	0.03099	0.03822	0.04545	0.05267	0.05990	0.06713	0.07435	0.08158	0.08881	0.09603	0.10326	0.11049	0.11771
9	0.02710	0.03430	0.04150	0.04870	0.05591	0.06311	0.07031	0.07751	0.08471	0.09191	0.09911	0.10632	0.11352
10	0.02322	0.03040	0.03758	0.04475	0.05193	0.05911	0.06628	0.07346	0.08063	0.08781	0.09499	0.10216	0.10934
11	0.01937	0.02652	0.03367	0.04082	0.04797	0.05512	0.06227	0.06942	0.07658	0.08373	0.09088	0.09803	0.10518
12	0.01552	0.02265	0.02978	0.03690	0.04403	0.05116	0.05828	0.06541	0.07254	0.07966	0.08679	0.09392	0.10104
13	0.01170	0.01880	0.02590	0.03300	0.04011	0.04721	0.05431	0.06141	0.06851	0.07562	0.08272	0.08982	0.09692
14	0.00789	0.01496	0.02204	0.02912	0.03620	0.04327	0.05035	0.05743	0.06451	0.07159	0.07866	0.08574	0.09282
15	0.00409	0.01114	0.01820	0.02525	0.03230	0.03936	0.04641	0.05346	0.06052	0.06757	0.07462	0.08168	0.08873
16	0.00031	0.00733	0.01436	0.02139	0.02842	0.03545	0.04248	0.04951	0.05654	0.06357	0.07060	0.07763	0.08466
17	–.00347	0.00354	0.01055	0.01755	0.02456	0.03156	0.03857	0.04557	0.05258	0.05958	0.06659	0.07360	0.08060
18	–.00723	–.00024	0.00674	0.01372	0.02070	0.02768	0.03467	0.04165	0.04863	0.05561	0.06260	0.06958	0.07656
19	–.01097	–.00401	0.00294	0.00990	0.01686	0.02382	0.03078	0.03774	0.04469	0.05165	0.05861	0.06557	0.07253
20	–.01471	–.00778	–.00084	0.00609	0.01303	0.01996	0.02690	0.03384	0.04077	0.04771	0.05464	0.06158	0.06851
21	–.01844	–.01153	–.00462	0.00230	0.00921	0.01612	0.02303	0.02994	0.03686	0.04377	0.05068	0.05759	0.06451
22	–.02216	–.01527	–.00838	–.00149	0.00540	0.01229	0.01918	0.02606	0.03295	0.03984	0.04673	0.05362	0.06051
23	–.02587	–.01901	–.01214	–.00527	0.00159	0.00846	0.01533	0.02219	0.02906	0.03593	0.04279	0.04966	0.05653
24	–.02958	–.02273	–.01589	–.00905	–.00220	0.00464	0.01149	0.01833	0.02517	0.03202	0.03886	0.04570	0.05255
25	–.03328	–.02646	–.01963	–.01281	–.00599	0.00083	0.00765	0.01447	0.02129	0.02811	0.03494	0.04176	0.04858
26	–.03697	–.03017	–.02337	–.01657	–.00978	–.00298	0.00382	0.01062	0.01742	0.02422	0.03102	0.03782	0.04462
27	–.04066	–.03388	–.02711	–.02033	–.01355	–.00678	–.00000	0.00678	0.01355	0.02033	0.02711	0.03388	0.04066
28	–.04435	–.03759	–.03084	–.02408	–.01733	–.01057	–.00382	0.00294	0.00969	0.01645	0.02320	0.02996	0.03671
29	–.04803	–.04130	–.03457	–.02783	–.02110	–.01437	–.00763	–.00090	0.00583	0.01256	0.01930	0.02603	0.03276
30	–.05172	–.04501	–.03829	–.03158	–.02487	–.01816	–.01145	–.00474	0.00197	0.00869	0.01540	0.02211	0.02882
31	–.05540	–.04871	–.04202	–.03533	–.02864	–.02195	–.01526	–.00857	–.00188	0.00481	0.01150	0.01819	0.02488
32	–.05908	–.05242	–.04575	–.03908	–.03241	–.02574	–.01907	–.01241	–.00574	0.00093	0.00760	0.01427	0.02094
33	–.06277	–.05612	–.04948	–.04283	–.03618	–.02953	–.02289	–.01624	–.00959	–.00295	0.00370	0.01035	0.01700
34	–.06646	–.05983	–.05321	–.04658	–.03995	–.03333	–.02670	–.02008	–.01345	–.00682	–.00020	0.00643	0.01305
35	–.07015	–.06355	–.05694	–.05034	–.04373	–.03713	–.03052	–.02392	–.01731	–.01071	–.00410	0.00251	0.00911
36	–.07385	–.06727	–.06068	–.05410	–.04751	–.04093	–.03434	–.02776	–.02118	–.01459	–.00801	–.00142	0.00516
37	–.07756	–.07099	–.06443	–.05786	–.05130	–.04474	–.03817	–.03161	–.02504	–.01848	–.01192	–.00535	0.00121
38	–.08127	–.07472	–.06818	–.06164	–.05509	–.04855	–.04201	–.03546	–.02892	–.02238	–.01583	–.00929	–.00275
39	–.08499	–.07847	–.07194	–.06542	–.05890	–.05237	–.04585	–.03933	–.03280	–.02628	–.01976	–.01324	–.00671
40	–.08872	–.08222	–.07571	–.06921	–.06271	–.05621	–.04970	–.04320	–.03670	–.03019	–.02369	–.01719	–.01069
41	–.09246	–.08598	–.07950	–.07301	–.06653	–.06005	–.05356	–.04708	–.04060	–.03412	–.02763	–.02115	–.01467

CORRECTION FOR UNIT DIFFERENCE IN COEFFICIENT OF EXPANSION (TABLES 3.27–3.30)

Corrections for unit difference in expansion constants are in kg/g/mg and are to be added to the mass of water measured in kg/g/mg as the unit of capacity is in $m^3/dm^3/cm^3$ respectively

Table 3.27 DEN = 8400 kg/m^3 REFERENCE TEMPERATURE = 27 °C

Temp	0.0	0.1	0.2	0.3	0.4	0.5	0.6	0.7	0.8	0.9
5	–21.9766	–21.8767	–21.7768	–21.6768	–21.5769	–21.4769	–21.3770	–21.2770	–21.1771	–21.0771
6	–20.9772	–20.8772	–20.7773	–20.6773	–20.5773	–20.4774	–20.3774	–20.2774	–20.1775	–20.0775
7	–19.9775	–19.8775	–19.7775	–19.6776	–19.5776	–19.4776	–19.3776	–19.2776	–19.1776	–19.0776
8	–18.9776	–18.8776	–18.7776	–18.6776	–18.5776	–18.4776	–18.3776	–18.2776	–18.1776	–18.0776
9	–17.9776	–17.8776	–17.7776	–17.6775	–17.5775	–17.4775	–17.3775	–17.2775	–17.1775	–17.0775
10	–16.9774	–16.8774	–16.7774	–16.6774	–16.5774	–16.4774	–16.3773	–16.2773	–16.1773	–16.0773
11	–15.9773	–15.8772	–15.7772	–15.6772	–15.5772	–15.4772	–15.3771	–15.2771	–15.1771	–15.0771
12	–14.9771	–14.8770	–14.7770	–14.6770	–14.5770	–14.4770	–14.3770	–14.2769	–14.1769	–14.0769
13	–13.9769	–13.8769	–13.7769	–13.6769	–13.5769	–13.4768	–13.3768	–13.2768	–13.1768	–13.0768
14	–12.9768	–12.8768	–12.7768	–12.6768	–12.5768	–12.4768	–12.3768	–12.2768	–12.1768	–12.0768
15	–11.9769	–11.8769	–11.7769	–11.6769	–11.5769	–11.4769	–11.3770	–11.2770	–11.1770	–11.0770
16	–10.9771	–10.8771	–10.7771	–10.6772	–10.5772	–10.4772	–10.3773	–10.2773	–10.1774	–10.0774
17	–9.9775	–9.8775	–9.7776	–9.6776	–9.5777	–9.4777	–9.3778	–9.2779	–9.1779	–9.0780
18	–8.9781	–8.8782	–8.7782	–8.6783	–8.5784	–8.4785	–8.3786	–8.2787	–8.1788	–8.0789
19	–7.9790	–7.8791	–7.7792	–7.6793	–7.5794	–7.4796	–7.3797	–7.2798	–7.1799	–7.0801
20	–6.9802	–6.8803	–6.7805	–6.6806	–6.5808	–6.4809	–6.3811	–6.2812	–6.1814	–6.0816
21	–5.9817	–5.8819	–5.7821	–5.6823	–5.5825	–5.4827	–5.3829	–5.2831	–5.1833	–5.0835
22	–4.9837	–4.8839	–4.7841	–4.6843	–4.5845	–4.4848	–4.3850	–4.2853	–4.1855	–4.0857
23	–3.9860	–3.8862	–3.7865	–3.6868	–3.5870	–3.4873	–3.3876	–3.2879	–3.1882	–3.0885
24	–2.9888	–2.8891	–2.7894	–2.6897	–2.5900	–2.4903	–2.3906	–2.2910	–2.1913	–2.0916
25	–1.9920	–1.8923	–1.7927	–1.6930	–1.5934	–1.4938	–1.3941	–1.2945	–1.1949	–1.0953
26	–0.9957	–0.8961	–0.7965	–0.6969	–0.5973	–0.4977	–0.3982	–0.2986	–0.1990	–0.0995
27	0.0001	0.0996	0.1992	0.2987	0.3982	0.4977	0.5973	0.6968	0.7963	0.8958
28	0.9953	1.0948	1.1943	1.2937	1.3932	1.4927	1.5921	1.6916	1.7910	1.8905
29	1.9899	2.0893	2.1888	2.2882	2.3876	2.4870	2.5864	2.6858	2.7852	2.8846
30	2.9839	3.0833	3.1827	3.2820	3.3814	3.4807	3.5801	3.6794	3.7787	3.8780
31	3.9773	4.0766	4.1759	4.2752	4.3745	4.4738	4.5731	4.6723	4.7716	4.8708
32	4.9701	5.0693	5.1685	5.2678	5.3670	5.4662	5.5654	5.6646	5.7638	5.8629
33	5.9621	6.0613	6.1604	6.2596	6.3587	6.4579	6.5570	6.6561	6.7552	6.8544
34	6.9535	7.0526	7.1516	7.2507	7.3498	7.4489	7.5479	7.6470	7.7460	7.8450
35	7.9441	8.0431	8.1421	8.2411	8.3401	8.4391	8.5381	8.6370	8.7360	8.8350
36	8.9339	9.0329	9.1318	9.2307	9.3297	9.4286	9.5275	9.6264	9.7253	9.8241
37	9.9230	10.0219	10.1207	10.2196	10.3184	10.4173	10.5161	10.6149	10.7137	10.8125
38	10.9113	11.0101	11.1089	11.2076	11.3064	11.4051	11.5039	11.6026	11.7013	11.8000
39	11.8988	11.9975	12.0962	12.1948	12.2935	12.3922	12.4908	12.5895	12.6881	12.7868
40	12.8854	12.9840	13.0826	13.1812	13.2798	13.3784	13.4770	13.5755	13.6741	13.7726
41	13.8712	13.9697	14.0682	14.1667	14.2652	14.3637	14.4622	14.5607	14.6591	14.7576

Corrections for unit difference in expansion coefficients are in kg/g/mg and are to be added to the mass of water measured in kg/g/mg as the unit of capacity is in m^3/dm^3/cm^3 respectively

Table 3.28 DEN = 8000 kg/m^3 REFERENCE TEMPERATURE = 27 °C

Temp	0.0	0.1	0.2	0.3	0.4	0.5	0.6	0.7	0.8	0.9
5	−21.9768	−21.8768	−21.7769	−21.6770	−21.5770	−21.4771	−21.3771	−21.2772	−21.1772	−21.0773
6	−20.9773	−20.8774	−20.7774	−20.6775	−20.5775	−20.4775	−20.3775	−20.2776	−20.1776	−20.0776
7	−19.9776	−19.8777	−19.7777	−19.6777	−19.5777	−19.4777	−19.3777	−19.2777	−19.1777	−19.0778
8	−18.9778	−18.8778	−18.7778	−18.6778	−18.5778	−18.4777	−18.3777	−18.2777	−18.1777	−18.0777
9	−17.9777	−17.8777	−17.7777	−17.6777	−17.5777	−17.4776	−17.3776	−17.2776	−17.1776	−17.0776
10	−16.9776	−16.8775	−16.7775	−16.6775	−16.5775	−16.4775	−16.3774	−16.2774	−16.1774	−16.0774
11	−15.9774	−15.8773	−15.7773	−15.6773	−15.5773	−15.4773	−15.3772	−15.2772	−15.1772	−15.0772
12	−14.9772	−14.8771	−14.7771	−14.6771	−14.5771	−14.4771	−14.3771	−14.2770	−14.1770	−14.0770
13	−13.9770	−13.8770	−13.7770	−13.6770	−13.5769	−13.4769	−13.3769	−13.2769	−13.1769	−13.0769
14	−12.9769	−12.8769	−12.7769	−12.6769	−12.5769	−12.4769	−12.3769	−12.2769	−12.1769	−12.0769
15	−11.9769	−11.8770	−11.7770	−11.6770	−11.5770	−11.4770	−11.3770	−11.2771	−11.1771	−11.0771
16	−10.9771	−10.8772	−10.7772	−10.6772	−10.5773	−10.4773	−10.3773	−10.2774	−10.1774	−10.0775
17	−9.9775	−9.8776	−9.7776	−9.6777	−9.5777	−9.4778	−9.3779	−9.2779	−9.1780	−9.0781
18	−8.9782	−8.8782	−8.7783	−8.6784	−8.5785	−8.4786	−8.3787	−8.2787	−8.1788	−8.0789
19	−7.9790	−7.8792	−7.7793	−7.6794	−7.5795	−7.4796	−7.3797	−7.2799	−7.1800	−7.0801
20	−6.9802	−6.8804	−6.7805	−6.6807	−6.5808	−6.4810	−6.3811	−6.2813	−6.1815	−6.0816
21	−5.9818	−5.8820	−5.7821	−5.6823	−5.5825	−5.4827	−5.3829	−5.2831	−5.1833	−5.0835
22	−4.9837	−4.8839	−4.7841	−4.6844	−4.5846	−4.4848	−4.3850	−4.2853	−4.1855	−4.0858
23	−3.9860	−3.8863	−3.7865	−3.6868	−3.5871	−3.4873	−3.3876	−3.2879	−3.1882	−3.0885
24	−2.9888	−2.8891	−2.7894	−2.6897	−2.5900	−2.4903	−2.3906	−2.2910	−2.1913	−2.0916
25	−1.9920	−1.8923	−1.7927	−1.6930	−1.5934	−1.4938	−1.3942	−1.2945	−1.1949	−1.0953
26	−0.9957	−0.8961	−0.7965	−0.6969	−0.5973	−0.4977	−0.3982	−0.2986	−0.1990	−0.0995
27	0.0001	0.0996	0.1992	0.2987	0.3982	0.4977	0.5973	0.6968	0.7963	0.8958
28	0.9953	1.0948	1.1943	1.2937	1.3932	1.4927	1.5921	1.6916	1.7910	1.8905
29	1.9899	2.0894	2.1888	2.2882	2.3876	2.4870	2.5864	2.6858	2.7852	2.8846
30	2.9840	3.0833	3.1827	3.2820	3.3814	3.4807	3.5801	3.6794	3.7787	3.8781
31	3.9774	4.0767	4.1760	4.2753	4.3745	4.4738	4.5731	4.6724	4.7716	4.8709
32	4.9701	5.0693	5.1686	5.2678	5.3670	5.4662	5.5654	5.6646	5.7638	5.8630
33	5.9622	6.0613	6.1605	6.2596	6.3588	6.4579	6.5571	6.6562	6.7553	6.8544
34	6.9535	7.0526	7.1517	7.2508	7.3498	7.4489	7.5480	7.6470	7.7461	7.8451
35	7.9441	8.0431	8.1422	8.2412	8.3402	8.4391	8.5381	8.6371	8.7361	8.8350
36	8.9340	9.0329	9.1319	9.2308	9.3297	9.4286	9.5275	9.6264	9.7253	9.8242
37	9.9231	10.0219	10.1208	10.2196	10.3185	10.4173	10.5161	10.6150	10.7138	10.8126
38	10.9114	11.0102	11.1089	11.2077	11.3065	11.4052	11.5040	11.6027	11.7014	11.8001
39	11.8988	11.9975	12.0962	12.1949	12.2936	12.3923	12.4909	12.5896	12.6882	12.7869
40	12.8855	12.9841	13.0827	13.1813	13.2799	13.3785	13.4770	13.5756	13.6742	13.7727
41	13.8713	13.9698	14.0683	14.1668	14.2653	14.3638	14.4623	14.5608	14.6592	14.7577

Corrections for unit difference in expansion constants are in kg/g/mg and are to be added to the mass of water measured in kg/g/mg as the unit of capacity is in m³/dm³/cm³ respectively

Table 3.29 DEN = 8400 kg/m³ REFERENCE TEMPERATURE = 20°C

Temp	0.0	0.1	0.2	0.3	0.4	0.5	0.6	0.7	0.8	0.9
5	−14.9841	−14.8841	−14.7842	−14.6843	−14.5844	−14.4844	−14.3845	−14.2846	−14.1847	−14.0847
6	−13.9848	−13.8849	−13.7849	−13.6850	−13.5850	−13.4851	−13.3852	−13.2852	−13.1853	−13.0853
7	−12.9854	−12.8854	−12.7855	−12.6855	−12.5856	−12.4856	−12.3857	−12.2857	−12.1858	−12.0858
8	−11.9859	−11.8859	−11.7860	−11.6860	−11.5860	−11.4861	−11.3861	−11.2862	−11.1862	−11.0863
9	−10.9863	−10.8863	−10.7864	−10.6864	−10.5865	−10.4865	−10.3866	−10.2866	−10.1866	−10.0867
10	−9.9867	−9.8868	−9.7868	−9.6869	−9.5869	−9.4870	−9.3870	−9.2871	−9.1871	−9.0872
11	−8.9872	−8.8873	−8.7873	−8.6874	−8.5874	−8.4875	−8.3875	−8.2876	−8.1876	−8.0877
12	7.9878	−7.8878	−7.7879	−7.6879	−7.5880	−7.4881	−7.3881	−7.2882	−7.1883	−7.0884
13	−6.9884	−6.8885	−6.7886	−6.6887	−6.5888	−6.4888	−6.3889	−6.2890	−6.1891	−6.0892
14	−5.9893	−5.8894	−5.7895	−5.6896	−5.5897	−5.4898	−5.3899	−5.2900	−5.1901	−5.0902
15	−4.9903	−4.8905	−4.7906	−4.6907	−4.5908	−4.4910	−4.3911	−4.2912	−4.1914	−4.0915
16	−3.9916	−3.8918	−3.7919	−3.6921	−3.5922	−3.4924	−3.3925	−3.2927	−3.1929	−3.0930
17	−2.9932	−2.8934	−2.7936	−2.6937	−2.5939	−2.4941	−2.3943	−2.2945	−2.1947	−2.0949
18	−1.9951	−1.8953	−1.7955	−1.6957	−1.5960	−1.4962	−1.3964	−1.2966	−1.1969	−1.0971
19	−0.9973	−0.8976	−0.7978	−0.6981	−0.5983	−0.4986	−0.3989	−0.2991	−0.1994	−0.0997
20	0.0000	0.0998	0.1995	0.2992	0.3989	0.4986	0.5983	0.6980	0.7977	0.8973
21	0.9970	1.0967	1.1964	1.2960	1.3957	1.4953	1.5950	1.6946	1.7943	1.8939
22	1.9935	2.0932	2.1928	2.2924	2.3920	2.4916	2.5912	2.6908	2.7904	2.8900
23	2.9896	3.0892	3.1887	3.2883	3.3879	3.4874	3.5870	3.6865	3.7861	3.8856
24	3.9851	4.0847	4.1842	4.2837	4.3832	4.4827	4.5822	4.6817	4.7812	4.8807
25	4.9801	5.0796	5.1791	5.2785	5.3780	5.4774	5.5769	5.6763	5.7758	5.8752
26	5.9746	6.0740	6.1734	6.2728	6.3722	6.4716	6.5710	6.6704	6.7697	6.8691
27	6.9685	7.0678	7.1672	7.2665	7.3658	7.4652	7.5645	7.6638	7.7631	7.8624
28	7.9617	8.0610	8.1603	8.2596	8.3588	8.4581	8.5574	8.6566	8.7559	8.8551
29	8.9543	9.0536	9.1528	9.2520	9.3512	9.4504	9.5496	9.6488	9.7479	9.8471
30	9.9463	10.0454	10.1446	10.2437	10.3429	10.4420	10.5411	10.6402	10.7393	10.8384
31	10.9375	11.0366	11.1357	11.2348	11.3338	11.4329	11.5319	11.6310	11.7300	11.8290
32	11.9281	12.0271	12.1261	12.2251	12.3241	12.4230	12.5220	12.6210	12.7199	12.8189
33	12.9178	13.0168	13.1157	13.2146	13.3135	13.4124	13.5113	13.6102	13.7091	13.8080
34	13.9069	14.0057	14.1046	14.2034	14.3022	14.4011	14.4999	14.5987	14.6975	14.7963
35	14.8951	14.9939	15.0926	15.1914	15.2902	15.3889	15.4876	15.5864	15.6851	15.7838
36	15.8825	15.9812	16.0799	16.1786	16.2772	16.3759	16.4746	16.5732	16.6719	16.7705
37	16.8691	16.9677	17.0663	17.1649	17.2635	17.3621	17.4607	17.5592	17.6578	17.7563
38	17.8548	17.9534	18.0519	18.1504	18.2489	18.3474	18.4459	18.5443	18.6428	18.7413
39	18.8397	18.9382	19.0366	19.1350	19.2334	19.3318	19.4302	19.5286	19.6270	19.7253
40	19.8237	19.9220	20.0204	20.1187	20.2170	20.3153	20.4136	20.5119	20.6102	20.7085
41	20.8068	20.9050	21.0033	21.1015	21.1997	21.2979	21.3961	21.4943	21.5925	21.6907

Corrections for unit difference in expansion constants are in kg/g/mg and are to be added to the mass of water measured in kg/g/mg as the unit of capacity is in $m^3/dm^3/cm^3$ respectively

Table 3.30 DEN = 8000 kg/m^3 REFERENCE TEMPERATURE = 20 °C

Temp	0.0	0.1	0.2	0.3	0.4	0.5	0.6	0.7	0.8	0.9
5	−14.9842	−14.8842	−14.7843	−14.6844	−14.5845	−14.4845	−14.3846	−14.2847	−14.1848	−14.0848
6	−13.9849	−13.8850	−13.7850	−13.6851	−13.5851	−13.4852	−13.3853	−13.2853	−13.1854	−13.0854
7	−12.9855	−12.8855	−12.7856	−12.6856	−12.5857	−12.4857	−12.3858	−12.2858	−12.1859	−12.0859
8	−11.9860	−11.8860	−11.7860	−11.6861	−11.5861	−11.4862	−11.3862	−11.2863	−11.1863	−11.0863
9	−10.9864	−10.8864	−10.7865	−10.6865	−10.5865	−10.4866	−10.3866	−10.2867	−10.1867	−10.0868
10	−9.9868	−9.8868	−9.7869	−9.6869	−9.5870	−9.4870	−9.3871	−9.2871	−9.1872	−9.0872
11	−8.9873	−8.8873	−8.7874	−8.6874	−8.5875	−8.4875	−8.3876	−8.2876	−8.1877	−8.0878
12	−7.9878	−7.8879	−7.7879	−7.6880	−7.5881	−7.4881	−7.3882	−7.2883	−7.1883	−7.0884
13	−6.9885	−6.8886	−6.7886	−6.6887	−6.5888	−6.4889	−6.3890	−6.2891	−6.1891	−6.0892
14	−5.9893	−5.8894	−5.7895	−5.6896	−5.5897	−5.4898	−5.3899	−5.2900	−5.1902	−5.0903
15	−4.9904	−4.8905	−4.7906	−4.6907	−4.5909	−4.4910	−4.3911	−4.2913	−4.1914	−4.0915
16	−3.9917	−3.8918	−3.7920	−3.6921	−3.5923	−3.4924	−3.3926	−3.2927	−3.1929	−3.0931
17	−2.9932	−2.8934	−2.7936	−2.6938	−2.5939	−2.4941	−2.3943	−2.2945	−2.1947	−2.0949
18	−1.9951	−1.8953	−1.7955	−1.6957	−1.5960	−1.4962	−1.3964	−1.2966	−1.1969	−1.0971
19	−0.9973	−0.8976	−0.7978	−0.6981	−0.5983	−0.4986	−0.3989	−0.2991	−0.1994	−0.0997
20	0.0000	0.0998	0.1995	0.2992	0.3989	0.4986	0.5983	0.6980	0.7977	0.8973
21	0.9970	1.0967	1.1964	1.2960	1.3957	1.4953	1.5950	1.6946	1.7943	1.8939
22	1.9936	2.0932	2.1928	2.2924	2.3920	2.4916	2.5912	2.6908	2.7904	2.8900
23	2.9896	3.0892	3.1888	3.2883	3.3879	3.4875	3.5870	3.6866	3.7861	3.8856
24	3.9852	4.0847	4.1842	4.2837	4.3832	4.4827	4.5822	4.6817	4.7812	4.8807
25	4.9802	5.0797	5.1791	5.2786	5.3780	5.4775	5.5769	5.6764	5.7758	5.8752
26	5.9746	6.0741	6.1735	6.2729	6.3723	6.4717	6.5710	6.6704	6.7698	6.8692
27	6.9685	7.0679	7.1672	7.2666	7.3659	7.4652	7.5645	7.6639	7.7632	7.8625
28	7.9618	8.0611	8.1604	8.2596	8.3589	8.4582	8.5574	8.6567	8.7559	8.8552
29	8.9544	9.0536	9.1528	9.2521	9.3513	9.4505	9.5496	9.6488	9.7480	9.8472
30	9.9463	10.0455	10.1447	10.2438	10.3429	10.4421	10.5412	10.6403	10.7394	10.8385
31	10.9376	11.0367	11.1358	11.2348	11.3339	11.4330	11.5320	11.6311	11.7301	11.8291
32	11.9281	12.0272	12.1262	12.2252	12.3242	12.4231	12.5221	12.6211	12.7200	12.8190
33	12.9179	13.0169	13.1158	13.2147	13.3136	13.4125	13.5114	13.6103	13.7092	13.8081
34	13.9070	14.0058	14.1047	14.2035	14.3023	14.4012	14.5000	14.5988	14.6976	14.7964
35	14.8952	14.9940	15.0927	15.1915	15.2903	15.3890	15.4877	15.5865	15.6852	15.7839
36	15.8826	15.9813	16.0800	16.1787	16.2774	16.3760	16.4747	16.5733	16.6720	16.7706
37	16.8692	16.9678	17.0664	17.1650	17.2636	17.3622	17.4608	17.5593	17.6579	17.7564
38	17.8550	17.9535	18.0520	18.1505	18.2490	18.3475	18.4460	18.5445	18.6429	18.7414
39	18.8398	18.9383	19.0367	19.1351	19.2335	19.3319	19.4303	19.5287	19.6271	19.7255
40	19.8238	19.9222	20.0205	20.1188	20.2172	20.3155	20.4138	20.5121	20.6104	20.7086
41	20.8069	20.9052	21.0034	21.1016	21.1999	21.2981	21.3963	21.4945	21.5927	21.6909

CORRECTION FACTOR WHEN MERCURY IS USED (TABLES 3.31 TO 3.46)

The factor K is multiplied to the mass of mercury delivered/contained in the measure in kg/g/mg to give its capacity at 20 °C in m³/dm³/cm³

Table 3.31 Reference Temperature = 20 °C Air density = 1.2 kg/m³, ALPHA = .00001/°C DEN = 8000 kg/m³

The values of 10^3K

Temp	0.0	0.1	0.2	0.3	0.4	0.5	0.6	0.7	0.8	0.9
5	0.073629	0.073631	0.073632	0.073633	0.073634	0.073636	0.073637	0.073638	0.073639	0.073641
6	0.073642	0.073643	0.073644	0.073646	0.073647	0.073648	0.073650	0.073651	0.073652	0.073653
7	0.073655	0.073656	0.073657	0.073658	0.073660	0.073661	0.073662	0.073663	0.073665	0.073666
8	0.073667	0.073668	0.073670	0.073671	0.073672	0.073674	0.073675	0.073676	0.073677	0.073679
9	0.073680	0.073681	0.073682	0.073684	0.073685	0.073686	0.073687	0.073689	0.073690	0.073691
10	0.073692	0.073694	0.073695	0.073696	0.073698	0.073699	0.073700	0.073701	0.073703	0.073704
11	0.073705	0.073706	0.073708	0.073709	0.073710	0.073711	0.073713	0.073714	0.073715	0.073716
12	0.073718	0.073719	0.073720	0.073722	0.073723	0.073724	0.073725	0.073727	0.073728	0.073729
13	0.073730	0.073732	0.073733	0.073734	0.073735	0.073737	0.073738	0.073739	0.073740	0.073742
14	0.073743	0.073744	0.073746	0.073747	0.073748	0.073749	0.073751	0.073752	0.073753	0.073754
15	0.073756	0.073757	0.073758	0.073759	0.073761	0.073762	0.073763	0.073764	0.073766	0.073767
16	0.073768	0.073770	0.073771	0.073772	0.073773	0.073775	0.073776	0.073777	0.073778	0.073780
17	0.073781	0.073782	0.073783	0.073785	0.073786	0.073787	0.073789	0.073790	0.073791	0.073792
18	0.073794	0.073795	0.073796	0.073797	0.073799	0.073800	0.073801	0.073802	0.073804	0.073805
19	0.073806	0.073807	0.073809	0.073810	0.073811	0.073813	0.073814	0.073815	0.073816	0.073818
20	0.073819	0.073820	0.073821	0.073823	0.073824	0.073825	0.073826	0.073828	0.073829	0.073830
21	0.073831	0.073833	0.073834	0.073835	0.073837	0.073838	0.073839	0.073840	0.073842	0.073843
22	0.073844	0.073845	0.073847	0.073848	0.073849	0.073850	0.073852	0.073853	0.073854	0.073855
23	0.073857	0.073858	0.073859	0.073861	0.073862	0.073863	0.073864	0.073866	0.073867	0.073868
24	0.073869	0.073871	0.073872	0.073873	0.073874	0.073876	0.073877	0.073878	0.073880	0.073881
25	0.073882	0.073883	0.073885	0.073886	0.073887	0.073888	0.073890	0.073891	0.073892	0.073893
26	0.073895	0.073896	0.073897	0.073898	0.073900	0.073901	0.073902	0.073904	0.073905	0.073906
27	0.073907	0.073909	0.073910	0.073911	0.073912	0.073914	0.073915	0.073916	0.073917	0.073919
28	0.073920	0.073921	0.073922	0.073924	0.073925	0.073926	0.073928	0.073929	0.073930	0.073931
29	0.073933	0.073934	0.073935	0.073936	0.073938	0.073939	0.073940	0.073941	0.073943	0.073944
30	0.073945	0.073947	0.073948	0.073949	0.073950	0.073952	0.073953	0.073954	0.073955	0.073957
31	0.073958	0.073959	0.073960	0.073962	0.073963	0.073964	0.073965	0.073967	0.073968	0.073969
32	0.073971	0.073972	0.073973	0.073974	0.073976	0.073977	0.073978	0.073979	0.073981	0.073982
33	0.073983	0.073984	0.073986	0.073987	0.073988	0.073990	0.073991	0.073992	0.073993	0.073995
34	0.073996	0.073997	0.073998	0.074000	0.074001	0.074002	0.074003	0.074005	0.074006	0.074007
35	0.074009	0.074010	0.074011	0.074012	0.074014	0.074015	0.074016	0.074017	0.074019	0.074020
36	0.074021	0.074022	0.074024	0.074025	0.074026	0.074027	0.074029	0.074030	0.074031	0.074033
37	0.074034	0.074035	0.074036	0.074038	0.074039	0.074040	0.074041	0.074043	0.074044	0.074045
38	0.074046	0.074048	0.074049	0.074050	0.074052	0.074053	0.074054	0.074055	0.074057	0.074058
39	0.074059	0.074060	0.074062	0.074063	0.074064	0.074065	0.074067	0.074068	0.074069	0.074071
40	0.074072	0.074073	0.074074	0.074076	0.074077	0.074078	0.074079	0.074081	0.074082	0.074083
41	0.074084	0.074086	0.074087	0.074088	0.074090	0.074091	0.074092	0.074093	0.074095	0.074096

The factor K is multiplied to the mass of mercury delivered/contained in the measure in kg/g/mg to give its capacity at 20 °C in $m^3/dm^3/cm^3$

Table 3.32 ReferenceTemperature = 20 °C Air density = 1.2 kg/m^3, ALPHA = .00015/°C DEN = 8000 kg/m^3

The values of 10^3K

Temp	0.0	0.1	0.2	0.3	0.4	0.5	0.6	0.7	0.8	0.9
5	0.073635	0.073636	0.073637	0.073639	0.073640	0.073641	0.073642	0.073643	0.073645	0.073646
6	0.073647	0.073648	0.073650	0.073651	0.073652	0.073653	0.073654	0.073656	0.073657	0.073658
7	0.073659	0.073661	0.073662	0.073663	0.073664	0.073666	0.073667	0.073668	0.073669	0.073670
8	0.073672	0.073673	0.073674	0.073675	0.073677	0.073678	0.073679	0.073680	0.073681	0.073683
9	0.073684	0.073685	0.073686	0.073688	0.073689	0.073690	0.073691	0.073692	0.073694	0.073695
10	0.073696	0.073697	0.073699	0.073700	0.073701	0.073702	0.073704	0.073705	0.073706	0.073707
11	0.073708	0.073710	0.073711	0.073712	0.073713	0.073715	0.073716	0.073717	0.073718	0.073719
12	0.073721	0.073722	0.073723	0.073724	0.073726	0.073727	0.073728	0.073729	0.073731	0.073732
13	0.073733	0.073734	0.073735	0.073737	0.073738	0.073739	0.073740	0.073742	0.073743	0.073744
14	0.073745	0.073746	0.073748	0.073749	0.073750	0.073751	0.073753	0.073754	0.073755	0.073756
15	0.073757	0.073759	0.073760	0.073761	0.073762	0.073764	0.073765	0.073766	0.073767	0.073769
16	0.073770	0.073771	0.073772	0.073773	0.073775	0.073776	0.073777	0.073778	0.073780	0.073781
17	0.073782	0.073783	0.073784	0.073786	0.073787	0.073788	0.073789	0.073791	0.073792	0.073793
18	0.073794	0.073796	0.073797	0.073798	0.073799	0.073800	0.073802	0.073803	0.073804	0.073805
19	0.073807	0.073808	0.073809	0.073810	0.073811	0.073813	0.073814	0.073815	0.073816	0.073818
20	0.073819	0.073820	0.073821	0.073823	0.073824	0.073825	0.073826	0.073827	0.073829	0.073830
21	0.073831	0.073832	0.073834	0.073835	0.073836	0.073837	0.073838	0.073840	0.073841	0.073842
22	0.073843	0.073845	0.073846	0.073847	0.073848	0.073849	0.073851	0.073852	0.073853	0.073854
23	0.073856	0.073857	0.073858	0.073859	0.073861	0.073862	0.073863	0.073864	0.073865	0.073867
24	0.073868	0.073869	0.073870	0.073872	0.073873	0.073874	0.073875	0.073877	0.073878	0.073879
25	0.073880	0.073881	0.073883	0.073884	0.073885	0.073886	0.073888	0.073889	0.073890	0.073891
26	0.073892	0.073894	0.073895	0.073896	0.073897	0.073899	0.073900	0.073901	0.073902	0.073904
27	0.073905	0.073906	0.073907	0.073908	0.073910	0.073911	0.073912	0.073913	0.073915	0.073916
28	0.073917	0.073918	0.073919	0.073921	0.073922	0.073923	0.073924	0.073926	0.073927	0.073928
29	0.073929	0.073931	0.073932	0.073933	0.073934	0.073935	0.073937	0.073938	0.073939	0.073940
30	0.073942	0.073943	0.073944	0.073945	0.073946	0.073948	0.073949	0.073950	0.073951	0.073953
31	0.073954	0.073955	0.073956	0.073958	0.073959	0.073960	0.073961	0.073962	0.073964	0.073965
32	0.073966	0.073967	0.073969	0.073970	0.073971	0.073972	0.073973	0.073975	0.073976	0.073977
33	0.073978	0.073980	0.073981	0.073982	0.073983	0.073985	0.073986	0.073987	0.073988	0.073989
34	0.073991	0.073992	0.073993	0.073994	0.073996	0.073997	0.073998	0.073999	0.074001	0.074002
35	0.074003	0.074004	0.074005	0.074007	0.074008	0.074009	0.074010	0.074012	0.074013	0.074014
36	0.074015	0.074016	0.074018	0.074019	0.074020	0.074021	0.074023	0.074024	0.074025	0.074026
37	0.074028	0.074029	0.074030	0.074031	0.074032	0.074034	0.074035	0.074036	0.074037	0.074039
38	0.074040	0.074041	0.074042	0.074043	0.074045	0.074046	0.074047	0.074048	0.074050	0.074051
39	0.074052	0.074053	0.074055	0.074056	0.074057	0.074058	0.074059	0.074061	0.074062	0.074063
40	0.074064	0.074066	0.074067	0.074068	0.074069	0.074071	0.074072	0.074073	0.074074	0.074075
41	0.074077	0.074078	0.074079	0.074080	0.074082	0.074083	0.074084	0.074085	0.074086	0.074088

The factor K is multiplied to the mass of mercury delivered/contained in the measure in kg/g/mg to give its capacity at 20 °C in m³/dm³/cm³

Table 3.33 ReferenceTemperature = 20 °C Air density = 1.2 kg/m³, ALPHA = .000025/°C DEN = 8000 kg/m³

The values of 10^3K

Temp	0.0	0.1	0.2	0.3	0.4	0.5	0.6	0.7	0.8	0.9
5	0.073646	0.073647	0.073648	0.073649	0.073651	0.073652	0.073653	0.073654	0.073655	0.073656
6	0.073657	0.073659	0.073660	0.073661	0.073662	0.073663	0.073664	0.073666	0.073667	0.073668
7	0.073669	0.073670	0.073671	0.073672	0.073674	0.073675	0.073676	0.073677	0.073678	0.073679
8	0.073680	0.073682	0.073683	0.073684	0.073685	0.073686	0.073687	0.073689	0.073690	0.073691
9	0.073692	0.073693	0.073694	0.073695	0.073697	0.073698	0.073699	0.073700	0.073701	0.073702
10	0.073704	0.073705	0.073706	0.073707	0.073708	0.073709	0.073710	0.073712	0.073713	0.073714
11	0.073715	0.073716	0.073717	0.073719	0.073720	0.073721	0.073722	0.073723	0.073724	0.073725
12	0.073727	0.073728	0.073729	0.073730	0.073731	0.073732	0.073734	0.073735	0.073736	0.073737
13	0.073738	0.073739	0.073740	0.073742	0.073743	0.073744	0.073745	0.073746	0.073747	0.073748
14	0.073750	0.073751	0.073752	0.073753	0.073754	0.073755	0.073757	0.073758	0.073759	0.073760
15	0.073761	0.073762	0.073763	0.073765	0.073766	0.073767	0.073768	0.073769	0.073770	0.073772
16	0.073773	0.073774	0.073775	0.073776	0.073777	0.073778	0.073780	0.073781	0.073782	0.073783
17	0.073784	0.073785	0.073787	0.073788	0.073789	0.073790	0.073791	0.073792	0.073793	0.073795
18	0.073796	0.073797	0.073798	0.073799	0.073800	0.073802	0.073803	0.073804	0.073805	0.073806
19	0.073807	0.073808	0.073810	0.073811	0.073812	0.073813	0.073814	0.073815	0.073817	0.073818
20	0.073819	0.073820	0.073821	0.073822	0.073823	0.073825	0.073826	0.073827	0.073828	0.073829
21	0.073830	0.073832	0.073833	0.073834	0.073835	0.073836	0.073837	0.073838	0.073840	0.073841
22	0.073842	0.073843	0.073844	0.073845	0.073847	0.073848	0.073849	0.073850	0.073851	0.073852
23	0.073853	0.073855	0.073856	0.073857	0.073858	0.073859	0.073860	0.073861	0.073863	0.073864
24	0.073865	0.073866	0.073867	0.073868	0.073870	0.073871	0.073872	0.073873	0.073874	0.073875
25	0.073876	0.073878	0.073879	0.073880	0.073881	0.073882	0.073883	0.073885	0.073886	0.073887
26	0.073888	0.073889	0.073890	0.073891	0.073893	0.073894	0.073895	0.073896	0.073897	0.073898
27	0.073900	0.073901	0.073902	0.073903	0.073904	0.073905	0.073906	0.073908	0.073909	0.073910
28	0.073911	0.073912	0.073913	0.073915	0.073916	0.073917	0.073918	0.073919	0.073920	0.073921
29	0.073923	0.073924	0.073925	0.073926	0.073927	0.073928	0.073930	0.073931	0.073932	0.073933
30	0.073934	0.073935	0.073936	0.073938	0.073939	0.073940	0.073941	0.073942	0.073943	0.073945
31	0.073946	0.073947	0.073948	0.073949	0.073950	0.073951	0.073953	0.073954	0.073955	0.073956
32	0.073957	0.073958	0.073960	0.073961	0.073962	0.073963	0.073964	0.073965	0.073966	0.073968
33	0.073969	0.073970	0.073971	0.073972	0.073973	0.073975	0.073976	0.073977	0.073978	0.073979
34	0.073980	0.073981	0.073983	0.073984	0.073985	0.073986	0.073987	0.073988	0.073990	0.073991
35	0.073992	0.073993	0.073994	0.073995	0.073996	0.073998	0.073999	0.074000	0.074001	0.074002
36	0.074003	0.074005	0.074006	0.074007	0.074008	0.074009	0.074010	0.074011	0.074013	0.074014
37	0.074015	0.074016	0.074017	0.074018	0.074020	0.074021	0.074022	0.074023	0.074024	0.074025
38	0.074026	0.074028	0.074029	0.074030	0.074031	0.074032	0.074033	0.074035	0.074036	0.074037
39	0.074038	0.074039	0.074040	0.074041	0.074043	0.074044	0.074045	0.074046	0.074047	0.074048
40	0.074050	0.074051	0.074052	0.074053	0.074054	0.074055	0.074056	0.074058	0.074059	0.074060
41	0.074061	0.074062	0.074063	0.074065	0.074066	0.074067	0.074068	0.074069	0.074070	0.074072

The factor K is multiplied to the mass of mercury delivered/contained in the measure in kg/g/mg to give its capacity at 20 °C in m³/dm³/cm³

Table 3.34 Reference Temperature = 20 °C Air density = 1.2 kg/m³, ALPHA = .00003/°C DEN = 8000 kg/m³

The values of 10^3K

Temp	0.0	0.1	0.2	0.3	0.4	0.5	0.6	0.7	0.8	0.9
5	0.073651	0.073653	0.073654	0.073655	0.073656	0.073657	0.073658	0.073659	0.073660	0.073661
6	0.073663	0.073664	0.073665	0.073666	0.073667	0.073668	0.073669	0.073670	0.073672	0.073673
7	0.073674	0.073675	0.073676	0.073677	0.073678	0.073679	0.073680	0.073682	0.073683	0.073684
8	0.073685	0.073686	0.073687	0.073688	0.073689	0.073690	0.073692	0.073693	0.073694	0.073695
9	0.073696	0.073697	0.073698	0.073699	0.073701	0.073702	0.073703	0.073704	0.073705	0.073706
10	0.073707	0.073708	0.073709	0.073711	0.073712	0.073713	0.073714	0.073715	0.073716	0.073717
11	0.073718	0.073720	0.073721	0.073722	0.073723	0.073724	0.073725	0.073726	0.073727	0.073728
12	0.073730	0.073731	0.073732	0.073733	0.073734	0.073735	0.073736	0.073737	0.073738	0.073740
13	0.073741	0.073742	0.073743	0.073744	0.073745	0.073746	0.073747	0.073749	0.073750	0.073751
14	0.073752	0.073753	0.073754	0.073755	0.073756	0.073757	0.073759	0.073760	0.073761	0.073762
15	0.073763	0.073764	0.073765	0.073766	0.073767	0.073769	0.073770	0.073771	0.073772	0.073773
16	0.073774	0.073775	0.073776	0.073778	0.073779	0.073780	0.073781	0.073782	0.073783	0.073784
17	0.073785	0.073786	0.073788	0.073789	0.073790	0.073791	0.073792	0.073793	0.073794	0.073795
18	0.073797	0.073798	0.073799	0.073800	0.073801	0.073802	0.073803	0.073804	0.073805	0.073807
19	0.073808	0.073809	0.073810	0.073811	0.073812	0.073813	0.073814	0.073815	0.073817	0.073818
20	0.073819	0.073820	0.073821	0.073822	0.073823	0.073824	0.073826	0.073827	0.073828	0.073829
21	0.073830	0.073831	0.073832	0.073833	0.073834	0.073836	0.073837	0.073838	0.073839	0.073840
22	0.073841	0.073842	0.073843	0.073844	0.073846	0.073847	0.073848	0.073849	0.073850	0.073851
23	0.073852	0.073853	0.073855	0.073856	0.073857	0.073858	0.073859	0.073860	0.073861	0.073862
24	0.073863	0.073865	0.073866	0.073867	0.073868	0.073869	0.073870	0.073871	0.073872	0.073874
25	0.073875	0.073876	0.073877	0.073878	0.073879	0.073880	0.073881	0.073882	0.073884	0.073885
26	0.073886	0.073887	0.073888	0.073889	0.073890	0.073891	0.073893	0.073894	0.073895	0.073896
27	0.073897	0.073898	0.073899	0.073900	0.073901	0.073903	0.073904	0.073905	0.073906	0.073907
28	0.073908	0.073909	0.073910	0.073911	0.073913	0.073914	0.073915	0.073916	0.073917	0.073918
29	0.073919	0.073920	0.073922	0.073923	0.073924	0.073925	0.073926	0.073927	0.073928	0.073929
30	0.073930	0.073932	0.073933	0.073934	0.073935	0.073936	0.073937	0.073938	0.073939	0.073941
31	0.073942	0.073943	0.073944	0.073945	0.073946	0.073947	0.073948	0.073949	0.073951	0.073952
32	0.073953	0.073954	0.073955	0.073956	0.073957	0.073958	0.073960	0.073961	0.073962	0.073963
33	0.073964	0.073965	0.073966	0.073967	0.073968	0.073970	0.073971	0.073972	0.073973	0.073974
34	0.073975	0.073976	0.073977	0.073978	0.073980	0.073981	0.073982	0.073983	0.073984	0.073985
35	0.073986	0.073987	0.073989	0.073990	0.073991	0.073992	0.073993	0.073994	0.073995	0.073996
36	0.073997	0.073999	0.074000	0.074001	0.074002	0.074003	0.074004	0.074005	0.074006	0.074008
37	0.074009	0.074010	0.074011	0.074012	0.074013	0.074014	0.074015	0.074016	0.074018	0.074019
38	0.074020	0.074021	0.074022	0.074023	0.074024	0.074025	0.074027	0.074028	0.074029	0.074030
39	0.074031	0.074032	0.074033	0.074034	0.074035	0.074037	0.074038	0.074039	0.074040	0.074041
40	0.074042	0.074043	0.074044	0.074046	0.074047	0.074048	0.074049	0.074050	0.074051	0.074052
41	0.074053	0.074054	0.074056	0.074057	0.074058	0.074059	0.074060	0.074061	0.074062	0.074063

The factor K is multiplied to the mass of mercury delivered/contained in the measure in kg/g/mg to give its capacity at 20 °C in $m^3/dm^3/cm^3$

Table 3.35 Reference Temperature = 20 °C Air density = 1.2 kg/m^3, ALPHA = .00001/°C DEN = 8400 kg/m^3

The values of 10^3K

Temp	0.0	0.1	0.2	0.3	0.4	0.5	0.6	0.7	0.8	0.9
5	0.073630	0.073631	0.073632	0.073634	0.073635	0.073636	0.073637	0.073639	0.073640	0.073641
6	0.073642	0.073644	0.073645	0.073646	0.073648	0.073649	0.073650	0.073651	0.073653	0.073654
7	0.073655	0.073656	0.073658	0.073659	0.073660	0.073661	0.073663	0.073664	0.073665	0.073666
8	0.073668	0.073669	0.073670	0.073672	0.073673	0.073674	0.073675	0.073677	0.073678	0.073679
9	0.073680	0.073682	0.073683	0.073684	0.073685	0.073687	0.073688	0.073689	0.073690	0.073692
10	0.073693	0.073694	0.073696	0.073697	0.073698	0.073699	0.073701	0.073702	0.073703	0.073704
11	0.073706	0.073707	0.073708	0.073709	0.073711	0.073712	0.073713	0.073714	0.073716	0.073717
12	0.073718	0.073720	0.073721	0.073722	0.073723	0.073725	0.073726	0.073727	0.073728	0.073730
13	0.073731	0.073732	0.073733	0.073735	0.073736	0.073737	0.073738	0.073740	0.073741	0.073742
14	0.073744	0.073745	0.073746	0.073747	0.073749	0.073750	0.073751	0.073752	0.073754	0.073755
15	0.073756	0.073757	0.073759	0.073760	0.073761	0.073762	0.073764	0.073765	0.073766	0.073768
16	0.073769	0.073770	0.073771	0.073773	0.073774	0.073775	0.073776	0.073778	0.073779	0.073780
17	0.073781	0.073783	0.073784	0.073785	0.073786	0.073788	0.073789	0.073790	0.073792	0.073793
18	0.073794	0.073795	0.073797	0.073798	0.073799	0.073800	0.073802	0.073803	0.073804	0.073805
19	0.073807	0.073808	0.073809	0.073811	0.073812	0.073813	0.073814	0.073816	0.073817	0.073818
20	0.073819	0.073821	0.073822	0.073823	0.073824	0.073826	0.073827	0.073828	0.073829	0.073831
21	0.073832	0.073833	0.073835	0.073836	0.073837	0.073838	0.073840	0.073841	0.073842	0.073843
22	0.073845	0.073846	0.073847	0.073848	0.073850	0.073851	0.073852	0.073853	0.073855	0.073856
23	0.073857	0.073859	0.073860	0.073861	0.073862	0.073864	0.073865	0.073866	0.073867	0.073869
24	0.073870	0.073871	0.073872	0.073874	0.073875	0.073876	0.073877	0.073879	0.073880	0.073881
25	0.073883	0.073884	0.073885	0.073886	0.073888	0.073889	0.073890	0.073891	0.073893	0.073894
26	0.073895	0.073896	0.073898	0.073899	0.073900	0.073902	0.073903	0.073904	0.073905	0.073907
27	0.073908	0.073909	0.073910	0.073912	0.073913	0.073914	0.073915	0.073917	0.073918	0.073919
28	0.073920	0.073922	0.073923	0.073924	0.073926	0.073927	0.073928	0.073929	0.073931	0.073932
29	0.073933	0.073934	0.073936	0.073937	0.073938	0.073939	0.073941	0.073942	0.073943	0.073945
30	0.073946	0.073947	0.073948	0.073950	0.073951	0.073952	0.073953	0.073955	0.073956	0.073957
31	0.073958	0.073960	0.073961	0.073962	0.073963	0.073965	0.073966	0.073967	0.073969	0.073970
32	0.073971	0.073972	0.073974	0.073975	0.073976	0.073977	0.073979	0.073980	0.073981	0.073982
33	0.073984	0.073985	0.073986	0.073988	0.073989	0.073990	0.073991	0.073993	0.073994	0.073995
34	0.073996	0.073998	0.073999	0.074000	0.074001	0.074003	0.074004	0.074005	0.074006	0.074008
35	0.074009	0.074010	0.074012	0.074013	0.074014	0.074015	0.074017	0.074018	0.074019	0.074020
36	0.074022	0.074023	0.074024	0.074025	0.074027	0.074028	0.074029	0.074031	0.074032	0.074033
37	0.074034	0.074036	0.074037	0.074038	0.074039	0.074041	0.074042	0.074043	0.074044	0.074046
38	0.074047	0.074048	0.074050	0.074051	0.074052	0.074053	0.074055	0.074056	0.074057	0.074058
39	0.074060	0.074061	0.074062	0.074063	0.074065	0.074066	0.074067	0.074069	0.074070	0.074071
40	0.074072	0.074074	0.074075	0.074076	0.074077	0.074079	0.074080	0.074081	0.074082	0.074084
41	0.074085	0.074086	0.074088	0.074089	0.074090	0.074091	0.074093	0.074094	0.074095	0.074096

The factor K is multiplied to the mass of mercury delivered/contained in the measure in kg/g/mg to give its capacity at 20 °C in $m^3/dm^3/cm^3$

Table 3.36 ReferenceTemperature = 20 °C Air density = 1.2 kg/m^3, ALPHA = .000015/°C DEN = 8400 kg/m^3

The values of 10^3K

Temp	0.0	0.1	0.2	0.3	0.4	0.5	0.6	0.7	0.8	0.9
5	0.073635	0.073637	0.073638	0.073639	0.073640	0.073642	0.073643	0.073644	0.073645	0.073646
6	0.073648	0.073649	0.073650	0.073651	0.073653	0.073654	0.073655	0.073656	0.073657	0.073659
7	0.073660	0.073661	0.073662	0.073664	0.073665	0.073666	0.073667	0.073668	0.073670	0.073671
8	0.073672	0.073673	0.073675	0.073676	0.073677	0.073678	0.073680	0.073681	0.073682	0.073683
9	0.073684	0.073686	0.073687	0.073688	0.073689	0.073691	0.073692	0.073693	0.073694	0.073695
10	0.073697	0.073698	0.073699	0.073700	0.073702	0.073703	0.073704	0.073705	0.073707	0.073708
11	0.073709	0.073710	0.073711	0.073713	0.073714	0.073715	0.073716	0.073718	0.073719	0.073720
12	0.073721	0.073722	0.073724	0.073725	0.073726	0.073727	0.073729	0.073730	0.073731	0.073732
13	0.073733	0.073735	0.073736	0.073737	0.073738	0.073740	0.073741	0.073742	0.073743	0.073745
14	0.073746	0.073747	0.073748	0.073749	0.073751	0.073752	0.073753	0.073754	0.073756	0.073757
15	0.073758	0.073759	0.073760	0.073762	0.073763	0.073764	0.073765	0.073767	0.073768	0.073769
16	0.073770	0.073772	0.073773	0.073774	0.073775	0.073776	0.073778	0.073779	0.073780	0.073781
17	0.073783	0.073784	0.073785	0.073786	0.073787	0.073789	0.073790	0.073791	0.073792	0.073794
18	0.073795	0.073796	0.073797	0.073798	0.073800	0.073801	0.073802	0.073803	0.073805	0.073806
19	0.073807	0.073808	0.073810	0.073811	0.073812	0.073813	0.073814	0.073816	0.073817	0.073818
20	0.073819	0.073821	0.073822	0.073823	0.073824	0.073825	0.073827	0.073828	0.073829	0.073830
21	0.073832	0.073833	0.073834	0.073835	0.073837	0.073838	0.073839	0.073840	0.073841	0.073843
22	0.073844	0.073845	0.073846	0.073848	0.073849	0.073850	0.073851	0.073852	0.073854	0.073855
23	0.073856	0.073857	0.073859	0.073860	0.073861	0.073862	0.073864	0.073865	0.073866	0.073867
24	0.073868	0.073870	0.073871	0.073872	0.073873	0.073875	0.073876	0.073877	0.073878	0.073879
25	0.073881	0.073882	0.073883	0.073884	0.073886	0.073887	0.073888	0.073889	0.073891	0.073892
26	0.073893	0.073894	0.073895	0.073897	0.073898	0.073899	0.073900	0.073902	0.073903	0.073904
27	0.073905	0.073906	0.073908	0.073909	0.073910	0.073911	0.073913	0.073914	0.073915	0.073916
28	0.073918	0.073919	0.073920	0.073921	0.073922	0.073924	0.073925	0.073926	0.073927	0.073929
29	0.073930	0.073931	0.073932	0.073933	0.073935	0.073936	0.073937	0.073938	0.073940	0.073941
30	0.073942	0.073943	0.073945	0.073946	0.073947	0.073948	0.073949	0.073951	0.073952	0.073953
31	0.073954	0.073956	0.073957	0.073958	0.073959	0.073961	0.073962	0.073963	0.073964	0.073965
32	0.073967	0.073968	0.073969	0.073970	0.073972	0.073973	0.073974	0.073975	0.073976	0.073978
33	0.073979	0.073980	0.073981	0.073983	0.073984	0.073985	0.073986	0.073988	0.073989	0.073990
34	0.073991	0.073992	0.073994	0.073995	0.073996	0.073997	0.073999	0.074000	0.074001	0.074002
35	0.074003	0.074005	0.074006	0.074007	0.074008	0.074010	0.074011	0.074012	0.074013	0.074015
36	0.074016	0.074017	0.074018	0.074019	0.074021	0.074022	0.074023	0.074024	0.074026	0.074027
37	0.074028	0.074029	0.074031	0.074032	0.074033	0.074034	0.074035	0.074037	0.074038	0.074039
38	0.074040	0.074042	0.074043	0.074044	0.074045	0.074046	0.074048	0.074049	0.074050	0.074051
39	0.074053	0.074054	0.074055	0.074056	0.074058	0.074059	0.074060	0.074061	0.074062	0.074064
40	0.074065	0.074066	0.074067	0.074069	0.074070	0.074071	0.074072	0.074074	0.074075	0.074076
41	0.074077	0.074078	0.074080	0.074081	0.074082	0.074083	0.074085	0.074086	0.074087	0.074088

The factor K is multiplied to the mass of mercury delivered/contained in the measure in kg/g/mg to give its capacity at 20 °C in m³/dm³/cm³

Table 3.37 Reference Temperature = 20 °C Air density = 1.2 kg/m³, ALPHA = .000025 DEN = 8400 kg/m³

The values of 10^3K

Temp	0.0	0.1	0.2	0.3	0.4	0.5	0.6	0.7	0.8	0.9
5	0.073646	0.073648	0.073649	0.073650	0.073651	0.073652	0.073653	0.073655	0.073656	0.073657
6	0.073658	0.073659	0.073660	0.073661	0.073663	0.073664	0.073665	0.073666	0.073667	0.073668
7	0.073669	0.073671	0.073672	0.073673	0.073674	0.073675	0.073676	0.073678	0.073679	0.073680
8	0.073681	0.073682	0.073683	0.073684	0.073686	0.073687	0.073688	0.073689	0.073690	0.073691
9	0.073693	0.073694	0.073695	0.073696	0.073697	0.073698	0.073699	0.073701	0.073702	0.073703
10	0.073704	0.073705	0.073706	0.073708	0.073709	0.073710	0.073711	0.073712	0.073713	0.073714
11	0.073716	0.073717	0.073718	0.073719	0.073720	0.073721	0.073723	0.073724	0.073725	0.073726
12	0.073727	0.073728	0.073729	0.073731	0.073732	0.073733	0.073734	0.073735	0.073736	0.073737
13	0.073739	0.073740	0.073741	0.073742	0.073743	0.073744	0.073746	0.073747	0.073748	0.073749
14	0.073750	0.073751	0.073752	0.073754	0.073755	0.073756	0.073757	0.073758	0.073759	0.073761
15	0.073762	0.073763	0.073764	0.073765	0.073766	0.073767	0.073769	0.073770	0.073771	0.073772
16	0.073773	0.073774	0.073776	0.073777	0.073778	0.073779	0.073780	0.073781	0.073782	0.073784
17	0.073785	0.073786	0.073787	0.073788	0.073789	0.073791	0.073792	0.073793	0.073794	0.073795
18	0.073796	0.073797	0.073799	0.073800	0.073801	0.073802	0.073803	0.073804	0.073806	0.073807
19	0.073808	0.073809	0.073810	0.073811	0.073812	0.073814	0.073815	0.073816	0.073817	0.073818
20	0.073819	0.073821	0.073822	0.073823	0.073824	0.073825	0.073826	0.073827	0.073829	0.073830
21	0.073831	0.073832	0.073833	0.073834	0.073835	0.073837	0.073838	0.073839	0.073840	0.073841
22	0.073842	0.073844	0.073845	0.073846	0.073847	0.073848	0.073849	0.073850	0.073852	0.073853
23	0.073854	0.073855	0.073856	0.073857	0.073859	0.073860	0.073861	0.073862	0.073863	0.073864
24	0.073865	0.073867	0.073868	0.073869	0.073870	0.073871	0.073872	0.073874	0.073875	0.073876
25	0.073877	0.073878	0.073879	0.073880	0.073882	0.073883	0.073884	0.073885	0.073886	0.073887
26	0.073889	0.073890	0.073891	0.073892	0.073893	0.073894	0.073895	0.073897	0.073898	0.073899
27	0.073900	0.073901	0.073902	0.073904	0.073905	0.073906	0.073907	0.073908	0.073909	0.073910
28	0.073912	0.073913	0.073914	0.073915	0.073916	0.073917	0.073919	0.073920	0.073921	0.073922
29	0.073923	0.073924	0.073925	0.073927	0.073928	0.073929	0.073930	0.073931	0.073932	0.073934
30	0.073935	0.073936	0.073937	0.073938	0.073939	0.073940	0.073942	0.073943	0.073944	0.073945
31	0.073946	0.073947	0.073949	0.073950	0.073951	0.073952	0.073953	0.073954	0.073955	0.073957
32	0.073958	0.073959	0.073960	0.073961	0.073962	0.073964	0.073965	0.073966	0.073967	0.073968
33	0.073969	0.073970	0.073972	0.073973	0.073974	0.073975	0.073976	0.073977	0.073979	0.073980
34	0.073981	0.073982	0.073983	0.073984	0.073985	0.073987	0.073988	0.073989	0.073990	0.073991
35	0.073992	0.073994	0.073995	0.073996	0.073997	0.073998	0.073999	0.074000	0.074002	0.074003
36	0.074004	0.074005	0.074006	0.074007	0.074009	0.074010	0.074011	0.074012	0.074013	0.074014
37	0.074015	0.074017	0.074018	0.074019	0.074020	0.074021	0.074022	0.074024	0.074025	0.074026
38	0.074027	0.074028	0.074029	0.074030	0.074032	0.074033	0.074034	0.074035	0.074036	0.074037
39	0.074039	0.074040	0.074041	0.074042	0.074043	0.074044	0.074045	0.074047	0.074048	0.074049
40	0.074050	0.074051	0.074052	0.074054	0.074055	0.074056	0.074057	0.074058	0.074059	0.074060
41	0.074062	0.074063	0.074064	0.074065	0.074066	0.074067	0.074069	0.074070	0.074071	0.074072

The factor K is multiplied to the mass of mercury delivered/contained in the measure in kg/g/mg to give its capacity at 20 °C in $m^3/dm^3/cm^3$

Table 3.38 Reference Temperature = 20 °C Air density = 1.2 kg/m^3, ALPHA = .00003/°C DEN = 8400 kg/m^3

The values of 10^3K

Temp	0.0	0.1	0.2	0.3	0.4	0.5	0.6	0.7	0.8	0.9
5	0.073652	0.073653	0.073654	0.073655	0.073656	0.073658	0.073659	0.073660	0.073661	0.073662
6	0.073663	0.073664	0.073665	0.073666	0.073668	0.073669	0.073670	0.073671	0.073672	0.073673
7	0.073674	0.073675	0.073677	0.073678	0.073679	0.073680	0.073681	0.073682	0.073683	0.073684
8	0.073685	0.073687	0.073688	0.073689	0.073690	0.073691	0.073692	0.073693	0.073694	0.073695
9	0.073697	0.073698	0.073699	0.073700	0.073701	0.073702	0.073703	0.073704	0.073706	0.073707
10	0.073708	0.073709	0.073710	0.073711	0.073712	0.073713	0.073714	0.073716	0.073717	0.073718
11	0.073719	0.073720	0.073721	0.073722	0.073723	0.073724	0.073726	0.073727	0.073728	0.073729
12	0.073730	0.073731	0.073732	0.073733	0.073735	0.073736	0.073737	0.073738	0.073739	0.073740
13	0.073741	0.073742	0.073743	0.073745	0.073746	0.073747	0.073748	0.073749	0.073750	0.073751
14	0.073752	0.073754	0.073755	0.073756	0.073757	0.073758	0.073759	0.073760	0.073761	0.073762
15	0.073764	0.073765	0.073766	0.073767	0.073768	0.073769	0.073770	0.073771	0.073772	0.073774
16	0.073775	0.073776	0.073777	0.073778	0.073779	0.073780	0.073781	0.073783	0.073784	0.073785
17	0.073786	0.073787	0.073788	0.073789	0.073790	0.073791	0.073793	0.073794	0.073795	0.073796
18	0.073797	0.073798	0.073799	0.073800	0.073801	0.073803	0.073804	0.073805	0.073806	0.073807
19	0.073808	0.073809	0.073810	0.073812	0.073813	0.073814	0.073815	0.073816	0.073817	0.073818
20	0.073819	0.073820	0.073822	0.073823	0.073824	0.073825	0.073826	0.073827	0.073828	0.073829
21	0.073831	0.073832	0.073833	0.073834	0.073835	0.073836	0.073837	0.073838	0.073839	0.073841
22	0.073842	0.073843	0.073844	0.073845	0.073846	0.073847	0.073848	0.073849	0.073851	0.073852
23	0.073853	0.073854	0.073855	0.073856	0.073857	0.073858	0.073860	0.073861	0.073862	0.073863
24	0.073864	0.073865	0.073866	0.073867	0.073868	0.073870	0.073871	0.073872	0.073873	0.073874
25	0.073875	0.073876	0.073877	0.073879	0.073880	0.073881	0.073882	0.073883	0.073884	0.073885
26	0.073886	0.073887	0.073889	0.073890	0.073891	0.073892	0.073893	0.073894	0.073895	0.073896
27	0.073897	0.073899	0.073900	0.073901	0.073902	0.073903	0.073904	0.073905	0.073906	0.073908
28	0.073909	0.073910	0.073911	0.073912	0.073913	0.073914	0.073915	0.073916	0.073918	0.073919
29	0.073920	0.073921	0.073922	0.073923	0.073924	0.073925	0.073927	0.073928	0.073929	0.073930
30	0.073931	0.073932	0.073933	0.073934	0.073935	0.073937	0.073938	0.073939	0.073940	0.073941
31	0.073942	0.073943	0.073944	0.073946	0.073947	0.073948	0.073949	0.073950	0.073951	0.073952
32	0.073953	0.073954	0.073956	0.073957	0.073958	0.073959	0.073960	0.073961	0.073962	0.073963
33	0.073964	0.073966	0.073967	0.073968	0.073969	0.073970	0.073971	0.073972	0.073973	0.073975
34	0.073976	0.073977	0.073978	0.073979	0.073980	0.073981	0.073982	0.073983	0.073985	0.073986
35	0.073987	0.073988	0.073989	0.073990	0.073991	0.073992	0.073994	0.073995	0.073996	0.073997
36	0.073998	0.073999	0.074000	0.074001	0.074002	0.074004	0.074005	0.074006	0.074007	0.074008
37	0.074009	0.074010	0.074011	0.074013	0.074014	0.074015	0.074016	0.074017	0.074018	0.074019
38	0.074020	0.074021	0.074023	0.074024	0.074025	0.074026	0.074027	0.074028	0.074029	0.074030
39	0.074032	0.074033	0.074034	0.074035	0.074036	0.074037	0.074038	0.074039	0.074040	0.074042
40	0.074043	0.074044	0.074045	0.074046	0.074047	0.074048	0.074049	0.074051	0.074052	0.074053
41	0.074054	0.074055	0.074056	0.074057	0.074058	0.074059	0.074061	0.074062	0.074063	0.074064

The factor K is multiplied to the mass of mercury delivered/contained in the measure in kg/g/mg to give its capacity at 20 °C in $m^3/dm^3/cm^3$

Table 3.39 Reference Temperature = 27 °C Air density = 1.1685 kg/m^3, ALPHA = .00001/°C DEN = 8400 kg/m^3

The values of 10^3K

Temp	0.0	0.1	0.2	0.3	0.4	0.5	0.6	0.7	0.8	0.9
5	0.073635	0.073636	0.073638	0.073639	0.073640	0.073641	0.073643	0.073644	0.073645	0.073646
6	0.073648	0.073649	0.073650	0.073651	0.073653	0.073654	0.073655	0.073656	0.073658	0.073659
7	0.073660	0.073662	0.073663	0.073664	0.073665	0.073667	0.073668	0.073669	0.073670	0.073672
8	0.073673	0.073674	0.073675	0.073677	0.073678	0.073679	0.073680	0.073682	0.073683	0.073684
9	0.073686	0.073687	0.073688	0.073689	0.073691	0.073692	0.073693	0.073694	0.073696	0.073697
10	0.073698	0.073699	0.073701	0.073702	0.073703	0.073704	0.073706	0.073707	0.073708	0.073710
11	0.073711	0.073712	0.073713	0.073715	0.073716	0.073717	0.073718	0.073720	0.073721	0.073722
12	0.073723	0.073725	0.073726	0.073727	0.073728	0.073730	0.073731	0.073732	0.073734	0.073735
13	0.073736	0.073737	0.073739	0.073740	0.073741	0.073742	0.073744	0.073745	0.073746	0.073747
14	0.073749	0.073750	0.073751	0.073752	0.073754	0.073755	0.073756	0.073758	0.073759	0.073760
15	0.073761	0.073763	0.073764	0.073765	0.073766	0.073768	0.073769	0.073770	0.073771	0.073773
16	0.073774	0.073775	0.073776	0.073778	0.073779	0.073780	0.073782	0.073783	0.073784	0.073785
17	0.073787	0.073788	0.073789	0.073790	0.073792	0.073793	0.073794	0.073795	0.073797	0.073798
18	0.073799	0.073801	0.073802	0.073803	0.073804	0.073806	0.073807	0.073808	0.073809	0.073811
19	0.073812	0.073813	0.073814	0.073816	0.073817	0.073818	0.073819	0.073821	0.073822	0.073823
20	0.073825	0.073826	0.073827	0.073828	0.073830	0.073831	0.073832	0.073833	0.073835	0.073836
21	0.073837	0.073838	0.073840	0.073841	0.073842	0.073843	0.073845	0.073846	0.073847	0.073849
22	0.073850	0.073851	0.073852	0.073854	0.073855	0.073856	0.073857	0.073859	0.073860	0.073861
23	0.073862	0.073864	0.073865	0.073866	0.073867	0.073869	0.073870	0.073871	0.073873	0.073874
24	0.073875	0.073876	0.073878	0.073879	0.073880	0.073881	0.073883	0.073884	0.073885	0.073886
25	0.073888	0.073889	0.073890	0.073892	0.073893	0.073894	0.073895	0.073897	0.073898	0.073899
26	0.073900	0.073902	0.073903	0.073904	0.073905	0.073907	0.073908	0.073909	0.073910	0.073912
27	0.073913	0.073914	0.073916	0.073917	0.073918	0.073919	0.073921	0.073922	0.073923	0.073924
28	0.073926	0.073927	0.073928	0.073929	0.073931	0.073932	0.073933	0.073935	0.073936	0.073937
29	0.073938	0.073940	0.073941	0.073942	0.073943	0.073945	0.073946	0.073947	0.073948	0.073950
30	0.073951	0.073952	0.073953	0.073955	0.073956	0.073957	0.073959	0.073960	0.073961	0.073962
31	0.073964	0.073965	0.073966	0.073967	0.073969	0.073970	0.073971	0.073972	0.073974	0.073975
32	0.073976	0.073978	0.073979	0.073980	0.073981	0.073983	0.073984	0.073985	0.073986	0.073988
33	0.073989	0.073990	0.073991	0.073993	0.073994	0.073995	0.073996	0.073998	0.073999	0.074000
34	0.074002	0.074003	0.074004	0.074005	0.074007	0.074008	0.074009	0.074010	0.074012	0.074013
35	0.074014	0.074015	0.074017	0.074018	0.074019	0.074021	0.074022	0.074023	0.074024	0.074026
36	0.074027	0.074028	0.074029	0.074031	0.074032	0.074033	0.074034	0.074036	0.074037	0.074038
37	0.074040	0.074041	0.074042	0.074043	0.074045	0.074046	0.074047	0.074048	0.074050	0.074051
38	0.074052	0.074053	0.074055	0.074056	0.074057	0.074059	0.074060	0.074061	0.074062	0.074064
39	0.074065	0.074066	0.074067	0.074069	0.074070	0.074071	0.074072	0.074074	0.074075	0.074076
40	0.074077	0.074079	0.074080	0.074081	0.074083	0.074084	0.074085	0.074086	0.074088	0.074089
41	0.074090	0.074091	0.074093	0.074094	0.074095	0.074096	0.074098	0.074099	0.074100	0.074102

The factor K is multiplied to the mass of mercury delivered/contained in the measure in kg/g/mg to give its capacity at 20 °C in $m^3/dm^3/cm^3$

Table 3.40 Reference Temperature = 27 °C Air density = 1.1685 kg/m^3, ALPHA = .000015/°C DEN = 8400 kg/m^3

The values of 10^3K

Temp	0.0	0.1	0.2	0.3	0.4	0.5	0.6	0.7	0.8	0.9
5	0.073643	0.073644	0.073646	0.073647	0.073648	0.073649	0.073650	0.073652	0.073653	0.073654
6	0.073655	0.073657	0.073658	0.073659	0.073660	0.073662	0.073663	0.073664	0.073665	0.073666
7	0.073668	0.073669	0.073670	0.073671	0.073673	0.073674	0.073675	0.073676	0.073677	0.073679
8	0.073680	0.073681	0.073682	0.073684	0.073685	0.073686	0.073687	0.073688	0.073690	0.073691
9	0.073692	0.073693	0.073695	0.073696	0.073697	0.073698	0.073700	0.073701	0.073702	0.073703
10	0.073704	0.073706	0.073707	0.073708	0.073709	0.073711	0.073712	0.073713	0.073714	0.073715
11	0.073717	0.073718	0.073719	0.073720	0.073722	0.073723	0.073724	0.073725	0.073727	0.073728
12	0.073729	0.073730	0.073731	0.073733	0.073734	0.073735	0.073736	0.073738	0.073739	0.073740
13	0.073741	0.073742	0.073744	0.073745	0.073746	0.073747	0.073749	0.073750	0.073751	0.073752
14	0.073753	0.073755	0.073756	0.073757	0.073758	0.073760	0.073761	0.073762	0.073763	0.073765
15	0.073766	0.073767	0.073768	0.073769	0.073771	0.073772	0.073773	0.073774	0.073776	0.073777
16	0.073778	0.073779	0.073780	0.073782	0.073783	0.073784	0.073785	0.073787	0.073788	0.073789
17	0.073790	0.073792	0.073793	0.073794	0.073795	0.073796	0.073798	0.073799	0.073800	0.073801
18	0.073803	0.073804	0.073805	0.073806	0.073807	0.073809	0.073810	0.073811	0.073812	0.073814
19	0.073815	0.073816	0.073817	0.073819	0.073820	0.073821	0.073822	0.073823	0.073825	0.073826
20	0.073827	0.073828	0.073830	0.073831	0.073832	0.073833	0.073834	0.073836	0.073837	0.073838
21	0.073839	0.073841	0.073842	0.073843	0.073844	0.073846	0.073847	0.073848	0.073849	0.073850
22	0.073852	0.073853	0.073854	0.073855	0.073857	0.073858	0.073859	0.073860	0.073861	0.073863
23	0.073864	0.073865	0.073866	0.073868	0.073869	0.073870	0.073871	0.073873	0.073874	0.073875
24	0.073876	0.073877	0.073879	0.073880	0.073881	0.073882	0.073884	0.073885	0.073886	0.073887
25	0.073888	0.073890	0.073891	0.073892	0.073893	0.073895	0.073896	0.073897	0.073898	0.073900
26	0.073901	0.073902	0.073903	0.073904	0.073906	0.073907	0.073908	0.073909	0.073911	0.073912
27	0.073913	0.073914	0.073915	0.073917	0.073918	0.073919	0.073920	0.073922	0.073923	0.073924
28	0.073925	0.073927	0.073928	0.073929	0.073930	0.073931	0.073933	0.073934	0.073935	0.073936
29	0.073938	0.073939	0.073940	0.073941	0.073942	0.073944	0.073945	0.073946	0.073947	0.073949
30	0.073950	0.073951	0.073952	0.073954	0.073955	0.073956	0.073957	0.073958	0.073960	0.073961
31	0.073962	0.073963	0.073965	0.073966	0.073967	0.073968	0.073969	0.073971	0.073972	0.073973
32	0.073974	0.073976	0.073977	0.073978	0.073979	0.073981	0.073982	0.073983	0.073984	0.073985
33	0.073987	0.073988	0.073989	0.073990	0.073992	0.073993	0.073994	0.073995	0.073997	0.073998
34	0.073999	0.074000	0.074001	0.074003	0.074004	0.074005	0.074006	0.074008	0.074009	0.074010
35	0.074011	0.074012	0.074014	0.074015	0.074016	0.074017	0.074019	0.074020	0.074021	0.074022
36	0.074024	0.074025	0.074026	0.074027	0.074028	0.074030	0.074031	0.074032	0.074033	0.074035
37	0.074036	0.074037	0.074038	0.074040	0.074041	0.074042	0.074043	0.074044	0.074046	0.074047
38	0.074048	0.074049	0.074051	0.074052	0.074053	0.074054	0.074055	0.074057	0.074058	0.074059
39	0.074060	0.074062	0.074063	0.074064	0.074065	0.074067	0.074068	0.074069	0.074070	0.074071
40	0.074073	0.074074	0.074075	0.074076	0.074078	0.074079	0.074080	0.074081	0.074083	0.074084
41	0.074085	0.074086	0.074087	0.074089	0.074090	0.074091	0.074092	0.074094	0.074095	0.074096

The factor K is multiplied to the mass of mercury delivered/contained in the measure in kg/g/mg to give its capacity at 20 °C in m³/dm³/cm³

Table 3.41 Reference Temperature = 27 °C Air density = 1.1685 kg/m³, ALPHA = .000025/°C DEN = 8400 kg/m³

The values of 10^3K

Temp	0.0	0.1	0.2	0.3	0.4	0.5	0.6	0.7	0.8	0.9
5	0.073659	0.073660	0.073662	0.073663	0.073664	0.073665	0.073666	0.073667	0.073669	0.073670
6	0.073671	0.073672	0.073673	0.073674	0.073675	0.073677	0.073678	0.073679	0.073680	0.073681
7	0.073682	0.073684	0.073685	0.073686	0.073687	0.073688	0.073689	0.073690	0.073692	0.073693
8	0.073694	0.073695	0.073696	0.073697	0.073699	0.073700	0.073701	0.073702	0.073703	0.073704
9	0.073705	0.073707	0.073708	0.073709	0.073710	0.073711	0.073712	0.073714	0.073715	0.073716
10	0.073717	0.073718	0.073719	0.073720	0.073722	0.073723	0.073724	0.073725	0.073726	0.073727
11	0.073729	0.073730	0.073731	0.073732	0.073733	0.073734	0.073735	0.073737	0.073738	0.073739
12	0.073740	0.073741	0.073742	0.073743	0.073745	0.073746	0.073747	0.073748	0.073749	0.073750
13	0.073752	0.073753	0.073754	0.073755	0.073756	0.073757	0.073758	0.073760	0.073761	0.073762
14	0.073763	0.073764	0.073765	0.073767	0.073768	0.073769	0.073770	0.073771	0.073772	0.073773
15	0.073775	0.073776	0.073777	0.073778	0.073779	0.073780	0.073782	0.073783	0.073784	0.073785
16	0.073786	0.073787	0.073788	0.073790	0.073791	0.073792	0.073793	0.073794	0.073795	0.073797
17	0.073798	0.073799	0.073800	0.073801	0.073802	0.073803	0.073805	0.073806	0.073807	0.073808
18	0.073809	0.073810	0.073812	0.073813	0.073814	0.073815	0.073816	0.073817	0.073818	0.073820
19	0.073821	0.073822	0.073823	0.073824	0.073825	0.073827	0.073828	0.073829	0.073830	0.073831
20	0.073832	0.073833	0.073835	0.073836	0.073837	0.073838	0.073839	0.073840	0.073841	0.073843
21	0.073844	0.073845	0.073846	0.073847	0.073848	0.073850	0.073851	0.073852	0.073853	0.073854
22	0.073855	0.073856	0.073858	0.073859	0.073860	0.073861	0.073862	0.073863	0.073865	0.073866
23	0.073867	0.073868	0.073869	0.073870	0.073871	0.073873	0.073874	0.073875	0.073876	0.073877
24	0.073878	0.073880	0.073881	0.073882	0.073883	0.073884	0.073885	0.073886	0.073888	0.073889
25	0.073890	0.073891	0.073892	0.073893	0.073895	0.073896	0.073897	0.073898	0.073899	0.073900
26	0.073901	0.073903	0.073904	0.073905	0.073906	0.073907	0.073908	0.073910	0.073911	0.073912
27	0.073913	0.073914	0.073915	0.073916	0.073918	0.073919	0.073920	0.073921	0.073922	0.073923
28	0.073925	0.073926	0.073927	0.073928	0.073929	0.073930	0.073931	0.073933	0.073934	0.073935
29	0.073936	0.073937	0.073938	0.073940	0.073941	0.073942	0.073943	0.073944	0.073945	0.073946
30	0.073948	0.073949	0.073950	0.073951	0.073952	0.073953	0.073955	0.073956	0.073957	0.073958
31	0.073959	0.073960	0.073961	0.073963	0.073964	0.073965	0.073966	0.073967	0.073968	0.073970
32	0.073971	0.073972	0.073973	0.073974	0.073975	0.073976	0.073978	0.073979	0.073980	0.073981
33	0.073982	0.073983	0.073985	0.073986	0.073987	0.073988	0.073989	0.073990	0.073991	0.073993
34	0.073994	0.073995	0.073996	0.073997	0.073998	0.074000	0.074001	0.074002	0.074003	0.074004
35	0.074005	0.074006	0.074008	0.074009	0.074010	0.074011	0.074012	0.074013	0.074015	0.074016
36	0.074017	0.074018	0.074019	0.074020	0.074021	0.074023	0.074024	0.074025	0.074026	0.074027
37	0.074028	0.074030	0.074031	0.074032	0.074033	0.074034	0.074035	0.074037	0.074038	0.074039
38	0.074040	0.074041	0.074042	0.074043	0.074045	0.074046	0.074047	0.074048	0.074049	0.074050
39	0.074052	0.074053	0.074054	0.074055	0.074056	0.074057	0.074058	0.074060	0.074061	0.074062
40	0.074063	0.074064	0.074065	0.074067	0.074068	0.074069	0.074070	0.074071	0.074072	0.074073
41	0.074075	0.074076	0.074077	0.074078	0.074079	0.074080	0.074082	0.074083	0.074084	0.074085

The factor K is multiplied to the mass of mercury delivered/contained in the measure in kg/g/mg to give its capacity at 20 °C in m³/dm³/cm³

Table 3.42 Reference Temperature = 27 °C Air density = 1.1685 kg/m³, ALPHA = .00003/°C DEN = 8400 kg/m³

The values of 10^3K

Temp	0.0	0.1	0.2	0.3	0.4	0.5	0.6	0.7	0.8	0.9
5	0.073667	0.073669	0.073670	0.073671	0.073672	0.073673	0.073674	0.073675	0.073676	0.073677
6	0.073679	0.073680	0.073681	0.073682	0.073683	0.073684	0.073685	0.073686	0.073688	0.073689
7	0.073690	0.073691	0.073692	0.073693	0.073694	0.073695	0.073696	0.073698	0.073699	0.073700
8	0.073701	0.073702	0.073703	0.073704	0.073705	0.073707	0.073708	0.073709	0.073710	0.073711
9	0.073712	0.073713	0.073714	0.073715	0.073717	0.073718	0.073719	0.073720	0.073721	0.073722
10	0.073723	0.073724	0.073725	0.073727	0.073728	0.073729	0.073730	0.073731	0.073732	0.073733
11	0.073734	0.073736	0.073737	0.073738	0.073739	0.073740	0.073741	0.073742	0.073743	0.073744
12	0.073746	0.073747	0.073748	0.073749	0.073750	0.073751	0.073752	0.073753	0.073754	0.073756
13	0.073757	0.073758	0.073759	0.073760	0.073761	0.073762	0.073763	0.073765	0.073766	0.073767
14	0.073768	0.073769	0.073770	0.073771	0.073772	0.073773	0.073775	0.073776	0.073777	0.073778
15	0.073779	0.073780	0.073781	0.073782	0.073784	0.073785	0.073786	0.073787	0.073788	0.073789
16	0.073790	0.073791	0.073792	0.073794	0.073795	0.073796	0.073797	0.073798	0.073799	0.073800
17	0.073801	0.073802	0.073804	0.073805	0.073806	0.073807	0.073808	0.073809	0.073810	0.073811
18	0.073813	0.073814	0.073815	0.073816	0.073817	0.073818	0.073819	0.073820	0.073821	0.073823
19	0.073824	0.073825	0.073826	0.073827	0.073828	0.073829	0.073830	0.073832	0.073833	0.073834
20	0.073835	0.073836	0.073837	0.073838	0.073839	0.073840	0.073842	0.073843	0.073844	0.073845
21	0.073846	0.073847	0.073848	0.073849	0.073850	0.073852	0.073853	0.073854	0.073855	0.073856
22	0.073857	0.073858	0.073859	0.073861	0.073862	0.073863	0.073864	0.073865	0.073866	0.073867
23	0.073868	0.073869	0.073871	0.073872	0.073873	0.073874	0.073875	0.073876	0.073877	0.073878
24	0.073880	0.073881	0.073882	0.073883	0.073884	0.073885	0.073886	0.073887	0.073888	0.073890
25	0.073891	0.073892	0.073893	0.073894	0.073895	0.073896	0.073897	0.073898	0.073900	0.073901
26	0.073902	0.073903	0.073904	0.073905	0.073906	0.073907	0.073909	0.073910	0.073911	0.073912
27	0.073913	0.073914	0.073915	0.073916	0.073917	0.073919	0.073920	0.073921	0.073922	0.073923
28	0.073924	0.073925	0.073926	0.073928	0.073929	0.073930	0.073931	0.073932	0.073933	0.073934
29	0.073935	0.073936	0.073938	0.073939	0.073940	0.073941	0.073942	0.073943	0.073944	0.073945
30	0.073947	0.073948	0.073949	0.073950	0.073951	0.073952	0.073953	0.073954	0.073955	0.073957
31	0.073958	0.073959	0.073960	0.073961	0.073962	0.073963	0.073964	0.073966	0.073967	0.073968
32	0.073969	0.073970	0.073971	0.073972	0.073973	0.073974	0.073976	0.073977	0.073978	0.073979
33	0.073980	0.073981	0.073982	0.073983	0.073984	0.073986	0.073987	0.073988	0.073989	0.073990
34	0.073991	0.073992	0.073993	0.073995	0.073996	0.073997	0.073998	0.073999	0.074000	0.074001
35	0.074002	0.074003	0.074005	0.074006	0.074007	0.074008	0.074009	0.074010	0.074011	0.074012
36	0.074014	0.074015	0.074016	0.074017	0.074018	0.074019	0.074020	0.074021	0.074022	0.074024
37	0.074025	0.074026	0.074027	0.074028	0.074029	0.074030	0.074031	0.074033	0.074034	0.074035
38	0.074036	0.074037	0.074038	0.074039	0.074040	0.074041	0.074043	0.074044	0.074045	0.074046
39	0.074047	0.074048	0.074049	0.074050	0.074052	0.074053	0.074054	0.074055	0.074056	0.074057
40	0.074058	0.074059	0.074060	0.074062	0.074063	0.074064	0.074065	0.074066	0.074067	0.074068
41	0.074069	0.074071	0.074072	0.074073	0.074074	0.074075	0.074076	0.074077	0.074078	0.074079

The factor K is multiplied to the mass of mercury delivered/contained in the measure in kg/g/mg to give its capacity at 20 °C in $m^3/dm^3/cm^3$

Table 3.43 Reference Temperature = 27 °C Air density = 1.1685 kg/m^3, ALPHA = .00001/°C DEN = 8000 kg/m^3

The values of 10^3K

Temp	0.0	0.1	0.2	0.3	0.4	0.5	0.6	0.7	0.8	0.9
5	0.073634	0.073636	0.073637	0.073638	0.073640	0.073641	0.073642	0.073643	0.073645	0.073646
6	0.073647	0.073648	0.073650	0.073651	0.073652	0.073653	0.073655	0.073656	0.073657	0.073658
7	0.073660	0.073661	0.073662	0.073664	0.073665	0.073666	0.073667	0.073669	0.073670	0.073671
8	0.073672	0.073674	0.073675	0.073676	0.073677	0.073679	0.073680	0.073681	0.073682	0.073684
9	0.073685	0.073686	0.073688	0.073689	0.073690	0.073691	0.073693	0.073694	0.073695	0.073696
10	0.073698	0.073699	0.073700	0.073701	0.073703	0.073704	0.073705	0.073706	0.073708	0.073709
11	0.073710	0.073712	0.073713	0.073714	0.073715	0.073717	0.073718	0.073719	0.073720	0.073722
12	0.073723	0.073724	0.073725	0.073727	0.073728	0.073729	0.073730	0.073732	0.073733	0.073734
13	0.073736	0.073737	0.073738	0.073739	0.073741	0.073742	0.073743	0.073744	0.073746	0.073747
14	0.073748	0.073749	0.073751	0.073752	0.073753	0.073754	0.073756	0.073757	0.073758	0.073760
15	0.073761	0.073762	0.073763	0.073765	0.073766	0.073767	0.073768	0.073770	0.073771	0.073772
16	0.073773	0.073775	0.073776	0.073777	0.073779	0.073780	0.073781	0.073782	0.073784	0.073785
17	0.073786	0.073787	0.073789	0.073790	0.073791	0.073792	0.073794	0.073795	0.073796	0.073797
18	0.073799	0.073800	0.073801	0.073803	0.073804	0.073805	0.073806	0.073808	0.073809	0.073810
19	0.073811	0.073813	0.073814	0.073815	0.073816	0.073818	0.073819	0.073820	0.073821	0.073823
20	0.073824	0.073825	0.073827	0.073828	0.073829	0.073830	0.073832	0.073833	0.073834	0.073835
21	0.073837	0.073838	0.073839	0.073840	0.073842	0.073843	0.073844	0.073845	0.073847	0.073848
22	0.073849	0.073851	0.073852	0.073853	0.073854	0.073856	0.073857	0.073858	0.073859	0.073861
23	0.073862	0.073863	0.073864	0.073866	0.073867	0.073868	0.073869	0.073871	0.073872	0.073873
24	0.073875	0.073876	0.073877	0.073878	0.073880	0.073881	0.073882	0.073883	0.073885	0.073886
25	0.073887	0.073888	0.073890	0.073891	0.073892	0.073894	0.073895	0.073896	0.073897	0.073899
26	0.073900	0.073901	0.073902	0.073904	0.073905	0.073906	0.073907	0.073909	0.073910	0.073911
27	0.073912	0.073914	0.073915	0.073916	0.073918	0.073919	0.073920	0.073921	0.073923	0.073924
28	0.073925	0.073926	0.073928	0.073929	0.073930	0.073931	0.073933	0.073934	0.073935	0.073937
29	0.073938	0.073939	0.073940	0.073942	0.073943	0.073944	0.073945	0.073947	0.073948	0.073949
30	0.073950	0.073952	0.073953	0.073954	0.073955	0.073957	0.073958	0.073959	0.073961	0.073962
31	0.073963	0.073964	0.073966	0.073967	0.073968	0.073969	0.073971	0.073972	0.073973	0.073974
32	0.073976	0.073977	0.073978	0.073980	0.073981	0.073982	0.073983	0.073985	0.073986	0.073987
33	0.073988	0.073990	0.073991	0.073992	0.073993	0.073995	0.073996	0.073997	0.073999	0.074000
34	0.074001	0.074002	0.074004	0.074005	0.074006	0.074007	0.074009	0.074010	0.074011	0.074012
35	0.074014	0.074015	0.074016	0.074017	0.074019	0.074020	0.074021	0.074023	0.074024	0.074025
36	0.074026	0.074028	0.074029	0.074030	0.074031	0.074033	0.074034	0.074035	0.074036	0.074038
37	0.074039	0.074040	0.074042	0.074043	0.074044	0.074045	0.074047	0.074048	0.074049	0.074050
38	0.074052	0.074053	0.074054	0.074055	0.074057	0.074058	0.074059	0.074061	0.074062	0.074063
39	0.074064	0.074066	0.074067	0.074068	0.074069	0.074071	0.074072	0.074073	0.074074	0.074076
40	0.074077	0.074078	0.074079	0.074081	0.074082	0.074083	0.074085	0.074086	0.074087	0.074088
41	0.074090	0.074091	0.074092	0.074093	0.074095	0.074096	0.074097	0.074098	0.074100	0.074101

The factor K is multiplied to the mass of mercury delivered/contained in the measure in kg/g/mg to give its capacity at 20 °C in $m^3/dm^3/cm^3$

Table 3.44 Reference Temperature = 27 °C Air density = 1.1685 kg/m³, ALPHA = .000015/°C DEN = 8000 kg/m³

The values of 10^3K

Temp	0.0	0.1	0.2	0.3	0.4	0.5	0.6	0.7	0.8	0.9
5	0.073643	0.073644	0.073645	0.073646	0.073648	0.073649	0.073650	0.073651	0.073652	0.073654
6	0.073655	0.073656	0.073657	0.073659	0.073660	0.073661	0.073662	0.073663	0.073665	0.073666
7	0.073667	0.073668	0.073670	0.073671	0.073672	0.073673	0.073674	0.073676	0.073677	0.073678
8	0.073679	0.073681	0.073682	0.073683	0.073684	0.073686	0.073687	0.073688	0.073689	0.073690
9	0.073692	0.073693	0.073694	0.073695	0.073697	0.073698	0.073699	0.073700	0.073701	0.073703
10	0.073704	0.073705	0.073706	0.073708	0.073709	0.073710	0.073711	0.073712	0.073714	0.073715
11	0.073716	0.073717	0.073719	0.073720	0.073721	0.073722	0.073724	0.073725	0.073726	0.073727
12	0.073728	0.073730	0.073731	0.073732	0.073733	0.073735	0.073736	0.073737	0.073738	0.073739
13	0.073741	0.073742	0.073743	0.073744	0.073746	0.073747	0.073748	0.073749	0.073751	0.073752
14	0.073753	0.073754	0.073755	0.073757	0.073758	0.073759	0.073760	0.073762	0.073763	0.073764
15	0.073765	0.073766	0.073768	0.073769	0.073770	0.073771	0.073773	0.073774	0.073775	0.073776
16	0.073778	0.073779	0.073780	0.073781	0.073782	0.073784	0.073785	0.073786	0.073787	0.073789
17	0.073790	0.073791	0.073792	0.073793	0.073795	0.073796	0.073797	0.073798	0.073800	0.073801
18	0.073802	0.073803	0.073804	0.073806	0.073807	0.073808	0.073809	0.073811	0.073812	0.073813
19	0.073814	0.073816	0.073817	0.073818	0.073819	0.073820	0.073822	0.073823	0.073824	0.073825
20	0.073827	0.073828	0.073829	0.073830	0.073831	0.073833	0.073834	0.073835	0.073836	0.073838
21	0.073839	0.073840	0.073841	0.073843	0.073844	0.073845	0.073846	0.073847	0.073849	0.073850
22	0.073851	0.073852	0.073854	0.073855	0.073856	0.073857	0.073858	0.073860	0.073861	0.073862
23	0.073863	0.073865	0.073866	0.073867	0.073868	0.073870	0.073871	0.073872	0.073873	0.073874
24	0.073876	0.073877	0.073878	0.073879	0.073881	0.073882	0.073883	0.073884	0.073885	0.073887
25	0.073888	0.073889	0.073890	0.073892	0.073893	0.073894	0.073895	0.073897	0.073898	0.073899
26	0.073900	0.073901	0.073903	0.073904	0.073905	0.073906	0.073908	0.073909	0.073910	0.073911
27	0.073912	0.073914	0.073915	0.073916	0.073917	0.073919	0.073920	0.073921	0.073922	0.073924
28	0.073925	0.073926	0.073927	0.073928	0.073930	0.073931	0.073932	0.073933	0.073935	0.073936
29	0.073937	0.073938	0.073939	0.073941	0.073942	0.073943	0.073944	0.073946	0.073947	0.073948
30	0.073949	0.073951	0.073952	0.073953	0.073954	0.073955	0.073957	0.073958	0.073959	0.073960
31	0.073962	0.073963	0.073964	0.073965	0.073967	0.073968	0.073969	0.073970	0.073971	0.073973
32	0.073974	0.073975	0.073976	0.073978	0.073979	0.073980	0.073981	0.073982	0.073984	0.073985
33	0.073986	0.073987	0.073989	0.073990	0.073991	0.073992	0.073994	0.073995	0.073996	0.073997
34	0.073998	0.074000	0.074001	0.074002	0.074003	0.074005	0.074006	0.074007	0.074008	0.074010
35	0.074011	0.074012	0.074013	0.074014	0.074016	0.074017	0.074018	0.074019	0.074021	0.074022
36	0.074023	0.074024	0.074025	0.074027	0.074028	0.074029	0.074030	0.074032	0.074033	0.074034
37	0.074035	0.074037	0.074038	0.074039	0.074040	0.074041	0.074043	0.074044	0.074045	0.074046
38	0.074048	0.074049	0.074050	0.074051	0.074052	0.074054	0.074055	0.074056	0.074057	0.074059
39	0.074060	0.074061	0.074062	0.074064	0.074065	0.074066	0.074067	0.074068	0.074070	0.074071
40	0.074072	0.074073	0.074075	0.074076	0.074077	0.074078	0.074080	0.074081	0.074082	0.074083
41	0.074084	0.074086	0.074087	0.074088	0.074089	0.074091	0.074092	0.074093	0.074094	0.074096

The factor K is multiplied to the mass of mercury delivered/contained in the measure in kg/g/mg to give its capacity at 20 °C in m³/dm³/cm³

Table 3.45 Reference Temperature = 27 °C Air density = 1.1685 kg/m³, ALPHA = .000025/°C DEN = 8000 kg/m³

The values of 10^3K

Temp	0.0	0.1	0.2	0.3	0.4	0.5	0.6	0.7	0.8	0.9
5	0.073659	0.073660	0.073661	0.073662	0.073663	0.073665	0.073666	0.073667	0.073668	0.073669
6	0.073670	0.073671	0.073673	0.073674	0.073675	0.073676	0.073677	0.073678	0.073680	0.073681
7	0.073682	0.073683	0.073684	0.073685	0.073686	0.073688	0.073689	0.073690	0.073691	0.073692
8	0.073693	0.073695	0.073696	0.073697	0.073698	0.073699	0.073700	0.073701	0.073703	0.073704
9	0.073705	0.073706	0.073707	0.073708	0.073710	0.073711	0.073712	0.073713	0.073714	0.073715
10	0.073716	0.073718	0.073719	0.073720	0.073721	0.073722	0.073723	0.073725	0.073726	0.073727
11	0.073728	0.073729	0.073730	0.073731	0.073733	0.073734	0.073735	0.073736	0.073737	0.073738
12	0.073739	0.073741	0.073742	0.073743	0.073744	0.073745	0.073746	0.073748	0.073749	0.073750
13	0.073751	0.073752	0.073753	0.073754	0.073756	0.073757	0.073758	0.073759	0.073760	0.073761
14	0.073763	0.073764	0.073765	0.073766	0.073767	0.073768	0.073769	0.073771	0.073772	0.073773
15	0.073774	0.073775	0.073776	0.073778	0.073779	0.073780	0.073781	0.073782	0.073783	0.073784
16	0.073786	0.073787	0.073788	0.073789	0.073790	0.073791	0.073793	0.073794	0.073795	0.073796
17	0.073797	0.073798	0.073799	0.073801	0.073802	0.073803	0.073804	0.073805	0.073806	0.073808
18	0.073809	0.073810	0.073811	0.073812	0.073813	0.073814	0.073816	0.073817	0.073818	0.073819
19	0.073820	0.073821	0.073823	0.073824	0.073825	0.073826	0.073827	0.073828	0.073829	0.073831
20	0.073832	0.073833	0.073834	0.073835	0.073836	0.073838	0.073839	0.073840	0.073841	0.073842
21	0.073843	0.073844	0.073846	0.073847	0.073848	0.073849	0.073850	0.073851	0.073853	0.073854
22	0.073855	0.073856	0.073857	0.073858	0.073859	0.073861	0.073862	0.073863	0.073864	0.073865
23	0.073866	0.073867	0.073869	0.073870	0.073871	0.073872	0.073873	0.073874	0.073876	0.073877
24	0.073878	0.073879	0.073880	0.073881	0.073882	0.073884	0.073885	0.073886	0.073887	0.073888
25	0.073889	0.073891	0.073892	0.073893	0.073894	0.073895	0.073896	0.073897	0.073899	0.073900
26	0.073901	0.073902	0.073903	0.073904	0.073906	0.073907	0.073908	0.073909	0.073910	0.073911
27	0.073912	0.073914	0.073915	0.073916	0.073917	0.073918	0.073919	0.073921	0.073922	0.073923
28	0.073924	0.073925	0.073926	0.073927	0.073929	0.073930	0.073931	0.073932	0.073933	0.073934
29	0.073936	0.073937	0.073938	0.073939	0.073940	0.073941	0.073942	0.073944	0.073945	0.073946
30	0.073947	0.073948	0.073949	0.073951	0.073952	0.073953	0.073954	0.073955	0.073956	0.073957
31	0.073959	0.073960	0.073961	0.073962	0.073963	0.073964	0.073966	0.073967	0.073968	0.073969
32	0.073970	0.073971	0.073972	0.073974	0.073975	0.073976	0.073977	0.073978	0.073979	0.073981
33	0.073982	0.073983	0.073984	0.073985	0.073986	0.073987	0.073989	0.073990	0.073991	0.073992
34	0.073993	0.073994	0.073996	0.073997	0.073998	0.073999	0.074000	0.074001	0.074002	0.074004
35	0.074005	0.074006	0.074007	0.074008	0.074009	0.074011	0.074012	0.074013	0.074014	0.074015
36	0.074016	0.074018	0.074019	0.074020	0.074021	0.074022	0.074023	0.074024	0.074026	0.074027
37	0.074028	0.074029	0.074030	0.074031	0.074033	0.074034	0.074035	0.074036	0.074037	0.074038
38	0.074039	0.074041	0.074042	0.074043	0.074044	0.074045	0.074046	0.074048	0.074049	0.074050
39	0.074051	0.074052	0.074053	0.074054	0.074056	0.074057	0.074058	0.074059	0.074060	0.074061
40	0.074063	0.074064	0.074065	0.074066	0.074067	0.074068	0.074069	0.074071	0.074072	0.074073
41	0.074074	0.074075	0.074076	0.074078	0.074079	0.074080	0.074081	0.074082	0.074083	0.074084

The factor K is multiplied to the mass of mercury delivered/contained in the measure in kg/g/mg to give its capacity at 20 °C in $m^3/dm^3/cm^3$

Table 3.46 Reference Temperature = 27 °C Air density = 1.1685 kg/m^3, ALPHA = .00003/°C DEN = 8000 kg/m^3

The values of 10^3K

Temp	0.0	0.1	0.2	0.3	0.4	0.5	0.6	0.7	0.8	0.9
5	0.073667	0.073668	0.073669	0.073670	0.073671	0.073672	0.073674	0.073675	0.073676	0.073677
6	0.073678	0.073679	0.073680	0.073681	0.073683	0.073684	0.073685	0.073686	0.073687	0.073688
7	0.073689	0.073690	0.073691	0.073693	0.073694	0.073695	0.073696	0.073697	0.073698	0.073699
8	0.073700	0.073702	0.073703	0.073704	0.073705	0.073706	0.073707	0.073708	0.073709	0.073710
9	0.073712	0.073713	0.073714	0.073715	0.073716	0.073717	0.073718	0.073719	0.073720	0.073722
10	0.073723	0.073724	0.073725	0.073726	0.073727	0.073728	0.073729	0.073731	0.073732	0.073733
11	0.073734	0.073735	0.073736	0.073737	0.073738	0.073739	0.073741	0.073742	0.073743	0.073744
12	0.073745	0.073746	0.073747	0.073748	0.073749	0.073751	0.073752	0.073753	0.073754	0.073755
13	0.073756	0.073757	0.073758	0.073760	0.073761	0.073762	0.073763	0.073764	0.073765	0.073766
14	0.073767	0.073768	0.073770	0.073771	0.073772	0.073773	0.073774	0.073775	0.073776	0.073777
15	0.073779	0.073780	0.073781	0.073782	0.073783	0.073784	0.073785	0.073786	0.073787	0.073789
16	0.073790	0.073791	0.073792	0.073793	0.073794	0.073795	0.073796	0.073797	0.073799	0.073800
17	0.073801	0.073802	0.073803	0.073804	0.073805	0.073806	0.073808	0.073809	0.073810	0.073811
18	0.073812	0.073813	0.073814	0.073815	0.073816	0.073818	0.073819	0.073820	0.073821	0.073822
19	0.073823	0.073824	0.073825	0.073827	0.073828	0.073829	0.073830	0.073831	0.073832	0.073833
20	0.073834	0.073835	0.073837	0.073838	0.073839	0.073840	0.073841	0.073842	0.073843	0.073844
21	0.073845	0.073847	0.073848	0.073849	0.073850	0.073851	0.073852	0.073853	0.073854	0.073856
22	0.073857	0.073858	0.073859	0.073860	0.073861	0.073862	0.073863	0.073864	0.073866	0.073867
23	0.073868	0.073869	0.073870	0.073871	0.073872	0.073873	0.073875	0.073876	0.073877	0.073878
24	0.073879	0.073880	0.073881	0.073882	0.073883	0.073885	0.073886	0.073887	0.073888	0.073889
25	0.073890	0.073891	0.073892	0.073894	0.073895	0.073896	0.073897	0.073898	0.073899	0.073900
26	0.073901	0.073902	0.073904	0.073905	0.073906	0.073907	0.073908	0.073909	0.073910	0.073911
27	0.073912	0.073914	0.073915	0.073916	0.073917	0.073918	0.073919	0.073920	0.073921	0.073923
28	0.073924	0.073925	0.073926	0.073927	0.073928	0.073929	0.073930	0.073931	0.073933	0.073934
29	0.073935	0.073936	0.073937	0.073938	0.073939	0.073940	0.073942	0.073943	0.073944	0.073945
30	0.073946	0.073947	0.073948	0.073949	0.073950	0.073952	0.073953	0.073954	0.073955	0.073956
31	0.073957	0.073958	0.073959	0.073961	0.073962	0.073963	0.073964	0.073965	0.073966	0.073967
32	0.073968	0.073969	0.073971	0.073972	0.073973	0.073974	0.073975	0.073976	0.073977	0.073978
33	0.073980	0.073981	0.073982	0.073983	0.073984	0.073985	0.073986	0.073987	0.073988	0.073990
34	0.073991	0.073992	0.073993	0.073994	0.073995	0.073996	0.073997	0.073998	0.074000	0.074001
35	0.074002	0.074003	0.074004	0.074005	0.074006	0.074007	0.074009	0.074010	0.074011	0.074012
36	0.074013	0.074014	0.074015	0.074016	0.074017	0.074019	0.074020	0.074021	0.074022	0.074023
37	0.074024	0.074025	0.074026	0.074028	0.074029	0.074030	0.074031	0.074032	0.074033	0.074034
38	0.074035	0.074036	0.074038	0.074039	0.074040	0.074041	0.074042	0.074043	0.074044	0.074045
39	0.074047	0.074048	0.074049	0.074050	0.074051	0.074052	0.074053	0.074054	0.074055	0.074057
40	0.074058	0.074059	0.074060	0.074061	0.074062	0.074063	0.074064	0.074066	0.074067	0.074068
41	0.074069	0.074070	0.074071	0.074072	0.074073	0.074074	0.074076	0.074077	0.074078	0.074079

Table 3.47 Density of Mercury in kg/m^3 Against Temperature in °C

Temp	0.0	0.1	0.2	0.3	0.4	0.5	0.6	0.7	0.8	0.9
0	13595.0763	4.8295	4.5826	4.3358	4.0889	3.8421	3.5953	3.3484	3.1016	2.8548
1	13592.6080	2.3612	2.1145	1.8677	1.6209	1.3741	1.1274	0.8807	0.6339	0.3872
2	13590.1405	9.8937	9.6470	9.4003	9.1536	8.9070	8.6603	8.4136	8.1669	7.9203
3	13587.6736	7.4270	7.1804	6.9337	6.6871	6.4405	6.1939	5.9473	5.7007	5.4541
4	13585.2075	4.9610	4.7144	4.4679	4.2213	3.9748	3.7283	3.4817	3.2352	2.9887
5	13582.7422	2.4957	2.2492	2.0027	1.7563	1.5098	1.2633	1.0169	0.7704	0.5240
6	13580.2776	0.0312	9.7847	9.5383	9.2919	9.0455	8.7992	8.5528	8.3064	8.0600
7	13577.8137	7.5673	7.3210	7.0747	6.8283	6.5820	6.3357	6.0894	5.8431	5.5968
8	13575.3505	5.1043	4.8580	4.6117	4.3655	4.1192	3.8730	3.6267	3.3805	3.1343
9	13572.8881	2.6419	2.3957	2.1495	1.9033	1.6571	1.4110	1.1648	0.9186	0.6725
10	13570.4263	0.1802	9.9341	9.6880	9.4419	9.1958	8.9497	8.7036	8.4575	8.2114
11	13567.9653	7.7193	7.4732	7.2272	6.9811	6.7351	6.4891	6.2430	5.9970	5.7510
12	13565.5050	5.2590	5.0131	4.7671	4.5211	4.2751	4.0292	3.7832	3.5373	3.2914
13	13563.0454	2.7995	2.5536	2.3077	2.0618	1.8159	1.5700	1.3241	1.0783	0.8324
14	13560.5866	0.3407	0.0949	9.8490	9.6032	9.3574	9.1116	8.8657	8.6199	8.3742
15	13558.1284	7.8826	7.6368	7.3911	7.1453	6.8995	6.6538	6.4081	6.1623	5.9166
16	13555.6709	5.4252	5.1795	4.9338	4.6881	4.4424	4.1967	3.9511	3.7054	3.4597
17	13553.2141	2.9685	2.7228	2.4772	2.2316	1.9860	1.7404	1.4948	1.2492	1.0036
18	13550.7580	0.5124	0.2669	0.0213	9.7758	9.5302	9.2847	9.0392	8.7936	8.5481
19	13548.3026	8.0571	7.8116	7.5661	7.3206	7.0752	6.8297	6.5842	6.3388	6.0933
20	13545.8479	5.6025	5.3570	5.1116	4.8662	4.6208	4.3754	4.1300	3.8846	3.6392
21	13543.3939	3.1485	2.9031	2.6578	2.4124	2.1671	1.9218	1.6765	1.4311	1.1858
22	13540.9405	0.6952	0.4499	0.2047	9.9594	9.7141	9.4688	9.2236	8.9783	8.7331
23	13538.4879	8.2426	7.9974	7.7522	7.5070	7.2618	7.0166	6.7714	6.5262	6.2810
24	13536.0359	5.7907	5.5455	5.3004	5.0553	4.8101	4.5650	4.3199	4.0747	3.8296
25	13533.5845	3.3394	3.0944	2.8493	2.6042	2.3591	2.1141	1.8690	1.6240	1.3789
26	13531.1339	0.8889	0.6438	0.3988	0.1538	9.9088	9.6638	9.4188	9.1738	8.9289
27	13528.6839	8.4389	8.1940	7.9490	7.7041	7.4591	7.2142	6.9693	6.7244	6.4795
28	13526.2346	5.9897	5.7448	5.4999	5.2550	5.0102	4.7653	4.5204	4.2756	4.0307
29	13523.7859	3.5411	3.2962	3.0514	2.8066	2.5618	2.3170	2.0722	1.8274	1.5827
30	13521.3379	1.0931	0.8484	0.6036	0.3589	0.1141	9.8694	9.6247	9.3799	9.1352
31	13518.8905	8.6458	8.4011	8.1564	7.9118	7.6671	7.4224	7.1778	6.9331	6.6885
32	13516.4438	6.1992	5.9546	5.7099	5.4653	5.2207	4.9761	4.7315	4.4870	4.2424
33	13513.9978	3.7532	3.5087	3.2641	3.0196	2.7750	2.5305	2.2860	2.0415	1.7969
34	13511.5524	1.3079	1.0634	0.8189	0.5745	0.3300	0.0855	9.8410	9.5966	9.3521
35	13509.1077	8.8633	8.6188	8.3744	8.1300	7.8856	7.6412	7.3968	7.1524	6.9080
36	13506.6636	6.4192	6.1749	5.9305	5.6861	5.4418	5.1974	4.9531	4.7088	4.4644
37	13504.2201	3.9758	3.7315	3.4872	3.2429	2.9986	2.7544	2.5101	2.2658	2.0215
38	13501.7773	1.5330	1.2888	1.0446	0.8003	0.5561	0.3119	0.0677	9.8235	9.5793
39	13499.3351	9.0909	8.8467	8.6025	8.3584	8.1142	7.8701	7.6259	7.3818	7.1376
40	13496.8935	6.6494	6.4053	6.1611	5.9170	5.6729	5.4288	5.1848	4.9407	4.6966
41	13494.4525	4.2085	3.9644	3.7204	3.4763	3.2323	2.9882	2.7442	2.5002	2.2562

REFERENCES

[1] Gupta S V, Practical Density Measurements and Hydrometry, 2002, Institute of Physics Publishing, Bristol and Philadelphia.

[2] Davis R S, 1992, Equation for Determination of Density of Moist Air (1981–1991), Metrologia, 29, 67–70

[3] OIML Recommendations R 117–2003.

[4] ISO 4787–1984, Use and Testing of Volumetric Glassware.

[5] Lida D R, 1997, CRC Handbook of Chemistry and Physics, 76th Edition, (London Chemical Rubber Company) pp 172.

[6] Beatti J A et al, 1941–Proc. Am. Acad. Arts Sci. 71, 71.

[7] Sommer K D and Poziemski J, 1993/1994, Density, Thermal Expansion and Compressibility of mercury, Metrologia, 30, 665–668.

VOLUMETRIC METHOD

4.1 APPLICABILITY OF VOLUMETRIC METHOD

When large number of measures of especially of high capacity is required to be calibrated and uncertainty requirements are not too stringent, the volumetric method is used. In this method the capacity of the under-test measure is compared with that of the standard of known capacity.

The volumetric method is applicable only when standard is of delivery type and measure under test is content type and vice-versa. That is, if the measure under test is delivery type, then the standard should be content type. Similarly, to test a content type measure, the standard measure of delivery type is taken. A working liquid, normally water, is delivered from the delivery measure, till the content type measure under test is full up to the specified mark. The volume of water delivered by the delivery measure and the capacity of the content measure is assumed to be equal. Therefore, the capacity of the standard measure should be either equal to or sub-multiple of the capacity of the measure under test.

Each capacity measure is kept in such a way that graduation marks are in horizontal plane or the axis of the delivery measure is vertical for over-flow and other non-graduated measures. As the marks on either measure are normal to their respective axis, so care should be taken that the content measure is kept on a horizontal ground and the delivery measure is in vertical position.

4.2 MULTIPLE AND ONE TO ONE TRANSFER METHODS

If the capacity of the measure under-test and that of the standard are equal then one to one transfer or direct comparison method is used.

If a content measure under-test is of larger capacity then the standard measure, then as stated above, a standard of delivery type, whose capacity is an integral sub-multiple of the capacity of the measure under-test, is used and multiple filling is carried out. Normally this procedure is used in situations, where standard is of delivery type and the measure under test is of content type.

4.3 CORRECTIONS APPLICABLE IN VOLUMETRIC METHOD

Corrections are applied due to (*i*) temperature of measurement is different from those of reference temperatures of the measures. In this case corrections are applied due to different coefficients of expansion of materials of the two measures and different reference temperatures to which the capacity of the measures are referring. (*ii*) Change in temperature of medium during its transfer from standard measure to under-test measure. For this, corrections in volume of water in the two measures are required.

The loss of water due to evaporation or spillage is one of the sources of error.

4.3.1 Temperature Correction in Volumetric Method

There are two possibilities (*i*) reference temperature is same for the two measures and (*ii*) the reference temperature for each measure is different.

4.3.1.1 Reference Temperatures are Equal

Let V, α, ρ and t respectively stands for volume, coefficient of expansion, density of water and reference temperature and subscripts s and u stand respectively for standard and under-test measures. In this case $t_{rs} = t_{ru} = t_r$.

If the standard measure is used n times to fill the measure under-test, then

$$V_{utu} = nV_{sts}$$

The temperature of water transferred to the measure under-test has changed from t_s to t_u. Assuming that there is no loss of water during transfer, irrespective of the fact that there is a change in volume of water, the mass of water transferred from standard measure to under-test measure remains unchanged. Then volumes of the two measures at different temperatures will be related to each other through the density of water at its temperatures in each measure.

Such that $$nV_{sts}\cdot \rho_{ts} = V_{utu}\cdot\rho_{tu} \qquad ...(1)$$

If V_{str} and V_{utr} are their respective capacities at reference temperature t_r, then

$$nV_{str}[1+\alpha_s(t_s-t_r)]\rho_{ts} = V_{utr}[1+\alpha_u(t_u-t_r)]\rho_{tu}$$

Giving $$V_{utr} = nV_{str}[1+\alpha_s(t_s-t_r)]\rho_{ts}/[1+\alpha_u(t_u-t_r)]\rho_{tu} \qquad ...(2)$$

If K is the factor such that $$V_{utr} = nK.V_{str} \qquad ...(3)$$

Then K is given by $$K = [1+\alpha_s(t_s-t_r)]\rho_{ts}/[1+\alpha_u(t_u-t_r)]\rho_{tu} \qquad ...(4)$$

As $\alpha_s(t_s - t_r)$ and $\alpha_u(t_u - t_r)$ are small in comparison to 1, using binomial expansion of the denominator and neglecting the terms containing higher powers of α_u, or $\alpha_u\ \alpha_s$ then K can be expressed as

$$K = [1+\alpha_s(t_s-t_r)-\alpha_u(t_u-t_r)]\rho_{ts}/\rho_{tu} \qquad ...(5)$$

By taking proper values of coefficients of cubical expansion of the materials used for the two measures and density of water at temperature of measurement, tables for K have been made with respect of the temperatures of the two measures. The value of α_u – the coefficient of expansion of material of the measure under test has been chosen to cover most of the materials used for manufacturing such measures.

The reference temperature for each measure is 20 °C in tables 4.1 and 4.2. We have taken ALPHAS (α_s)= 27.10^{-6}/°C and ALPHAU (α_u) = 51.10^{-6}/°C for Table 4.1, but in table 4.2 ALPHAS (α_s) and ALPHAU(α_u) are equal and each is equal to 51×10^{-6}/°C. For tables 4.3 to 4.8, the value of ALPHAS (α_s) is 54×10^{-6}/°C and reference temperature is 27 °C and ALPHAU respectively takes values of 54×10^{-6}/°C, 33×10^{-6}/°C, 30×10^{-6}/°C, 25×10^{-6}/°C, 15×10^{-6}/°C and 10×10^{-6}/°C.

It may be mentioned that coefficient of cubical expansion of various materials used are
10×10^{-6}/°C for Borosilicate glass
15×10^{-6} /°C for neutral glass
25×10^{-6}/°C to 30×10^{-6} for soda glass,
33×10^{-6} /°C is for galvanised iron sheet for stainless steel.
54×10^{-6} /°C for admiralty bronze
For coefficients of expansions of other materials, please refer [1]

4.3.1.2 Reference Temperatures are Different

Let t_{rs} and t_{ru} be respectively the reference temperatures of standard and under-test measures. If we respectively replace t_r by t_{rs} and t_r by t_{ru} for reference temperatures for standard and under-test measures then also equation (2) is satisfied. So following logic of section of 4.3.1.1, the new equation for the value of K will be as follows

$$K = [1 + \alpha_s(t_s - t_{rs}) - a_u(t_u - t_{ru})]\rho_{ts}/\rho_{tu} \quad ...(6)$$

Here large number of permutations of different reference temperatures and coefficients of expansion are possible, moreover expression is quite simple in calculation, so it is advisable not to construct special tables but use directly the equation (6) to calculate the multiplying factor K.

4.4 USE OF A VOLUMETRIC MEASURE AT A TEMPERATURE OTHER THAN ITS STANDARD TEMPERATURE

Let the temperature of the measure, which was calibrated at 27 °C is filled with water at t°C. If V_n be the nominal capacity, then the actual capacity of the measure at t°C is given by

$$V_n(1 + \alpha(t - 27))$$

Here α is the coefficient of cubical expansion of the material of the measure.

So volume of water V_w at that instant, assuming temperature of the measure and water is same, is given by

$$V_w = V_n(1 + \alpha(t - 27)) \quad ...(7)$$

If V_{27w} is the volume of water at 27 °C, then

$$V_w = V_{27w}\,(1 + \gamma(t - 27)), \text{ giving}$$

$$V_{27w} = V_n\{1 + \alpha(t - 27)\}/\{1 - \gamma(t - 27)\} = V_n[1 - (\gamma - \alpha)(t - 27)]$$

If C be correction to be added to V_n to obtain the volume of water at reference temperature, then C is given by the expression

$$C = -V_n\,(\gamma - \alpha)(t - 27) \quad ...(8)$$

However in case of water, the values of its density at various temperatures are better known. So expressing the volumes of water at various temperatures in terms of its density, we get:

$$V_w/V_{27w} = d_{27w}/d_w$$

Substitution of these in the equation (7) gives us

$$V_w = V_{27w}.d_{27w}/d_w = V_n(1 + \alpha(t - 27)), \text{ giving } V_{27w} \text{ as}$$

$$V_{27w} = (d_w/d_{27w})[V_n(1 + \alpha(t - 27))]$$

Writing

$$V_{27w} = V_n + C \text{ giving}$$

$$C = V_n[(d_w - d_{27w})/d_{27w} + d_w\,\alpha(t - 27)/d_{27w}] \quad ...(9)$$

The values of the correction C (in grams) against temperature for $V_n = 1000$ cm^3 for different values of coefficients of expansion are given in the Tables from 4.9 to 4.15. Here also if C and mass of water is measured in kilograms then capacity is 1000 dm^3, similarly if C and mass of water is in milligrams then the capacity is in mm^3.

If the glass measure is used at temperatures other than the standard temperature of 27°C, then the aforesaid correction is added to the nominal capacity to give the volume of water at 27°C. Conversely, by subtracting the correction from the nominal value gives the volume of water, which must measure at temperature t°C to obtain nominal volume at 27°C.

4.5 VOLUMETRIC METHOD

4.5.1 From a Delivery Measure to a Content Measure

Keep the standard delivery measure and the content measure under test, together with water to be used as medium, at least for 24 hours in the same air-conditioned room so that temperatures of both the measures and water becomes same.

1. Set-up the delivery-measure in vertical position. Its height must be such that the measure under test can be taken out and put underneath easily.
2. Fill the delivery measure from below with water under gravity. The water jar or reservoir must be a few metres above the delivery measure, so that water is filled under gravity in reasonable time. Time of filling should be equal to the delivery time of the measure.
3. Clean the content measure and dry it perfectly, and place it under the delivery measure, with a glass rod inside it, which is resting on its wall. Water is delivered so that the water falls first on the rod without splashing and then trickles down to the content measure. The observer may hold a small measure in his hand in slightly inclined position so that water falls on its wall without splashing, in that case separate rod will not be necessary. The tip of the delivery measure may be quite close to the walls but should never touch them.
4. Fix a thermometer at the outlet of the delivery measure and a thermometer in the content measure under-test, the temperature of water at the delivery point should not differ from that of water inside the measure by more than 0.1°C.
5. Start filling the measure and stop when 75% volume of water has been delivered, remove the rod and thermometer and start filling the rest of water. Till the measure is full, there are two possibilities (1) the capacity of the measure under test is smaller than that of the delivery measure or (2) its capacity is larger than that of the standard.

(1) **The capacity of the measure under-test is smaller than that of the standard measure:** Fill the measure under test up to the graduation mark or up to the brim in case of un-graduated measures. Test the measure under test by sliding the glass plate on its top, ensure no water overflows and there is no air gap due to short filling, in latter case fill the measure from the delivery measure. Take the delivery of remaining water in a graduated cylinder with appropriate graduation mark. The measuring cylinder should be graduated so that difference between the consecutive graduation marks is smaller than the 1/3 of the maximum permissible error of the measure under-test. Measure the volume of water in the measuring cylinder and if the volume of water is larger than the maximum permissible error in deficiency then the measure under test is short in capacity and should be rejected.

(2) **The capacity of the measure under-test is larger than that of the standard measure:** In this case, deliver all the water till zero of the delivery measure and fill the rest with water using a graduated pipette or a burette. The volume of extra water filled should be less than the MPE of the measure under test.

In case the content measure has a capacity, which is an integral multiple of the capacity of the delivery measure, then multiple filling is carried out. The measure is tested for deficiency or excess, with the last fill, in the manner as discussed above.

Alternative method is to take the standard of delivery type with graduated delivery tube. The indication of nominal capacity is in the centre of the scale and graduations are such that these cover the maximum permissible errors for the measures to be tested. Volume of water delivered to fill the measure under test is directly obtained from the graduated scale.

4.5.2 Calibration of Content to Content Measure (working standard capacity measures)

In some countries like India, secondary as well working standard capacity measures are both content type, so method in section 4.5.1 cannot be used as such. There are two possibilities; one is to calibrate the measure under-tests through Gravimetric method using Secondary Standard Weights. This, however, requires distilled water and makes Secondary Standard Capacity measures redundant. So the method for verifying the working standard capacity measures is by one to one volumetric method. The method is enumerated as follows:

1. Clean both the measures properly. The equivalent secondary measure is place on a levelled table. The level of the table is seen with a spirit level.
2. Fill the Secondary Standard Capacity measure with clean and preferably distilled water slightly below the edge of the measure. Remove any air bubble sticking to its walls with the help of the clean glass rod. Note the temperature of water. Slide the Striking glass carefully across the rim of the measure until the glass covers the measure leaving about 2 cm distance uncovered. The measure is now slowly filled and at the same time slide the glass across the rim, until it is completely full. Make sure that there is no air bubble between water and striking glass.
3. Now the water from the secondary measure is to be carefully transferred to the measure under-test. For this slide back the striking glass by very small amount, use a pipette to take out water from the Secondary measure and deliver it into the measure under-test, till sufficient gap has occurred between water surface and striking glass. The pipette should be previously wetted. Wait for a few seconds so that water, which was touching the striking glass, trickles down to the measure and a clear air gap between water-surface and striking glass is visible. Remove the striking glass completely, taking care that no water is spill over or remain sticking to it while it is drawn out.
4. Use a glass wetted siphon to transfer bulk of water to the measure under-test.

Again two possibilities arise

(1) **The capacity of the measure under-test is more than that of standard:** Transfer last part of the liquid in to the measure under-test by tilting the standard measure and bringing it to bottom up position. Note the temperature of water. The temperature should not be different from that taken in step 2. Slide the glass of the measure under-test carefully till a small distance remains uncovered. Finally add water by a previously wetted graduated pipette, till no air bubble is visible on drawing in the striking glass completely. If air bubble is still visible then drop a few drops of water from the graduated pipette into the cavity of the striking glass. By pressing the glass repeatedly water will get in and air will come out filling the measure under-test completely. So error in excess of the measure under-test is estimated by the amount of water delivered by the graduated pipette.

(2) **The capacity of the measure under-test is less than that of standard:** The water from the standard measure is siphoned till the measure under-test becomes full, which is seen by drawing in the striking glass completely. The water left out in the standard measure is measured with the help of a measuring cylinder. Alternatively, we may reverse the roles of the two measures. That is fill the standard measure with the help of measure under test and apply the method described in (1) above.

Here it may be noticed that if materials of the measure under-test and standard measure are different then volume correction as given in the appropriate Table from 4.9 to 4.14 is to be applied. Similarly if the reference temperatures of the measure under test and standard are not the same, correction factor K is also to be applied from appropriate Tables from 4.1 to 4.8.

4.6 ERROR DUE TO EVAPORATION AND SPILLAGE

In volumetric measurements, basic source of additional errors come from the spillage and evaporation of liquid (water) used. In each of the two methods gravimetric or volume transfer, water is exposed to atmosphere, so is liable to evaporate and cause loss in volume. Evaporation of water causes loss in volume of water due to two counts, (1) the actual volume of water evaporated and (2) loss in volume due to fall in temperature, as evaporation causes cooling.

The fall in temperature due to evaporation may be calculated as follows: If P% by weight of water is evaporated at temperature t °C, and then fall in temperature t_f is given as

$P536. = (100 - P).\ t_f$ 536 is the latent heat of evaporation of water in calories.

Giving

$t_f = 536P/(100 - P) = 5.36\,P$ approximately if $P << 100$

It may be seen from above that even 0.1% of evaporation (P = 0.1), the temperature change is about 0.5°C, which will affect the capacity determination by 0.01 %.

We know, rate of evaporation increases

- with rise in air temperature
- for liquids of lower boiling point
- with increased temperature of the liquid
- with increase in exposed surface area of liquid
- with increase in surface area of measure under test
- with increase in air speed
- with decrease in atmospheric (environmental) pressure
- with decrease in relative humidity.

So it is advisable to make volumetric measurements in air-conditioned room with water as transfer liquid. Water has fairly high boiling point and vapour pressure so evaporation is less. Relative humidity of air should be around 50% to reduce water evaporation and condensation of water vapours on the surface of the measure under-test.

Though I am not able to arrive at a reasonable formula, which we can use in case of volumetric method of calibration of large capacity measures. I have collected some formulae connected with evaporation of water with varying parameters like wind velocity, exposed surface area, ambient temperature etc. I sincerely hope that somebody will take up the work to arrive at a formula on theoretical basis. Alternatively one may establish a relation of water evaporation during filling on experimental basis. The experiments are to be carried out in a laboratory with well-controlled environmental conditions, which can be varied at will.

4.6.1 Collected Formulae

1. In a steady flow of air with horizontal (wind) velocity w

$$V = K\sqrt{w} \quad ...(10)$$

$$V = A.\ r^{3/2}, \text{ where } r \text{ is the radius of the tube, area } 250\ m^2 - 10\ cm^2 \quad ...(11)$$

In gentle breeze 25 m/s to 1 cm/s. V is the rate of evaporation.

2. [2] General form of rate of fall of level with respect of time (dE/dt) is given by

$$dE/dt = (A + BW)(P_s - P_d) \quad ...(12)$$

Where P_s and P_d are saturation vapour pressure at environmental temperature and at dew point respectively, A and B are constants, dE/dt is the rate of change of water level.

A typical formula perhaps on empirical basic is given as

$$E(\text{in mm}) = 0.425(p_s - p_d)(1 + 0.805W) \quad ...(13)$$

A similar formula used in Chemical Engineering is given as

$$M = 0.02(p_s - p_d), \quad ...(14)$$

3. [2] Evaporation from pans in air current is given by

$$M = (0.031 + 0.0135W)(p_s - p_d)(p_o/p_1). \quad ...(15)$$

Where p_0 is 760 mm of Hg and p_1 is actual pressure, W is air velocity from 0.5 m/s to 4 m/s, M is hourly loss of mass of water in kg per m^2 i.e. M is in (kg/m^{2}h), W is wind velocity in m/s, valid temperature range is 20 –70 °C.

It may be calculated from the above formulae that at 50 °C, the evaporation of water in air current of 2.5 m/s is 2.8 times that of still air and will become 3.8 if air velocity W changes to 5 m/s.

Comment: It appears to be a reasonably good formula but for stationary water. For flowing water, W may be taken as a relative speed of water with respect to air. That W will be the sum of air and water flow velocities.

4. Rate of evaporation of water is proportional to p_s, temperature range is (B_p – 15)°C, where B_p is boiling point of water [9]

4.6.2 Miscellaneous Statements

1. Evaporation of seawater is 5% less than fresh water.
2. Number of gram molecule of a liquid evaporated per unit time per unit surface area is proportional to its vapour pressure.
3. Evaporation from large areas like lakes is about 2/3rd of from small pans.
4. Evaporation of ocean is almost 820 mm per year.
5. Rate of evaporation of water is proportional to its vapour pressure in the temperature range (t – 15) °C [9], where t is the boiling point in °C.

Other litterateur collected during the study are sited from [3] to [8].

4.6.3 Spillage

Spillage is a process in which small droplets are spilled over. Some of them get evaporated. Some of them sit outside the measures especially in crevices of steps on the measure. In gravimetric method these add to the mass of water required to fill the measure up to the certain graduation mark for content measures and subtract from the mass of water in case of a delivery measure, thus have opposite errors in the two cases.

Table 4.1 The Values of *K* Factor for ALPHAS = 27 × 10^{-6}/°C ALPHAU 51 × 10^{-6}/°C

Ref. temp. of standard = 20 °C; Ref. temp. of under-test = 20 °C

t_s	t_u °C										
°C	15	16	17	18	19	20	21	22	23	24	25
15	1.00012	1.00023	1.00035	1.00048	1.00062	1.00077					
16	0.99999	1.00010	1.00022	1.00034	1.00048	1.00063	1.00079				
17	0.99984	0.99995	1.0007	1.00020	1.00034	1.00049	1.00065	1.00082			
18	0.99969	0.99980	0.99992	1.00005	1.00019	1.00034	1.00049	1.00065	1.00082		
19	0.99953	0.99964	0.99978	0.99988	1.00002	1.00017	1.00033	1.00050	1.00069	1.00088	
20	0.99935	0.99946	0.99958	0.99971	0.99985	1.00000	1.00016	1.00033	1.00052	1.00071	1.00091
21		0.99928	0.99940	0.99953	0.99967	0.99982	0.99998	1.00015	1.00033	1.00052	1.00072
22			0.99921	0.99933	0.99947	0.99962	0.99978	0.99995	1.00014	1.00033	1.00053
23				0.99912	0.99926	0.99941	0.99957	0.99974	0.99993	1.00012	1.00032
24					0.99905	0.99920	0.99936	0.99952	0.99971	0.99990	1.00010
25						0.99897	0.99913	0.99930	0.99949	0.99968	0.99988

Table 4.2 The Values of *K* Factor for ALPHAS = 51 × 10^{-6}/°C ALPHAU = 51 × 10^{-6}/°C

Ref. temp. of standard = 20 °C Ref. temp. of under-test = 20 °C

t_s	t_u °C										
°C	15	16	17	18	19	20	21	22	23	24	25
15	1.00000	1.00011	1.00023	1.00036	1.00050	1.00065					
16	0.99989	1.00000	1.00012	1.00025	1.00039	1.00054	1.00070				
17	0.99977	0.99988	1.00000	1.00013	1.00027	1.00042	1.00058	1.00075			
18	0.99964	0.99975	0.99987	1.00000	1.00014	1.00029	1.00045	1.00062	1.00081		
19	0.99950	0.99961	0.99973	0.99986	1.00000	1.00015	1.00031	1.00048	1.00067	1.00085	
20	0.99935	0.99946	0.99958	0.99971	0.99985	1.00000	1.00016	1.00033	1.00052	1.00071	1.00091
21		0.99930	0.99942	0.99955	0.99969	0.99984	1.00000	1.00017	1.00036	1.00055	1.00075
22			0.99925	0.99938	0.99952	0.99967	0.99983	1.00000	1.00019	1.00038	1.00058
23				0.99919	0.99933	0.99948	0.99964	0.99981	1.00000	1.00019	1.00039
24					0.99914	0.99929	0.99945	0.99962	0.99981	1.00000	1.00020
25						0.99909	0.99925	0.99942	0.99961	0.99980	1.00000

Table 4.3 The Values of *K* Factor for ALPHAS = 54 × 10^{-6}/°C ALPHAU = 54 × 10^{-6} /°C

Ref. temp. of standard = 27 °C Ref. temp. of under-test = 27 °C

t_s	t_u °C										
°C	15	16	17	18	19	20	21	22	23	24	25
15	1.00000	1.00010	1.00021	1.00034	1.00048	1.00062	1.00078	1.00095	1.00113	1.00132	1.00152
16	0.99989	1.00000	1.00011	1.00024	1.00037	1.00052	1.00068	1.00085	1.00103	1.00121	1.00141
17	0.99978	0.99988	1.00000	1.00012	1.00026	1.00041	1.0006	1.00073	1.00091	1.00110	1.00130
18	0.99965	0.99976	0.99987	1.00000	1.00013	1.00028	1.00044	1.00061	1.00079	1.00097	1.00117
19	0.99952	0.99962	0.99973	0.99986	1.00000	1.00015	1.00031	1.00047	1.00065	1.00084	1.00104
20	0.99937	0.99947	0.99959	0.99972	0.99985	1.00000	1.00015	1.00032	1.00050	1.00069	1.00089
21	0.99921	0.99932	0.99943	0.99960	0.99969	0.99984	1.00000	1.00017	1.00035	1.00054	1.00073
22	0.99905	0.99915	0.99926	0.99938	0.99952	0.99967	0.99983	1.00000	1.00018	1.00037	1.00057
23	0.99887	0.99897	0.99908	0.99921	0.99935	0.99950	0.99965	0.99982	1.00000	1.00019	1.00039
24	0.99868	0.99878	0.99890	0.99902	0.99916	0.99931	0.99946	0.99963	0.99981	1.00000	1.00020
25	0.99848	0.99858	0.99870	0.99882	0.99896	0.99911	0.99927	0.99943	0.99961	0.99980	1.00000

Table 4.4 The Values of *K* Factor for ALPHAS= 54 x10^{-6} /°C ALPHAU= 33 x10^{-6}/°C

Ref. temp. of standard = 27 °C Ref. temp. of under-test = 27 °C

T_s \ T_u	22	23	24	25	26	27	28	29	30	31	32
22	0.99992	1.00013	1.00035	1.00057	1.00081	1.00106	1.00132	1.00158	1.00186	1.00214	1.00244
23	0.99973	0.99994	1.00016	1.00039	1.00062	1.00087	1.00113	1.00140	1.00167	1.00195	1.00225
24	0.99953	0.99974	0.99996	1.00019	1.00043	1.00067	1.00093	1.00120	1.00147	1.00176	1.00205
25	0.99932	0.99953	0.99975	0.99998	1.00022	1.00047	1.00072	1.00099	1.00126	1.00155	1.00184
26	0.99910	0.99931	0.99953	0.99976	1.00000	1.00025	1.00050	1.00077	1.00105	1.00133	1.00162
27	0.99888	0.99909	0.99931	0.99953	0.99977	1.00002	1.00028	1.00054	1.00082	1.00110	1.00140
28	0.99864	0.99885	0.99907	0.99930	0.99954	0.99979	1.00004	1.00031	1.00058	1.00087	1.00116
29	0.99840	0.99861	0.99883	0.99905	0.99929	0.99954	0.99980	1.00006	1.00034	1.00062	1.00091
30	0.99814	0.99835	0.99857	0.99880	0.99904	0.99929	0.99954	0.99981	1.00008	1.00037	1.00066
31	0.99788	0.99809	0.99831	0.99854	0.99878	0.99902	0.99928	0.99955	0.99982	1.00011	1.00040
32	0.99761	0.99782	0.99804	0.99827	0.99851	0.99875	0.99901	0.99928	0.99955	0.99983	1.00013

Table 4.5 The Values of *K* Factor for ALPHAS = 54×10^{-6}/°C ALPHAU = 30×10^{-6}/°C

Ref. temp. of standard = 27 °C Ref. temp. of under-test = 27 °C

T_s \ T_u	22	23	24	25	26	27	28	29	30	31	32
22	0.99990	1.00012	1.00034	1.00057	1.00081	1.00106	1.00132	1.00159	1.00187	1.00216	1.00245
23	0.99972	0.99993	1.00015	1.00038	1.00062	1.00088	1.00114	1.00140	1.00168	1.00197	1.00227
24	0.99952	0.99973	0.99995	1.00018	1.00043	1.00068	1.00094	1.00121	1.00148	1.00177	1.00207
25	0.99931	0.99952	0.99974	0.99998	1.00022	1.00047	1.00073	1.00100	1.00128	1.00156	1.00186
26	0.99909	0.99930	0.99953	0.99976	1.00000	1.00025	1.00051	1.00078	1.00106	1.00134	1.00164
27	0.99887	0.99908	0.99930	0.99953	0.99977	1.00002	1.00028	1.00055	1.00083	1.00112	1.00141
28	0.99863	0.99884	0.99906	0.99930	0.99954	0.99979	1.00005	1.00032	1.00059	1.00088	1.00118
29	0.99839	0.99860	0.99882	0.99905	0.99929	0.99954	0.99980	1.00007	1.00035	1.00064	1.00093
30	0.99813	0.99834	0.99857	0.99880	0.99904	0.99929	0.99955	0.99982	1.00010	1.00038	1.00068
31	0.99787	0.99808	0.99830	0.99854	0.99878	0.99903	0.99929	0.99956	0.99983	1.00012	1.00042
32	0.99760	0.99781	0.99803	0.99827	0.99851	0.99876	0.99902	0.99928	0.99956	0.99985	1.00014

Table 4.6 The Values of *K* Factor for ALPHAS = 54×10^{-6} /°C ALPHAU = 25×10^{-6} /°C

Ref. temp. of standard = 27 °C Ref. temp. of under-test = 27 °C

T_s \ T_u	22	23	24	25	26	27	28	29	30	31	32
22	0.99988	1.00010	1.00033	1.00057	1.00081	1.00107	1.00133	1.00161	1.00189	1.00218	1.00248
23	0.99970	0.99991	1.00014	1.00038	1.00062	1.00088	1.00115	1.00142	1.00170	1.00199	1.00230
24	0.99950	0.99971	0.99994	1.00018	1.00043	1.00068	1.00095	1.00122	1.00150	1.00180	1.00210
25	0.99929	0.99951	0.99973	0.99997	1.00022	1.00047	1.00074	1.00101	1.00130	1.00159	1.00189
26	0.99907	0.99929	0.99952	0.99975	1.00000	1.00026	1.00052	1.00079	1.00108	1.00137	1.00167
27	0.99885	0.99906	0.99929	0.99953	0.99977	1.00003	1.00029	1.00057	1.00085	1.00114	1.00144
28	0.99861	0.99883	0.99905	0.99929	0.99954	0.99979	1.00006	1.00033	1.00061	1.00091	1.00121
29	0.99837	0.99858	0.99881	0.99905	0.99929	0.99955	0.99981	1.00009	1.00037	1.00066	1.00096
30	0.99811	0.99833	0.99856	0.99879	0.99904	0.99929	0.99956	0.99983	1.00012	1.00041	1.00071
31	0.99785	0.99807	0.99829	0.99853	0.99878	0.99903	0.99930	0.99957	0.99985	1.00014	1.00045
32	0.99758	0.99780	0.99802	0.99826	0.99851	0.99876	0.99903	0.99930	0.99958	0.99987	1.00017

Table 4.7 The Values of *K* Factor for ALPHAS= 54 x10^{-6} /ºC ALPHAU= 15 x10^{-6} /ºC

Ref. temp. of standard = 27 °C Ref. temp. of under-test = 27 °C

T_s \ T_u	22	23	24	25	26	27	28	29	30	31	32
22	0.99984	1.00007	1.00031	1.00056	1.00081	1.00108	1.00135	1.00164	1.00193	1.00223	1.00254
23	0.99966	0.99988	1.00012	1.00037	1.00062	1.00089	1.00117	1.00145	1.00174	1.00204	1.00236
24	0.99946	0.99968	0.99992	1.00017	1.00043	1.00069	1.00097	1.00125	1.00154	1.00185	1.00216
25	0.99925	0.99948	0.99971	0.99996	1.00022	1.00048	1.00076	1.00104	1.00134	1.00164	1.00195
26	0.99903	0.99926	0.99950	0.99974	1.00000	1.00027	1.00054	1.00082	1.00112	1.00142	1.00173
27	0.99881	0.99903	0.99927	0.99952	0.99977	1.00004	1.00031	1.00060	1.00089	1.00119	1.00150
28	0.99857	0.99880	0.99903	0.99928	0.99954	0.99980	1.00008	1.00036	1.00065	1.00096	1.00127
29	0.99833	0.99855	0.99879	0.99904	0.99929	0.99956	0.99983	1.00012	1.00041	1.00071	1.00102
30	0.99807	0.99830	0.99854	0.99878	0.99904	0.99930	0.99958	0.99986	1.00016	1.00046	1.00077
31	0.99781	0.99804	0.99827	0.99852	0.99878	0.99904	0.99932	0.99960	0.99989	1.00020	1.00051
32	0.99754	0.99777	0.99800	0.99825	0.99851	0.99877	0.99905	0.99933	0.99962	0.99992	1.00023

Table 4.8 The Values of *K* Factor for ALPHAS = 54 × 10^{-6}/°C ALPHAU = 10 × 10^{-6}/ºC

Ref. temp. of standard = 27 °C Ref. temp. of under-test = 27 °C

T_s \ T_u	22	23	24	25	26	27	28	29	30	31	32
22	0.99982	1.00006	1.00030	1.00055	1.00081	1.00108	1.00136	1.00165	1.00195	1.00226	1.00257
23	0.99964	0.99987	1.00011	1.00036	1.00062	1.00090	1.00118	1.00146	1.00176	1.00207	1.00239
24	0.99944	0.99967	0.99991	1.00016	1.00043	1.00070	1.00098	1.00127	1.00156	1.00187	1.00219
25	0.99923	0.99946	0.99970	0.99996	1.00022	1.00049	1.00077	1.00106	1.00136	1.00166	1.00198
26	0.99901	0.99924	0.99949	0.99974	1.00000	1.00027	1.00055	1.00084	1.00114	1.00144	1.00176
27	0.99879	0.99902	0.99926	0.99951	0.99977	1.00004	1.00032	1.00061	1.00091	1.00122	1.00153
28	0.99855	0.99878	0.99902	0.99928	0.99954	0.99981	1.00009	1.00038	1.00067	1.00098	1.00130
29	0.99831	0.99854	0.99878	0.99903	0.99929	0.99956	0.99984	1.00013	1.00043	1.00074	1.00105
30	0.99805	0.99828	0.99853	0.99878	0.99904	0.99931	0.99959	0.99988	1.00018	1.00048	1.00080
31	0.99779	0.99802	0.99826	0.99852	0.99878	0.99905	0.99933	0.99962	0.99991	1.00022	1.00054
32	0.99752	0.99775	0.99799	0.99825	0.99851	0.99878	0.99906	0.99934	0.99964	0.99995	1.00026

Table 4.9 Correction in cm^3

Add to nominal capacity of the measure to obtain the volume of water at reference temperature when a 1 dm^3 measure is used at a temperature other than its reference temperature 27 °C ALPHA (α) = 54×10^{-6} °C

Temp.	0.0	0.1	0.2	0.3	0.4	0.5	0.6	0.7	0.8	0.9
5	2.2709	2.2746	2.2782	2.2817	2.2850	2.2881	2.2910	2.2939	2.2965	2.2991
6	2.3015	2.3037	2.3057	2.3077	2.3094	2.3110	2.3125	2.3139	2.3151	2.3161
7	2.3170	2.3177	2.3183	2.3187	2.3191	2.3192	2.3193	2.3191	2.3188	2.3184
8	2.3179	2.3172	2.3163	2.3153	2.3142	2.3129	2.3115	2.3100	2.3083	2.3065
9	2.3045	2.3024	2.3001	2.2978	2.2953	2.2927	2.2899	2.2869	2.2839	2.2807
10	2.2773	2.2739	2.2703	2.2665	2.2627	2.2587	2.2546	2.2503	2.2459	2.2414
11	2.2367	2.2319	2.2270	2.2219	2.2168	2.2115	2.2060	2.2005	2.1947	2.1889
12	2.1830	2.1769	2.1707	2.1644	2.1579	2.1513	2.1446	2.1377	2.1308	2.1237
13	2.1165	2.1092	2.1017	2.0941	2.0864	2.0785	2.0706	2.0625	2.0543	2.0460
14	2.0375	2.0290	2.0202	2.0115	2.0025	1.9935	1.9843	1.9750	1.9656	1.9560
15	1.9464	1.9366	1.9267	1.9168	1.9067	1.8964	1.8861	1.8756	1.8650	1.8543
16	1.8435	1.8325	1.8215	1.8103	1.7991	1.7877	1.7762	1.7645	1.7528	1.7409
17	1.7290	1.7169	1.7047	1.6924	1.6800	1.6674	1.6548	1.6421	1.6292	1.6162
18	1.6032	1.5900	1.5766	1.5632	1.5497	1.5361	1.5224	1.5085	1.4945	1.4804
19	1.4663	1.4520	1.4376	1.4231	1.4085	1.3938	1.3790	1.3640	1.3490	1.3338
20	1.3186	1.3032	1.2878	1.2722	1.2566	1.2408	1.2249	1.2089	1.1928	1.1766
21	1.1603	1.1439	1.1274	1.1108	1.0941	1.0773	1.0604	1.0434	1.0262	1.0090
22	0.9917	0.9743	0.9568	0.9391	0.9214	0.9035	0.8856	0.8676	0.8495	0.8312
23	0.8129	0.7945	0.7759	0.7573	0.7386	0.7198	0.7009	0.6818	0.6627	0.6435
24	0.6241	0.6047	0.5852	0.5656	0.5459	0.5261	0.5062	0.4862	0.4661	0.4459
25	0.4256	0.4052	0.3848	0.3642	0.3435	0.3228	0.3019	0.2810	0.2599	0.2387
26	0.2175	0.1962	0.1748	0.1532	0.1316	0.1100	0.0882	0.0662	0.0443	0.0222
27	0.0000	–0.0223	–0.0446	–0.0670	–0.0896	–0.1122	–0.1349	–0.1578	–0.1807	–0.2037
28	–0.2268	–0.2499	–0.2732	–0.2966	–0.3200	–0.3435	–0.3672	–0.3909	–0.4147	–0.4386
29	–0.4626	–0.4867	–0.5108	–0.5351	–0.5595	–0.5839	–0.6084	–0.6330	–0.6577	–0.6825
30	–0.7074	–0.7324	–0.7574	–0.7825	–0.8078	–0.8331	–0.8585	–0.8840	–0.9096	–0.9352
31	–0.9610	–0.9868	–1.0127	–1.0387	–1.0648	–1.0910	–1.1172	–1.1436	–1.1700	–1.1966
32	–1.2231	–1.2499	–1.2766	–1.3035	–1.3304	–1.3574	–1.3846	–1.4118	–1.4391	–1.4664
33	–1.4938	–1.5214	–1.5490	–1.5767	–1.6045	–1.6324	–1.6603	–1.6883	–1.7164	–1.7446
34	–1.7729	–1.8013	–1.8297	–1.8582	–1.8868	–1.9155	–1.9443	–1.9731	–2.0021	–2.0311
35	–2.0602	–2.0893	–2.1186	–2.1479	–2.1774	–2.2068	–2.2364	–2.2661	–2.2958	–2.3256
36	–2.3555	–2.3855	–2.4155	–2.4457	–2.4759	–2.5062	–2.5365	–2.5670	–2.5975	–2.6282
37	–2.6588	–2.6896	–2.7204	–2.7513	–2.7824	–2.8134	–2.8446	–2.8759	–2.9071	–2.9385
38	–2.9700	–3.0015	–3.0331	–3.0648	–3.0966	–3.1285	–3.1603	–3.1923	–3.2244	–3.2566
39	–3.2888	–3.3211	–3.3535	–3.3859	–3.4185	–3.4511	–3.4837	–3.5165	–3.5493	–3.5822
40	–3.6152	–3.6482	–3.6814	–3.7146	–3.7478	–3.7812	–3.8146	–3.8481	–3.8817	–3.9153
41	–3.9490	–3.9828	–4.0166	–4.0506	–4.0846	–4.1187	–4.1528	–4.1870	–4.2213	–4.2557

Table 4.10 Correction in cm^3

Add to nominal capacity of the measure to obtain the volume of water at reference temperature when a 1 dm^3 measure is used at a temperature other than its reference temperature 27 °C ALHPA (α)= 33×10^{-6}/°C

Temp.	0.0	0.1	0.2	0.3	0.4	0.5	0.6	0.7	0.8	0.9
5	2.7345	2.7361	2.7376	2.7390	2.7401	2.7411	2.7420	2.7427	2.7433	2.7437
6	2.7440	2.7441	2.7440	2.7439	2.7435	2.7430	2.7424	2.7416	2.7407	2.7396
7	2.7384	2.7370	2.7355	2.7338	2.7321	2.7301	2.7280	2.7258	2.7234	2.7209
8	2.7182	2.7154	2.7125	2.7093	2.7061	2.7027	2.6992	2.6955	2.6917	2.6878
9	2.6837	2.6796	2.6752	2.6707	2.6661	2.6614	2.6565	2.6514	2.6462	2.6409
10	2.6355	2.6299	2.6242	2.6184	2.6124	2.6063	2.6001	2.5936	2.5872	2.5806
11	2.5738	2.5669	2.5598	2.5527	2.5454	2.5380	2.5304	2.5227	2.5149	2.5070
12	2.4989	2.4907	2.4824	2.4740	2.4654	2.4567	2.4478	2.4389	2.4299	2.4206
13	2.4113	2.4019	2.3923	2.3826	2.3728	2.3628	2.3527	2.3426	2.3322	2.3218
14	2.3112	2.3006	2.2898	2.2789	2.2679	2.2567	2.2454	2.2340	2.2225	2.2108
15	2.1991	2.1872	2.1752	2.1631	2.1509	2.1385	2.1261	2.1135	2.1008	2.0879
16	2.0750	2.0620	2.0488	2.0355	2.0222	2.0087	1.9951	1.9813	1.9675	1.9535
17	1.9394	1.9252	1.9110	1.8966	1.8820	1.8674	1.8526	1.8378	1.8228	1.8077
18	1.7925	1.7772	1.7618	1.7463	1.7306	1.7149	1.6991	1.6831	1.6670	1.6509
19	1.6346	1.6182	1.6017	1.5851	1.5684	1.5515	1.5346	1.5176	1.5005	1.4832
20	1.4659	1.4484	1.4308	1.4131	1.3954	1.3775	1.3595	1.3414	1.3232	1.3049
21	1.2865	1.2680	1.2494	1.2307	1.2118	1.1929	1.1739	1.1548	1.1356	1.1162
22	1.0968	1.0773	1.0577	1.0379	1.0181	0.9981	0.9781	0.9580	0.9378	0.9174
23	0.8970	0.8764	0.8558	0.8351	0.8143	0.7933	0.7723	0.7512	0.7299	0.7086
24	0.6872	0.6657	0.6440	0.6224	0.6006	0.5786	0.5566	0.5345	0.5123	0.4900
25	0.4676	0.4451	0.4226	0.3999	0.3771	0.3543	0.3313	0.3083	0.2851	0.2618
26	0.2385	0.2151	0.1916	0.1679	0.1442	0.1205	0.0966	0.0725	0.0485	0.0243
27	0.0000	–0.0244	–0.0488	–0.0733	–0.0980	–0.1227	–0.1475	–0.1725	–0.1975	–0.2226
28	–0.2478	–0.2730	–0.2984	–0.3239	–0.3494	–0.3750	–0.4008	–0.4266	–0.4525	–0.4785
29	–0.5046	–0.5308	–0.5570	–0.5834	–0.6098	–0.6364	–0.6630	–0.6897	–0.7165	–0.7434
30	–0.7703	–0.7974	–0.8246	–0.8518	–0.8791	–0.9065	–0.9340	–0.9616	–0.9893	–1.0170
31	–1.0449	–1.0728	–1.1008	–1.1289	–1.1571	–1.1854	–1.2137	–1.2422	–1.2707	–1.2993
32	–1.3280	–1.3568	–1.3857	–1.4146	–1.4436	–1.4728	–1.5020	–1.5313	–1.5606	–1.5901
33	–1.6196	–1.6493	–1.6789	–1.7087	–1.7386	–1.7686	–1.7986	–1.8287	–1.8589	–1.8892
34	–1.9196	–1.9500	–1.9806	–2.0112	–2.0418	–2.0727	–2.1035	–2.1345	–2.1655	–2.1966
35	–2.2278	–2.2590	–2.2903	–2.3218	–2.3533	–2.3849	–2.4166	–2.4483	–2.4801	–2.5120
36	–2.5440	–2.5761	–2.6082	–2.6404	–2.6727	–2.7051	–2.7375	–2.7701	–2.8027	–2.8354
37	–2.8682	–2.9010	–2.9339	–2.9669	–3.0000	–3.0332	–3.0664	–3.0998	–3.1331	–3.1666
38	–3.2001	–3.2338	–3.2675	–3.3012	–3.3351	–3.3690	–3.4030	–3.4371	–3.4713	–3.5055
39	–3.5398	–3.5742	–3.6087	–3.6432	–3.6778	–3.7125	–3.7472	–3.7821	–3.8170	–3.8519
40	–3.8870	–3.9221	–3.9573	–3.9927	–4.0280	–4.0634	–4.0989	–4.1345	–4.1701	–4.2059
41	–4.2416	–4.2775	–4.3134	–4.3495	–4.3855	–4.4217	–4.4579	–4.4942	–4.5306	–4.5670

Table 4.11 Correction in cm^3

Add to nominal capacity of the measure to obtain the volume of water at reference temperature when a 1 dm3 measure is used at a temperature other than its reference temperature = 27 °C ALHPA (α) =30 × 10^{-6}/C

Temp.	0.0	0.1	0.2	0.3	0.4	0.5	0.6	0.7	0.8	0.9
5	2.8007	2.8021	2.8032	2.8043	2.8052	2.8058	2.8064	2.8068	2.8071	2.8072
6	2.8072	2.8070	2.8067	2.8062	2.8055	2.8047	2.8038	2.8027	2.8015	2.8001
7	2.7986	2.7969	2.7951	2.7931	2.7911	2.7888	2.7864	2.7839	2.7812	2.7784
8	2.7754	2.7723	2.7690	2.7656	2.7621	2.7584	2.7546	2.7506	2.7465	2.7423
9	2.7379	2.7334	2.7287	2.7240	2.7190	2.7140	2.7088	2.7035	2.6980	2.6924
10	2.6866	2.6808	2.6748	2.6686	2.6623	2.6560	2.6494	2.6427	2.6359	2.6290
11	2.6219	2.6147	2.6074	2.5999	2.5923	2.5846	2.5767	2.5688	2.5606	2.5524
12	2.5441	2.5356	2.5270	2.5182	2.5093	2.5003	2.4912	2.4819	2.4726	2.4630
13	2.4534	2.4437	2.4338	2.4238	2.4137	2.4035	2.3931	2.3826	2.3720	2.3612
14	2.3504	2.3394	2.3283	2.3171	2.3058	2.2943	2.2827	2.2710	2.2592	2.2472
15	2.2351	2.2230	2.2107	2.1983	2.1858	2.1731	2.1604	2.1475	2.1345	2.1213
16	2.1081	2.0948	2.0813	2.0677	2.0541	2.0402	2.0264	2.0123	1.9981	1.9839
17	1.9695	1.9550	1.9404	1.9257	1.9109	1.8959	1.8809	1.8658	1.8505	1.8351
18	1.8196	1.8040	1.7883	1.7725	1.7565	1.7405	1.7244	1.7081	1.6917	1.6752
19	1.6586	1.6420	1.6251	1.6082	1.5912	1.5741	1.5569	1.5396	1.5221	1.5045
20	1.4869	1.4691	1.4513	1.4333	1.4152	1.3970	1.3787	1.3604	1.3419	1.3233
21	1.3045	1.2857	1.2668	1.2478	1.2287	1.2094	1.1902	1.1707	1.1512	1.1316
22	1.1118	1.0920	1.0721	1.0520	1.0319	1.0117	0.9913	0.9709	0.9504	0.9297
23	0.9090	0.8882	0.8672	0.8462	0.8251	0.8038	0.7825	0.7611	0.7395	0.7179
24	0.6962	0.6744	0.6525	0.6305	0.6084	0.5861	0.5638	0.5414	0.5189	0.4963
25	0.4736	0.4508	0.4280	0.4050	0.3819	0.3588	0.3355	0.3122	0.2887	0.2652
26	0.2415	0.2178	0.1940	0.1700	0.1460	0.1220	0.0978	0.0734	0.0491	0.0246
27	0.0000	–0.0247	–0.0494	–0.0742	–0.0992	–0.1242	–0.1493	–0.1746	–0.1999	–0.2253
28	–0.2508	–0.2763	–0.3020	–0.3278	–0.3536	–0.3795	–0.4056	–0.4317	–0.4579	–0.4842
29	–0.5106	–0.5371	–0.5636	–0.5903	–0.6170	–0.6439	–0.6708	–0.6978	–0.7249	–0.7521
30	–0.7793	–0.8067	–0.8341	–0.8617	–0.8893	–0.9170	–0.9448	–0.9727	–1.0007	–1.0287
31	–1.0568	–1.0851	–1.1134	–1.1418	–1.1703	–1.1989	–1.2275	–1.2563	–1.2851	–1.3140
32	–1.3430	–1.3721	–1.4012	–1.4305	–1.4598	–1.4892	–1.5188	–1.5484	–1.5780	–1.6077
33	–1.6376	–1.6675	–1.6975	–1.7276	–1.7578	–1.7881	–1.8184	–1.8488	–1.8793	–1.9099
34	–1.9405	–1.9713	–2.0021	–2.0330	–2.0640	–2.0951	–2.1263	–2.1575	–2.1888	–2.2202
35	–2.2517	–2.2833	–2.3149	–2.3466	–2.3784	–2.4103	–2.4423	–2.4743	–2.5065	–2.5386
36	–2.5709	–2.6033	–2.6357	–2.6682	–2.7008	–2.7335	–2.7662	–2.7991	–2.8320	–2.8650
37	–2.8981	–2.9312	–2.9644	–2.9977	–3.0311	–3.0646	–3.0981	–3.1318	–3.1654	–3.1992
38	–3.2330	–3.2670	–3.3009	–3.3350	–3.3692	–3.4034	–3.4377	–3.4721	–3.5065	–3.5411
39	–3.5757	–3.6103	–3.6451	–3.6799	–3.7149	–3.7499	–3.7849	–3.8200	–3.8552	–3.8905
40	–3.9258	–3.9613	–3.9968	–4.0324	–4.0680	–4.1037	–4.1395	–4.1754	–4.2113	–4.2474
41	–4.2834	–4.3196	–4.3558	–4.3922	–4.4285	–4.4650	–4.5015	–4.5381	–4.5747	–4.6115

Table 4.12 Correction in cm^3

Add to nominal capacity of the measure to obtain the volume of water at reference temperature when a 1 dm^3 measure is used at a temperature other than its reference temperature 27 °C ALHPA (α) = 25 × 10^{-6}/°C

Temp.	0.0	0.1	0.2	0.3	0.4	0.5	0.6	0.7	0.8	0.9
5	2.9111	2.9119	2.9126	2.9132	2.9135	2.9137	2.9138	2.9137	2.9135	2.9131
6	2.9125	2.9119	2.9110	2.9100	2.9089	2.9076	2.9062	2.9046	2.9029	2.9010
7	2.8989	2.8967	2.8945	2.8920	2.8894	2.8866	2.8838	2.8807	2.8775	2.8742
8	2.8707	2.8671	2.8634	2.8594	2.8554	2.8512	2.8469	2.8424	2.8378	2.8331
9	2.8282	2.8232	2.8180	2.8128	2.8073	2.8018	2.7961	2.7902	2.7843	2.7782
10	2.7719	2.7655	2.7591	2.7524	2.7456	2.7387	2.7317	2.7245	2.7172	2.7098
11	2.7022	2.6944	2.6866	2.6786	2.6706	2.6624	2.6540	2.6455	2.6369	2.6281
12	2.6193	2.6103	2.6012	2.5919	2.5826	2.5731	2.5634	2.5536	2.5438	2.5337
13	2.5236	2.5134	2.5030	2.4925	2.4819	2.4711	2.4602	2.4493	2.4381	2.4269
14	2.4155	2.4041	2.3924	2.3807	2.3689	2.3569	2.3449	2.3327	2.3204	2.3079
15	2.2953	2.2826	2.2698	2.2570	2.2439	2.2307	2.2175	2.2041	2.1906	2.1770
16	2.1632	2.1494	2.1354	2.1214	2.1072	2.0929	2.0785	2.0639	2.0493	2.0345
17	2.0196	2.0046	1.9895	1.9743	1.9590	1.9435	1.9280	1.9124	1.8966	1.8807
18	1.8647	1.8486	1.8324	1.8161	1.7996	1.7830	1.7664	1.7497	1.7328	1.7158
19	1.6987	1.6815	1.6642	1.6468	1.6293	1.6116	1.5939	1.5761	1.5582	1.5401
20	1.5220	1.5036	1.4853	1.4668	1.4483	1.4296	1.4108	1.3919	1.3729	1.3538
21	1.3346	1.3153	1.2958	1.2764	1.2567	1.2370	1.2172	1.1973	1.1772	1.1571
22	1.1369	1.1165	1.0961	1.0755	1.0550	1.0342	1.0134	0.9924	0.9714	0.9503
23	0.9290	0.9077	0.8862	0.8647	0.8431	0.8214	0.7995	0.7776	0.7555	0.7334
24	0.7112	0.6889	0.6665	0.6440	0.6214	0.5986	0.5758	0.5529	0.5300	0.5069
25	0.4836	0.4604	0.4370	0.4135	0.3899	0.3663	0.3425	0.3187	0.2947	0.2707
26	0.2465	0.2223	0.1980	0.1735	0.1490	0.1245	0.0998	0.0749	0.0501	0.0251
27	0.0000	−0.0252	−0.0504	−0.0757	−0.1012	−0.1267	−0.1523	−0.1781	−0.2039	−0.2298
28	−0.2558	−0.2818	−0.3080	−0.3343	−0.3606	−0.3870	−0.4136	−0.4402	−0.4669	−0.4937
29	−0.5206	−0.5476	−0.5746	−0.6018	−0.6290	−0.6564	−0.6837	−0.7113	−0.7388	−0.7665
30	−0.7943	−0.8222	−0.8501	−0.8781	−0.9063	−0.9345	−0.9628	−0.9911	−1.0196	−1.0482
31	−1.0768	−1.1056	−1.1344	−1.1632	−1.1922	−1.2213	−1.2504	−1.2797	−1.3090	−1.3385
32	−1.3679	−1.3975	−1.4272	−1.4569	−1.4868	−1.5167	−1.5467	−1.5768	−1.6070	−1.6372
33	−1.6675	−1.6980	−1.7284	−1.7590	−1.7897	−1.8205	−1.8513	−1.8822	−1.9132	−1.9443
34	−1.9754	−2.0067	−2.0380	−2.0695	−2.1009	−2.1325	−2.1642	−2.1959	−2.2277	−2.2596
35	−2.2916	−2.3237	−2.3558	−2.3880	−2.4203	−2.4527	−2.4852	−2.5177	−2.5503	−2.5830
36	−2.6158	−2.6487	−2.6816	−2.7146	−2.7477	−2.7809	−2.8141	−2.8474	−2.8809	−2.9144
37	−2.9479	−2.9816	−3.0153	−3.0491	−3.0830	−3.1169	−3.1509	−3.1851	−3.2192	−3.2535
38	−3.2878	−3.3223	−3.3567	−3.3913	−3.4259	−3.4607	−3.4955	−3.5303	−3.5653	−3.6004
39	−3.6354	−3.6706	−3.7059	−3.7412	−3.7766	−3.8121	−3.8476	−3.8833	−3.9190	−3.9547
40	−3.9906	−4.0265	−4.0625	−4.0986	−4.1347	−4.1709	−4.2072	−4.2436	−4.2800	−4.3165
41	−4.3531	−4.3897	−4.4265	−4.4633	−4.5002	−4.5371	−4.5741	−4.6112	−4.6484	−4.6856

Table 4.13 Correction in cm^3

Added to nominal capacity of the measure to obtain the volume of water at reference temperature when a 1 dm3 measure is used at a temperature other than its reference temperature = 27 °C ALHPA (α) = 15 × 10^{-6}/°C

Temp.	0.0	0.1	0.2	0.3	0.4	0.5	0.6	0.7	0.8	0.9
5	3.1319	3.1317	3.1314	3.1309	3.1303	3.1295	3.1285	3.1274	3.1262	3.1248
6	3.1233	3.1216	3.1197	3.1177	3.1156	3.1133	3.1109	3.1083	3.1056	3.1026
7	3.0996	3.0964	3.0931	3.0896	3.0860	3.0823	3.0784	3.0743	3.0701	3.0658
8	3.0613	3.0567	3.0520	3.0471	3.0420	3.0368	3.0315	3.0260	3.0204	3.0147
9	3.0088	3.0028	2.9966	2.9903	2.9839	2.9774	2.9707	2.9638	2.9568	2.9497
10	2.9425	2.9351	2.9276	2.9199	2.9121	2.9042	2.8962	2.8880	2.8797	2.8713
11	2.8627	2.8539	2.8451	2.8361	2.8270	2.8178	2.8084	2.7990	2.7893	2.7796
12	2.7697	2.7597	2.7496	2.7394	2.7290	2.7185	2.7078	2.6971	2.6862	2.6751
13	2.6640	2.6528	2.6414	2.6299	2.6182	2.6065	2.5946	2.5827	2.5705	2.5583
14	2.5459	2.5334	2.5208	2.5081	2.4953	2.4823	2.4692	2.4560	2.4427	2.4292
15	2.4156	2.4019	2.3881	2.3743	2.3602	2.3460	2.3318	2.3174	2.3029	2.2882
16	2.2735	2.2587	2.2437	2.2286	2.2135	2.1981	2.1827	2.1671	2.1515	2.1357
17	2.1198	2.1038	2.0878	2.0716	2.0552	2.0387	2.0222	2.0056	1.9887	1.9719
18	1.9549	1.9378	1.9205	1.9032	1.8858	1.8682	1.8506	1.8328	1.8149	1.7969
19	1.7788	1.7607	1.7424	1.7239	1.7054	1.6868	1.6681	1.6492	1.6303	1.6112
20	1.5921	1.5728	1.5534	1.5339	1.5144	1.4947	1.4749	1.4550	1.4350	1.4149
21	1.3947	1.3744	1.3539	1.3334	1.3128	1.2920	1.2713	1.2503	1.2293	1.2082
22	1.1869	1.1656	1.1442	1.1226	1.1010	1.0792	1.0574	1.0355	1.0134	0.9913
23	0.9690	0.9467	0.9243	0.9018	0.8791	0.8564	0.8336	0.8106	0.7876	0.7645
24	0.7412	0.7179	0.6945	0.6710	0.6474	0.6237	0.5999	0.5759	0.5520	0.5279
25	0.5036	0.4794	0.4550	0.4305	0.4060	0.3813	0.3565	0.3317	0.3067	0.2817
26	0.2565	0.2313	0.2060	0.1805	0.1550	0.1295	0.1038	0.0779	0.0521	0.0261
27	0.0000	–0.0262	–0.0524	–0.0787	–0.1052	–0.1317	–0.1583	–0.1851	–0.2119	–0.2388
28	–0.2658	–0.2928	–0.3200	–0.3473	–0.3746	–0.4020	–0.4296	–0.4572	–0.4849	–0.5127
29	–0.5405	–0.5686	–0.5966	–0.6247	–0.6530	–0.6813	–0.7097	–0.7382	–0.7668	–0.7955
30	–0.8243	–0.8532	–0.8821	–0.9111	–0.9402	–0.9694	–0.9987	–1.0281	–1.0576	–1.0872
31	–1.1168	–1.1465	–1.1763	–1.2062	–1.2362	–1.2663	–1.2964	–1.3267	–1.3570	–1.3874
32	–1.4179	–1.4485	–1.4791	–1.5098	–1.5407	–1.5716	–1.6026	–1.6337	–1.6649	–1.6961
33	–1.7274	–1.7589	–1.7903	–1.8219	–1.8536	–1.8854	–1.9172	–1.9491	–1.9811	–2.0132
34	–2.0453	–2.0775	–2.1099	–2.1423	–2.1747	–2.2074	–2.2400	–2.2727	–2.3055	–2.3384
35	–2.3714	–2.4045	–2.4376	–2.4708	–2.5041	–2.5375	–2.5709	–2.6045	–2.6381	–2.6718
36	–2.7055	–2.7394	–2.7733	–2.8073	–2.8414	–2.8756	–2.9098	–2.9441	–2.9785	–3.0131
37	–3.0476	–3.0822	–3.1169	–3.1517	–3.1866	–3.2216	–3.2566	–3.2917	–3.3269	–3.3621
38	–3.3974	–3.4329	–3.4683	–3.5039	–3.5395	–3.5753	–3.6110	–3.6469	–3.6828	–3.7189
39	–3.7549	–3.7911	–3.8274	–3.8637	–3.9001	–3.9366	–3.9731	–4.0097	–4.0464	–4.0832
40	–4.1200	–4.1569	–4.1939	–4.2310	–4.2681	–4.3053	–4.3426	–4.3800	–4.4174	–4.4549
41	–4.4924	–4.5301	–4.5678	–4.6056	–4.6435	–4.6814	–4.7194	–4.7575	–4.7956	–4.8339

Table 4.14 Correction in cm^3

Add to nominal capacity of the measure to obtain the volume of water at reference temperature when a 1000 cm^3 measure is used at a temperature other than its reference temperature = 27 °C ALHPA (α) = 10×10^{-6}/°C

Temp.	0.0	0.1	0.2	0.3	0.4	0.5	0.6	0.7	0.8	0.9
5	3.2422	3.2416	3.2407	3.2398	3.2387	3.2373	3.2359	3.2343	3.2326	3.2307
6	3.2286	3.2264	3.2241	3.2216	3.2189	3.2161	3.2132	3.2101	3.2069	3.2035
7	3.2000	3.1962	3.1925	3.1885	3.1844	3.1801	3.1757	3.1712	3.1665	3.1617
8	3.1567	3.1515	3.1463	3.1409	3.1353	3.1296	3.1238	3.1178	3.1117	3.1055
9	3.0991	3.0926	3.0859	3.0791	3.0722	3.0652	3.0580	3.0506	3.0431	3.0355
10	3.0277	3.0199	3.0119	3.0037	2.9954	2.9870	2.9785	2.9697	2.9609	2.9520
11	2.9429	2.9337	2.9243	2.9149	2.9053	2.8956	2.8857	2.8757	2.8656	2.8553
12	2.8450	2.8345	2.8238	2.8131	2.8022	2.7912	2.7800	2.7688	2.7574	2.7458
13	2.7342	2.7225	2.7106	2.6986	2.6864	2.6742	2.6618	2.6493	2.6367	2.6240
14	2.6111	2.5981	2.5850	2.5718	2.5584	2.5449	2.5313	2.5177	2.5038	2.4898
15	2.4758	2.4616	2.4473	2.4329	2.4184	2.4037	2.3889	2.3740	2.3590	2.3439
16	2.3286	2.3133	2.2978	2.2822	2.2666	2.2507	2.2348	2.2188	2.2026	2.1863
17	2.1700	2.1534	2.1369	2.1202	2.1033	2.0863	2.0693	2.0522	2.0348	2.0175
18	2.0000	1.9824	1.9646	1.9468	1.9288	1.9108	1.8927	1.8744	1.8560	1.8375
19	1.8189	1.8003	1.7814	1.7625	1.7435	1.7243	1.7051	1.6858	1.6664	1.6468
20	1.6271	1.6073	1.5875	1.5675	1.5474	1.5272	1.5069	1.4866	1.4661	1.4454
21	1.4247	1.4039	1.3830	1.3620	1.3408	1.3196	1.2983	1.2769	1.2553	1.2337
22	1.2120	1.1901	1.1682	1.1461	1.1240	1.1018	1.0794	1.0570	1.0345	1.0118
23	0.9891	0.9662	0.9433	0.9203	0.8971	0.8739	0.8506	0.8271	0.8036	0.7800
24	0.7562	0.7324	0.7085	0.6845	0.6604	0.6362	0.6119	0.5875	0.5630	0.5384
25	0.5136	0.4889	0.4640	0.4390	0.4140	0.3888	0.3635	0.3382	0.3127	0.2872
26	0.2615	0.2358	0.2100	0.1840	0.1580	0.1320	0.1058	0.0794	0.0531	0.0266
27	0.0000	–.0267	–.0534	–.0802	–.1072	–.1342	–.1613	–.1886	–.2159	–.2433
28	–.2708	–.2983	–.3260	–.3538	–.3816	–.4095	–.4376	–.4657	–.4939	–.5222
29	–.5505	–.5791	–.6076	–.6362	–.6650	–.6938	–.7227	–.7517	–.7808	–.8100
30	–.8393	–.8686	–.8981	–.9276	–.9572	–.9869	–1.0167	–1.0466	–1.0766	–1.1066
31	–1.1368	–1.1670	–1.1973	–1.2277	–1.2581	–1.2887	–1.3194	–1.3501	–1.3809	–1.4118
32	–1.4428	–1.4739	–1.5051	–1.5363	–1.5676	–1.5990	–1.6306	–1.6622	–1.6938	–1.7255
33	–1.7574	–1.7893	–1.8213	–1.8533	–1.8855	–1.9178	–1.9501	–1.9825	–2.0150	–2.0476
34	–2.0802	–2.1130	–2.1458	–2.1787	–2.2117	–2.2448	–2.2779	–2.3111	–2.3444	–2.3778
35	–2.4113	–2.4448	–2.4785	–2.5122	–2.5460	–2.5798	–2.6138	–2.6479	–2.6820	–2.7161
36	–2.7504	–2.7848	–2.8192	–2.8537	–2.8883	–2.9229	–2.9577	–2.9925	–3.0274	–3.0624
37	–3.0974	–3.1326	–3.1678	–3.2030	–3.2384	–3.2739	–3.3094	–3.3450	–3.3807	–3.4164
38	–3.4522	–3.4882	–3.5241	–3.5602	–3.5963	–3.6326	–3.6688	–3.7052	–3.7416	–3.7782
39	–3.8147	–3.8514	–3.8881	–3.9249	–3.9618	–3.9988	–4.0358	–4.0730	–4.1101	–4.1474
40	–4.1847	–4.2221	–4.2596	–4.2972	–4.3348	–4.3725	–4.4103	–4.4481	–4.4860	–4.5241
41	–4.5621	–4.6002	–4.6385	–4.6768	–4.7151	–4.7536	–4.7920	–4.8306	–4.8692	–4.9080

REFERENCES

[1] CRC, Handbook of Physics and Chemistry, 1996-97, **12**, 172.
[2] International Critical Tables vol 5, page 54.
[3] Hill and Hargood Ash, 1919, NBS Scientific papers, 5B, 90,438.
[4] Thomas Ferguson, 1917, Proc of Royal Society Edinburgh **3**, 34, 308.
[5] Proc. Roy. Soc. London A, 1908, 506, **36**, 24.
[6] Sutton, 1907, Proc. Phys. Soc. London, **117**, 11, 137.
[7] Marvin, 1909, Proc. Russian Physico Chemical Society, 506, **37**: 57.
[8] Himus and Hinchy, 1924, Chemistry and Industry, **43**, 840.
[9] Becker, 1917, Phil Magazine and Journal of Science, Lond. **17**, 241, 23.

5
CHAPTER

VOLUMETRIC GLASSWARE

5.1 INTRODUCTION

Very large amount of glassware is used in Pharmaceutical research and Chemical technology. Volumetric glassware is extensively used also in chemical industry and any industry involved with chemical analysis. It is equally important for those involved in technological research doing physical measurement of volume and capacity, education and training for understanding and practising the volumetric analysis and titration. Glassware consists of pipettes, burettes, one mark volumetric flask, measuring cylinders, and micropipettes. It is important that such apparatus is of assured accuracy. In order to achieve consistent results, all glassware should be accurate, or at least its inaccuracy must be known. Every National Metrology Laboratory, like in India we have National Physical Laboratory at New Delhi caters this important need of industry, by way of calibrating all types of glassware as per National and International Standards Specifications. In formative days of volumetric industry, NPL also rendered the service of providing reference standards of capacity and volume. For example 25 dm^3 measures were fabricated, adjusted, calibrated and were supplied to many user industrial concerns. NPL also used to fabricate, graduate and supply some special instruments and standard equipment like automatic pipettes, which deliver automatically a pre-assigned volume.

5.1.1 Facilities at NPL for Calibration of Volumetric Glassware

NPL has facilities to calibrate volumetric measures from 25 dm^3 down to a few μ*l*. Triple distilled water is used for determination of capacity of all but for micropipettes and micro-volumetric measures. For micropipettes and like, freshly distilled mercury is used for determination of capacity. All types of glass measures like burettes, pipettes, flasks, measuring cylinders, graduated pipettes and butyrometers are received for calibration. Graduating of special type of glass measures, like 0.3125 cm^3 automatic pipettes used in the calibration of butyrometers is also undertaken. Uncertainty in measurement is 0.01%. NPL has its own primary standard in the form of quartz sphere and several spheres in zerodur. These spheres are also used to establish the density of de-ionised water taken from the tapes in its premises.

5.1.2 Special Volumetric Equal-arm Balances

Special balances with double platform pans are used for calibrating the volumetric measures. These balances have two pans one above the other on each side. The measure to be calibrated is placed on the lower right-hand pan and standard weights on the upper pan. A similar measure to compensate for the variation in environmental conditions is kept on the lower left-hand pan, while counterpoise weights are kept on the upper left-hand pan. Sufficient vertical distance is kept in between the two pans, so that even tall measuring cylinders received for calibration, can be easily accommodated on the lower pan. These balances have the advantage of keeping the weights and volumetric measures centrally on the pans with ease and keep the weights out of contact with water. The balances are checked periodically for their continued satisfactory performance and for the mass value of the smallest graduated interval on the scale. The weights used are calibrated periodically against NPL standards of mass.

5.2 VOLUMETRIC GLASSWARE

In the previous chapter, we have discussed the classification of volumetric measures. Measures are of two types namely (*i*) Content type and (*ii*) Delivery type. Delivery type measures must be capable to deliver the same amount of liquid every time the measure is used under the specified conditions.

Necessary condition for obtaining consistent results with a delivery measure is that when measure is emptied, its interior surface remains wetted with uniformly distributed film of liquid. The liquid, in no condition, should collect together to form a drop.

The measures considered in this and next chapters are:

A. One mark flasks
B. One mark pipettes
C. Graduated pipettes
D. Serological pipettes
E. 44.7 µl content type pipettes
F. Piston operated pipettes and burette in mm^3 range
G. Disposable glass micro pipettes
H. Micro-volumetric vessels, flasks and centrifuge tubes
I. Burettes
J. Micro-burettes
K. Measuring cylinders

5.3 CLEANING OF VOLUMETRIC GLASSWARE

In order to achieve the full efficacy of a volumetric measure, its cleaning and keeping it in cleaned condition is vital [1, 2]. Normal source of un-cleanliness is minute amount of grease, which sticks to inside the measure and is very difficult to remove. The use of specific cleaning agents depends upon the end use of the measure. Methods of cleaning and cleaning agents are as follows:

1. Obvious loose contamination is removed mechanically by brushing and shaking the measure.
2. A good cleaning may be achieved by using aqueous solutions of soap-less detergent. The measure is nearly filled with it and well shaken.

3. If the measure is not required for immediate use, it is left for overnight filled with hot mixture of sulphuric acid and saturated solution of Potassium bi- chromate. Next day morning, the measure is emptied and rinsed with distilled water. The emptied mixture is kept safely for future use. The mixture is highly corrosive in nature so all precautions are required for handling.
4. Alcoholic solution of caustic soda may also be similarly employed.
5. Freshly prepared potassium permanganate solution in sulphuric acid is used for quicker results.
6. Sometimes fuming sulphuric acid is also used for rapid results.
7. A good rapid method of cleaning is to shake vigorously a little absolute alcohol and then empty it, allow a small time to drain off, then shake a little strong nitric acid in the measure and wash it thoroughly with water.
8. For obstinate stains, strong soap solution, hot if necessary has proved to be good alternative cleaning agent. Modern day detergents may also be profitably used. But only problem with these detergents is that even after repeated washings, minute traces of detergent have been found in the measure. The main effect of these micro-residues is lowering of the surface tension of water and hence reducing the meniscus volume.
9. Measures used with mercury, develop black stains, which are difficult to remove by ordinary cleaning agents. Zinc dust and dilute hydrochloric acid when shaken reduces the stain to mercury, which then is removed by dissolving it in nitric acid.
10. Filling partially the measure with water and a number of tiny pieces of filter paper and shaking the measure vigorously does a good mechanical cleaning.
11. For plastic ware, do not use any cleaning agent, which may attack, discolour or swell it.

5.3.1 Precautions in use of Cleaning Agents

All the aforesaid cleaning agents like sulphuric acid, hydrochloric acid and nitric acid are highly corrosive in nature and burn the skin deep if comes in its contact. Even their fumes are pungent and may create giddiness and headache. All precautions are taken to avoid their direct contact with body or inhalation. Always go close to them with surgical or rubber gloves and apparels. Proper ventilation of the room or use of fume chamber is also necessary. Maintaining a small first aid kit is recommended.

5.3.2 Cleaning of Small Volumetric Glassware

For cleaning of small articles such as pipettes, micropipettes etc., it is easier to fill them with a cleaning agent by suction. Suction is done by using of a water vacuum pump or a rubber bulb but never by mouth. Cleaning agent should be passed through it several times until the inside surface is evenly coated with the cleaning agent. For cleaning flask, pour cleaning agent while rotating the flask slowly so that a film of cleaning agent covers entirely the inside surface. For filling a burette, it should be fitted in vertical position and filled by pouring the cleaning solution through a funnel from above. Due care should be taken that its stopcock is closed and cleaning agent does not spill over. Open the stopcock to drain the cleaning solution. Repeat the process several times, till the inside surface is uniformly coated with the cleaning agent. If necessary, the small articles after filling with the required cleaning solution may be left over night in a long cylinder, containing the cleaning agent. One advantage of this method is that the outer surface of ware also gets cleaned. This is useful for the content type volumetric ware. As to calibrate content measure whole measure is weighed, so if outer surface is clean then only the variation due to change in environmental conditions and handling will have minimal effect. To remove a cleaning agent, thoroughly rinse with tap water several times and finally with distilled water.

Plastic volumetric ware is also similarly cleaned with proper cleaning agents taking in precautions mentioned in the point 11 of section 5.3.

5.3.3 Delivery Measure kept filled with Distilled Water

If the measure is for delivery, then after cleaning and thorough washing and final rinsing with distilled water, the measure is kept ready for use after filling it with distilled water. All delivery measures should be kept filled with distilled water even if they are not in use.

5.3.4 Drying of a Content Measure

All content measures are dried before use. A very quick method is to rinse the measure with pure absolute alcohol and then with ether. The cold dry air may be drawn through the measure. The air should be clean and dry, so air is made to pass through a proper filter to ensure absence of oil or dirt. If cost of alcohol prohibits its use then Acetone is its substitute. The outer surface of the measure should also be properly cleaned and dried.

5.3.5 Test of Cleanliness

In order to get a good idea about the cleanliness of a delivery type measure like burette or pipette, clamp it vertically. Fill it slowly with distilled water through its delivery jet. A well-formed meniscus should be visible, which should be rising at a constant speed without any deformation of its shape. A change in the shape of the meniscus at a point indicates dirt or not cleaned portion of the measure at that point. One can also see clearly the movement of a thin film of water travelling ahead of the main water surface. In a perfect clean measure, the front edge of this film appears to be rising at the same rate as the main surface of water i.e. keeping the same constant distance in front of it. Should the measure be slightly dirty at any point, the front edge of the film gets retarded and water meniscus overtakes it and front edge of the film crinkles. Similar phenomenon is observed when water is passed through the contaminated surface. A pipette or burette when filled up through its delivery jet, and in which the front edge of the film keeps on advancing with constant speed and without any deformation may be relied upon for its cleanliness. It also ensures that when filled with water and emptied, it will have a uniform film of water left over through out its surface. The advantage of knowing for certainty about the cleanliness of a delivery measure is that it may be emptied without fear of error due to irregular wetting of the walls.

5.4 READING AND SETTING THE LEVEL OF MENISCUS

5.4.1 Convention for Reading

Universally adopted convention is that in case of all transparent liquids the tangent at the lower edge of the meniscus is taken as reference. For opaque liquids making a convex meniscus it is the tangent at the upper edge of the meniscus and those liquids making concave meniscus but are not transparent like $KMnO_4$ and milk, it is top rim of the meniscus, which are taken as reference. However in regard to which part of the graduation mark, the meniscus should touch, in general, there are two methods of setting:

1. Set out the lowest part of concave meniscus tangential to upper most part of the graduation mark and upper most part of the convex meniscus to the lower most edge

of the graduation mark. In case of opaque liquids having concave meniscus, set it out such that rim of the meniscus is in the central part of the graduation mark.

2. The position of the lowest point of the meniscus with reference to the graduation mark is such that it is in the plane of the middle of the graduation mark. The position of the meniscus is obtained by making the setting in the centre of the ellipse formed by the front and back portions of the graduation mark as observed by having the eye slightly below the plane of the graduation mark.

The difference between the two methods of setting the graduation marks as horizontal tangent to the meniscus amounts to a difference in volume equal to the product of the area of cross-section at the air liquid interface and half the thickness of graduation mark. Normally the thickness of the graduation marks is one fifth of the height of the cylinder at the air liquid interface having a volume equal to the maximum permissible error. So the difference due to two methods of setting would be only one tenth of the maximum permissible error.

For opaque liquids, readings are taken where the liquid meets the wall of the tube i.e. at the contact circle. Necessary correction due to surface tension is applied for opaque liquids.

5.4.2 Method of Reading

When water or similar meniscus is viewed in daylight or in ordinary illuminations, reflection and refraction at the glass water surfaces render the exact location of the lowest point of the meniscus rather difficult. A very simple device is used to overcome this problem.

A strip of black tough paper is folded round the neck of the measure. The upper edge of the strip is cut clean and straight and is kept a little below say 1 mm below the meniscus. The strip is held in such a way that the top edges of the two ends of strip where they meet after encircling the neck is exactly on the mark. The strip then is held in position through a gem clip. When so shaded meniscus is viewed against the white background, the bottom of the meniscus becomes quite dark and its outline is sharply defined against the white background. The Figure 5.1A shows the water meniscus viewed against a white background without any supplementary device. The Figure 5.1B shows the same meniscus shaded by a device described above. The difference in clarity of the outline of the meniscus is evident from the two figures and becomes more evident in actual practice.

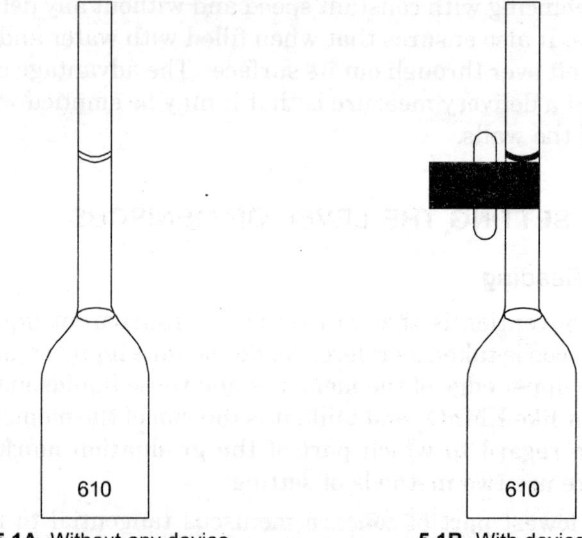

5.1A Without any device **5.1B** With device

Figure 5.1. Advantage of shading device

The above-mentioned device gives better and highly satisfactory results than placing a black and white screen behind the measure.

For mercury or for any convex meniscus, the lowest edge of the black strip is kept about 1 mm above the meniscus.

This device becomes extra useful in cases where the graduation marks are not complete circles, for example all the marks on a class B burette are not complete circles. In those cases, placing the eye, so that the top edge of the strip on the front of the burette coincides with the top edge of the strip at back of the burette, eliminates practically all the error due to parallax.

5.4.3 Error due to Meniscus Setting

Let the diameter of the cylindrical tube at the meniscus be d mm. If δh is the error in setting, then δV the volume difference due to setting is given by

$$\delta V = \pi d^2 \delta h/4 \text{ mm}^3$$

Taking δh as 0.2 mm, the error in volume δV is given by

$$\delta V = (3.1416).d^2(0.2)/4 = 0.15\,708 d^2 \text{ mm}^3$$

As the error in volume is proportional to the square of the diameter, so the diameter at the graduation mark for all volumetric ware is specified.

The error in volume corresponding to the neck diameters of the flask suggested in ISO [1] are given in the table 5.1.

Table 5.1 Error in Volume mm^3

Neck dia. in mm	6	8	10	12	14	16	18	20	25	30	35	40
δh in mm	0.2	0.2	0.2	0.2	0.2	0.3	0.3	0.5	0.5	0.5	1	1
δV mm^3	5.65	10.1	15.7	22.60	30.80	42.0	50.9	62.8	98.2	141	192	251
Capacity cm^3	10	25	50	100	250	500	1000	1500	2000	3000	4000	5000
% Error	0.06	0.04	0.03	0.02	0.01	.012	.075	.01	.012	.062	.025	.025

5.5 FACTORS INFLUENCING THE CAPACITY OF A MEASURE

5.5.1 Temperature

Due to expansion property of material, the capacity of a volumetric measure varies with the temperature. Therefore, it is necessary to define the temperature at which its capacity is intended to be correct. In India the capacity of a measure is referred to at a temperature of 27 °C while in Britain the reference temperature is 20 °C. For a change of 1 °C, for vessels of soda glass, the change in capacity is 30 parts per million while in the case of borosilicate glass, the change is only 10 parts per million.

5.5.2 Delivery Time and Drainage Time

The delivery time is defined, as the time required in delivering the total quantity of water for which a measure is calibrated when it is emptied in a specified manner. When the meniscus is set at the graduation line and liquid is allowed to flow, the instant it starts is the beginning of the delivery time. If the motion of the liquid surface down the delivery tube of a pipette is observed it will be seen that the liquid surface comes to rest a little above the bottom end of the of the delivery tube. The instant at which the liquid surface comes to rest can be noted fairly definitely and is taken as the end of delivery time and beginning of the drainage time. The time elapsed from this instant to the instant, receiving vessel is taken away from the pipette or the stopcock of the burette is closed, is the drainage time.

In general, when the contents of a delivery vessel are discharged, a residual film of the liquid adheres to the surface of the vessel. As a result, the volume of liquid delivered is less than the volume contained, by an amount equal to the volume of the liquid film, which remains adhered to the walls. The volume of this film, which we may call as V_w, naturally depends upon the delivery time. The volume of the film decreases up to a certain limit, as the time of delivery is increased. So if the delivery time is increased delivered volume will increase only up to certain limit, after which the increase in delivery time will not increase the volume delivered. The delivery time, for which volume delivered reaches its maximum, let us call it limiting delivery time (LDT). Further it has been observed that volume delivered will be consistent if the delivery time is more than a certain minimum time (MDT). That is type A error will be less. So for a delivery measure minimum and maximum delivery times are specified.

Quite often the difference in volume delivered in MDT and LDT will be of the order of maximum permissible error allowed on that delivery measure.

5.5.3 Delivery Time and Drainage Volume for a Burette

A study [5,6] was performed from 1920 to 1923 by taking a 50-cm^3 burette with scale length of 534 mm and detachable stopcocks with varying orifice diameter so that delivery time for water in seconds was 206, 152, 106, 74, 56, 37, and 20.

Delivery time of a burette is defined as the time taken for the delivery of water from the zero mark to the full capacity mark of the burette under free flow conditions i.e. stopcock is fully open and the burette is in vertical position.

Experimental Observations

The burette was mounted in front of a reading telescope with a micrometer eye-piece having a moveable horizontal cross wire. The telescope was focused on 50 cm^3 mark in such a way that some part of the burette was visible. The burette was in vertical position and each time filled from below with distilled water up to the level a few mm above zero graduation line. Water was run out from the burette very slowly till the lowest part of the meniscus touched the zero line. The burette was allowed to empty freely until water meniscus reached in between 49.9 cm^3 and 50 cm^3 marks and delivery time was noted and the stopcock was turned off as soon as the level reaches 50 cm^3 mark. The moving cross wire, just after the outflow stopped, was set at the lowest point of the meniscus as soon possible. The time taken was generally within 5 to 10 seconds. The micrometer reading was taken at frequent intervals over a period of about 30 minutes.

The rise of the meniscus level equivalent to 0.1 cm^3 was equivalent to 650 divisions of the micrometer scale. Settings were repeatable within a few divisions so that a very small change in the position of the meniscus could be accurately observed.

For better visibility of the meniscus, a cleanly cut black paper was wrapped round the burette a few mm below the 50 cm^3 mark. With this arrangement, it was found that the measured rise of meniscus was in excess of 0.01 cm^3 than actual rise at maximum drainage volume of 0.24 cm^3. This was because of certain distance between bottom of the meniscus and edge of the paper. Necessary corrections were applied to all the measurements made.

The results of observations plotted are shown in Figure 5.2. The amount of water drained from the walls is on Y-axis and the time in seconds is on the horizontal axis. The drained time is reckoned from the instant the water flow is ceased. The delivery time given by the jets used for each set of observations is shown against each mean curve.

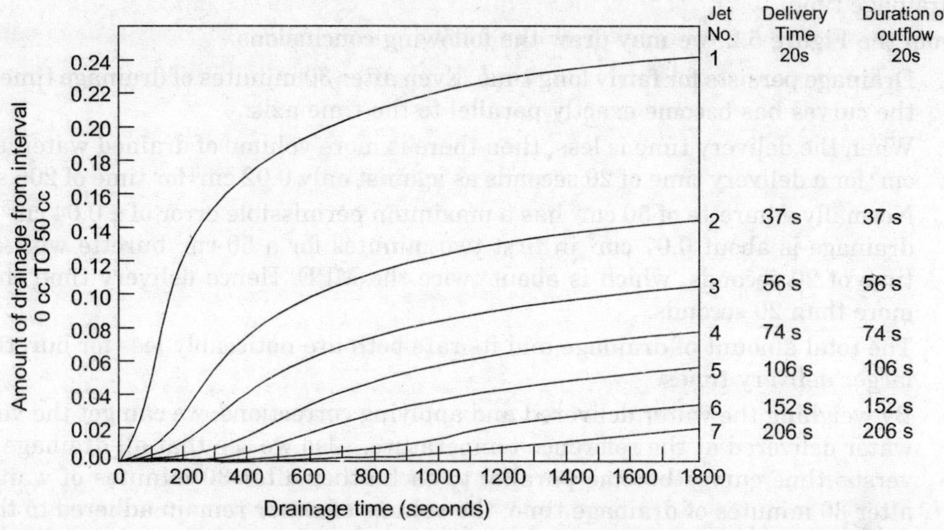

Figure 5.2 Drained volume versus drainage time for a burette

The curves were smooth line drawn through the plotted observations. Because of the large number of observations 619 in number the actual observations were not marked on the curves. The Table 2 [5] gives a fairly good idea about the accuracy obtained.

Table 5.2

Delivery Time seconds	*Number of Sets of Observations*	*Number of Observations in each set*	*Greatest departure From the mean curve in cm³*	*Greatest departure From mean curve at the end of 5 minutes in cm³*
1	2	3	4	5
20	3	150	0.008	0.004
37	3	115	0.006	0.005
56	2	76	0.005	0.002
74	2	61	0.003	0.0008
106	2	41	0.005	0.001
152	2	43	0.003	0.000
206	2	33	0.002	0.0005

The value given in the 4th column represents to the greatest difference between an actual observed value and the corresponding value on the mean curve. It was also reported that, in all cases greatest departure of the observations had occurred at the extreme right hand of the curves, that is for about thirty minutes drainage.

The values in the 5th column represent the greatest departure of the observational curves from the mean curves at the ordinate of 5 minutes interval. These values are considerably less than those given in the 4th column. This is because the lines joining actual observations in a given set formed fairly smooth curve and curves for the independent sets for any particular delivery time agreed closely over their initial portions and diverged somewhat toward the end of the drainage time.

From the Figure 5.2, we may draw the following conclusions:

1. Drainage persists for fairly long time. Even after 30 minutes of drainage time none of the curves has become exactly parallel to the time axis.
2. When the delivery time is less, then there is more volume of drained water, say 0.24 cm^3 for a delivery time of 20 seconds as against only 0.02 cm^3 for time of 206 seconds.
3. Normally a burette of 50 cm^3 has a maximum permissible error of ± 0.04 cm^3 but the drainage is about 0.07 cm^3 in first two minutes for a 50-cm^3 burette with delivery time of 20 seconds, which is about twice the MPE. Hence delivery time should be more than 20 seconds.
4. The total amount of drainage and its rate both are noticeably less for burettes with larger delivery times.
5. By weighing the water delivered and applying corrections, we can get the volume of water delivered at the reference temperature. Also we see that all drainage volume versus time curves become parallel to each other after 30 minutes of waiting, i.e. after 30 minutes of drainage time, the volume of water remain adhered to the walls of the burette is same. Hence volume of water delivered plus the drained volume for corresponding delivery times should be constant. So the difference between the ordinates of any two curves at 30 minutes (sufficiently long time) is taken as difference of volume of water initially left on the walls of the burette at the beginning of the drainage time for the two curves in question. For example the difference in ordinates of the curves marked 20 seconds and 206 seconds respectively after thirty minutes of drainage is 0.22_1 cm^3. In other words when a burette is emptied in 20 seconds then, 0.22_1 cm^3 more water remain adhered to the walls of the burette than when emptied in 206 seconds. Obviously, therefore, when the burette is emptied in 20 seconds 0.22_1 cm^3 of less water would be delivered than when the same burette is emptied in 206 seconds. If we find out volume of water delivered by the burette with 206 seconds of delivery time, then we can deduce the volume delivered by the same burette having delivery time of 20 seconds. Similarly we can deduce the volume of water delivered by the same burette with any other delivery time. We can also determine the volume of water delivered by the same burette with different delivery times directly by using the gravimetric method. So we have two sets of values of volume delivered (1) by measurement and (2) the other by deduction, these values have been plotted in the graph shown in Figure 5.3. Crosses represent measured volume of water delivered while circles represent the deduced values of water volume, which the burette would have delivered with the same delivery time. Small differences between the corresponding values ensure the efficacy of the two methods.
6. From the curve in Figure 5.3, one can find out the minimum and maximum values of delivery time so that the burette can delivery the volume of water well with in the prescribed tolerance limits. Similar exercise could be carried out with burettes of different capacities and scale length to arrive at logically derived values of delivery times. This work was done long back at the National Physical Laboratory, U.K.

So we have seen that, on opening the stopcock of a burette, besides normal flow of liquid, some liquid remain adhering to the walls of the burette, which slides down the walls. So with increase in drainage time the volume delivered also increases. In the initial period of drainage time the delivered volume will increase more rapidly and becomes constant for all delivery times if the drainage times increases to 5 minutes (300 s) or more. The quantity of liquid adhered to walls depends upon the delivery time. It has been observed that the volume of liquid delivered is less for smaller delivery time but the adhered volume increases with the decrease in delivery time. So due to the drainage, liquid slides down the walls of the burette, the reading obtained on its scale increases with time elapsed after the stopcock is closed. Naturally the volume of liquid adhered will depend upon the viscosity of the liquid.

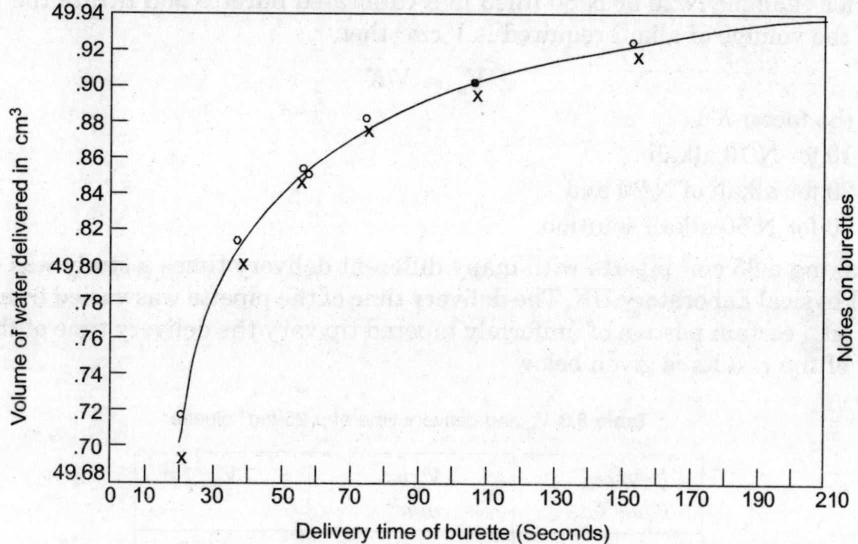

Figure 5.3 Volume delivered versus delivery time

5.5.4 Volume Delivered and Delivery Time of Pipettes

To take full advantage of a pipette, one has to take in account,

(1) the difference in delivery time inscribed and actual time of delivery,

(2) last drop, which remains attached to the tip of the jet of pipette and

(3) finally the meniscus adjustment on the graduated line.

The point (3)-meniscus adjustment will not pose problem if the graduated line, encircles the suction tube, is thin enough and properly coloured, and a proper reading method with a suitable device is used.

For point (2)-last drop of liquid, the convention, in general, is to take it as a part of volume delivered. To take the last drop, the receiving vessel is touched at an angle to the tip. One should not blow out air to take the last drop or keep the pipette in touch with the liquid in the receiving vessel during drainage time.

As regards point (1), the volume of the fluid remaining adhered to the walls depends upon several factors like:

- The delivery time

- The diameter of the pipette and
- Length of the pipette from tip to the graduation line.

To establish an empirical relation between V_w and above factors, a good deal of work has been done in the early part of the last century [6].

5.5.4.1 Determination of Adhered Volume

To determine the volume V_w of water remain adhered to walls of the pipette, easiest way is to use N/1(normal) hydrochloric acid and let it be delivered under the stated conditions. Remove the last drop of acid from its tip by touching it with the wall of the receiving vessel. Make the pipette horizontal and rinse it at least five times with double distilled water, collect all the rinsed water with out loosing any drop of it. Take carbon dioxide free alkali of strength N/10 or even less for example N/20 or N/50 filled in a calibrated burette and titrate the rinsed water with it. If the volume of alkali required is V cm^3 then

$$V_{w'} = V/K$$

the factor K is
10 for N/10 alkali
20 for alkali of N/20 and
50 for N/50 alkali solution.

By taking a 25 cm^3 pipette with many different delivery times a study was conducted at National Physical Laboratory UK. The delivery time of the pipette was varied from 2 s to 360 s. Chipping off a certain portion of uniformly tapered tip vary the delivery time of the pipette. A summary of the results is given below

Table 5.3 V_w and delivery time of a 25-cm^3 pipette

Delivery Time(T)s	V_w *in mm*^{3}	$V_w \sqrt{T}$
1.4	410	485
5.0	268	598
29.0	118	635
360.0	31	588

From the third column of the table, it is clear that the product of the volume V_w and square root of the delivery time is practically the same. This concludes that volume of water remained adhered to the walls of the pipette is inversely proportional to the square root of the delivery time, giving.

$$V_w = F/\sqrt{T}$$

F is a constant of proportionality

The following was also shown in the same study [6]

1. That standard deviation of the volume delivered σ_w was about $V_w/20$
2. Also by using several pipettes of different capacities, it was also shown that standard deviation of the delivered volume was

   (*a*) $\sigma_w = 0.13\sqrt{V}$, where σ_w and V both are in mm^3.

   (*b*) $\sigma_w = k/\sqrt{T}$ and

(*c*) $\sigma_w = 2.2\ V_w/(\sqrt{tD})$, where *D* is diameter and *t* is the mean time in seconds for a meniscus fall of 1 cm.

3. σ_w is practically independent of temperature variation. It was shown that in the temperature range 6°C to 31°C, the variation for one degree change in temperature was only 0.0006% of σ_w.

5.5.5 Relation between V_w and Parameters of a Delivery Measure

Effect of delivery time *T* on the wall fluid V_w, was studied simultaneously for 50 ml burette and two tubes of internal diameter 0.290 and 0.105. A piece of extra tubing was attached to one end by a short rubber band and the end of this tube was drawn out to simulate the dead space of the ordinary burette. To determine the volume of fluid left on the walls, the method of rinsing and titration of the rinsed water was used. However to get a better accuracy in the volume of the wall fluid, alkali of N/200 was used with a calibrated burette.

Q is the wall fluid in mm³ per cm² area of the wall, *t* is time in seconds the meniscus takes to descend per cm and *D* is internal diameter of burette or tubes.

Table 5.4

D in cm	V_w in mm³	*Q*	Time per cm *t*	$Q\sqrt{t}$	$Q.\sqrt{t}/\sqrt{D}$
I. Burette 50 ml with scale of 52.3 cm					
1.11	349	1.93	0.35	1.14	1.09
	234	1.29	0.72	1.09	1.04
	199	1.10	1.27	1.24	1.18
	144	0.80	2.15	1.17	1.11
	104	0.51	4.12	1.04	0.99
II. 16.0 cm long simple tube with extra tube one end of which was drawn out					
0.290	27.3	2.05	0.110	0.68	1.26
	11.4	0.86	0.453	0.58	1.07
	7.4	0.56	0.876	0.52	0.96
	4.9	0.37	2.670	0.61	1.13
III. 27.3 cm long simple tube with extra tube one end of which was drawn out					
0.105	14.2	1.58	0.048	0.35	1.07
	6.0	0.67	0.233	0.32	0.99
	3.1	0.35	0.630	0.28	0.86
IV. Same as in III above except the tube is held in horizontal position end drawn out					
0.105	3.0	0.33	0.845	0.30	0.93
	9.9	1.10	0.092	0.33	1.02

In case of burette acid was filled from below exactly up to the zero graduation mark and taken out. To find the volume of acid remaining adhered to the walls, the stopcock was closed and burette was rinsed with double distilled water and was inverted to collect rinsed water. This method avoided the acid remained adhered in the stopcock to be included with V_w.

From the scale length and internal diameter of the burette, area is calculated so wall fluid Q in mm^3 per cm^2 is given in the third column. Delivery times divided by scale length for burette and each tube are indicated in column four. To show the results, the products of Q and square root of t and the products of Q and square root of time t divided by square root of internal diameter are respectively shown in columns 5th and 6th.

From the last column one can easily conclude that all values of $Q.\sqrt{(t/D)}$ are practically constant giving

$Q = V_w/S = 1.1.\sqrt{(D/t)}$ 1.1 is the mean value of the constant.

Where V_w is fluid volume in mm^3 and S is internal surface area of burette/tube. One can easily see that $S = \pi$ DL, giving us

$$V_w = \pi \text{DL}\, 1.1\sqrt{(D/t)}$$

$$= 1.1\,(\pi \text{DL})^{3/2}\,(\sqrt{1/T}),$$ where T is the delivery time.

5.6 FACTORS INFLUENCING THE DETERMINATION OF CAPACITY

5.6.1 Meniscus Setting

For precise measurement of capacity correct setting of meniscus is very important. When water is used for calibration, the upper edge of graduation line is set tangentially to the lowest point of the meniscus. To obtain better contrast, a black and white background is chosen in such a way that dividing line be almost at the level of meniscus of water. An inverted image of meniscus is formed, the eye level is adjusted in such a way that meniscus of the water and its image appears to touch each other. A black paper strip hold at the level of meniscus also gives a well-defined image of the lowest part of the meniscus of water.

In the case of mercury, the highest point of the meniscus is set to the lower edge of the graduation line.

To make a precise setting, the lighting conditions are so arranged that the meniscus appears dark and clear in background. This is achieved by cutting undesirable illumination by folding a suitable strip of black paper round the vessel and allowing a gap of not more than 1 mm between it and the level of setting. A water meniscus appears quite black and its outline is sharply defined against the white background. The water meniscus is shaded from below and a mercury meniscus from above. Line of sight is kept normal to the water column.

5.6.2 Surface Tension

When the flask is filled slowly, the water surface rises in its bulb and becomes contaminated by slight traces of dirt. When the flask is filled up to the graduation mark, the water surface becomes more and more contaminated as all washed off contamination is collected in a comparatively smaller area. This reduces surface tension considerably. Osborne and Veazey [7] have observed that reduction of surface tension to half is quite normal. Porter [8] investigated the effect of surface tension on the meniscus volume for tubes of different diameters. The results quoted in [7] are reproduced in Table 5.5.

5.6.3 Effect of Change in Surface Tension

Let us consider a 1000 cm³ flask having neck diameter 20 mm. From the Table 5.5 the error in determination of volume may be 0.1 cm³ owing to decrease in surface tension from 70 mN/m to 50 mN/m. In order to avoid changes in surface tension due to contamination, the measure should be thoroughly cleaned before being calibrated.

Table 5.5 Volume of Meniscus of Water in cm³

Surface Tension of water mN/m	*Internal diameter of tube in millimetres* 5	10	15	20	25
70	0.015	0.084	0.21	0.36	0.50
60	0.015	0.080	0.19	0.32	0.44
50	0.014	0.075	0.18	0.28	0.38
40	0.014	0.069	0.16	0.23	0.31

To keep the surface tension unchanged, the contamination of water due to dirt is avoided. For this purpose, the measure under test is overfilled and then water is withdrawn to the desired graduation mark. In this method water washes away contamination left on the wall up to a level well above the graduation mark. The wetting of the wall above the final position of the meniscus also ensures that angle of contact between the water surface and the wall is close to zero. Thus in this way, the shape of the meniscus, as well as the meniscus volume in any tube of given diameter remains constant well within consistency required.

5.6.4 The Error in Meniscus Volume when Surface Tension is Reduced to Half

Table 5.6

S.No.	*Capacity cm³*	*Neck internal diameter mm*	*Error cm³*	*Neck internal diameter mm*	*Error cm³*
1	50	10	0.018	6	0.003
2	100	12	0.032	8	0.008
3	200	13	0.041	9	0.012
4	300	15	0.058	10	0.018
5	500	18	0.063	12	0.032
6	1000	20	0.105	14	0.049
7	2000	25	0.155	18	0.065

5.6.5 Use of Liquids other than Water

The difference in volume contained by the measure when filled by some other liquid is simply the difference between two volumes of the meniscus. The Table 5.7 due to Osborne and Veazey, which gives difference in volumes of water and liquids of varying capillary constant 'a', is reproduced below:

Where a^2 is defined as

$$a^2 = 2T/g\rho, \text{ where } \rho \text{ is the density of the liquid}$$

5.6.6 Correction in Volume in mm³ (0.001 cm³) against Capillary Constants and Tube Diameters

The correction is to be subtracted from the capacity as determined with water

Table 5.7

S. No.	Tube diameter	Capillary constant in mm²										
		14	13	12	11	10	9	8	7	6	5	4
1	4	0	0	0	0	0	0	1	1	1	1	1
2	5	0	0	0	1	1	1	1	2	2	3	3
3	6	0	1	1	1	2	2	3	4	4	5	7
4	7	0	1	1	2	3	4	5	6	7	9	12
5	8	1	1	2	3	5	6	8	10	12	15	18
6	9	1	2	4	6	7	10	12	15	18	23	27
7	10	1	4	6	8	11	14	18	22	27	32	—
8	11	2	5	8	12	16	20	24	30	36	—	—
9	12	3	7	11	16	21	25	32	40	—	—	—
10	13	4	9	14	20	26	33	41	—	—	—	—
11	14	5	11	17	25	33	42	—	—	—	—	—
12	15	6	13	22	30	41	—	—	—	—	—	—
13	16	7	16	27	37	—	—	—	—	—	—	—

5.6.7 Non-uniformity of Temperature

If the temperature of the measure under test and medium used are not the same, then there will be some error in measurement. A difference of 0.1°C makes an error of 25 parts per million.

Non-uniformity in temperature may be due to

1. Temperature of the working medium and the measure under test not being same.
2. The temperature from one point to another within the medium may be different and
3. The temperature of the medium during the set of observation may be varying.

These errors may be reduced to minimum if the medium used and the under test measures are kept together for long time so that the temperature equilibrium is attained, and also the ambient temperature of the room is not allowed to vary by more than 0.5 °C/ hour.

5.7 INFLUENCE PARAMETERS AND THEIR CONTRIBUTION TO FRACTIONAL UNCERTAINTY

Errors committed in influence parameters and their effects on measured volume are given in Table 5.8 [9]

Table 5.8

Parameter	*Error committed in Parameter*	*Fractional error in volume*
Water temperature	± 0.5°C	$\pm 10^{-4}$
Air pressure	± 8 mbar or 10.8 kPa	$\pm 10^{-5}$
Air temperature	± 2.5°C	$\pm 10^{-5}$
Relative humidity	± 10%	$\pm 10^{-6}$
Density of standard weights	± 0.6 g/cm^3	$\pm 10^{-5}$

5.8 FILLING A MEASURE

5.8.1 Filling the Content Measure

A Flask

A content measure is cleaned and dried along with its stopper/striking glass if any. Take a proper size funnel, such that water is discharged below the stopper portion. Pour water in the funnel from a beaker and manipulate the funnel so that the entire neck of the flask below the stopper mark is wetted. Stop filling as soon as the water level is just below the graduation mark. Wait for two minutes to allow walls of flask to drain and use a burette or a glass tube with a jet to fill necessary water from the beaker so that meniscus touches the upper edge of the graduation mark. Alternatively, the measure is slight overfilled and final adjustment and setting of the lowest point of the meniscus to the upper edge of the graduation mark is carried out by removing water bit by bit with an ash-less filter paper or with the help of the glass tube drawn in to a jet. Care is taken that water does not splash on the walls of the flask.

A Non-graduated Measure

In case of a non-graduated measure with a striking glass, overfill it slightly and slide the glass over the rim of the measure keeping it always in horizontal position, so that there is no air bubble between the striking glass and water surface inside the measure. Clean the measure from the out side including the upper surface of the striking glass before reweighing the measure. Alternatively the method described in 4.5.2 may be used to fill the measure under-test.

5.8.2 Filling of a Delivery Measure

The delivery measure is clamped in vertical position and water is filled against gravity through its stopcock. The measure, in this case also, is filled to a few mm above the graduation mark to be tested and the water remaining on the outside of the jet is removed. Running out the surplus water through the jet by manipulating finely the stopcock, makes the meniscus setting to the required mark. Any drop of water, adhering to the jet, is removed by bringing a clean wetted glass surface into contact with the tip of the jet. Water is delivered into the tarred weighing flask with unrestricted flow of water till a few mm above the desired graduation mark and then the stopcock is manipulated so that water meniscus just touches the upper edge of the required graduation line. Specified drainage/waiting time is then allowed.

5.9 DETERMINATION OF THE CAPACITY WITH MERCURY AS MEDIUM

Mercury is the only metal element, which is found in liquid state at ordinary temperatures. It has a high density of the order of 13560 kg/m^3. So mass of its small volume is quite large. For small capacity measures, say 100 mm^3, mass of mercury required will be 1.35 gram. Moreover it may be easily purified and its density is known very accurately. So for calibration of small capacity measures like micro-burettes or micropipettes, mercury is used instead of water employing gravimetric method. Only drawback of mercury is its toxicity. It is a poison if swallowed orally. The surface of the standards weights adsorbs the mercury vapours causing a change in their mass values. Special care is to be taken not to allow any spilling over of mercury or prolonged exposure to it. For this a fume chamber with a special working table is recommended. The table should have a grove along it sides and collecting port in one corner connected to reservoir of used mercury. So the table is brushed quite frequently to get all mercury collected in the reservoir. The mercury to be used should be freshly distilled, and filtered through ash-less filter paper of very fine bore. The process of filtering is necessary to remove the minute dust particles. It is better that the worker uses surgical mask, to avoid mercury vapours. No gold ornament, even the ring should be worn while working with mercury, otherwise mercury vapours will amalgamate the gold. Besides all these precautions mercury vapours, being heavy, are injurious to health.

If m is the mass of mercury required to be filled or is delivered by the measure, as determined by weighing mercury in air of density σ g/cm^3 against weights of density D g/cm^3, the capacity of the measure V_{27} at 27°C is given by

$$m(1 - \sigma/D) = [V_{27}(1 + \alpha(t - 27)](\rho - \sigma) \text{ giving}$$

$$V_{27} = (m/\rho)\{1 - \sigma(1/D - 1/\rho)\}\{1 - \alpha(t - 27)\}$$

Where α is the cubical expansion coefficient of glass per degree Celsius.

The term $\sigma(1/D - 1/\rho)$ is so small that density D of weights; σ the density of air and ρ density of water are taken as constants. To calculate the aforesaid term, following values of σ, D and ρ are taken.

$$\sigma = 1.17 \text{ g/dm}^3$$

$$D = 8000 \text{ kg/m}^3 \text{ and}$$

$$\rho = 13560 \text{ kg, giving}$$

$$\sigma(1/D - 1/\rho) = 0.000\,05967 \text{ and hence}$$

$$V_{27} = m(0.999\,94033/\rho)\{1 - \alpha(t - 27)\}$$

Values of $(0.999\,94033/\rho)\{1 - \alpha(t - 27)\}$ have been tabulated for temperature range of 5 to 40°C in steps of 0.1°C in IS: 1991[10]. The value of α for coefficient of expansion of glass was taken 25.10^{-6}/ °C in the specification.

However the author has calculated the values of $\{(1/\rho).(1 - \sigma/D)\}/\{(\rho - \sigma).(1 + \alpha(t - 27)\}$ for more accurate work for different parameters like reference temperature and corresponding density of air, density of standard weights, coefficients of volume expansion and their combination. Temperature range chosen is 5 °C to 41 °C in steps of 0.1 °C and their values are given in Chapter 3 Tables 3.31 onward.

5.10 CRITERION FOR FIXING MAXIMUM PERMISSIBLE ERRORS

Maximum permissible error should naturally depend upon the capability of observing the instrument under question within reasonable repeatability. Basically there are two errors one is the capability of eye and the other error, which can reasonably occur, is due to parallax.

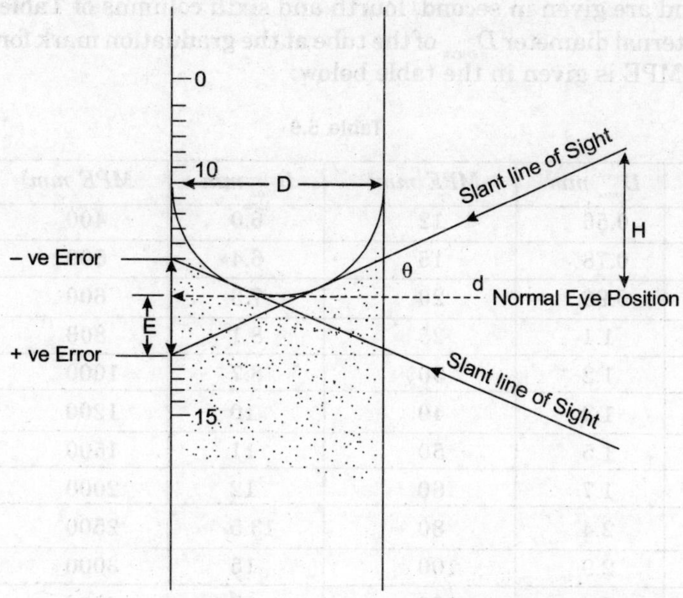

Figure 5.4 Error due to line of sight

The best eye can estimate is 0.4 mm even for very small-bore tube, so this we can take inherent error due to eye. If the line of sight of the observer is not in the horizontal plane, tangential at the lowest point of the meniscus, then the error due to parallax will occur. The error denoted by E mm is given by the following relation:

$$\tan\theta = E/(D/2) = H/(d + D/2)$$

Where E is error in observing the graduation line in mm

D is diameter of the tube or neck, as the case may be, where the mark is graduated in mm H is the vertical offset of the of the eye from the horizontal plane in mm and d is distance of the eye along the tangential horizontal plane at the lowest point of the meniscus.

In normal circumstances d is 200 mm, D would vary from 1 mm to 100 mm and H about 5 mm.

Giving

$$E = HD/(2d + D)$$

E may vary for different values of D. But normally D varies from 1 mm to 100 mm

So $E = 5D/(400 + 1)$

$E = 0.0125\,D$ for $D = 1$ mm

and $E = 5D/(400 + 100) = 0.01D$, for $D = 100$ mm

So E may be taken uniformly for 0.01 D.

Hence minimum error in volume due to capability of eye and by displaced line of sight even for very good observer V_E is given as

$$V_E = (\pi D^2/4) \times (0.4 + 0.01D) \text{ in mm}^3$$

Hence Maximum Permissible error allowed should in no case be less than V_E given above or for closer tolerance D should be made small accordingly. So for given maximum permissible errors indicated in the first, third and fifth columns, maximum values of tube diameter D have been calculated and are given in second, fourth and sixth columns of Table 5.9.

Maximum internal diameter D_{max} of the tube at the graduation mark for selected maximum permissible error MPE is given in the table below:

Table 5.9

MPE mm³	D_{max} *mm*	*MPE mm³*	D_{max} *mm*	*MPE mm³*	D_{max} *mm*
0.1	0.56	12	6.0	400	27
0.2	0.78	15	6.4	500	29
0.3	0.96	20	7.3	600	32
0.4	1.1	25	8.1	800	36
0.5	1.2	30	8.7	1000	40
0.6	1.3	40	10	1200	44
0.8	1.5	50	11	1500	47
1	1.7	60	12	2000	52
2	2.4	80	13.5	2500	57
3	2.9	100	15	3000	61
4	3.4	120	17	4000	68
5	3.8	150	18	5000	74
6	4.2	200	20	6000	80
8	4.7	250	23	8000	83
10	5.3	300	25	10000	96

REFERENCES

[1] ISO 4787-1984:1984 Use and Testing of Capacity of Volumetric Glassware.
[2] Notes on Applied Science No 6 Volumetric Glassware, 1957, Her Majesty Stationary Office, London.
[3] ASTM standard E-542: 1979 Standard Practice for Calibration of Volumetric Ware.
[4] ISO 1042:1983 One-mark Volumetric Flasks.
[5] Stott V, "Notes on Burettes", 1923, Trans. Soc. Glass Tech. **7**, 169-198.
[6] Stott V. "Notes on pipettes", 1921, Trans. Soc. Glass Tech **5**, 307-325.
[7] Osborne and Veazey, 1908 National Bureau of Standards Bulletin, 567-574.
[8] Porter A. W. 1934 On the Volume of the Meniscus at the Surface of a Liquid, **17**, 511.
[9] ISO 4787:1984 Use and Testing of Capacity of Volumetric Glassware.
[10] IS:1991 Calibration Tables for Water and Mercury for Laboratory Glassware.

6
CHAPTER

CALIBRATION OF GLASS WARE

6.1 BURETTE

A burette [7, 8, 9, 10, 11 and 12] is essentially a tube of uniform diameter. Stopcock for controlling the flow of water with a jet of such dimension so that the delivery period of the burette lies in between specified delivery times. Upper end is open and rim is bevelled. The tube is marked with equi-spaced graduation lines indicating the volume, which the burette will deliver from the zero line to that line, length of the graduation lines are according to the class and specification of the burette. Some blank space for mandatory inscription like capacity, reference temperature and the letter 'D' or 'Ex' indicating that the burette is for delivery, is left out. Some burettes may have separate filling and delivering tube. In such cases three-way stopcock is used. A typical burette with a three-way stopcock is shown in the Figure 6.1.

Capacity: The burettes are available in capacity of 10 cm^3, 25 cm^3, 50 cm^3 and 100 cm^3.

6.1.1 Jets for Stopcock of Burettes

The National Bureau of Standards, USA, at one time suggested that to avoid splashing, the jet of the burette should be curved. Many chemists prefer to have straight jet, which delivers directly into the liquid. Burette jets, which taper rapidly or have a sudden constriction at the orifice, caused by reducing the size of the opening by heating the one end of the jet in a flame after cutting off drawn out portion are objectionable. They are more likely to cause splashing than a jet with gradual taper. Moreover slight damage to tip may change the delivery time considerably. A jet with larger gradual taper, therefore, is better. The jet with constricted tip is difficult to clean. The internal diameter of the tubing for jet should be equal to the diameter of the hole drilled through the tap. Also in sealing the jet to the barrel of the tap, it should not enlarge appreciably. If the internal diameter of the jet was such that it is not completely filled with liquid when the burette is in use, then there would be errors of varying nature. As the amount of the portion remained unfilled will vary, so will be the error in volume of liquid delivered. Different jets are shown in Figure 6.2.

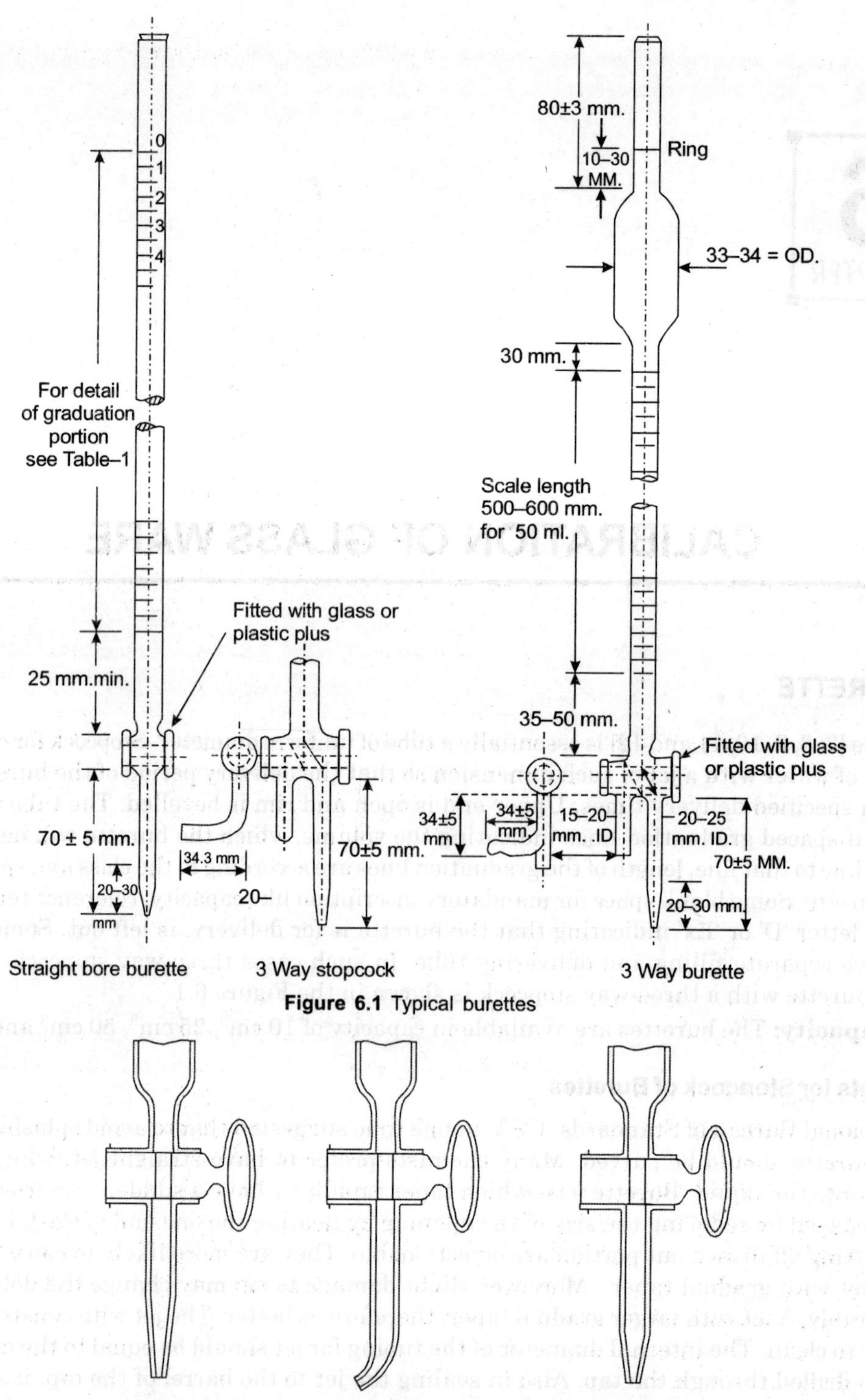

Figure 6.1 Typical burettes

Figure 6.2 Jets for stopcock of a burette

6.1.2 Burette-key

The key of the burette tap should be ground into its barrel so that two have a good fit. Leakage should be prevented by the goodness of the fit of the key in the barrel and not by using too much of grease.

6.1.3 Graduations on a Burette

General style for graduation lines is common to all the burettes, for example these should be permanently engraved or printed with indelible ink, however every national standard specification specifies slightly different requirements. For example ASTM [7] requirements are given in Figure 6.3.

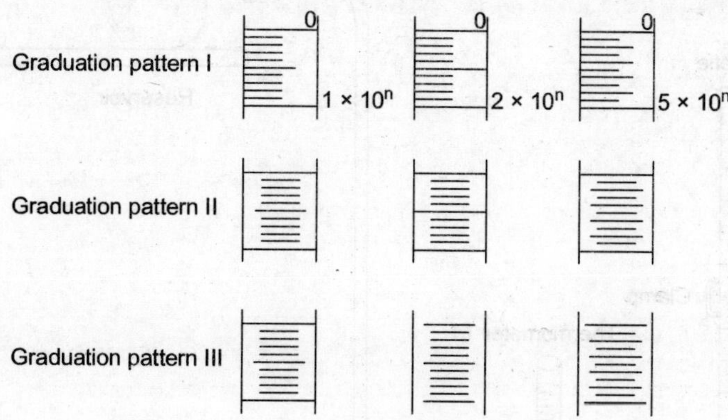

6.3A Linear

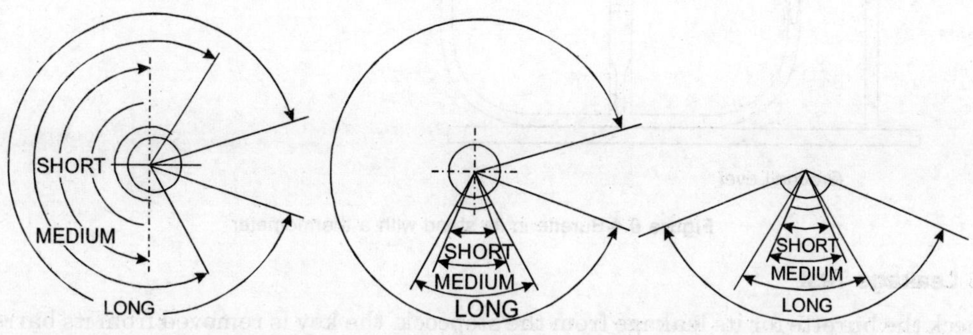

6.3B Angular

Figure 6.3 Scheme of graduation lines

6.1.4 Setting up a Burette

For use, the burette is clamped vertically on a support stand. If the burette itself is not large enough to hold a thermometer for recording temperature of water, then it is clamped in a T section tube fitted in the rubber tube carrying the water as shown in Figure 6.4.

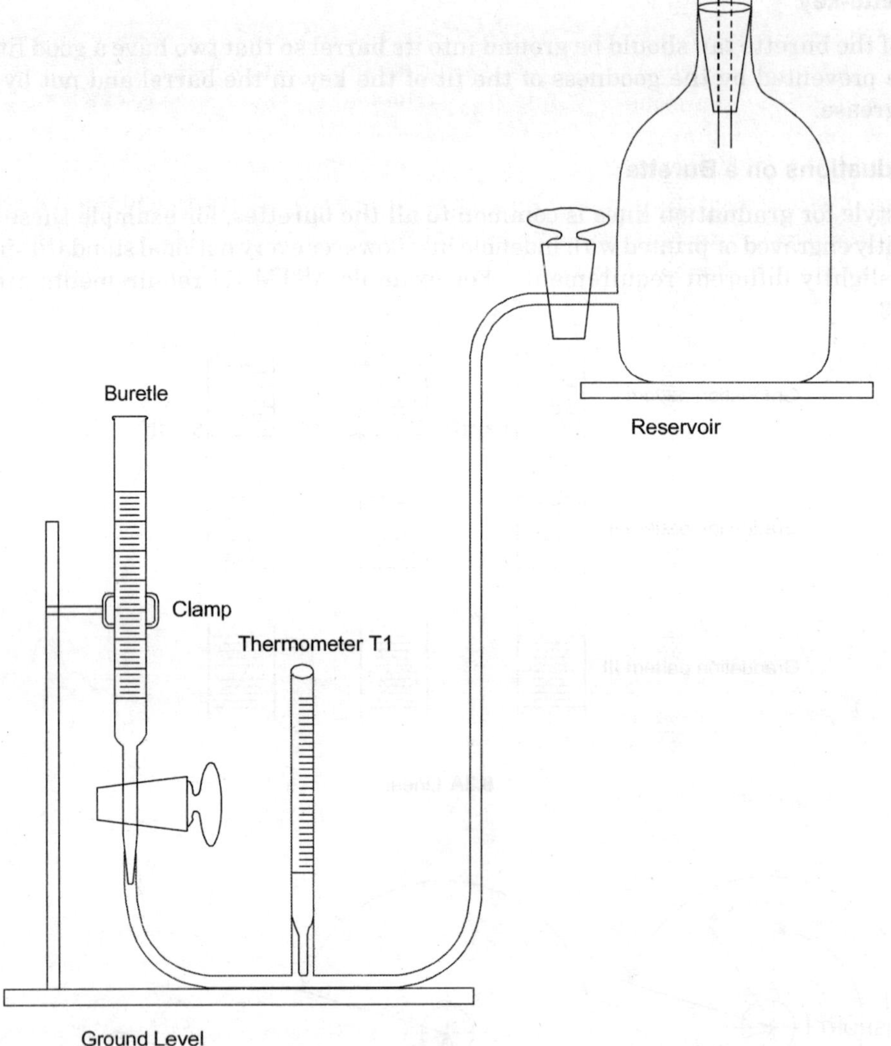

Figure 6.4 Burette in its stand with a thermometer

6.1.5 Leakage Test

To check the burette for its leakage from the stopcock, the key is removed from its barrel. Both key and barrel are thoroughly cleaned with alcohol to remove any traces of grease. The key is dipped in water and fitted in the barrel of the stopcock. Fix the burette in a vertical stand and fill it with water from a reservoir or storage flask in which water has reached thermal equilibrium with room temperature. Set the meniscus at zero cm^3 mark. The burette is left for about 30 minutes in one shut off position. Note the time elapsed and the reading of the water meniscus level. Turn the stopcock by 180° and repeat the procedure. Take the mean value of the readings of water meniscus and calculate the fall of meniscus level per minute. Normally it should not exceed 0.1 mm per minute. Some national specifications give permissible limits of the volume, which could leak in certain specified time. If the specified period for leakage test is more than 8 hours, the test may start in the evening and continue next day. During this period the room temperature should not change by more than 5°C. Set the meniscus to the zero graduation line, close the stopcock properly and start the stopwatch or note the time from a wristwatch. Leave the burette undisturbed till the specified time or for time planned for leakage test, note the time. Observe the meniscus reading. This is the leakage in volume in the time allowed for leakage, dividing the volume by time gives the rate of leakage, see if it is within the prescribed limits, if not reject the burette. No further tests should be carried out on such a burette.

Caution: To perform the leakage test, tap is never greased. Quite bad taps can be made to withstand leakage for an appreciable time if these are sufficiently greased and not turned during the time for which test is conducted. Such stopcocks, of course, fail in actual use when the taps are repeatedly turned on and off.

6.1.6 Delivery Time

Drain water from the burette with stopcock fully open and record the time water level takes from zero mark to full capacity mark of the burette. Delivery time is defined as the time taken by the unrestricted flow of water from the zero line to the lowest graduation line with the stopcock fully open. Time may be recorded in seconds or in terms of 0.5 s and three such delivery periods are taken and their mean value is reported as its delivery time. Maximum difference between any two delivery-periods should not be more than 1 second.

6.1.7 Calibration of Burette

Refill the burette to approximately 10 mm above the zero line, and record the temperature. Set the meniscus at the zero graduation line; use the stopcock of the burette to lower the level of water. Once the meniscus is set to the desired graduation line, touch the tip of the stopcock with a wetted wall of a beaker to remove excess water. Ensure the meniscus setting is unchanged. Record the temperature T1 from thermometer Figure 6.4. Take a weighing flask with a tight stopper and ensure that outer surface is dry and clean, then weigh it empty and observe and record mass of weights. Certificate correction to weights is applied if necessary. Bring the weighing flask under the burette so that it covers fully the tip of the burette and the tip of the stopcock touches its inclined wall. Fully open the stopcock until the water level reaches a few mm above the graduation line being tested. The stream is slowed down so as to make an accurate setting at the desired line, move the weighing flask horizontally breaking the contact with the burette. Recheck for the proper setting at the desired line. Stopper the weighing flask and weigh it, observe and record the mass of weights. Apply certificate corrections to weights removed/added. Check and record the temperature T2. Mean of the two temperatures is taken as the actual temperature of water delivered. The difference between the two apparent masses gives the apparent mass of water at the mean temperature of measurement. Find the correction from the suitable tables from 3.1 to 3.24 corresponding to the mean temperature. The selection of table will depend upon

- Density of weights used
- Reference temperature
- And coefficient of expansion of the glass and
- Reference temperature.

Find the correction, for the nominal value of the graduation line, and add it to the apparent mass of water to get the volume of water, which the burette will deliver at the reference temperature. The value of the graduation line minus volume of water so calculated gives the error at that graduation line. Repeat the process for another interval; always start from the zero line to the graduation line to be tested. Four such intervals are taken, which must include an interval from zero line to the last graduation line i.e. full capacity of the burette. The burette is tested also for the accuracy of volume of water delivered between two consecutive lines. The maximum difference in any two errors should not exceed the maximum permissible error prescribed for the burette.

For the burettes with specified waiting time, after adjustment at the zero graduation, open the stopcock fully till the meniscus reaches a few mm above the desired graduation line, Wait for the specified time and then adjust the meniscus to the desired graduation line by manipulating the stopcock.

Delivery time would naturally depend on the scale length, so scale length is also measured for compliance to the desired specification. A typical set of delivery time is given in Table 6.1. Depending upon the accuracy, the burettes are divided in two accuracy classes. The two accuracy classes are normally designated as class A and class B. The delivery times as per scale length of the burette as adopted at NPL U.K. are also indicated in Table 6.1.

6.1.8 Delivery Time of Burettes in Seconds–A Comparison

Table 6.1 [3]

Length of	*NBS*		*NPL Class A Burette*		*NPL class B Burette*	
Scale mm	*Minimum*	*Maximum*	*Minimum*	*Maximum*	*Minimum*	*Maximum*
150	30 s	180 s	30 s	60 s	20 s	60 s
200	35 s	180 s	40 s	80 s	30 s	80 s
250	40 s	180 s	50 s	100 s	35 s	100 s
300	50 s	180 s	60 s	120 s	45 s	120 s
350	60 s	180 s	70 s	140 s	50 s	140 s
400	70 s	180 s	80 s	160 s	55 s	160 s
450	80 s	180 s	90 s	180 s	60 s	180 s
500	90 s	180 s	100 s	200 s	70 s	200 s
550	105 s	180 s	110 s	220 s	75 s	220 s
600	120 s	180 s	120 s	240 s	80 s	240 s
650	140 s	180 s	130 s	260 s	85 s	260 s
700	160 s	180 s	140 s	280 s	90 s	280 s
750	—		150 s	300 s	100 s	300 s

The maximum difference allowed between the actual and inscribed times is 8% up to 200 s and 20 s for a time period of 300 s.

6.1.9 MPE (Tolerance) / Basic Dimensions of Burettes

6.1.9.1 Maximum Permissible Error

Tolerance on capacity is defined as the maximum error allowed at any point of scale and also maximum difference allowed between the errors at any two points of the scale.

These, as adopted in India are given in Table 6.2.

Table 6.2 [3]

Total Capacity	*Maximum Permissible Error* ± *cm*^{3}	
cm^3	*Class A*	*Class B*
2	0.01	0.015
10	0.02	0.035
30	0.03	0.05
50	0.04	0.07
75	0.06	0.10
100	0.08	0.14
200	0.15	0.25

A burette conforming to the above requirements guarantee a reasonable accuracy and repeatability. While emptying the burette the tap should be kept fully open. Even small quantities at a time may be drawn by sharply turning the tap fully on and then off again. With class A burettes the sum of volumes obtained by emptying a given interval in successive stages differs only slightly from the volume delivered when whole interval is emptied at a time.

It is of course not feasible to keep the tap fully open while in the last stages of adding a few drops of the liquid say in the process of titration. The last 1 cm^3 or so must be added slowly. Moreover a similar condition appears while calibrating as in that case also rate of out flow of water is to be reduced considerably to exactly reach the graduation line i.e. touching of lowest part of meniscus with the upper part of the graduation line. In effect, it simply amounts to an increase in drainage time, which has been shown not to introduce serious error, provided the delivery time is within the prescribed limits. The final reading therefore may be taken at the user's convenience within reasonable time without giving any allowance for drainage time.

6.1.9.2 Basic Requirements for Burettes

Table 6.3 [7] Basic Requirements for Burettes

Capacity	*I*	*S*	*L1*	*Numbering*	*MPE*
10 cm^3	0.05	350 to 450	30 to 75	0.5	0.02
20 cm^3	0.1	350 to 450	30 to 75	1	0.03
50 cm^3	0.1	500 to 600	40 to 100	1	0.05
100 cm^3	0.2	550 to 650	40 to 100	2	0.10

I is value of smallest sub-division in cm^3
S scale length in mm
L1 Distance from top to the zero graduation line in cm
Numbering at every cm^3
MPE Maximum permissible error in cm^3

6.2 GRADUATED MEASURING CYLINDERS

6.2.1 Types of Measuring Cylinders

Measuring cylinders [13, 14, 15, and 16] are made from a colourless glass with no special tint. The glass of a good measuring cylinder is free from any visible defect and is reasonably free from internal strain. As regards to alkalinity, the glass may belong to any of three categories, which meet the requirements of the relevant national standard for example of IS: 2303- 1963. The measuring cylinders are available with stopper having grounded neck to receive the stopper. A typical content type measuring cylinder is shown in Figure 6.5 A. The measuring cylinders are also available without stopper having a lip for delivering the liquid. A typical delivery type cylinder is shown in Figure 6.5B.

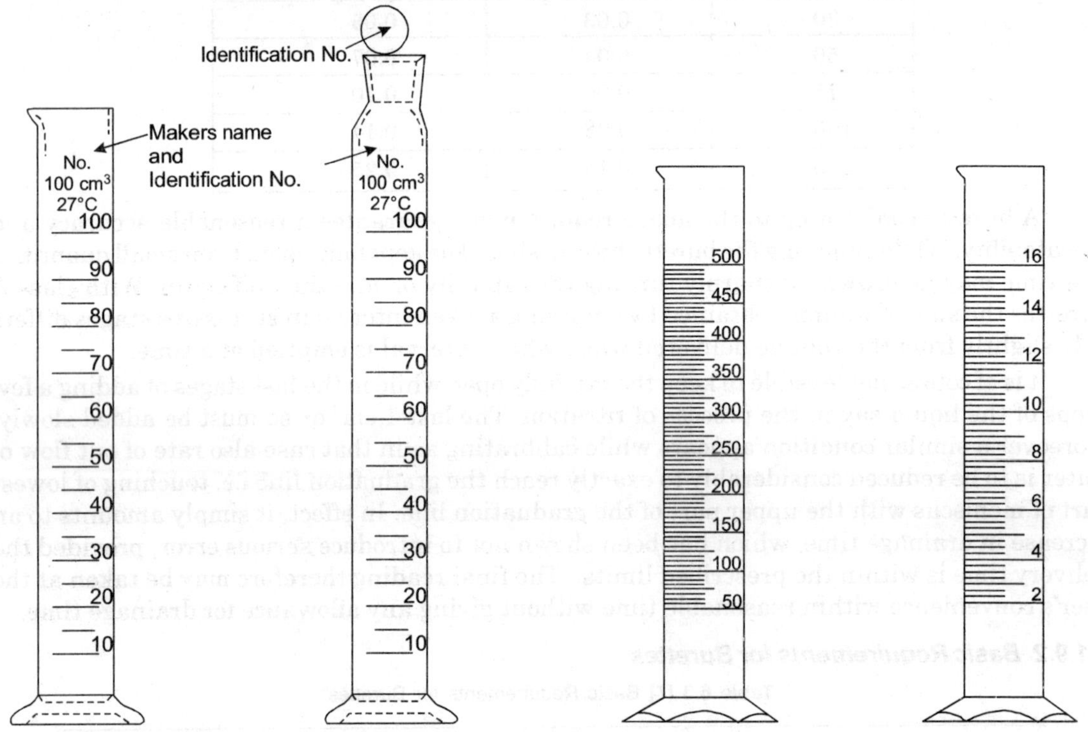

Figure 6.5A Content type **Figure 6.5B** Delivery type measuring cylinders

As these are available both as delivery and content types, so there should have inscription about the type of the cylinder for example 'D' or 'Ex' for delivery and 'C' or 'In' for content. Similarly the reference temperature and capacity of the cylinder must be permanently marked on the cylinder. Graduated cylinders should have base, big enough, to stand firmly not only on the level surface but be able to stand without toppling on an inclined plane at 15° with the horizontal.

The internal diameter of cylindrical measures, especially when these are mould blown, is liable to be of considerable variation. So graduations, from the bottom to a distance of one tenth of the total scale length, are omitted. The scale graduation lines should be etched and filled with indelible ink, the lines should be, equally spaced and at right angles to the axis of the cylinder. For easy reading lines should be in three sizes namely long, medium and short. The length of the longer, medium and shorter lines should respectively at least be one quarter, one sixth and one eighth of the circumference of the cylinder. The lines should be symmetrically placed about some imaginary central line. Sequence of different lines is prescribed by a national standard specification for example prescribed sequence of lines in India, as per Indian standard [13] is shown in Figure 6.6.

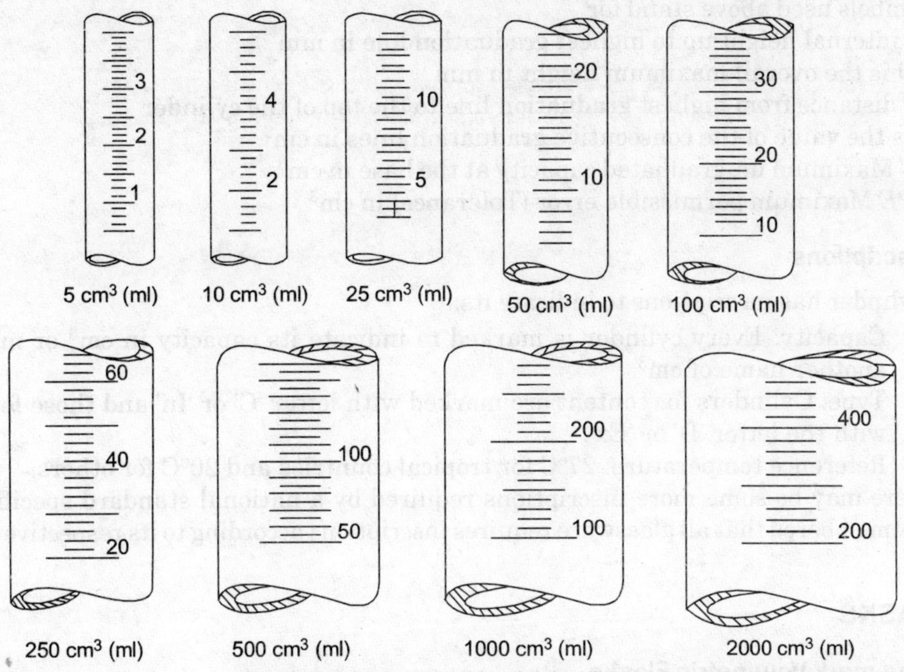

Figure 6.6 Sequence of graduation lines

When calibrated for content, the mass of water required to fill up to the desired graduation line is determined and necessary correction for corresponding capacity, type of glass used and temperature etc. is applied to give its capacity at the reference temperature. These cylinders can also be calibrated by volumetric filling method by employing delivery type automatic pipettes.

Measures meant for delivery are necessarily provided with lips. While calibrating, this is emptied by gradually inclining it until the continuous stream of water ceases; this is made nearly vertical with bottom up. It should be maintained in this position for a drainage time of 30 seconds and the lip is then stroked gently against the inside of the receiving vessel to remove any drop of water adhering to the lip.

The Maximum permissible errors and dimensions given in the Table 6.4 are as per Indian standard [13]

Table 6.4 [13] Maximum Permissible Errors and Mandatory Dimensions

Capacity	*HI*	*HO*	*LT*	*D value*	*UC at base*	*MPE*
5	55	115	20	0.1	0.5	± 0.1
10	70	140	20	0.2	1.0	± 0.2
25	90	170	25	0.5	3	± 0.5
50	115	200	30	1	5	± 1
100	145	260	35	1	10	± 1
250	200	335	40	2	20	± 2
500	250	390	45	5	50	± 5
1000	315	470	50	10	100	± 10
2000	400	570	50	20	200	± 20

Symbols used above stand for
HI internal height up to highest graduation line in mm
HO is the overall maximum height in mm
LT distance from highest graduation line to the top of the cylinder
D is the value of the consecutive graduation lines in cm^3
UC Maximum un-graduated capacity at the base in cm^3
MPE Maximum permissible error (Tolerance) in cm^3

6.2.2 Inscriptions

Every cylinder has inscriptions to indicate its:

- Capacity: Every cylinder is marked to indicate its capacity in cm^3 or ml as ml is another name of cm^3.
- Type: Cylinders for content are marked with letter 'C' or 'In' and those for delivery with the letter 'D' or 'Ex'.
- Reference temperature: 27°C for tropical countries and 20°C for others.

There may be some more inscriptions required by a national standard specification. It may be remembered that all glassware requires inscriptions according to its respective standard.

6.3 FLASKS

6.3.1 One-mark Volumetric Flasks

Most common measure, among all the glass volumetric measures is a one mark volumetric flask [17, 18, 19, 20, 21, 22, 23, and 24]. These are made of colourless glass not having any pronounced tint. They are normally made of glass free from any visible defects such as seeds, bubbles and stones. As regards alkalinity, the glass should meet the requirements of a relevant national standard such as IS 2303-1963 in India or ASTM [60] in US. Normally two types of glass are used in fabrication of volumetric one-mark flasks, namely (1) Borosilicate and (2) Soda glass. Any other neutral glass having cubical thermal coefficients of 10.10^{-6}/ °C, 15.10^{-6}/°C, 25.10^{-6}/ °C or 30.10^{-6}/ °C may also be used. The flasks are available both with stopper and without it. Typical flasks are shown in Figure 6.7.

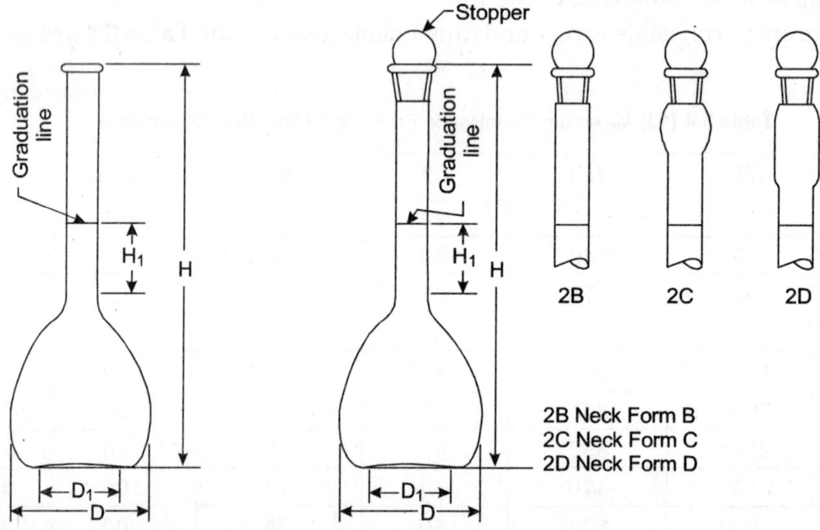

Figure 6.7 One mark flask without and with stopper

Capacity: One mark flasks are available in capacities of 2000, 1000, 500, 200, 100, 50, 20, 10 cm^3 (ml).

Shape: Main body (the bulb) is a frustum of a cone or pear shaped with a flat base, surmounted with a long cylindrical tube. These are of single capacity type measures. A permanent complete circular line on its neck defines the capacity. The line should be square to the axis of the flask and should be horizontal when flask is standing on a level ground. The flask may be with a stopper or without it. The dimension of the base (the bulb) visa-vis other dimensions is such that it remains stable, without toppling, when placed on an inclined plane of 15° with horizontal. The circular line should not be close to the joint of neck with body and also should be well below the top of the neck. Important dimensions are given in various national standards. The essential dimensions and maximum permissible error as per Indian Standard are given in Table 6.5.

Table 6.5 [18]

Capacity	*ID*	$H1_{min}$	*H*	*Maximum permissible Error*	
				Class A	*Class B*
5 cm^3	6 to 8	5	70	±0.025	±0.05
10 cm^3	6 to 8	5	90	±0.025	±0.05
25 cm^3	8 to 10	5	110	±0.04	±0.08
50 cm^3	10 to 12	10	140	±0.06	±0.12
100 cm^3	12 to 14	10	170	±0.10	±0.20
200 cm^3	14 to 17	10	210	±0.15	±0.30
250 cm^3	14 to 17	10	220	±0.15	±0.30
500 cm^3	17 to 21	15	260	±0.25	±0.50
1000 cm^3	21 to 25	15	300	±0.40	±0.80
2000 cm^3	25 to 30	15	370	±0.60	±0.120

Where

$H1_{min}$ is the distance of graduation line from the point of change of internal diameter of neck in mm

H is the overall height of a flask without stopper in mm

ID is the internal diameter.

6.3.1.1 *Basic Requirements of Flasks as per ASTM E 288 [17]*

Table 6.6 [17]

Capacity	*MPE*	*ID*	*Position of line*		*MCA style*		*Stopper size*
			Lt	*Lb*			
5	±0.02	6 to 8	22	5	—	—	8 or 9
10	±0.02	7 to 8	28	7	—	—	9
25	±0.03	7 to 8	35	7	—	—	9
50	±0.05	8 to10	40	8	10	8	9
100	±0.08	10 to12	40	10	10	10	13
300	±0.10	11 to14	45	10	10	10	13 or 16
250	±0.12	12 to15	45	10	10	10	16
500	±0.20	15 to18	60	15	10	10	16 or 19
1000	±0.30	16 to 20	60	15	10	10	22
2000	±0.50	21 to 25	60	15	10	15	27

Where Lt is the distance of the graduation line from the top;

Lb is the distance of the graduation line from main body from where neck is of almost uniform diameter.

The neck should be cylindrical above the graduation line and should not show any marked taper towards the top. The internal diameter of the neck should not exceed the values given in the table 6.6.

6.3.1.2 Calibration of Flasks by Volumetric Method

One-mark flasks are used in large numbers, so a simple, quick but accurate method of their initial calibration is very important especially at the premises of manufacturers. Calibration of flasks by gravimetric method requires lot of time and thus poses a problem to manufacturers. Volumetric calibration method is the answer to their problem. A pipette with automatic zero setting and graduated delivery tube, as shown in the Figure 6.8 is used for the purpose. The tube at the top of the graduated pipette is drawn into a fine jet, which is grounded off, and polished smooth. The pipette is filled through the side tube and stopcock A. The stopcock A is partially opened together with stopcock B, so that water fills completely the lower delivery jet. To ensure it a small amount of water is allowed to run out. The stopcock B is closed and water is allowed to enter the main pipette. The stopcock A is fully opened till the water overflows through the top jet of the pipette. A small flask C with the side overflow tube D is inverted over the jet through a rubber cork. There is small hole at the top of flask C to have communication with outside air. When the water is overflowing freely, the stopcock A is gradually closed. The pipette is made full from top to the delivery jet.

The flask to be tested is placed on a rotating Table E, which can be easily raised and lowered. The rotating Table E should always remain horizontal. The flask is raised until the tip of the jet is just above the graduation line and almost touches it. The stopcock B is fully opened and the slightly curved jet directs the outflow water on to the inside of the flask just above the graduation line. This avoids splashing and the formation of air bubbles. The stopcock B is kept fully open until the water level comes a few mm below the graduated line. The flask is continuously rotated, so that neck is fully wet. The stopcock B is then closed. The filling is completed by manipulation of the stopcock B so that only small amount of water comes to the flask at a time until the water meniscus just touches the upper part of the graduation line after the last drop of water adhering to the jet is taken in the flask by touching the jet to its wall. The reading at the pipette gives the capacity of the flask under test. The same apparatus may be used for graduating the flask instead of calibrating it. In that case stopcock B is very nearly closed till the level of water in the pipette comes exactly at 100-cm^3-mark. The flask is pre-coated with a very thin but uniform layer of bees wax. The circular line is graduated by a fine stylus, which itself is fixed through a spring-loaded holder.

6.3.1.3 Calibration of Flask by Gravimetric Method

After cleaning and drying, place the empty flask including its stopper in one, say right hand, pan of a two-pan balance and keep a similar flask on the other pan. Put standard weights at the rate of one gram per cm^3 of its capacity.

Put similar amount of load on the other pan to counter balance it. Take the observations, note and record the total mass value of the standard weights. The measure is filled to well above the graduation line say to a distance of few mm. Final adjustment is done by withdrawing of water with ash-less filter paper or a small glass tube drawn into a jet.

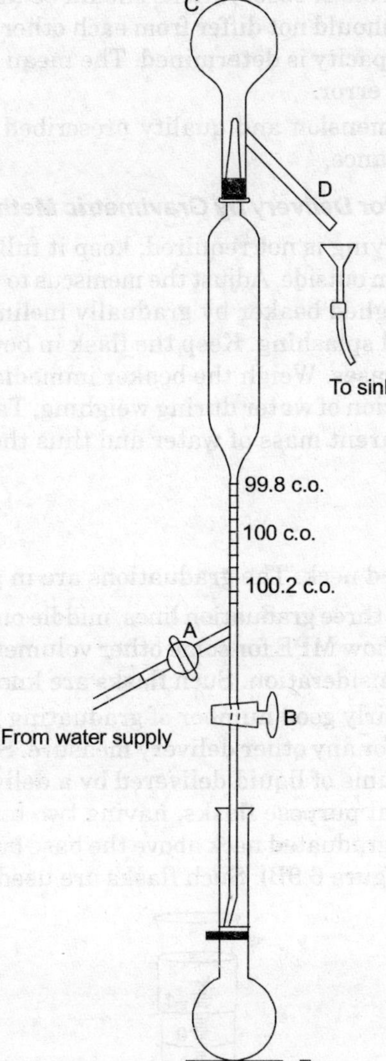

Figure 6.8 Arrangement for calibration of a flask

Alternatively, the wall of the measure is wetted for a considerable distance above the graduation line to be tested. The measure is filled to a few mm below the graduation line by running water down the wetted wall of the neck. Two minutes drainage time is allowed. The final setting is then made by discharging the required liquid against the wall about 1 cm above the graduation line and rotating the measure to wet the wall uniformly. The flask is placed in the pan of the balance remove the standard (weights equivalent to mass of water filled) to restore equilibrium. Thus water is substituted by the standard weights i.e. substitution weighing is used. Record the observations and determine the apparent mass of water. Temperature of water is measured in the water beaker before filling and after weighing the flask and emptying in to the same beaker. Appropriate correction table from Tables 3.1 to 3.24 of chapter 3, may be consulted to find the value of correction to be added to give the capacity of the flask at the reference temperature. Three sets of observations should be taken, giving rise to three values of capacity of the flask, which should not differ from each other by more than 1/3rd of the MPE. The mean of three values of capacity is determined. The mean value of the capacity should lie within the prescribed limits of error.

Any other mandatory dimension and quality prescribed in the concerned specification should be examined for compliance.

6.3.1.4 Calibration of a Flask for Delivery by Gravimetric Method

In case of flask for delivery, drying is not required; keep it full at least up to a few mm above the graduation line. Clean it from outside. Adjust the meniscus to the upper edge of the graduation line and empty it in a pre-weighed beaker by gradually inclining the flask from the vertical position. Care is taken to avoid splashing. Keep the flask in bottom up position for 30 seconds after the main flow of water ceases. Weigh the beaker immediately. The care is taken so that there is no significant evaporation of water during weighing. Take the temperature of water in the beaker. Calculate the apparent mass of water and thus the volume of water delivered by the flask.

6.3.2 Graduated Neck Flask

There are flasks with graduated neck. The graduations are in general three types

1. The flask having only three graduation lines, middle one indicates its nominal capacity and other two lines show MPE for some other volumetric ware, which can be verified by the flask under consideration. Such flasks are known as tolerance flasks.
2. The flask having a fairly good number of graduating lines (Figure 6.9A), which may be used as standard for any other delivery measure. So the flask is of content type. It will measure the volume of liquid delivered by a delivery measure under test.

Yet there are other special purpose flasks, having two bulbs, which will hold the major portion of the volume. There is graduated neck above the base bulb and another fully graduated neck above the second bulb (Figure 6.9B). Such flasks are used in some special reactions.

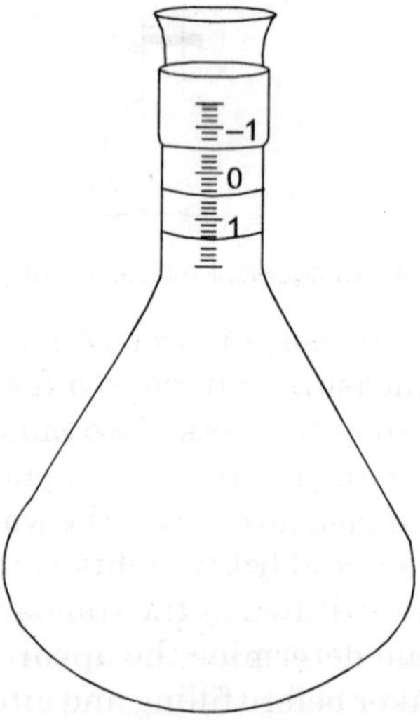

Figure 6.9A Graduated neck

F Maximum overall height including stopper's height in mm

G size number of the stopper

MPE Maximum permissible error (tolerance) in cm^3

6.4 PIPETTES

6.4.1 One Mark Bulb Pipette

6.4.1.1 Construction

The pipette [25 to 35] consists of four main parts namely suction tube through which liquid is sucked, the bulb, which constitutes its main capacity, the delivery tube with a jet from which the liquid flows out and finally a graduation line to mark its capacity.

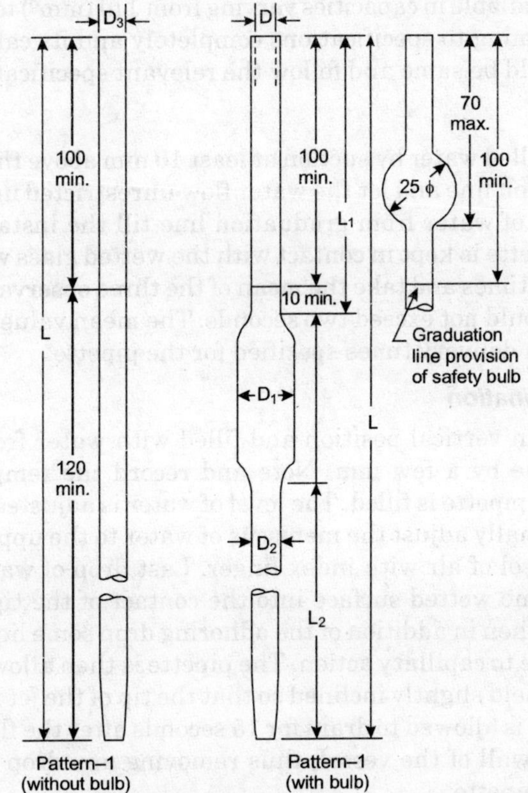

Figure 6.11 One mark bulb and straight pipettes

The bulb, suction tube and delivery tube should have a common axis. The graduation line is made by fine clean line completely round the suction tube. The circular line is square of the axis of the pipette. The top of the suction tube is square to the axis of the pipette and well ground. The ground surface must be smooth. The delivery jet should have a gradual taper. Sudden constriction at the orifice of the jet is not allowed. Size of the well grounded off jet must be such that the delivery time lies in between the specified limits.

One-mark pipettes are without bulb for capacity up to 1 cm^3 for class A pipettes and up to 2 cm^3 for class B pipettes.

6.4.1.2 Inscriptions

One-mark bulb pipette should have the following inscriptions:

- The delivery and the drainage time on the bulb of the pipette for class A pipettes only.
- Delivery time.
- Reference temperature.
- "Ex" or 'D' to indicate that the pipette is for delivery.
- The letter 'A 'or 'B' to indicate the class of accuracy of the pipette.

6.4.1.3 Capacity

One mark pipettes are available in capacities varying from 1 μl (mm^3) to 100 cm^3. Its effectiveness depends upon its conforming to specifications completely and its calibration. The methods of calibration and use should be same and follow the relevant specification.

6.4.1.4 Delivery Time

Fill the pipette with distilled water by suction at least 10 mm above the graduation line. Set the meniscus to the graduation line and let the water flow unrestricted under gravity. Observe the time taken for free flow of water from graduation line till the instant at which regular flow ceases. The tip of the pipette is kept in contact with the wetted glass wall of an inclined beaker. Repeat the process three times and take the mean of the three observations. Difference between any two observations should not exceed two seconds. The mean value should lie in between the maximum and minimum delivery times specified for the pipette.

6.4.1.5 Capacity Determination

The pipette is clamped in vertical position and filled with water from below through its jet, above the graduated line by a few mm. Note and record the temperature of water in the reservoir from which the pipette is filled. The level of water is adjusted by manipulating airflow with the index finger. Finally adjust the meniscus of water to the upper edge of the graduation line with still finer control of air with index finger. Last drop of water adhering to the jet is removed by bringing some wetted surface into the contact of the tip. If the jet is brought in contact of a dry surface, then in addition of the adhering drop some liquid from the inside of the jet may also come out due to capillary action. The pipette is then allowed to deliver into a clean and pre-weighed vessel held slightly inclined so that the tip of the jet is in contact with the side of the vessel. The pipette is allowed to drain for 15 seconds after the flow of water stops. The jet is still in contact of the wall of the vessel, thus removing any drop of water adhering to the outside of the jet of the pipette.

The difference between two weighing will give the apparent mass of water delivered. Instead of controlling airflow with index finger, a rubber bulb with air valve may be used. Note the temperature of water coming out of the pipette. Mean of two observed temperatures is used to apply corrections from the suitable Tables 3.1 to 3.24 corresponding to the mean temperature of water and other parameters. Addition of correction gives the corrected volume of water, which the pipette will deliver at the reference temperature. Three independent sets of values of water delivered by the pipette are taken and the mean value is calculated. The subtraction of the mean value of volume from the nominal capacity of the pipette gives the error. The error should not exceed the maximum permissible error as prescribed in the specification. No calculated value of water delivered at reference temperature by the pipette should differ by more than one third of the maximum permissible error.

For the purpose of counting the drainage time, the motion of the water down the delivery tube is observed, and the delivery time is supposed to be complete when the water meniscus comes to stop slightly above the end of the delivery jet. 15 seconds of drainage time is counted from this moment.

In some pipettes drainage time is not allowed.

To properly assess the position of the water meniscus with respect of graduation line, a black paper folded round the burette is held with a gem clip about a mm below the graduation line the meniscus then appears darker and with sharper contrast.

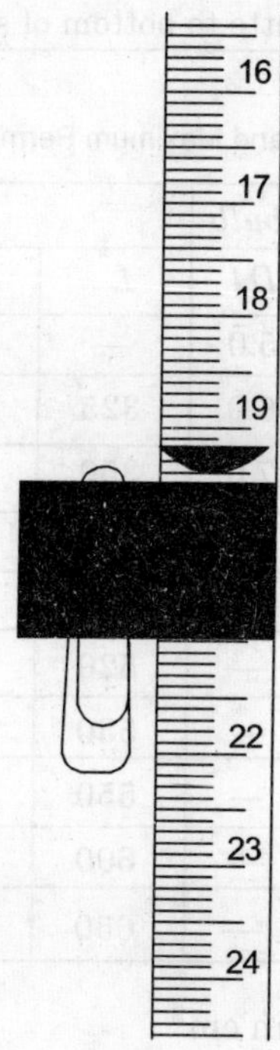

Figure 6.12 Use of black paper for meniscus setting

6.4.1.6 Mandatory Dimensions

Mandatory dimensions should also be measured with required accuracy and seen for compliance of the requirement of the specification. Length of the suction tube above the graduation line is necessary from the safety point of view. Total length of a pipette is important from packing point of view.

In some cases larger number of dimensions is to be checked for compliance, in that case put the requirements in a tabular form and tick mark for each fulfilled requirement. The table should include every clause of the specification, which requires compliance. The table may consist of only three columns. In the first column write the requirement of the specification, in the second column, the clause number of the specification and in the third column put the tick mark (✓) if complies otherwise put a cross (x).

Basic Common Dimensions for One-mark Pipettes [27]

Dimensions	*mm*
Minimum distance of graduation line from top of the pipette	100
Minimum Distances from graduation line to top of bulb,	10
Minimum Distance from graduation to tip of the delivery jet	120
Minimum Wall thickness	0.7
For 1.0 and 2 cm^3 having bulb	
For other pipettes	1.0
Minimum Diameter of the safety bulb if provided	25
Maximum Distance from top of pipette to bottom of safety bulb	70

Table 6.8 [27] Important Dimensions and Maximum Permissible Error for One mark-Pipettes

Capacity	*MPE*		*Without bulb*		*With bulb*					
C	*A*	*B*	*L*	*D4*	*L*	*L1*	*L2*	*D1*	*D2*	*D3*±1.5
0.1	± 0.005	± 0.01	280	5.0	—	—	—	—	—	—
1	± 0.007	± 0.015	280	6.0	325	150	110	3.0	9	5.5
2	± 0.010	± 0.02	280	7.0	350	150	125	3.5	9	5.5
5	± 0.015	± 0.03	—	—	410	150	145	4.0	12	6.5
10	± 0.02	± 0.04	—	—	450	160	160	4.5	16	6.5
20	± 0.03	± 0.06	—	—	520	170	210	5.5	22	7.0
25	± 0.03	± 0.06	—	—	530	170	220	5.5	24	7.0
50	± 0.05	± 0.10	—	—	550	170	230	6.0	30	7.5
100	± 0.08	± 0.16	—	—	600	170	240	7.5	38	8
200	± 0.10	± 0.20	—	—	650	170	240	8.5	49	9

MPE Maximum permissible error in cm^3
L Maximum Overall length in mm
D4 Maximum External diameter of tube in mm
L2 Minimum Length of suction tube in mm
L3 Minimum Length of delivery tube in mm
D1 Maximum Internal diameter of suction tube in mm
D2 External diameter of bulb in mm
D3 External diameter of delivery tube
C Capacity in cm^3
Min is minimum delivery time in seconds
Max is maximum delivery time in seconds
Diff Maximum permissible difference between observed and inscribed times in seconds.

6.4.1.7 Delivery Time of Pipettes Versus Capacity

Table 6.9 [27]

C	0.5	1.0	2	5	10	20	25	50	100	200
Class A pipettes										
Min	10 s	10	10	10	15	15	25	30	40	50
Max	20	20	25	30	40	50	50	60	60	70
Class B pipettes										
Min	7	7	7	10	10	20	20	20	30	40
Max	20	20	25	30	40	50	50	60	60	70
Diff	2	2	2	3	3	4	4	5	5	5

6.4.2 Graduated Pipettes

Graduated pipettes [31 to 35] consist of a graduated cylindrical tube drawn out to a delivery jet. In other words it is a burette without a stopcock or may be seen as cylindrical pipette with a graduated scale. So all the considerations, of delivery and drainage time, volume delivered and proper graduated scale, are equally applicable to these pipettes. Similarly requirements as applicable to burettes in respect of delivery time, graduated scale, and maximum permissible error are applicable to these pipettes. There are two types of graduated pipettes, namely type 1 pipettes and type 2 pipettes. Type 1 pipettes are those in which volume is defined between the lowest graduated line and the other graduation lines. For delivering the liquid, this type of pipettes are filled up to the graduation line and then allowed to deliver by free fall only about 1 cm above the required graduation line, from where the movement of liquid is controlled again, so that it just comes up to the top edge of the lowest graduation line. That is in such pipettes, delivery of the liquid is manipulated in a similar way as in the case of burette, so these may be considered as burettes without stopcock.

Type 2 pipettes are those in which volume delivered is defined between the tip of the jet to the graduated line. Such pipettes are not to be manipulated while delivering the volume from a specific mark and thus behave as a simple pipette.

Each of the two types of pipettes are available in class A and class B accuracy.

In addition to these there are three more classes of graduated pipettes, in which scale is graduated just like burette but capacity is from graduation line to the tip of the jet, in one no drainage time is allowed, in second 15 seconds drainage time is to elapse and in third, last drop is blown out by mouth. The last category of pipettes is also called as blown out pipettes.

Formally we may define these pipettes as:

1. Graduated pipettes adjusted for delivery of a liquid from zero line at the top to any graduation line. The lowest graduation line defines the nominal capacity. Accuracy wise pipettes of class A as well as of class B are permitted. No waiting time is required.

   These pipettes work as a burette. Graduations of class A pipettes are similar to the pipette 1 in Figure 6.13, while graduations of class B pipettes are similar to 2 of Figure 6.13. These are in fact type 1 pipettes.

2. Graduated pipettes adjusted for delivery of a liquid from any graduation line down to the jet. Nominal capacity is represented by upper most graduation line. Accuracy wise pipettes of class A as well as of class B are permitted. No waiting time is required.

   These pipettes work as an ordinary pipette described as type 2 above. Graduations of for class A pipette is shown Figure 6.13 pipette 3. Class B pipettes of this category is shown as pipette 4.
3. Graduated pipettes adjusted for delivery of a liquid from zero line at the top to any graduation line. Nominal capacity is in between the jet to the highest graduation line. Accuracy wise pipettes of class B only are permitted. No waiting time is required. These pipettes work as a burette. Graduations are shown in Figure 13.3 (5).
4. Graduated pipettes adjusted for delivery of a liquid from zero line to any graduated line. Nominal capacity is in between the jet to the lowest graduation line. Accuracy wise pipettes of class A as well as of class B are permitted. Waiting time of 15 seconds is required. These pipettes work as a burette.
5. Graduated pipettes adjusted for delivery of a liquid from any graduation line down to the jet. Last drop of liquid is taken out by blowing. These pipettes are sometimes known as blown out pipettes. Accuracy wise pipettes of class B only are permitted.

The pipettes defined at Serial number 3 to 5 work as an ordinary pipette, except the method of taking last drop, in 3 no drainage time, in 4 drainage time of 15 seconds and in 5 last drop of water is blown out.

The scale graduations of all types of pipettes are shown in Figure 6.13.

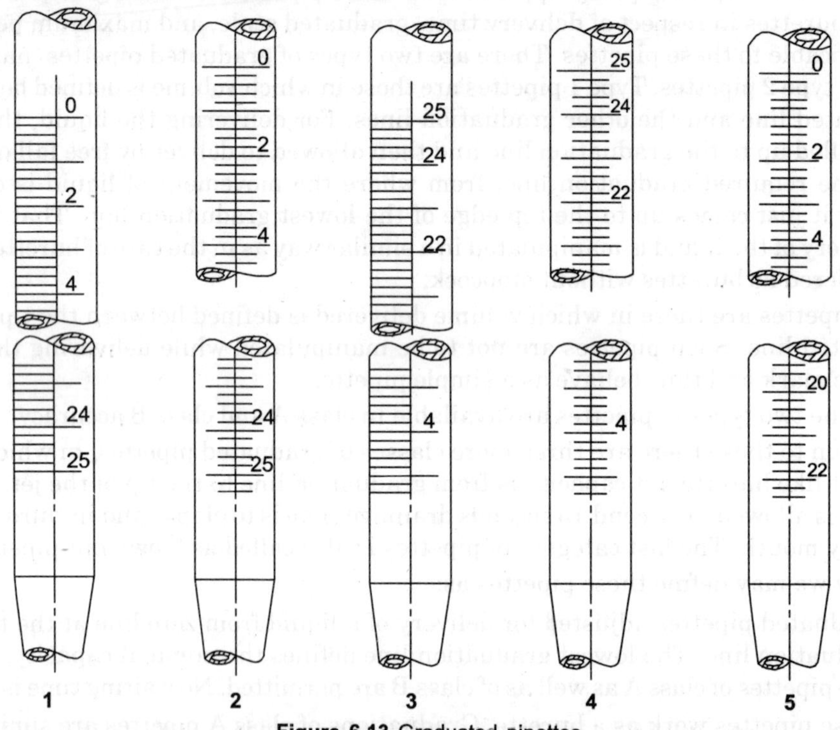

Figure 6.13 Graduated pipettes

6.4.2.1 Delivery Time, MPE for Graduated Pipettes

Delivery time and maximum permissible errors are indicated in Table 6.10.

Table 6.10

Delivery time and maximum permissible errors

Nominal Capacity cm³	*Delivery time type 1*				*Delivery time type 2*				*Delivery time type 3*				*Maximum Permissible error*	
	Class A		*Class B*		*Class A*		*Class B*		*Class A*		*Class B*		A	B
0.5													0.005	—
1	7	10	2	10	5	7	2	10	—	—	2	10	0.006	0.01
2	8	12	2	12	6	9	2	12	—	—	2	12	0.01	0.02
5	10	14	5	14	8	11	5	14	—	—	5	14	0.03	0.05
10	13	17	5	17	10	13	5	17	—	—	5	17	0.05	0.1
25	15	21	9	21	11	16	9	21	—	—	9	21	0.1	—
25													0.1	0.2

6.5 MICRO-PIPETTES

6.5.1 Capacity and Colour Code

Micropipettes of capacities of 1, 2, 3, 4, 5, 6, 7, 8, 9, 10, 15, 20, 25, 35, 50, 60, 75, 100, 150, 200, 250, 300, 400, 500, 1000 mm³ (µl) are available [37].

Micro-pipettes of capacities 0.2, 0.5, 1, 2 and 3 cm³ are shown in Figure 6.14. While the pipettes, of 0.1 ml and 0.2 ml shown in Figure 6.15, are content type micropipettes. A wash out micropipette is shown in Figure 6.16. Micro-weighing pipettes are shown in Figure 6.17 and 6.18 and are used for density determination of liquids available in very small quantity. Pipettes in capacity of 1 mm³ to 1 cm³ (1µl to 1000 µl), shown in Figures 6.19 and 6.20, are known as micro-litre pipettes.

It is difficult to tell the capacity of a micropipette by simple vision especially for micropipettes of smaller capacity. To facilitate in assessing the capacity a micropipette, a certain colour code is followed. The band or bands of a certain colour will straight away tell the nominal capacity; As per Indian standards [36] it is as follows:

Colour	Nominal capacity of a micropipette
White	5 µl
Orange	10 µl
Black	20 µl
2 white bands,	25 µl
Green	47.5 µl and 50 µl
Blue	100 µl
Red	200 µl

6.5.2 Nomenclature of Micropipettes

In order to differentiate among various pipettes, one way of classification is according to the use they are put, for example pipettes shown in Figures 6.14, 6.15, and 6.16 are specially designed for use in biological and clinical chemistry, while pipettes shown in Figure 6.17 are also a micropipette but are used for determination of density of liquids available in smaller quantity and are known as micro-weighing pipettes or pycnometers. So micro-pipettes are named as

Measuring micropipettes for delivery (Figure 6.14)
Folin's type micropipettes for content (Figure 6.15)
Washed out micropipettes for content (Figure 6.16)
Micro-weighing pipettes density type for content (Figure 6.17 and 6.18)
Micro-litre pipettes for content (Figures 6.19 and 6.20)

6.5.3 Measuring Micropipettes

These have been shown in Figure 6.14 and are available in five capacities, namely 0.2 cm^3, 0.5 cm^3, 1 cm^3, 2 cm^3 and 3 cm^3. All pipettes have similar tips with glazed ends.

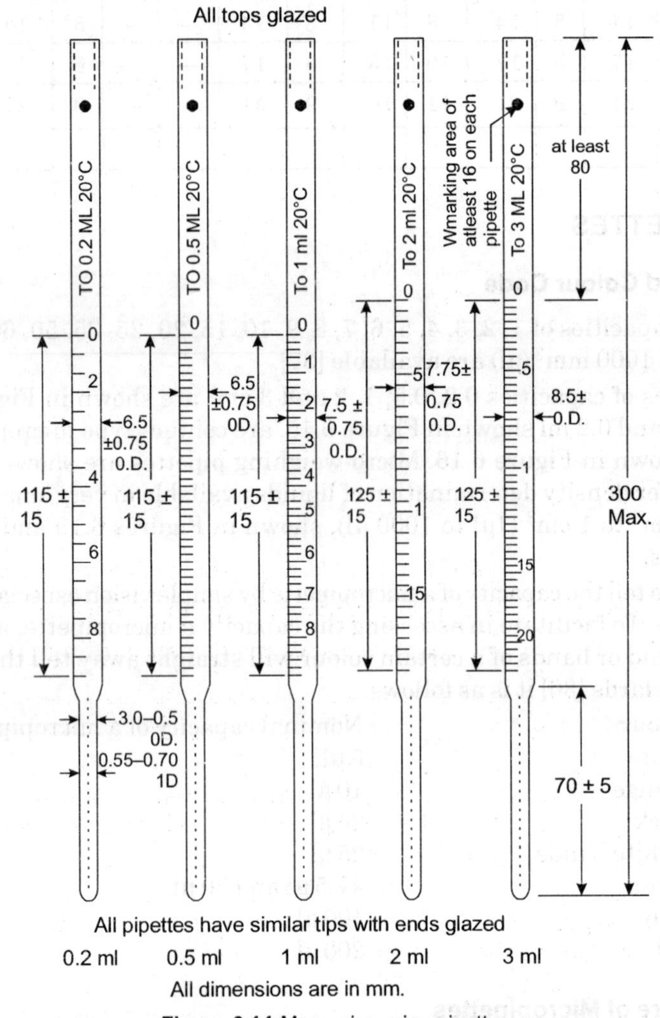

Figure 6.14 Measuring micropipette

Requirements pertaining to graduations are given in the Table 6.11 with the notations defined as follows:

C: Capacity in cm^3
S: Subdivisions in cm^3
G: Graduated interval from 0 to G in cm^3

Fm: Full circular graduation line at every in cm^3
Hm: Half ring at each mark in cm^3
No: Numbered at zero and every cm^3
MPE: Maximum permissible error (Tolerances) cm^3

Table 6.11[37] Basic Dimensions and MPE of Measuring Micropipettes

Capacity	*S*	*G*	*Fm*	*Hm*	*No*	*MPE*
0.2 cm^3	0.01	0.18	0.02	0.01	0.02	0.005
0.5 cm^3	0.01	0.45	0.05	0.01	0.1	0.01
1 cm^3	0.02	0.90	0.1	0.02	0.1	0.02
2 cm^3	0.05	1.75	0.25	0.05	0.5	0.04
3 cm^3	0.05	2.70	0.25	0.05	0.5	0.06

Basic dimensions are indicated in the Figure 6.14.

6.5.4 Folin's Type Micropipettes

These are available in two capacities namely 0.1 and 0.2 cm^3. The delivery tips are ground and bevelled. All graduations should be at least three fourth of the circle. All numbered graduations should be full circle. Wall thickness shall nowhere be less than 0.5 mm. Basic dimensions of these pipettes are indicated in Figure 6.15.

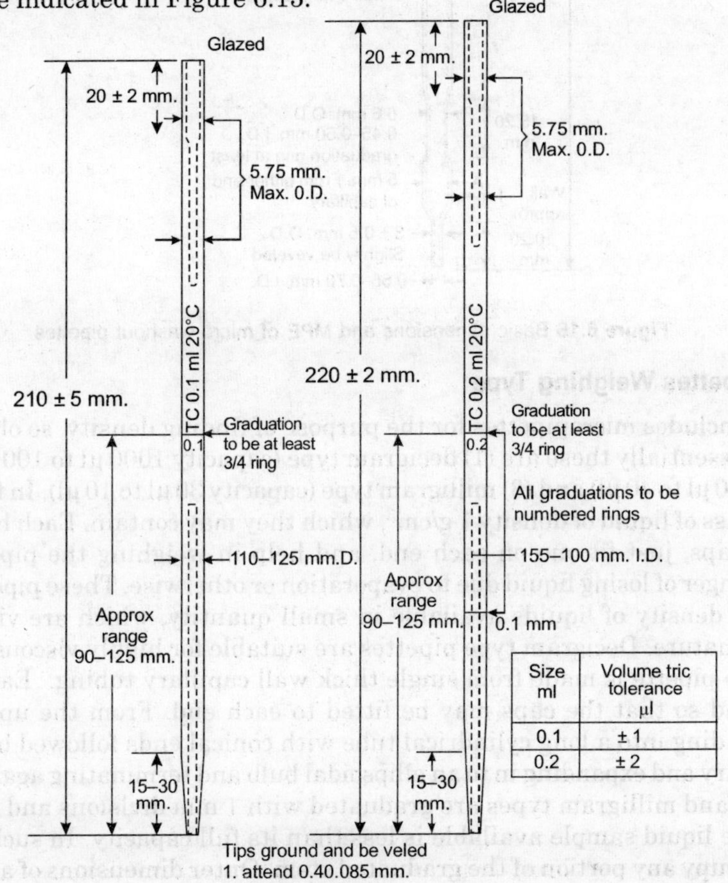

Figure 6.15 Basic dimensions and MPE of Folin's type micropipettes

6.5.5 Micro Washout Pipettes

Micro washout pipettes are of four capacities namely 0.1 cm^3, 0.2 cm^3, 0.5 cm^3 and 1.0 cm^3. Quantitative delivery of the volume indicated by the graduation line is obtained by rinsing out the contents with wash liquids added from the top of the pipette. The name of washout pipette comes from this requirement. The delivery tip is slightly bevelled. A typical micro-washout pipette is shown in Figure 6.16.

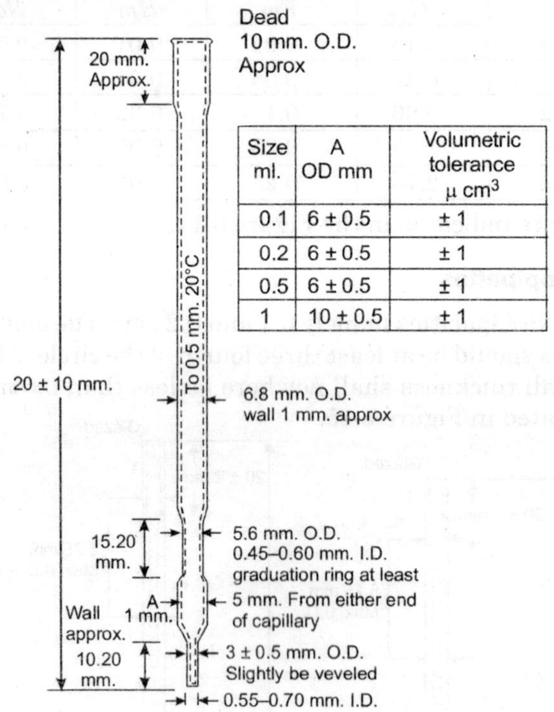

Size ml.	A OD mm	Volumetric tolerance µ cm^3
0.1	6 ± 0.5	± 1
0.2	6 ± 0.5	± 1
0.5	6 ± 0.5	± 1
1	10 ± 0.5	± 1

Figure 6.16 Basic dimensions and MPE of micro-washout pipettes

6.5.6 Micro Pipettes Weighing Type

This category includes micropipettes for the purpose of finding density, so obviously these are content type. Essentially these are (1) decigram type (capacity 1000 µl to 100 µl), (2) centigram type (capacity 80 µl to 40 µl) and (3) milligram type (capacity 30 µl to 10 µl). In fact nomenclature is as per the mass of liquid of density 1 g/cm^3, which they may contain. Each has a ground glass cap. The two caps, just fit one on each end, and help in weighing the pipette in a balance without any danger of losing liquid due to evaporation or otherwise. These pipettes are designed to find out the density of liquids available in small quantity, which are viscous, volatile or hygroscopic in nature. Decigram type pipettes are suitable for highly viscous liquids.

The whole pipette is made from single thick wall capillary tubing. Each end is tapered and well ground so that the caps may be fitted to each end. From the upper end, it has a capillary expanding into a long cylindrical tube with conical ends followed by about the same length of capillary and expanding in to an ellipsoidal bulb and terminating again into a capillary. The centigram and milligram types are graduated with 1 mm divisions and can, therefore, be used even if the liquid sample available is less than its full capacity. In such cases the liquid sample may occupy any portion of the graduated stem. Outer dimensions of all micro-weighing pipettes are same.

A typical micro-weighing pipette of decigram type is shown in Figure 6.17A, and that of centigram and milligram types in Figure 6.17B. Dimensions as per ASTM [37] are given below.

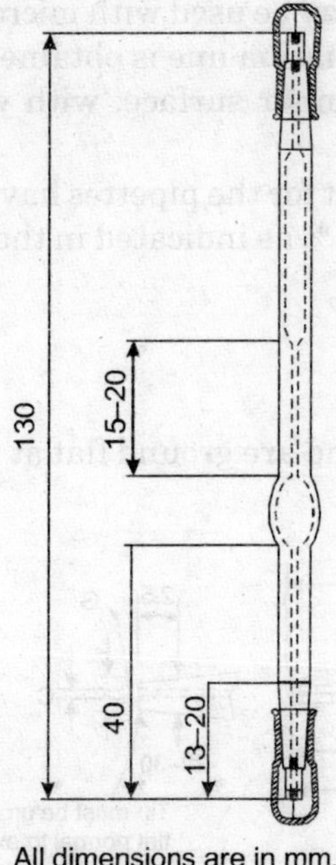

All dimensions are in mm.

Figure 6.17A Decigram type

All dimensions are in mm.

Figure 6.17B Centigram and milligram type

Overall length from one to another end faces without taking thickness of cap into account 130 ±5 mm

Length of tapered portion including cylindrical tip at each end 13 to 20 mm

Length of cylindrical tip 4 to 6 mm

Length of capillary between two bulbs 15 to 20 mm

Distance from bottom face to highest graduation mark 100±1 mm

Outer diameter of the capillary tube 5.0 to 5.5 mm

Maximum outer diameter at the ellipsoidal bulb 7 mm

The first bulb from the tip 40 mm

Cap of the Pipette

The cap, which fits at each end, is shown in Figure 6.18. Its total length is 18 to 20 mm and width is 7 to 8 mm.

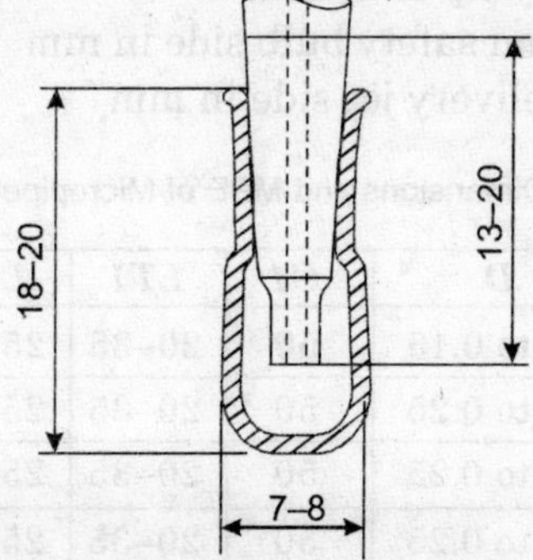

All dimensions are in mm.

Figure 6.18 Cap of micro weighing pipette

6.5.7 Micro-litre Pipettes of Content Type

These micropipettes are available in 25 capacities from 1 μl to 4 μl in steps of 1 μl and 5 μl to 1000 μl. The dimensions of these pipettes are such that they may be used with micro volumetric flasks. Complete delivery of the volume indicated at the graduation line is obtained by rinsing out several times, to remove any solution adhering to the inner surface, with wash liquid, drawn up from its tip.

Maximum permissible error MPE (Tolerance) is 1 percent for the pipettes having capacity from 1 μl to 4 μl. For larger pipettes it varies from 0.5 % to 0.2 % as indicated in the Table 6.13.

6.5.8 Micro-litre Pipettes

6.5.8.1 Micropipettes of Capacity from 1 μl to 4 μl

A typical micropipette is shown in Figure 6.19. Tips at each end are ground flat at right angles to its axis and glazed.

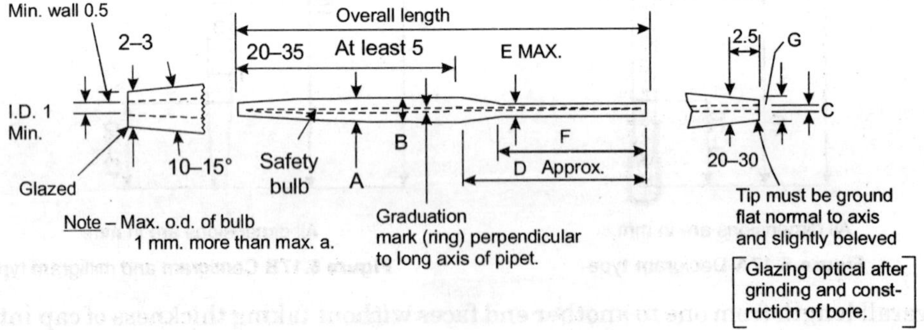

Figure 6.19 Micro-litre pipettes (1μl to 4 μl)

Other dimensions are given in the Table 6.12 with the following notations:

C Capacity in μl
L1 Overall length in mm
ID Internal diameter of the tubing in mm
OD Minimum diameter at the ends in mm
C1 Minimum capacity of safety bulb SB in ml
MPE Maximum Permissible Error (Tolerance) in percentage
LT1 Length of taper on safety bulb side in mm
LT2 Length of taper on delivery tip side in mm
WT1 Minimum wall thickness on safety bulb side in mm
WT2 Minimum thickness on delivery jet side in mm

Table 6.12 [37] Basic Dimensions and MPE of Micropipettes (1 μl to 4 μl)

C	*L1*	*ID*	*D*	*C1*	*LT1*	*LT2*	*WT1*	*WT2*	*MPE*
1	140±5	0.12 to 0.16	0.10 to 0.15	50	20–35	25–40	0.5	0.5–0.75	1
2	140±5	0.16 to 0.25	0.15 to 0.25	50	20–35	25–40	0.5	0.5–0.75	1
3	140±5	0.20 to 0.28	0.15 to 0.25	50	20–35	25–40	0.5	0.5–0.75	1
4	140±5	0.24 to 0.32	0.15 to 0.25	50	20–35	25–40	0,5	0.5–0.75	1

6.5.8.2 *Micropipettes from* 5 µl *to* 1000 µl

Dimensions and maximum permissible errors of micropipettes are given below:

C Capacity of the micro-litre pipette in µl
L1 over all length in mm
OD outer diameter of tubing in mm
D Maximum outer diameter of tubing
ID Minimum internal diameter of tubing in mm
ID1 inner diameter at ends in mm
J Approximate length of the delivery jet in mm
L3 minimum length of tapered portion of the deliver jet in mm
W Minimum wall thickness at end in mm
C1 minimum capacity of the safety bulb SB in µl
MPE maximum permissible error (Tolerance) of the pipette in percentage

Some dimensions to all pipettes between 5 µl to 1000 µl are common which are shown in the Figure 6.20.

Table 6.13[37] Basic dimensions and MPE of micropipettes (5 µl to 1000 µl)

C	*L1*	*OD*	*ID*	*ID1*	*J*	*D*	*L3*	*W*	*C1*	*MPE*
5	140±5	5 to 6	0.18 to 0.25	0.15 to 0.25	60	4	55	0.5–0.7	50	0.5
6	140±5	5 to 6	0.18 to 0.25	0.15 to 0.25	65	4	55	0.5–0.7	50	0.5
7	140±5	5 to 6	0.18 to 0.25	0.15 to 0.25	65	4	55	0.5–0.7	50	0.5
8	140±5	5 to 6	0.18 to 0.25	0.15 to 0.25	65	4	55	0.5–0.7	50	0.5
9	140±5	5 to 6	0.18 to 0.25	0.15 to 0.25	65	4	55	0.5–0.7	50	0.5
10	140±5	5 to 6	0.20 to 0.35	0.15 to 0.35	65	4	55	0.5–0.7	50	0.5
15	140±5	5 to 6	0.25 to 0.40	0.15 to 0.40	65	4	55	0.5–0.7	50	0.5
20	140±5	5 to 6	0.35 to 0.50	0.25 to 0.50	65	4	55	0.5–0.7	50	0.5
25	140±5	5 to 6	0.35 to 0.50	0.25 to 0.50	65	4	55	0.5–0.7	50	0.5
35	140±5	5 to 6	0.35 to 0.50	0.25 to 0.50	65	4	55	0.5–0.7	50	0.3
50	140±5	5 to 6	0.35 to 0.50	0.25 to 0.50	65	4	55	0.5–0.7	50	0.3
60	140±5	5 to 6	0.40 to 0.55	0.30 to 0.55	65	4	55	0.5–0.7	50	0.3
75	140±5	5 to 6	0.40 to 0.60	0.30 to 0.50	65	4	55	0.5–0.7	75	0.3
100	140±5	5 to 6	0.50 to 0.75	0.30 to 0.50	65	4	55	0.5–0.7	75	0.3
150	140±5	5 to 6	0.75 to 1.00	0.40 to 0.60	65	4	55	0.5–0.7	100	0.3
200	145±10	5 to 6	0.75 to 1.00	0.40 to 0.60	65	4	55	0.6–0.8	100	0.2
250	145±10	5 to 6	0.75 to 1.00	0.40 to 0.60	65	4	55	0.6–0.8	100	0.2
300	145±10	5 to 6	0.75 to 1.00	0.40 to 0.70	65	4	55	0.6–0.8	200	0.2
400	150±10	6 to 67	1.00 to 1.25	0.40 to 0.70	70	6	60	0.6–0.8	200	0.2
500	160±10	6 to 7	1.25 to 1.50	0.40 to 0.70	70	6	60	0.6–0.8	200	0.2
1000	170±10	7 to 8	2.00 to 2.25	0.40 to 0.70	80	7	60	0.6–0.8	300	0.2

Note: All dimensions are accordance of ASTM [37].

There are, disposable glass micropipettes for various purposes [43, 56], blood collection pipettes [44, 45], and disposable Pasteur type pipettes [46, 47]

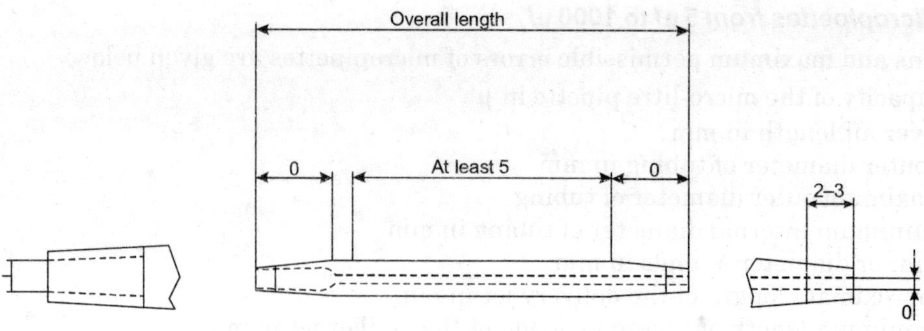

Figure 6.20 Micro-litre pipettes (5 μl to 1000 μl)

6.6 SPECIAL PURPOSE GLASS PIPETTES

6.6.1 Disposable Serological Pipettes

Capacity

Serological pipettes are available in capacities of 0.1 cm^3, 0.2 cm^3, 0.3 cm^3, 0.5 cm^3, 1 cm^3, 2 cm^3, 5 cm^3 and 10 cm^3 [48 to 51]. These are made from either Borosilicate or soda glass conforming to the requirements of ASTM 714 or any other national standard specification.

Construction

It is straight and one piece-construction. Its cross section at any point perpendicular to its longitudinal axis is circular. The delivery tip is made with gradual taper of 10 mm to 20 mm for capacities up to 2 cm^3 and 15 mm to 30 mm for 5 cm^3 and 10 cm^3 capacities. Mouthpiece of 10 cm^3 pipette is tooled to a diameter of 7 mm to 9 mm and is of length 15 mm to 25 mm. Alternatively mouthpiece is of unreduced diameter of the pipette with a constriction located 15 to 25 mm from the top. Each mouthpiece end is fired when tooled or constricted, the mouthpiece should be suitable for plugging with filtering material.

Fine lines of thickness 0.2 to 0.5 mm are graduated in a plane perpendicular to the longitudinal axis of the pipette and are parallel to each other. Main graduation lines are extended to at least three fifth of the way around the pipette. All these main graduation lines are numbered. Intermediate graduation lines are at least one fifth of the circle round the pipette and the smallest graduation lines are of at least one seventh of the circle round the pipette. Zero graduation line must be at least 90 mm below the top.

The pipette is marked with the reference temperature, capacity and symbol to indicate that pipette is for delivery.

Dimensions

Dimensions, delivery time and maximum permissible error are given in Table 6.14 with the following notations:

C nominal capacity in cm^3
S is least value i.e. volume between successive graduation lines
No numbered at graduations lines indicating the capacity
G Minimum range of graduations from 0 to in cm^3
D Outside diameter of the graduated portion of the tube
T is delivery time in seconds

W Minimum wall thickness in mm
A is maximum permissible error in percentage
A in this case is defined as percentage deviations of the mean X_{mean} value from the stated capacity *C* i.e.,

$$A = 100(X - X_{mean})/C$$

V Coefficient of variation = 100.SD/Mean value, where *SD* is the standard deviation from the mean = $\sqrt{[\Sigma(X - X_{mean})^2/(n-1)]}$

Normally MPE *A* and coefficient of variation *CV* is determined by measuring the capacity of thirty randomly selected pipettes.

Table 6.14[50] Basic dimensions and MPE of Serological pipettes

C	*S*	*No*	*G*	*D*	T_{Min}	T_{Max}	*W*	% *A*	% *V*
0.1	0.01	0.01	0.09	3.5 to 4.0	0.5	3	1.0	± 7	≤ 2.5
0.2	0.01	0.02	0.18	3.5 to 4.5	0.5	3	1.0	± 6	≤ 2.0
0.2	0.01	0.05	0.18	3.5 to 4.5	0.5	3	1.0	± 3	≤ 1.5
0.5	0.01	0.1	0.4	4.25 to 4.75	0.5	3	0.8	± 3	≤ 1.5
1.0	0.01	0.1	0.9	4.25 to 4.75	0.5	3	0.8	± 3	≤ 1.5
1.0	0.1	0.1	0.9	4.25 to 4.75	0.5	3	0.8	± 3	≤ 1.5
2.0	0.01	0.1	1.9	5.5 to 6.0	0.5	5	0.8	± 3	≤ 1.5
5.0	0.1	0.1	4.5	7.5 to 8.25	3.0	10	0.8	± 3	≤ 1.5
10.0	0.1	01	9.0	9.5 to 11.25	4.5	15	0.8	± 3	≤ 1.5

6.6.2 Piston Operated Volumetric Instrument

In this section piston operated instrument or pipettor with pipette tips [57 to 64] are being discussed.

Definitions

1. Deficiency of the pipettor is the ratio of the difference between the mean value of the volume delivered and its nominal capacity.
2. Coefficient of variation of a pipettor is the percentage ratio of the standard deviation from the mean to the mean of the volume delivered.
3. Micro-litre (μl) volume is any volume in between one micro-litre (1 μl) and one thousand micro-litres (1000 μl).

The Piston operated apparatus (pipettes /burettes) are of 'two types:

Type I –Air displacement (type A)

The volume of liquid is drawn into or dispensed from the apparatus tip by a measured volume of air. The precise movement of a close fitting airtight piston in a cylinder determines the volume of air. Liquid does not come in contact with the piston as only the pipette tip is dipped in liquid.

Type II- Positive displacement (type D)

The volume of liquid is drawn into or dispensed from its tip by mechanical action, which displaces measured liquid within the tip. The precise movement of a close fitting piston within the pipette determines the volume of liquid. Liquid comes in contact both with the pipette tip and piston.

6.6.2.1 Piston Operated Pipettes

Piston operated pipettes of single channel and multi channel pipettes are shown in Figures 6.21.

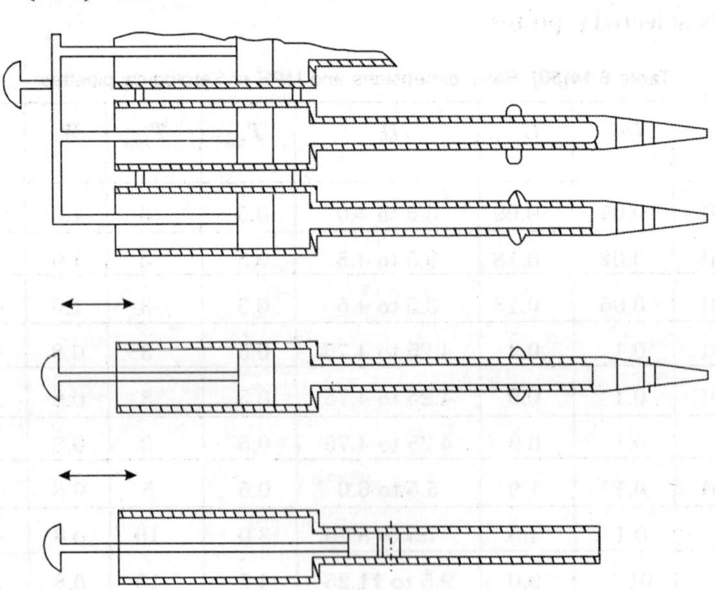

Figure 6.21 Piston operated pipettes (Top one is multi-channel)

Principle of Operation

The tip of the pipette made of plastic or glass is attached to the piston pipette. With piston in lower operation limit, the tip is dipped into the liquid to be dispensed as a measured volume. When moved to the upper aspiration limit, the piston aspirates the liquid. Moving the piston downward expel the liquid volume. Some air displacement pipettes type A have an extra air volume, which can be used to expel the last drop of the liquid.

Pipette tips are available in two types (1) Replaceable pipette tip i.e. which can be used repeatedly and (2) Disposable pipette tip, which is to be used only once. In case of type D pipettes either the plunger or the capillary or both may be reusable (type D1) or disposable (type D2).

Design

The piston pipettes is of fixed volume, designed and adjusted by the manufacturer to dispense only the specified volume.

These are also designed in such a way that they dispense selectable volumes within certain range by the user, for example between 10 mm^3 to 100 mm^3 (10 µl to 100 µl).

Testing of a Piston Pipette

Air displacement pipettes of capacity 1 µl to 9 µl are tested by gravimetric method using water as standard and a balance having uncertainty of not more than 1 µg. 30 deliveries of volume from the pipette are taken. Mean value of the water delivered at the reference temperature is calculated. Standard deviation from mean is also calculated. Efficiency should be better than the prescribed and coefficient of variation should be less than the prescribed limits given in Table 6.15. Balance for 10 µl to 1000 µl pipettes may be of uncertainty better than 10 µg. Built-in weights in the balance should be pre-calibrated with commensurate but known uncertainty.

Adjustment of the pipettes refers to a temperature of 20 °C with 50% relative humidity and air pressure of 101 kPa.

Positive displacement pipette with capacities 1 µl to 9 µl may be tested with water or with tripled distilled mercury as medium. A pipette for more viscous liquids than water should be tested with oils of known viscosity and density at reference temperature and pressure.

Table 6.15 [57] MPE and Coefficients of Pipettor

Capacity in mm³	*Accuracy %*	*Coefficient of variation %*
1 to 9	± 4.0	± 4.0
10 to 99	± 3.0	± 3.0
100 to 200	± 2.0	± 2.0
200 to 1000	± 1.0	± 1.0

Accuracy and coefficients of variation given in Table 6.15 are taken from ASTM [57] for hand held pipettor with pipette tips.

6.6.2.2 Piston Burettes

Piston burettes are used for the accurate delivery of liquids. In contrast to piston pipettes, dispensers and dilutors which are designed for accurately prescribed volumes, piston burettes are required to dispense volumes of liquids until external criteria as pH or conductivity are met, at which point it is necessary to know accurate volume dispensed.

The piston can be operated manually, or by electronic means. The drive, the piston and the cylinder can be one unit or modular to permit the use of different pistons and cylinders with the same drive.

Prior to delivery, the piston system is charged by aspiration of liquid from reservoir. After air free filling of the liquid, movement of the piston in one direction dispenses the liquid whose volume is to be measured; movement in the other direction recharges the system with liquid from the reservoir, please refer to Figure 6.22.

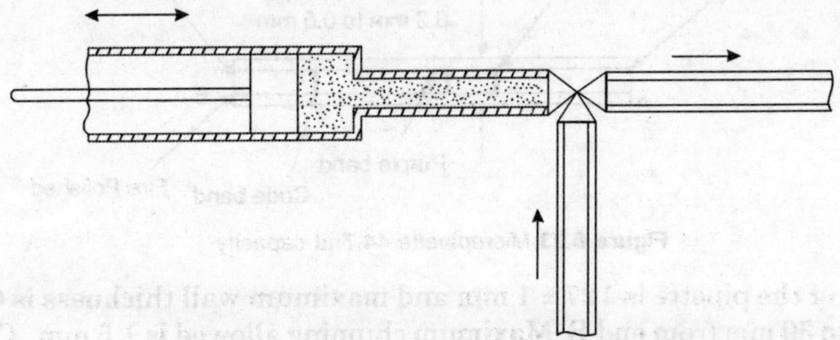

Figure 6.22 Schematic drawing of a piston burette

6.6.3 Special Purpose Micro-pipette (44.7 μl capacity)

These pipettes are content measures and are of two types:

Type I: Coated with heparin

Heparin of Sodium salt isolated from the intestinal mucosa of hog origin. The heparin potency should be 1 mg of sodium heparin compound and is equal to 100 United States Pharmacopoeia (USP) units.

Type II: Uncoated

The pipettes of type I are fabricated from borosilicate glass and those of type II can be made of soda glass also.

The pipettes are of one piece in glass with circular section.

Testing of a pipette

Using mercury: Allow dry pipette and triple distilled mercury to stand at room temperature for at least 2 hours so that temperature equilibrium is reached between mercury, standard weights of the balance or separate weights required for weighing and the pipette. Mercury is filled cautiously with the help of an air bulb. Set the meniscus so that lower edge of the graduation line forms a horizontal tangent at the highest point of the meniscus. To get it in a convenient and sure way is to put a black paper with a fine edge a little above the meniscus. This will make the profile of the meniscus dark against a light background. Discharge the mercury in a clean and pre-weighed dish; weigh it and obtain the apparent mass of mercury discharged, from the temperature reading of the mercury in container find out the factor for multiplication and get the volume of mercury delivered by the pipette. Repeat such observations 30 times and find the average, standard deviation from the mean and the coefficient of variation to assess the deficiency of the pipette. Difference between the mean of discharge from the stated value gives the error, which in no case should exceed the prescribed MPE.

Some literature suggested that the meniscus may be so set that horizontal plane through the middle of the graduation line is tangent to the highest point of the meniscus. The difference in volumes by the two methods will be equal to volume of a cylinder of height equal to half the width of the graduation line and area of cross-section of the pipette at the line of graduation. This difference, for nicely made pipettes will be around 0.4 percent.

Using water: When distilled water is used then balance used must have an over all uncertainty not worse than 1 μg and it is at the lowest point of the meniscus at which the upper edge or the middle of the graduation line is tangent.

Shape: A typical 44.7 μl pipette, as per ASTM [52] requirements, is shown in Figure 6.23.

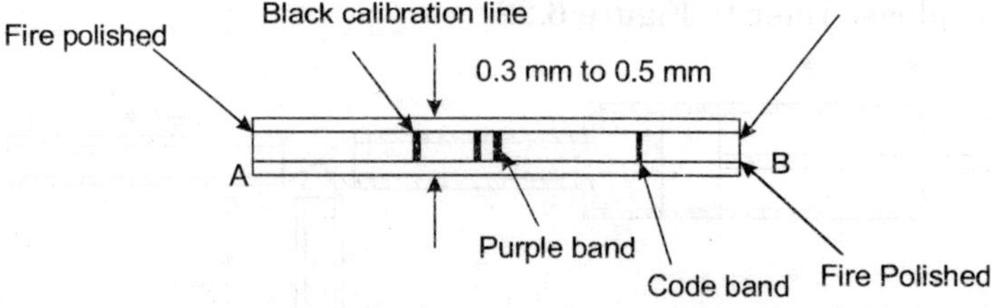

Figure 6.23 Micropipette 44.7 μl capacity

Total length of the pipette is 127 ± 1 mm and maximum wall thickness is 0.5. Code band should be 15 mm to 30 mm from end B. Maximum chipping allowed is 1.5 mm. Calibration line should be 50 to 90 mm from end A.

6.7 AUTOMATIC PIPETTE

6.7.1 Automatic Pipettes in Micro-litre Range

The pipette was first designed and used in National Physical Laboratory, U.K. [66], we at NPL, India, also made a similar pipette to deliver 0.3125 $cm^3 \pm 0.0003$ cm^3 at 27°C. This pipette proved to be useful for testing of the scale of a butyrometer representing the fat content in milk.

Construction

It has three basic components, a filling BCD tube in the form of "U" have a funnel attached to the longer arm at B with short high-pressure tubing. The U tube C to D is a capillary at D it is attached to another vertical tube DHE with a short high-pressure tubing. The tube DHE consist of a stopcock S1 bifurcating at M in two tubes, the straight vertical tube from H to E, which serves as measuring tube and have a very fine capillary from H to G, opening in to a bulb GF and again terminating to a capillary tube up to E. The automatic zero action is achieved by closing of the top of the tube FE by the inner surface of a spherical glass cap. The upper end of FE is ground to a truncated cone with a tip. The tip is finely ground flat, which is perpendicular to the axis of tube FE. Mercury flowing up the tube FE displaces air, which escapes between the flat end at E and the spherical cap, and when the mercury reaches the top of the tube, does not lift the cap. This is achieved by maintaining the equilibrium between pressure exerted by the level difference between the level of mercury in the funnel and the end E of the cap and the weight of the spherical cap. The flow of mercury ceases immediately the rising mercury surface reaches the top of the tube FE and this gives a very consistent cut-off of the inflowing mercury and the automatic zero. An inverted U shape tube with a stopcock S2 terminating into a fine capillary, works as a delivery system. The inverted U tube has a wider tube from stopcock S2 to K a point close to the highest point after that it turns into a capillary of fine bore. The tip of the inverted U tube is a few mm above the horizontal plane passing through H. The end O is ground polished. The delivery tube LO is parallel to tube EH. Function, of the two rubber tubes, is to join the respective components, so that apparatus becomes easy to handle and is easily dismantled for cleaning. Whole apparatus can be easily mounted on a flat board, which then is held in a vertical plane. The whole pipette is attached to the board in such a way that the tube EF is vertical. The pipette along with dimensions is shown in Figure 6.24. Except the relative diameters of capillary GH and delivery limb terminating at O, all other dimensions are for guidance only.

Working

Initially both stopcocks are kept closed and mercury is poured in the funnel, some mercury may enter in BCD but capillary portion CD remains empty. The stopcock S1 is very slowly opened so that mercury starts filling the pipette, then Stopcock S2 is opened and mercury fills the whole of inverted U tube and starts flowing out from orifice O. So a small beaker may be placed under it to avoid spilling of mercury. Close the stopcock S2, ensure that no air bubble is entrapped and mercury stands at E under pressure of spherical cap, which slides over the tube. Now if the stopcock S2 is opened and S1 is kept closed, then mercury will flow out, from the orifice O, to H the bottom of capillary GH. The point H is a few mm below the point O. When towards the end of delivery, the mercury surface enters GH the momentum of the flowing mercury causes initially to fall below the level of O at least to the level H. The hydrostatic pressure due to mercury level difference between O and H and the excess pressure due to the curvature of meniscus at O will restore mercury surface in GH above O, but the capillary depression in tube GH prevent this and the level of mercury comes to rest at the bottom of GH where the tube begins to open out. Thus a closely reproducible end-point is ensured. A drop of mercury usually remains pendant from the orifice O, which is without internal taper. The drop is detached and taken as the part of delivered volume. Such variations as occur in the precise level where the mercury detaches itself at the orifice are not large enough due to very small diameter of capillary to affect significantly the volume delivered. Same thing is true is respect of capillary GH. It may be pointed out that although the capillary GH opens out almost imperceptibly into the bulb above it, small air bubbles are liable to become trapped at the junction of the capillary and bulb if the rate of filling is two great. The rate of filling is thus adequately controlled by the narrow part of CD of the filling tube. The pipette takes about 15 seconds to fill up.

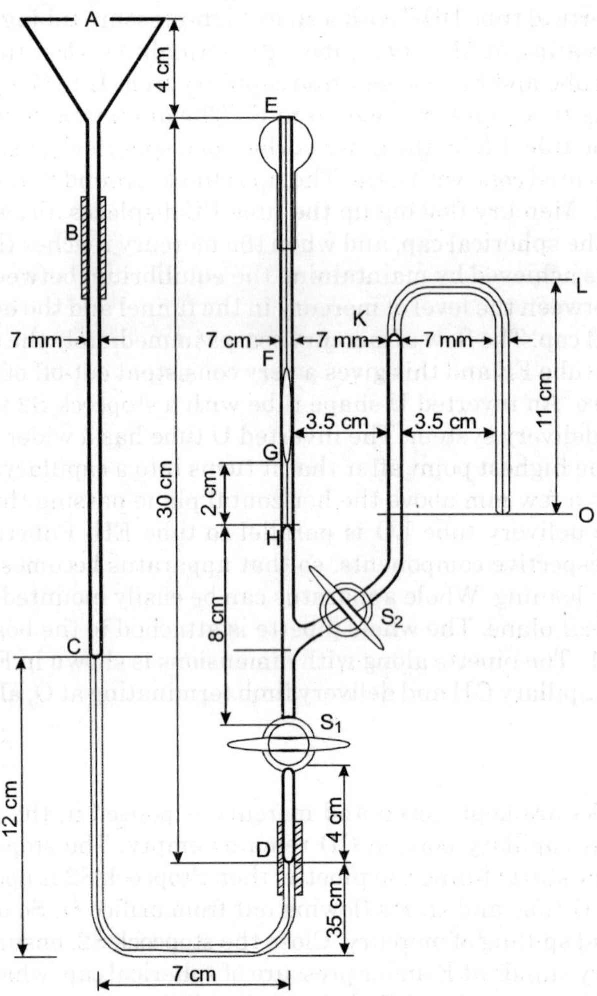

Figure 6.24 Automatic pipette of 0.3125 cm^3

Positions of stopcocks are also noteworthy. Their positions are such that stopcocks are always full of mercury and the flow is always upward through each of them, so that should any air bubble is trapped during initial filling they are always carried away by the mercury stream. Moreover, it is not possible for air to enter the stopcocks, as the pressure inside is more than the outside pressure. For this reason, there is also a less tendency of lubricant in the stopcock to works its way into mercury, and contaminate it. The volume of mercury delivered is not significantly affected by small variation in the height of orifice O relative to H the bottom of capillary. It is therefore sufficient if the pipette is set up with EH tube vertical as judged by eye.

If after one delivery, the stopcock S2 is closed and stopcock S1 is opened so that mercury from the funnel again fills the tube HE. The surface area of the funnel is so large at the level of mercury such that there will be no appreciable fall in its level so mercury will terminate at E again. If we close the stopcock S1 and open the S2, always the mercury from E to H will flow out from O. This action will be repeated as many times as one wishes to do.

Volume delivered will be equal to the volume of mercury in the tube HE. Bulk of volume is in bulb GF, so altering the size of the bulb GF, the pipette capable of delivering another volume may be constructed.

Automatic pipettes of capacities say from 5 cm^3 to 10 dm^3 are commercially available [67]. A typical automatic pipette is described below.

6.7.2 Automatic Pipettes (5 cm^3 to 5 dm^3)

By defining the capacity of pipette by overflow of liquid from the top, automatic pipettes from 5 cm^3 to 5 dm^3 are being made. In this case there is no need of adjusting the level of the liquid upto a certain specified graduation mark. One such automatic pipette is shown in Figure 6.25. It has

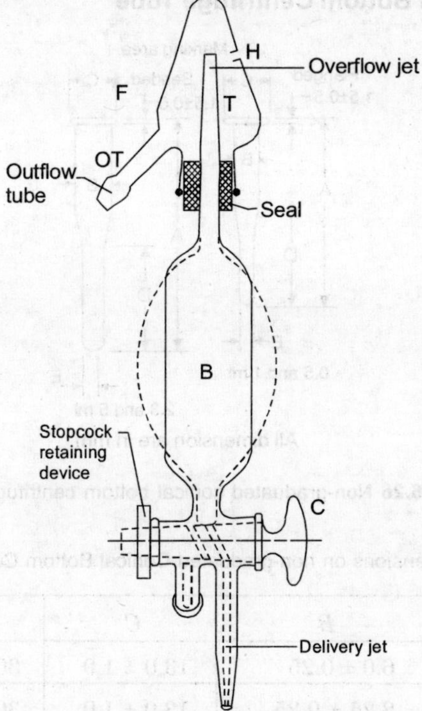

Figure 6.25 Automatic pipette

a 3-way stopcock C with a separate tube for filling the pipette with water or any other liquid by the action of gravity. In Figure 6.25, the position of the stopcock is such that delivery jet is connected to the main body, i.e. the pipette is in delivery mode. If we turn the stopcock through 180°, then body gets connected to the input tube, so that liquid can be filled. Bulk of the liquid is accommodated in the cylindrical tube B. The upper tube T is of much smaller diameter and the liquid overflows from the tip of this tube to have a fixed volume of liquid. The tip of the tube is fire polished and properly bevelled. The tip of the tube is shown vertical but quite often it is bent to about 60° to vertical. F serves a cap to the pipette and has an outflow tube OT. The tube OT is connected to sink so that over flown liquid is collected. H is a hole made in the overflow bulb so that no excessive pressure is generated, while the liquid flows out of the tip of tube T. The cap is sealed to the pipette through a rubber cork.

6.8 CENTRIFUGE TUBES

Centrifuge tubes are of four types. Before the centrifugation starts, it is better to remove the stopper if provided. The four types of centrifuge tube described below are:

- Non-graduated conical bottom centrifuge tube without stopper
- Non-graduated conical bottom centrifuge tube with stopper
- Graduated cylindrical bottom centrifuge tube with stopper
- Non-Graduated cylindrical bottom centrifuge tube without stopper

All dimensions for centrifuge tubes given Tables 6.16 and 6.17 are as per ASTM E237 [65]

6.8.1 Non-graduated Conical Bottom Centrifuge Tube

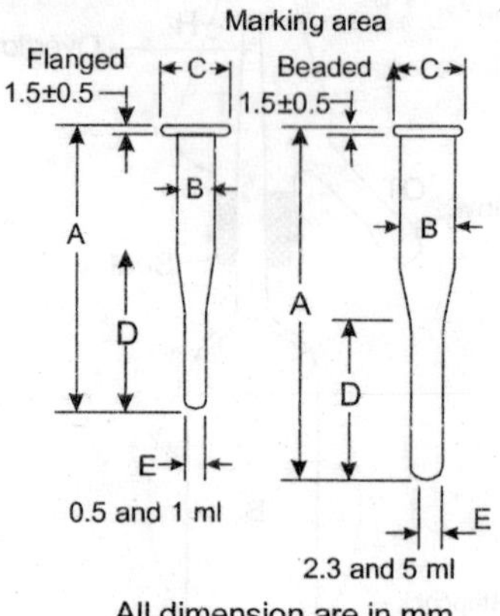

Figure 6.26 Non-graduated conical bottom centrifuge tube

Table 6.16 Dimensions on non-graduated Conical Bottom Centrifuge Tubes

Capacity	*A*	*B*	*C*	*D*	*E*
0.5 cm^3	58 ± 2	6.0 ± 0.25	13.0 ± 1.0	30 ± 2	3.5 ± 0.5
1 cm^3	61 ± 2	8.25 ± 0.25	13.0 ± 1.0	30 ± 2	3.5 ± 0.5
2 cm^3	66 ± 2	10.75 ± 0.25	13.5 ± 1.0	30 ± 2	4.0 ± 0.5
3 cm^3	74 ± 2	10.75 ± 0.25	13.5 ± 1.0	30 ± 2	4.0 ± 0.5
5 cm^3	101 ± 2	13.00 ± 0.50	16.25 ± 0.75	40 ± 2	4.0 ± 0.5

Where A is height of the tube in mm
B Outer diameter of cylindrical portion in mm
C Outer diameter of top in mm
D Length of the tapered portion in mm
E Outer diameter of the bottom in mm
Capacity in cm^3

6.8.2 Non-graduated Conical Bottom Centrifuge Tube with Stopper

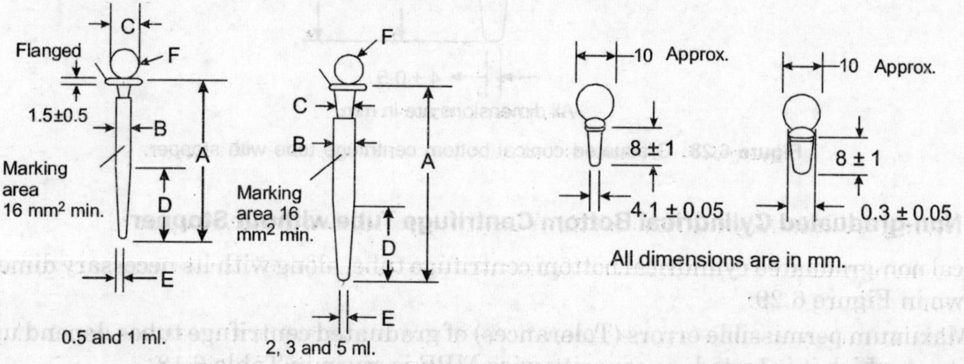

Figure 6.27 Non-graduated Conical bottom centrifuge tube with stopper

Dimensions are given in the Table 6.17 with the following symbols
A Height of the tube in mm
B Outer diameter of the cylindrical tube in mm
C Outer diameter of top in mm
D Length of tapered portion of the tube in mm
E Outer diameter of bottom in mm
F Stopper No
G and H are the details of neck and stopper, which are shown separately.

Table 6.17 Non-graduated Conical bottom Centrifuge Tube with Stopper

Capacity	*A*	*B*	*C*	*D*	*E*	*F*
0.5	66 ± 2	6.0 ± 0.25	13.0 ± 1.0	30 ± 2	3.5 ±0.5	Detail G
1	69 ± 2	8.25 ± 0.25	13.0 ± 1.0	30 ± 2	3.5 ± 0.5	Detail G
2	80 ± 2	10.75 ± 0.25	13.5 ± 1.0	30 ± 2	4.0 ± 0.5	Detail H
3	88 ± 2	10.75 ± 0.25	13.5 ± 1.0	30 ± 2	4.0± 0.5	Detail H
5	115 ± 2	13.00 ± 0.5	13.5 ± 1.0	40 ± 2	4.0 ± 0.5	Detail H

6.8.3 Graduated Conical Centrifuge Tube with Stopper

A typical graduated conical bottom centrifuge tube along with it dimensions is shown in Figure 6.28.

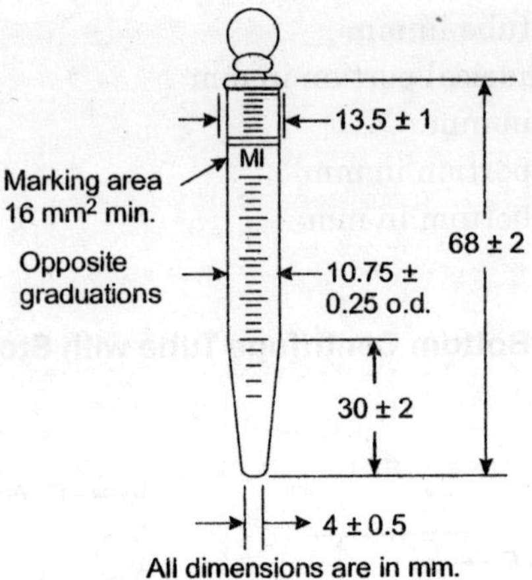

Figure 6.28 Graduated conical bottom centrifuge tube with stopper

6.8.4 Non-graduated Cylindrical Bottom Centrifuge Tube without Stopper

A typical non-graduated cylindrical bottom centrifuge tube, along with its necessary dimensions, is shown in Figure 6.29.

Maximum permissible errors (Tolerances) of graduated centrifuge tubes depend upon the capacity at which it is tested, so capacity wise MPE is given in Table 6.18.

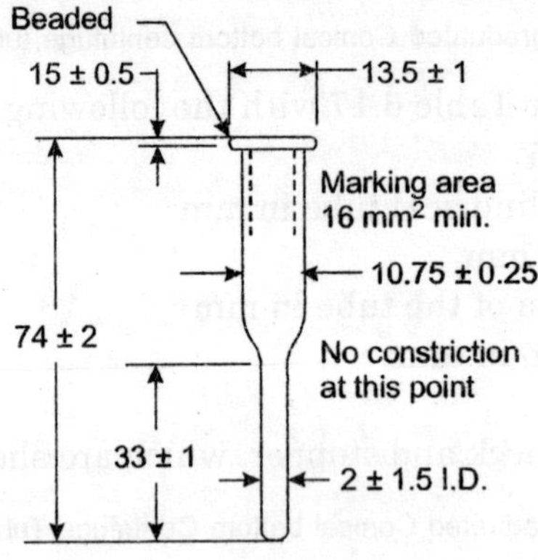

Figure 6.29 Non-graduated cylindrical bottom centrifuge tube without stopper

Table 6.18[65] Maximum Permissible Errors of Graduated Centrifuge Tubes

Capacity mark	*Maximum permissible error*
At 0.1 ml line	±0.01 ml
At 0.2 ml line	±0.02 ml
At 0.3 ml line	±0.03 ml
At 0.4 ml line	±0.05 ml
Above 0.4 ml line	±0.075 ml

6.9 USE OF A VOLUMETRIC MEASURE AT A TEMPERATURE OTHER THAN ITS STANDARD TEMPERATURE

Let the temperature of the measure, which was calibrated at 27 °C is filled with water at t °C. If V_n is the nominal capacity, then to obtain the actual volume of water at t °C, certain correction C given in Tables 4.10 to 4.14 Chapter 4 is to be applied.

If the glass measure is used at temperatures other than the standard temperature 27 °C, then the aforesaid correction is added to the nominal capacity to give the volume of water at 27 °C. Conversely, by subtracting the correction from the nominal value gives the volume of water, which must measure at temperature t °C to obtain nominal volume at 2 °C.

The difference between the values of the volume delivered by the burette with the nominal value of the graduation gives the error of the burette at that graduation line.

6.10 EFFECTIVE VOLUME OF REAGENTS USED IN VOLUMETRIC ANALYSIS

An important study carried out by W. Schlosser [72], showed that the following solutions, normally used in volumetric analysis, could be used with a pipette calibrated with water. The error in volume is not expected to be more than 0.01%.

Nitric acid, Sulphuric acid, Hydrochloric acid, Oxalic acid, Sodium hydroxide, Potassium hydroxide, Ammonium hydroxide, Barium chloride, Potassium bi-chromate, Ammonium sulpha-cyanide each of one normal strength and Sodium carbonate of N/2, Sodium Thio-sulphate, Sodium Chloride, Potassium permanganate, Silver nitrate and Iodine each of N/10 strength, sugar solution of 1%, Ferric chloride 0.012 g Fe per cm^3, Indigo solution, Mercuric nitrate may also be used without causing excessive errors.

However, the volume of absolute alcohol, Conc. Sulphuric acid, Conc. Potassium Hydroxide and milk will be in error of 0.1% to 0.44%. So, the pipette should be calibrated for these liquids separately.

6.11 EXAMPLES OF CALIBRATION

6.11.1 Calibration of a Burette

Alpha 25 × 10^{-6}/°C, Reference Temp 20 °C, density of standard weights used 8000 kg/m^3

Date

Burette:	Capacity	Type of glass	Markings
	50 cm^3	ALPHA 25 × 10^{-6}/°C	0.1 cm^3 divisions, D 20 °C

Observer:

Environmental parameters	Start	Finish
Time:	11.15 h	12.20 h
Air temp	26.0	26.0
Barometer	755 mm/Hg	750 mm/Hg

Balance particulars

Capacity	200 g
Type:	Single pan with optical projection
Optical scale	Equivalent to 100 mg with 1 mg divisions
Readability	0.1 mg with optical vernier
Mass standards used:	Stainless steel weights of Class A (F1) density 8000 kg/m^3

From cm³	*To cm³*	*Temp °C*	*Indication Dial g*	*Indication Scale mg*	*Mass of water g*	*Temp °C*	*Correction g*	*Volume water cm³*
0	0	25.9	10.5	54.5	–	25.9	–	–
0	50	25.9	60.3	82.5	49.828	25.9	0.2045	50.0325
0	10	25.9	20.45	58.3	9.9538	25.9	0.0409	9.9949
0	20	25.9	30.45	45.2	19.9407	25.9	0.0818	20.0225
0	30	25.9	40.36	76.7	29.8822	25.9	0.1227	30.0049
0	40	25.9	50.35	61.8	39.8573	25.9	0.1636	40.0209

6.11.2 Calibration of a Micropipette

About 100 cm³ of mercury is taken out in beaker and kept for at least overnight to acquire the temperature of the air-conditioned room. Mercury is withdrawn from this beaker only. Temperature of the mercury in the beaker is taken as the temperature of mercury inside the micro-pipette under test. We assume that temperature of mercury inside the under-test pipette remains constant and is same as that of the mercury in the beaker. A clean dry weighting-tube is weighed empty and weighed again with mercury delivered by the pipette. The diference of the two weighing results will give the mass of mercury. Mass of mercury is expressed in mg so that when multiplied by the correction factor, the result is in cm³ at refrence temperature.

Date sheet

Date

Micropipette:	Capacity 1 cm³	Type of glass ALPHA 25 × 10^{-6}/°C	Markings D 20 °C

Observer:

Environmental parameters	Start	Finish
Time:	10.15 h	10.30 h
Air temp	24.4	24.4
Barometer	755 mm/Hg	750 mm/Hg

Balance particulars

Capacity: 200 g

Type: Single pan with optical projection

Optical scale: Equivalent to 100 mg with 1 mg divisions

Readability: 0.1 mg with optical vernier

Mass standards used: Stainless steel weights of Class A (F1) density 8400 kg/m³

Observation and calculations

In the pan	*Temp*	*Dial reading g*	*Scale reading*	*Mass of mercury*	*Temp mg*	*10³ Factor*	*Capacity cm³*
Tube only	24.3	10.5	54.6		24.3		
Tube +Hg	24.3	24.0	90.6	13536.0	24.3	0.073883	1.000 08
Tube only	24.3	10.5	54.8				
Tube +Hg	24.3	24.0	91.3	13536.5	24.3	0.073883	1.000 12
						Mean	1.000 1

REFERENCES

General

[1] National Physical Laboratory U. K. Notes on Applied Science No. 6 Volumetric glassware, 1957, Her Majesty Stationary Office, London.

[2] ASTM E 691–1979 Standards Practice for Conducting an Inter-laboratory Test Program to Determine Precision of Test Method.

[3] Glaze brook Sir R (ed), 1950 A dictionary of Applied Physics, Volume 3, Volume measurement, pp783– 813, (New York) Smith and reprinted with arrangement with Macmillan.

[4] ASTM E 671–1979 Standards Specification for Maximum Permissible Residual Stress in Annealed Glass Laboratory Apparatus.

[5] ASTM E 784–1980 Standards Specification for Clamps, Utility, laboratory, and holders, burettes and clamps.

[6] IS 1058: 1960 Metric measures.

Burettes

[7] ASTM E 287–1976 Standard Specifications for Burettes.

[8] IS 1997:1992 Indian Standard Specifications for Burettes.

[9] ISO 385:1984 Laboratory glassware Burettes Part 1 General requirements.

[10] BS- 846:1985 Specifications for Burettes.

[11] ASTM E-694-1979 Standards Specification for volumetric ware
Volumetric flask;
Measuring cylinder;
Straight Burettes;
Transfer pipettes.

[12] Stott V, "Notes on Burettes", 1923, Trans. Soc. Glass Tech. **7**, 169-198.

Measuring cylinders

[13] IS 878:1975 (1991) Graduated measuring cylinders.

[14] IS10073:1982 Graduated plastic measuring cylinders.

[15] ISO 4788:1993 Laboratory glassware Graduated measuring cylinders.

[16] BS 604:1982 Specifications on measuring cylinders.

Flask

[17] ASTM E 288-1976 Standards Specification for flasks.

[18] IS 915: 1975 (1991), One-mark volumetric flask.

[19] ISO 1042- 1983, One- mark flask.

[20] BS 1792:1982- One- mark flask.

[21] BSENISO 1042: 1992 One mark flask.

[22] BS 676: 2002 Plastic graduated neck flask.

[23] ASTM E-694-1979 Standards Specification for volumetric ware
Volumetric flask;

[24] ASTM E 237-1980 Standards Specification for Micro-volumetric vessels. Volumetric flask and Centrifuge tubes.

Pipettes

[25] Stott V. "Notes on pipettes", 1921, Trans. Soc. Glass Tech **5**, 307–325.

[26] ISO 648-1983 Laboratory glassware: One mark pipettes.

[27] IS 1117:1975 (2000) One mark pipettes.

[28] ISO 835 part 4-1983 Blow out pipettes.

[29] BS 700 part 3, 1982 Blow out pipettes.

[30] BS 1583:2000 One mark pipette.

[31] ISO 835 Part 1: Laboratory glassware Graduated pipettes–General requirements

[32] ISO 835 Part 2: Laboratory glassware: Graduated pipettes– Pipettes for which no waiting time is specified.

[33] ISO 835 Part 3: Laboratory glassware: Graduated pipettes for which a waiting time of 15 second is specified.

[34] IS 4162:1993 Graduated pipettes.

[35] BS 700:1993 Graduated pipettes; 1-general requirements; 2-pipettes with no waiting time.

Micro-glassware -pipettes/burettes/ flasks

[36] IS 11383:1985 Capillary pipettes (1 μl to 250 μl).

[37] ASTM E 193-76 Standard specifications of micropipettes.
Measuring micropipettes for delivery
Folin's type micropipettes for content type
Washed out micropipettes for content
Micro-weighing pipettes density type for content
Micro-litre pipettes for content

[38] BS 2058:1992 Lung ray weighing pipette.

[39] BS 1428-D1 1993 Burettes with pressure device and automatic zero.

[40] BS 1428-D2 1963 Washout pipettes.

[41] BS 1428-D4:1993, Capillary pipettes (0.1, 0.2 and 0.5 ml, with 0.005 to 0.05 ml readability).

[42] BS 1428-D5:2002 Syringe pattern micropipettes.

[43] ASTM E 672-1978 Standards Specification for disposable glass micropipettes.

[44] ASTM E 787-1980 Standards Specification for Disposable glass micro-blood Collection pipette.

[45] IS 4087:1980 Haemoglobin and blood pipettes.

[46] ASTM E 732-1980 Standards Specification for disposable pasture type pipettes.

[47] IS 14284-1995 Laboratory glassware pasture disposable pipettes.

[48] BS/ISO 12771: 1997Disposable plastic serological pipettes.

[49] BS 6706:1992 Glass serological pipettes.

[50] ASTM E 714-80 Standard Specifications for Serological pipettes.

[51] IS 4364:1967 Serological pipettes..

[52] ASTM E 733-80 Standards Specification for 44.7 μl disposable micropipette.

[53] BS 6674:1997 Disposable plastic pasture pipettes.

[54] BS 5732:1997 Disposable glass pasture pipettes.

[55] IS 6543-1972 Disposable glass pipettes for artificial insemination for cattle.

[56] IS 7179 1982 Disposable plastic pipettes for artificial insemination for cattle.

Piston operated pipettes/burettes

[57] ASTM E 735-1980 Standards Specification for Minimum Performance Standards for hand held Pipettor with pipette tips.

[58] BSEN/ISO 8655-2,3:2002 Piston volumetric pipettes, and burettes.

[59] ISO 86655:2002(E)–1Piston-operated volumetric apparatus- General terminology, General requirements and user recommendations.

[60] ISO 86655:2002(E)–2 Piston-operated volumetric apparatus- Piston pipettes.

[61] ISO 86655:2002(E)–3 Piston-operated volumetric apparatus- Piston burettes.

[62] ISO 86655:2002(E)–4 Piston-operated volumetric apparatus.

[63] ISO 86655:2002(E)–5 Piston-operated volumetric apparatus.

[64] ISO 86655:2002(E)–6 Piston-operated volumetric apparatus- Gravimetric methods for the determination of measurement error.

[65] ASTM E 237-80 Standards Specification for Micro-volumetric vessels Centrifuge tubes.

Automatic volumetric pipettes

[66] Bigg P. H. "An accurate automatic mercury pipette" 1936, J. Sc. Instum.1936, **13**, 156–157.

[67] BS 1132:2003 Automatic pipettes (5 ml to 10 l).

Butyrometer

[68] IS 1223-2001 Apparatus for determination of milk fat by Gerber Method.

Pycnometers

[69] ISO 3507-1976 Pycnometers.

[70] BS 733 Part 1, 1983 Pycnometers.

[71] IS 5717:1991 Pycnometers.

[72] W. Schlosser, Chemiker Zeitung 1906, **XXX**, 1701.

EFFECT OF SURFACE TENSION ON MENISCUS VOLUME

7.1 INTRODUCTION

If a tube is dipped in a liquid, then the level of the liquid outside and inside of the tube will not be the same, also the air-liquid-interface will not be a plane, it will be either concave upward or concave downward.

The shape of the interface depends upon the relative strengths of adhesive and cohesive forces. Adhesive forces are in between interacting molecules of different substances (like liquid and those of the solid of which the tube is made off). Cohesive forces are in between the molecules of the same substance e.g. liquid here. For a liquid molecule well within the liquid, there are molecules of the same liquid surrounding it completely, so the net resultant force is zero and molecule is able to move in any direction. But the molecules, which are on the surface of the air-liquid interface, are attracted by liquid-molecules and on one-side and air molecules on the other. Air molecules are lighter than the liquid molecules so liquid molecules on the surface of the liquid are constantly attracted downward. Due to these cohesive forces, the surface of the liquid is stretched and makes a concave surface. Curving of surface causes increase in its area hence the surface gets stretched. So the molecules at the air-liquid interface are always in tension. Three types of molecules, namely molecules of the liquid, air molecules and molecules of the solid of which the tube is made, interact with each other. Molecules of the material of the tube, near the walls of the tube attract liquid molecules. If solid molecules are heavier than those of liquid molecules, then the net force will be upward, so the liquid surface at the air-liquid interface will be concave upward. The centre of curvature of the surface at any point will be above the liquid surface. Examples of liquids having this type of interface are water, milk, petroleum liquids, alcohol and all aqueous solutions contained in a glass tube. Reverse will be the case if the liquid molecules are heavier than those of solid molecules, say mercury. In this case, the centre of curvature at any point of the liquid surface will be below the surface and hence the air-liquid surface will be concave downward. If the air-liquid interface is concave upward the liquid inside the tube will rise, while it will fall if interface is convex (concave downward). These two situations are shown in Figures 7.1A, 7.1B.

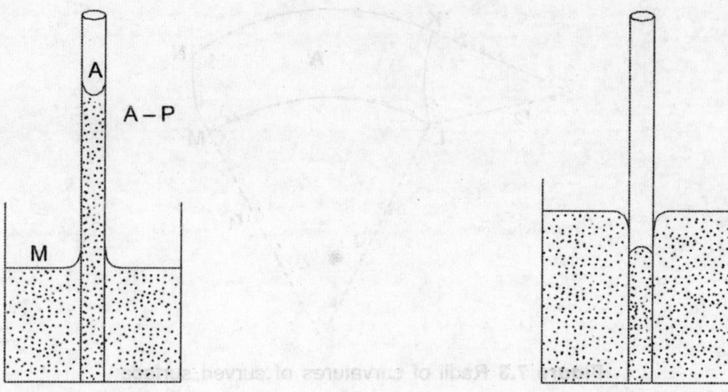

Figure 7.1A Concave interface **Figure 7.1B** Convex interface

Considering the vertical section of the interface and calling the point where air-liquid interface meets the wall of the containing tube as point of contact, then angle of contact is, which the tangent at the point of the contact makes with the vertical wall of the containing tube, the angle is measured from the wall and taken positive in the anticlockwise direction. So angle of contact is acute for concave interface and obtuse for the convex interface as shown in Figure 7.2A and 7.2B respectively.

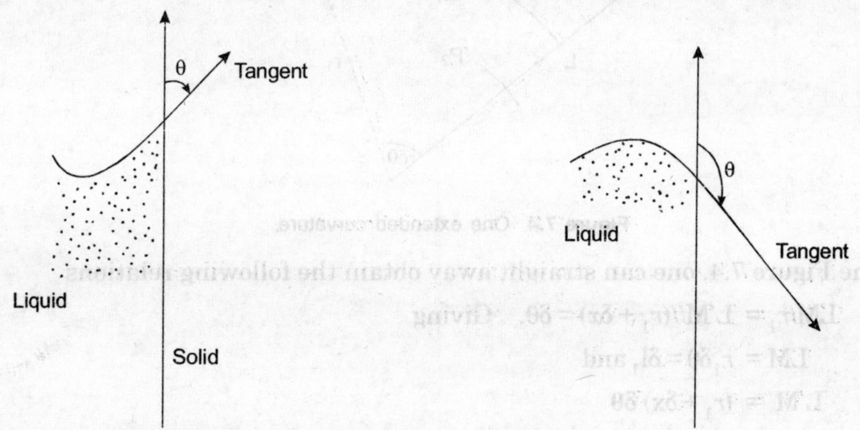

Figure 7.2A Acute angle of contact **Figure 7.2B** Obtuse angle of contact

7.2 EXCESS OF PRESSURE ON CONCAVE SIDE OF AIR-LIQUID INTERFACE

Consider an interface separating two fluids. As explained above this surface will be curved. Around any point *A*, let there be an elementary area KLMN, with LM = δl_1 and KL = δl_2. Further let r_1 and r_2 be the respective radii of curvatures of LM and KL, without loosing the generality, we may assume that their centres of curvature are on the same side of the surface, refer Figure 7.3.

If the pressure on upper side is P_1 and lower side is P_2, then due to difference in pressure, the elementary area will be stretched and its new dimensions L′M′ and K′L′ will become $\delta l_1 + \alpha$, and $\delta l_2 + \beta$ respectively.

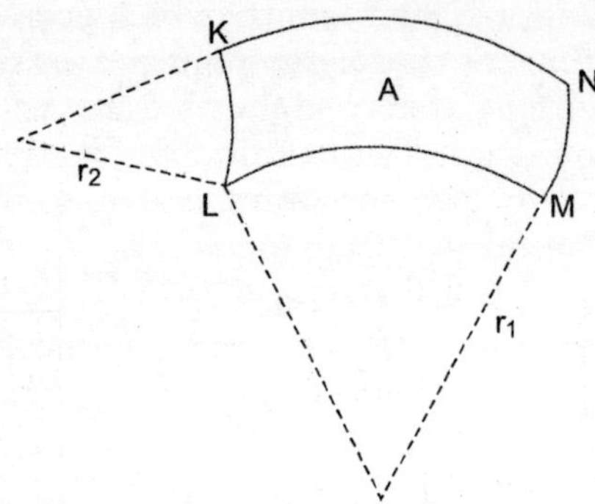

Figure 7.3 Radii of curvatures of curved surface

Old and new positions of the elementary arc LM are shown in Figure 7.4. It may be noted that the arc LM has moved a distance normal to surface LMNK by δx to its new position L′ M′.

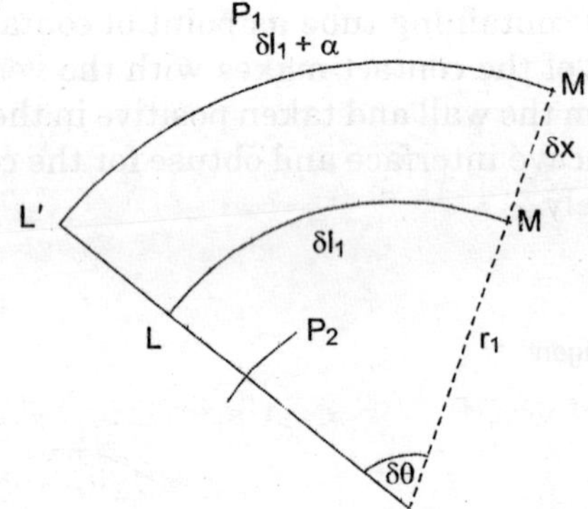

Figure 7.4 One extended curvature

From the Figure 7.4, one can straight away obtain the following relations

$$LM/r_1 = L'M'/(r_1 + \delta x) = \delta\theta. \quad \text{Giving}$$

$$LM = r_1\delta\theta = \delta l_1 \text{ and} \qquad ...(1)$$

$$L'M' = (r_1 + \delta x)\, \delta\theta$$

Here δx is the distance normal to the surface to which it has moved to occupy the new position.

So δx is the increase in its radius of curvature. Giving us

$$\delta l_1 + \alpha = (r_1 + \delta x)\, \delta\theta, \text{ giving}$$

$$\alpha = \delta x\delta\theta = \delta x\delta l_1/r_1 \qquad ...(2)$$

Similarly

$$\beta = \delta x\delta l_2/r_2$$

New area of the stretched surface will be

$(\delta l_1 + \alpha)(\delta l_2 + \beta) = (\delta l_1\delta l_2 + \delta l_1\beta + \delta l_2\alpha + \alpha\beta)$, neglecting the term $\alpha\beta$,

New area of stretched surface = $(\delta l_1\delta l_2 + \delta l_2\alpha + \delta l_1\beta)$

Substituting the values of α and β, we get the change in area as

$$\delta x\delta l_2\delta l_1(1/r_1 + 1/r_2) \qquad ...(3)$$

T- the surface tension is force per unit length, which is due to stretching of surface, but the surface tension T is also defined as the energy required in stretching a surface of a film by unit area under isothermal conditions. It may be clarified that the work divided by area has the same dimensions as force per unit length. Two definitions given above are, therefore, equivalent. Assuming that the surface, due to excess of pressure, is stretched under isothermal conditions, energy required in expanding the surface, from (3), is given by

$$T.\delta x \delta l_2 \delta l_1 (1/r_1 + 1/r_2) \quad ...(4)$$

Now the effective pressure acting is $(P_2 - P_1)$ and is normal to the surface, so force acting normal to the surface is $(P_2 - P_1)\delta l_2 \delta l_1$. Hence work done by this force is given by

$$(P_2 - P_1)\delta l_2 \delta l_1 \delta x. \quad ...(5)$$

This work has been utilized in expanding the surface of the film.

Thus equating the work required to stretch the film surface, from (4), to the work done by excess of pressure, from (5), gives us

$$T\delta x \delta l_2 \delta l_1 (1/r_1 + 1/r_2) = (P_2 - P_1)\delta l_2 \delta l_1 \delta x$$

Giving us

$$P_2 - P_1 = T(1/r_1 + 1/r_2) \quad ...(6)$$

Since right hand side is positive, so the pressure P_2 on the lower (concave side) of the film will be larger than P_1. In general we can write $P_2 - P_1 = P$ excess of pressure on concave side of the air-liquid interface as

$$P = T(1/r_1 + 1/r_2) \quad ...(7)$$

If the centre of curvature of side KL is on the opposite side as that of LM, then

$$P = T(1/r_1 - 1/r_2) \quad ...(7A)$$

7.3 DIFFERENTIAL EQUATION OF THE INTERFACE SURFACE

Let LM be the curve of the principal section of the interface separating the two incompressible fluids in contact and be shown in Figure 7.5. Lighter fluid is resting on the heavier. Take origin O much below the interface with axes Ox and Oz. Let the interface undergo an elementary virtual displacement in which every element of the surface moves from its initial position to its final position along the normal to itself.

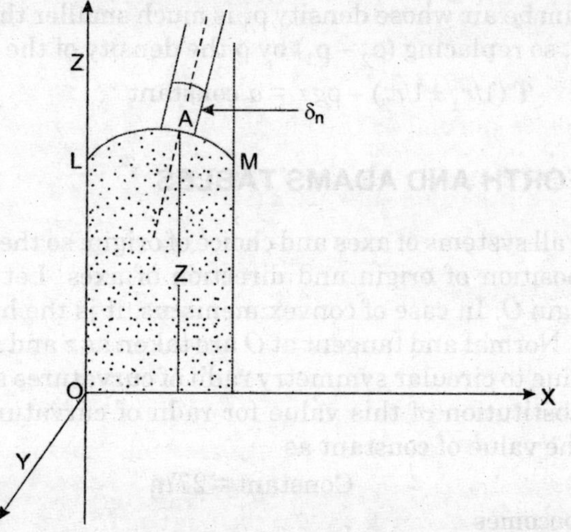

Figure 7.5 Air-liquid interface

Consider an elementary surface around a point A, whose sides are δl_1 and δl_2. It is displaced normal to itself by a distance dn. The area of the elementary surface δS is given by

$$\delta S = \delta l_1 \delta l_2$$

The excess pressure on the concave side is P, then the force acting on the elementary area is $P\delta S$ and virtual work done is

$\delta W = P\delta S\delta n$, but $\delta S\delta n$ is elementary volume δV around the point A, so

$\delta W = P\delta V$

Substituting the value of P from (6), we get

$$\delta W = T(1/r_1 + 1/r_2)\delta V \quad ...(8)$$

Here r_1 and r_2 are the principal radii of curvature of the surface at the point A. Due to this displacement, heavier liquid of volume δV and density ρ_2 has been removed and replaced by the same volume of the lighter liquid of density ρ_1. If vertical ordinate of the elementary area is z, then change in potential energy is $(\rho_2 - \rho_1)gz\delta V$. Taking the summation over entire surface and applying the principle of virtual work, we get

$$\iiint [T(1/r_1 + 1/r_2)]dV = \iiint [(\rho_2 - \rho_1)gz]\,dV$$

$$\iiint [T(1/r_1 + 1/r_2) - (\rho_2 - \rho_1)gz]dV = 0$$

As the liquids are incompressible and $\iiint dV$ represent the effective change in volume, so

$$\iiint dV = 0.$$

Hence the integrand must be a constant, giving us

$$T(1/r_1 + 1/r_2) - (\rho_2 - \rho_1)gz = a \text{ constant} \quad ...(9)$$

The above reasoning holds good, even if one of the media is incompressible and the other is a gas, as in that case also $\iiint dV$ will be zero.

As r_1, and r_2 are the radii of curvatures of the surface at a point $A(x, z)$, so these are functions of first and second order differential coefficients of y with respect to x and z. Hence the equation (9) is a second order non-linear differential equation in three variables.

Let the other medium be air whose density ρ_1 is much smaller than that of liquid ρ_2, then it can be neglected in (9), so replacing $(\rho_2 - \rho_1)$ by ρ the density of the liquid and gives

$$T(1/r_1 + 1/r_2) - \rho gz = a \text{ constant} \quad ...(10)$$

7.4 BASIS OF BASHFORTH AND ADAMS TABLES

Equation (10) is valid for all systems of axes and choice of origin, so the equation (10) holds good for any change of the position of origin and direction of axes. Let the lowest point of the meniscus is taken as origin O. In case of convex meniscus, it is the highest point of meniscus which is taken as origin. Normal and tangent at O are taken as z and x axes. As O is the lowest point of meniscus, then due to circular symmetry radii of curvatures at O will be equal to each other, let each be η. Substitution of this value for radii of curvatures and putting $z = 0$ in equation (10), gives us the value of constant as

$$\text{Constant} = 2T/\eta$$

So (10) in general becomes

$$T(1/r_1 + 1/r_2) = gz\rho + 2T/\eta \quad ...(11)$$

Dividing by T/η both sides, we get

$$(1/r_1 + 1/r_2)/(1/\eta) = g\rho z/T/\eta + 2$$

In a circular tube, air-liquid interface is a surface of revolution about the axis of the tube, then one of the radius of curvature at any point will be $x/\sin\psi$, where ψ is the angle which the tangent at any point P makes with the x-axis or the normal makes with z-axis–the axis of the tube.

Expressing all linear dimensions in terms of η, we write the above equation as

$$1/(r_1/\eta) + \sin\psi/(x/\eta) = 2 + (g\rho/T)\eta^2.\ z/\eta \qquad ...(12)$$

Writing as $T/g\rho$ as a^2, (12) becomes

$$1/(r_1/\eta) + \sin\psi/(x/\eta) = 2 + (\eta^2/a^2)(z/\eta) \qquad ...(13)$$

In equation (13), all terms are dimensionless, hence the above equation will hold good for all systems of measurement.

For the purpose of brevity write $\eta^2/a^2 = \beta$

Then (13) becomes

$$1/(r_1/\eta) + \sin\psi/(x/\eta) = 2 + \beta(z/\eta) \qquad ...(14)$$

r_1 and $\sin\psi$ can be expressed in terms of differential coefficients of z with respect of x, giving us a non linear differential equation for the vertical section of the interface, which as such is not integrable, so method of numerical solution is to be applied. Bashforth and Adams [1] have used the numerical method successfully.

For different values of β from 0.1 to 100 in steps of 1, they calculated the values of x/η and z/η for values of ψ from 0° to 180° in steps of 5°. To use the tables [1], the values of ψ and β are chosen and the tables give the corresponding values of x/η and z/η.

7.5 EQUILIBRIUM EQUATION OF A LIQUID COLUMN RAISED DUE TO CAPILLARITY

The volume of the liquid, which rises in the tube comprises of a straight cylindrical portion surmounted by a curved surface concave upward for acute angle of contact. The portion of the liquid, bounded by the air-liquid interface and the horizontal plane tangential to the curved surface at its lowest point, is called meniscus. The total upward force, due to surface tension T, balances the gravitational force due to mass of liquid of the cylindrical portion and that of the meniscus.

Giving

$$2\pi r\, T\cos\theta = (V_c + V_m)\rho g \qquad ...(15)$$

V_c and V_m are respectively the volumes of the cylindrical portion and meniscus of the liquid and θ is the angle of contact between the liquid and the vertical walls of the circular tube.

Let h be the height of the cylindrical portion of the liquid inside the tube of radius r. The volume of the cylindrical portion will be $\pi r^2 h$, than (15) becomes

$$2\pi r T\cos\theta = (\pi r^2 h + V_m)\rho g$$

or

$$2\pi T\cos\theta = (\pi r h + V_m/r)\rho g \qquad ...(16)$$

The height h of the cylindrical portion of the liquid, in terms of surface tension and radii of curvature at the lowest point of the air-liquid interface, may be expressed as follows:

Tube has circular symmetry about its axis, so radii of curvatures, in two mutually perpendicular planes, at the lowest point of the air-liquid interface, will be equal to each other.

That is

$$r_1 = r_2$$

So the excess of pressure P at the lowest point of the air-liquid interface from (7) will be equal to

$$P = T(1/r_1 + 1/r_1) = 2T/r_1 \quad ...(17)$$

Considering the Figure 7.6, pressure at the centre of the interface just above it is atmospheric. Let us denote it by A, then the pressure immediately below it in the liquid will be $A - P$.

The pressure at the point M on the air-liquid interface in the trough in which the tube is dipped will be $A + h\sigma g$. If N is another point in the horizontal plane passing through M but inside the tube then pressure at N will also be $A + h\sigma g$. Equating this pressure equal to the hydrostatic pressure of liquid column of height h plus the pressure at the point just inside the air-liquid interface, gives us

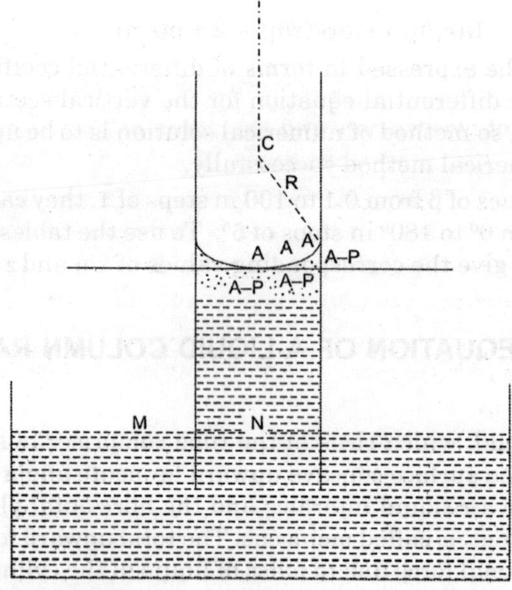

Figure 7.6 Air-liquid interface

$$A + h\sigma g = A - P + h\rho g$$

Giving

$$P = h(\rho - \sigma)g.$$

Neglecting σ in comparison of ρ

$P = h\rho g$, using the value of P from (7) we get

$$2T/r_1 = h\rho g \quad ...(18)$$

So (16) becomes

$$2\pi r T\cos\theta = 2\pi r^2 T/r_1 + V_m \rho g \quad ...(19)$$

Writing

$$T/\rho g = a^2$$

Equation (19) is re-written as

$$V_m/\pi = 2ra^2\,[\cos\theta - r/r_1] \qquad ...(20)$$

If we divide both sides by r^3, we get

$$V_m/\pi r^3 = (2a^2/r^2)[\cos\theta - (r/a)(a/r_1)] \qquad ...(21)$$

Similarly if we divide both sides of (19) by a^3, we get

$$V_m/\pi a^3 = 2r/a[\cos\theta - (r/a).(a/r_1)] \qquad ...(22)$$

The height of the cylindrical portion h of the liquid from (18) is inversely proportional to radius of curvature at the lowest point of the meniscus and is given as

$$2a^2/r_1 = h$$

$$r_1 h = 2a^2 \qquad ...(23)$$

This relation will be quite often used for air-liquid interface.

Equation (23) may be rewritten as

$$a/r_1 = h/2a$$

Using this value in (21), we get

$$V_m/\pi a^3 = r/a[2\cos\theta - (r/a).(h/a)] \qquad ...(24)$$

For given values of r/a, we can find the value of $V_m/\pi a^3$ either knowing the radius of curvature at the lowest point of the meniscus and using Equation (22) or the value of h-height of the cylindrical portion and using equation (24).

The Equation (22) is used for calculating $V_m/\pi a^3$ for smaller bore tubes say for r/a from 0 to 3. This method is known radius of curvature method.

For larger values of r/a say greater than 6, we use Equation (24) for calculating $V_m/\pi a^3$ and the method is called as height of cylindrical portion method.

It may be noticed that for larger diameter tubes, h is very small, so from (24), one may conclude that

$$V_m/\pi a^3 \leq 2r/a \text{ or}$$

$V_m/\pi a^3$ asymptotically approaches to $2r/a$.

Meniscus is a basic entity due to cohesive and adhesive forces and its shape depends upon their relative strengths. We will show later that for liquids, for which angle of contact is zero, the shape of the meniscus will be an ellipsoidal. The shape of the meniscus will never be a perfect hemisphere, except in the limiting case when r-radius of the tube approaches zero.

In general, meniscus will have a finite radius of curvature at its lowest point. For finite value of radius of curvature there will be a finite value of h-height of the cylindrical portion, hence irrespective of the diameter of the tube, both cylindrical portion as well as meniscus will co-exist.

7.6 RISE OF LIQUID IN NARROW CIRCULAR TUBE

Lord Rayleigh's Approach [2]

When a narrow tube is dipped in a vessel containing liquid, liquid rises in it. While doing the theoretical calculation, the height of liquid is reckoned from the free plane surface. But in actual practice, the liquid itself is contained in another tube or vessel of certain diameter, the air-liquid interface in this vessel will not be a perfect horizontal plane, hence the datum line for measuring the height of different points of the air-liquid interface for the narrow tube will not be the same. Lord Rayleigh, in his theoretical calculations, used a variable u to counteract it.

Let axis of the tube is taken as z-axis and the vertical plane passing through it as z-x plane. Because of symmetry of the circular tube about the z-axis, air-liquid interface will also be symmetrical about z-axis and may be taken as the part of the surface of the sphere of radius c, such that the interface meets the wall of the narrow tube at an angle equal to the angle of contact.

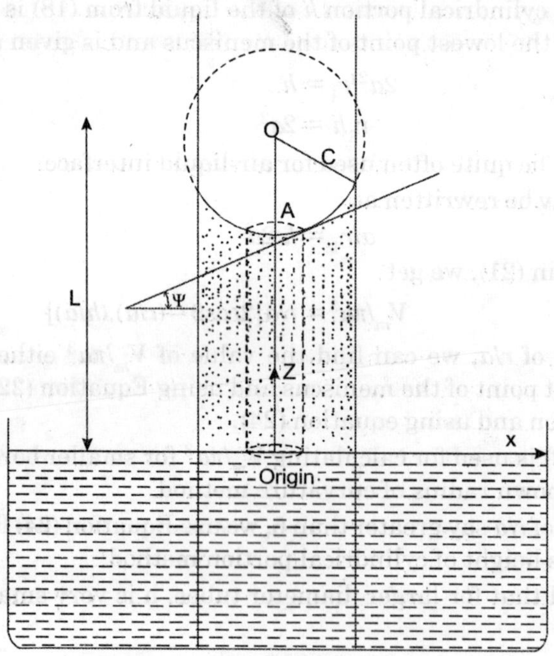

Figure 7.7 Rayleigh's approach

Consider the equilibrium of the liquid cylinder of radius x. The density of the liquid is ρ and tangent at the periphery of the cylinder makes an angle ψ with the x-axis. If T is the surface tension of the liquid then

$$2\pi x T \sin\psi = \int 2\pi x \,.\, z\rho g \,.\, dx$$

Limit of integration is $x = 0$ to $x = x$

$$x \sin\psi = (\rho g/T)\int zxdx$$

$$x \sin\psi = 1/a^2 \int zxdx \qquad ...(25)$$

z the length of the cylinder is given as

$$z = L - \sqrt{(c^2 - x^2)} + u \qquad ...(26)$$

Where u is a variable correction assigned by Lord Rayleigh. L is the height of the centre of spherical surface forming the air liquid interface from the datum.

If θ is the angle of contact and ψ_w the angle which the tangent makes at the intersection of the air-liquid interface with the wall of the tube, then

$$\psi_w = \pi/2 - \theta, \text{ and}$$

$$L = c + h$$

Where h is the height of the lowest point of meniscus from the datum line.

So equation (25) may be written as

$$a^2 r\cos\theta = \int[(c+h) - \sqrt{(c^2 - x^2)} + u]x\,dx \qquad ...(27)$$

The limits of integration are $x = 0$ to $x = r$ (radius of the tube)

7.6.1 Case I $u = 0$

The equation (27) gives

$$a^2 r\cos\theta = [(c+h)x^2/2 - (c^2 - x^2)^{3/2}(-1/2)/(3/2)], \text{ limit of } x \text{ are from 0 to } r, \text{ giving us}$$

$$a^2 r^2/c = (c+h)r^2/2 + (1/3)\{(c^2 - r^2)^{3/2} - c^3\}, \text{ as } \cos\theta = r/c \qquad ...(28)$$

Taking a particular case when angle of contact is zero, as is the case of water and quite a number of other liquids, so in this case

$\cos\theta = 1$ i.e. $r = c$, which implies that air liquid interface is a semi-sphere whose radius is equal to that of the tube.

putting $c = r$ in (28) gives us

$$a^2 r = (r+h)r^2/2 - r^3/3$$

$$2a^2 r = hr^2 + r^3/3$$

Multiplying both sides by π and writing a^2 as $T/\rho g$, we get

$$2\pi r T/\rho g = \pi r^2 h + \pi r^3/3$$

$$2\pi r T = (\pi r^2 h + \pi r^3/3)\rho g \qquad ...(29)$$

Comparing it with (16)

We see the volume of meniscus in this simple case is $\pi r^3/3$. The air liquid interface, in a narrow tube of circular section, if the liquid is wetting the tube ($\theta = 0$), is semi-spherical with radius equal to that of the tube. However, we will see in section 7.8.1 that surface of the interface can never be spherical.

7.6.2 Case II $u \neq 0$ but du/dx is small

If ψ is the angle which the tangent, at any point, makes with x-axis, then

$$\sin\psi = \sin\psi/\cos\psi.\sec\psi = \tan\psi/(1 + \tan^2\psi)^{1/2}$$

$$\sin\psi = dz/dx/\{1 + (dz/dx)^2\}^{1/2} \qquad ...(30)$$

But dz/dx from (26)

$$dz/dx = x/(c^2 - x^2)^{1/2} + du/dx \qquad ...(30a)$$

Squaring both sides gives

$$(dz/dx)^2 = x^2/(c^2 - x^2) + (du/dx)^2 + 2x(du/dx)/\{(c^2 - x^2)^{1/2}\}$$

Adding 1 to both sides, we get

$$1 + (dz/dx)^2 = 1 + x^2/(c^2 - x^2) + (du/dx)^2 + 2x(du/dx)/\{(c^2 - x^2)^{1/2}\}$$

$$= c^2/(c^2 - x^2) + (du/dx)^2 + 2xdu/dx)/\{(c^2 - x^2)^{1/2}\}$$

Taking square root of both side and considering only positive value, we get

$$\{1 + (dz/dx)^2\}^{1/2} = c/(c^2 - x^2)^{1/2}\,[1 + du/dx\{2x(c^2 - x^2)^{1/2}/c^2 + (c^2 - x^2)(du/dx)/c^2\}]^{1/2} \qquad ...(30b)$$

Combining (30), (30a) and (30b) give

$$x\sin\psi = xdz/dx/\{1 + (dz/dx)^2\}^{1/2} = [\{x^2/(c^2 - x^2)^{1/2} + xdu/dx\}]\,\{c/(c^2 - x^2)^{1/2}\}$$

$$[1 + du/dx\{2x(c^2 - x^2)^{1/2}/c^2 + (c^2 - x^2)\,du/dx/c^2\}]^{-1/2} \qquad ...(31)$$

As du/dx is small, right hand side of (31) may be expanded by Binomial theorem. Neglecting terms containing cubes and higher powers of du/dx, we get

$$x\sin\psi = 1 + du/dx(c^2 - x^2)^{3/2}/xc^2 - 3(c^2 - x^2)^2(du/dx)^2/2c^4$$

Substituting the value of z from (26) in equation (25) and integrating, we get

$$1 + du/dx(c^2 - x^2)^{3/2}/xc^2 - 3(c^2 - x^2)^2 (du/dx)^2/2c^4 = c/a^2x^2 \, [Lx^2/2 + \{(c^2 - x^2)^{3/2} - c^3\}/3 + \int uxdx] \quad ...(31a)$$

The limits of x in the integral are from 0 to x.

As du/dx is small for the first approximation $(du/dx)^2$ and $\int uxdx$ are neglected, giving us

$$du/dx = (cL/2a^2 - 1).\, c^2x(c^2 - x^2)^{-3/2} - c^6/\{3a^2x(c^2 - x^2)^{3/2}\} + c^3/3a^2x \quad ...(32)$$

Integrating both sides of (32) with respect of x, we get

$$u = c^2 (cL/2a^2 - c^2/3a^2 - 1)(c^2 - x^2)^{-1/2} + (c^3/3a^2) \log[c + (c^2 - x^2)^{1/2}] + C$$

C is the constant of integration. ...(33)

To avoid u becoming infinite at $x = c = r$ the multiplying factor in the first term should be zero. So putting

$(cL/2a^2 - c^2/3a^2 - 1) = 0$, u from (33) becomes

$$u = (c^3/3a^2)\log[c + (c^2 - x^2)^{1/2}] + C \quad ..(34)$$

Differentiating both sides of (34) with respect of x

$$du/dx = (c^3/3a^2)\{(c^2 - x^2)^{1/2} - c\}/x(c^2 - x^2)^{1/2} \quad ...(35)$$

To determine c the radius of the sphere, part of whose surface forms the air-liquid interface. Apply the condition that at wall of the tube ψ is complement of the angle of contact. That is

$$\cot\theta = (dz/dx)_{x=r} = r/(c^2 - r^2)^{1/2} + (du/dx)_{x=r}, \text{ giving}$$

$$\cot\theta = \{r/(c^2 - r^2)^{1/2}\}[1 - (c^3/3a^2)\{c - (c^2 - r^2)^{1/2}\}/r^2] \quad ...(36)$$

Equation (36) gives c in terms of θ and r.

However c can be explicitly expressed as:

$$c = r/\cos\theta - (r^3/3a^2)\{\sin^2\theta/\cos^3\theta\,(1 + \sin\theta)\} \quad ...(37)$$

The height h of the cylindrical portion of the liquid column at the lowest point of the air-liquid interface ($x = 0$) is given by

$h = L - c + u_{x=0}$, put $x = 0$ in equation (34) to get $u_{x=0}$, so h becomes

$$h = L - c + C + (c^3/3a^2)\log(2c) \quad ..(38)$$

Using this value for h in (27) and integrating it for $x = 0$ to $x = r$ we get

$$a^2r\cos\theta = (r^2/2)(L + C) + \{(c^2 - r^2)^{3/2} - c^3 +\}/3 + \int(u - C)xdx \quad ...(39)$$

Limits of x in the integral are 0 to r.

Substituting the value of $u - C$ from (34) and integrating, we get

$$a^2r\cos\theta = (h + c)r^2/2 + \{(c^2 - r^2)^{3/2} - c^3\}/3 + (c^3/3a^2)[(r^2/2)\log\{(c + (c^2 - r^2)^{1/2})/2c\} + 0.25\{c + (c^2 - r^2)^{1/2}\}^2 - c(c^2 - r^2)^{1/2}] \quad ...(40)$$

For liquids wetting the wall of the tube, put $\theta = 0$ and $c = r$ in (40), and divide both sides of (40) by r, we get

$$a^2 = r(h + r/3)/2 - (r^4/6a^2)(\log 2 - 0.5) \quad ...(41)$$

Replacing h from the relation (23)

$$2a^2 = rh$$

Equation (41) becomes

$$a^2 = (r/2)[\, h + r/3 - (2r^2/3h)(\log 2 - 0.5)] \quad ...(42)$$

Lord Rayleigh derived the above relation independently in 1915 [2].

But Poisson, long back, gave a similar expression in 1831 [3].

Mathieu's [4] objections to above relation were

- du/dx becomes infinite at $x = r$
- $(du/dx)\,(c^2 - x^2)^{1/2}$ and $(du/dx)^2\,(c^2 - x^2)^2$ both vanishes at $x = r$

To circumvent these objections Lord Rayleigh [2] took another approximation by not neglecting the term containing $(du/dx)^2$ in (31a) but taking its approximate value from (35) as

$$(c^2 - x^2)^2\,(du/dx)^2 = (c^6/9a^4x^2)\,\{c^2\,(c^2 - x^2) + (c^2 - x^2)^2 - 2c(c^2 - x^2)^{3/2}\} \quad ...(43)$$

Substituting it in the equation (31a) back, we get

$$\int uxdx = (C/2)x^2 + (c^3/6a^2)[a^2\log\{c + (c^2 - x^2)^{1/2}\} + c^2/2 - c(c^2 - x^2)^{1/2} + (c^2 - x^2)/2] \quad ...(44)$$

Thus giving

$$du/dx = \{c(L + C)/2a^2 - 1\}\,c^2x/(c^2 - x^2)^{3/2} - c^6/3a^2x(c^2 - x^2)^{3/2} + c^3/3a^2x + (c^4/6xa^4(c^2 - x^2)^{3/2})[3c^2(c^2 - x^2)/2 + (c^2 - x^2)^2 - 2c(c^2 - x^2)^{3/2} + c^2x^2\log\{c + (c^2 - x^2)^{1/2}\} + c^4/2 - c^3(c^2 - x^2)^{1/2}] \quad ...(45)$$

On integration, (45) gives the following expression for u

$$u = \{c(L + C)/2a^2 - c^2/3a^2 - 1\}c^2/(c^2 - x^2)^{1/2} + (c^3/3a^2)\log\{c + (c^2 - x^2)^{1/2}\} + (c^5/6a^4)[-2\log\{c + (c^2 - x^2)^{1/2}\} + (c^2 - x^2)^{1/2}/c - 1 + c/2(c^2 - x^2)^{1/2} + (c/(c^2 - x^2)^{1/2})\log\{c + (c^2 - x^2)^{1/2}\}] + C' \text{ (a constant of integration)} \quad ...(46)$$

So putting $c = r$ in (45), we get

$$(du/dx)_{x=r}\,(r^2 - x^2)/r^2 = (r/(r^2 - x^2)^{1/2})\{r(L + C)/2a^2 - 1 - r^2/3a^2 + (r^4/6a^4)(\log r + 0.5)\} - r^4/6a^4$$
$+$ other vanishing terms at $x = r$, ...(47)

We have to choose $L + C$ in such a way that for liquids which wets the tube i.e. for the case of $c = r$, the product of $(du/dx)_{x=r}$ and $(c^2 - x^2)$ remain a small quantity. For satisfying this condition

The term within curly brackets of the first term in (47) should be zero at $r = c$, giving us

$$c(L + C)/2a^2 - 1 - c^2/3a^2 + (c^4/6a^4)\{\log c + 0.5\} = 0, \quad ...(47a)$$

Using this condition in (46), u becomes

$$u = (c^3/3a^2)\,(1 - c^2/a^2)\log\{(c + (c^2 - x^2)^{1/2})/c\} + (c^5/6a^4)[((c^2 - x^2)^{1/2} - c)/c) + (c/(c^2 - x^2)^{1/2})\log\{(c + (c^2 - x^2)^{1/2})/c\}] + C' \quad ...(48)$$

It may be noted that with the aforesaid conditions, u does not become infinite when $c = r$ and $x = r$, which was the main objection to the earlier derivation. In fact, under these conditions, for $c = r$, u is given by

$$u = -r^5/6a^4 + r^5/6a^4\,[r(r^2 - x^2)^{-1/2}\log\{1 + ((c^2 - x^2)^{1/2}/r\}] + C' \quad ...(49)$$

putting $x = r$ at the walls, u_r becomes C'

The second term within square brackets becomes 1 as

$$\{\log(1+X)\}/X = [X - X^2/2 + X^3/3 + ...]/X = 1 + \text{terms containing } X \text{ and its higher powers}$$

so $\{\log(1+X)\}/X = 1$ for $X = 0$

For general value of c, the value of u_0 is obtained by putting $x=0$ in (48), which is given by

$$u_0 = (c^3/3a^2)(1 - c^2/2a^2)\log 2 + C' \quad ...(50)$$

From (26), h the height of the liquid column at the lowest point of the meniscus is obtained by putting $x = 0$ and taking $u = u_0$, giving us

$$h = L - c + u_0 = L - c + C' + (c^3/3a^2)(1 - c^2/2a^2)\log(2) \quad ...(51)$$

So from (27) we get

$$ra^2\cos\theta = (L + C')\,r^2/2 + \{(c^2 - r^2)^{3/2} - c^3\}/3 + \int(u - C')dx \quad ...(52)$$

Substituting the value of $L + C'$ in terms of h and c from (51), we get

$$ra^2\cos\theta = (r^2/2)\,\{h + c - (c^3/3a^2)(1 - c^2/2a^2)\log 2\} + \{(c^2 - r^2)^{3/2} - c^3\}/3 + \int(u - C')dx \quad ...(53)$$

The limits of x in the integral is from 0 to r and is equal to

$$\text{But } \int(u - C')x\,dx = (C^3/3a^2)(1 - c^2/a^2)[(r^2/2)\log[\{c + (c^2 - r^2)^{1/2}\}/c] + c^2/4 - (c/2)(c^2 - r^2)^{1/2} + (c^2 - r^2)/4] + (C^5/6a^4)\,[\{c^3 - (c^2 - r^2)^{3/2}\}/3c - r^2/2 - c\{c + (c^2 - r^2)^{1/2}\}\{\log(c + (c^2 - r^2)^{1/2})/c - 1\} + 2c^2(\log 2 - 1)] \quad ...(54)$$

It may be noted that it is difficult to explicitly write $ra^2\cos\theta$ or θ in terms of c or $\cot\theta$. AT $x = r$, $\cot\theta$ is given by

$$\cot\theta = r/(c^2 - r^2)^{1/2} + (du/dx)_{x=r}.$$

But for the liquids wetting the wall of the tube i.e. for $\theta = 0$ and $c = r$ (54) becomes

$$\int(u - C')x\,dx = (r^5/12a^2) + (r^7/6a^4)(2\log 2 - 5/3) \quad ...(55)$$

Substituting this value of integral from (55) in (53) and putting $c = r$, we obtain

$$a^2 = r(h + r/3)/2 - (r^4/6a^2)(\log 2 - 0.5) + (5r^6/36a^4)(3\log 2 - 2) \quad ...(56)$$

Expressing a^2 and a^4 in terms of h in (56), we get

$$a^2 = (r/2)[\,h + r/3 - 2r^2/3h(\log 2 - 0.5) + (r^3/9h^2)(30\log 2 - 20)]$$
$$= (r/2)[h + r/3 - 0.1288\,r^2/h + 0.08826\,r^3/h^2] \quad ...(57)$$

multiplying both sides of (56) by $2\pi r$

$$2\pi rT/\rho g = \pi r^2(h + r/3) - \pi r^5/3a^2(\log 2 - 0.5) + 5\pi r^7/18a^4(3\log 2 - 2)$$
$$= \pi r^2 h + \pi r^3[1/3 - r^2/3a^2(\log 2 - 0.5) + 5r^4/18a^4(3\log 2 - 2)] \quad ...(58)$$

So by definition volume of the meniscus V_m is the second term in Equation (58) and is given by

$$V_m = \pi r^3(1/3 - 0.0644\,r^2/a^2 + 0.02206r^4/a^4) \quad ...(59)$$

or

$$V_m/\pi r^3 = 1/3 - 0.0644r^2/a^2 + 0.02206r^4/a^4. \quad ...(60)$$

7.7 RISE OF LIQUID IN WIDER TUBE

7.7.1 Rayleigh Formula

Again considering equation (25) and writing $\sin\psi$ in terms of dz/dx, we get

$$x\sin\psi = x\,dz/dx\,\{1 + (dz/dx)^2\}^{-1/2} = 1/a^2\int xz\,dx$$

Differentiating the above equation, we get

$$d^2z/dx^2 + (1/x)(dz/dx)\{1 + (dz/dx)^2\} = z/a^2\{1 + (dz/dx)^2\}^{3/2} \quad ...(61)$$

For wider tubes it has been observed that the height of the meniscus is small. Hence for wider tubes dz/dx should be small, so expanding the right hand side of equation (61) and neglecting terms containing cube of dz/dx and its higher powers, we get

$$d^2z/dx^2 + (1/x)(dz/dx) - z/a^2 = 3(z^2/2a^2)(dz/dx)^2 - (1/x)(dz/dx)^3 \quad ...(62)$$

Neglecting further the square and higher powers of dz/dx gives

$$d^2z/dx^2 + (1/x)(dz/dx) - z/a^2 = 0$$

This is the Bessel equation of zero order so its solution is

$$z = h\,J_0(ix/a) = h_0\,I_0(x/a) = h_0\,\{1 + x^2/2^2.a^2 + x^4/2^2.4^2.a^4 + x^6/2^2.4^2.6^2.a^6 + \text{-----}\} \quad ...(63)$$

J_0 is Bessel function or Fourier Function I_0 of zero order. h_0 is the elevation at the axis of the tube above the free and plane level surface. For wide tubes h_0 is very small so that it can be neglected in the experiment. Eventually dz/dx should also be small for sufficiently large value of x/a.

But dz/dx from (63) is given as

$$dz/dx = h_0/a\,I_1\,(x/a) \quad ...(64)$$

For large values of x/a, the Fourier function of order one can also be expressed as

$$I_0(x) = I_1(x) = h_0(1/2\pi)\,(x/a)^{-1/2}\,\mathrm{Exp}(x/a),$$

But $dz/dx = \tan\psi$, for small values of ψ, $\tan(\psi) = \psi$, giving

$$\psi = h_0\,(1/2\pi)(x/a)^{-1/2}\,\mathrm{Exp}\,(x/a) \quad ...(65)$$

In order to follow the curve further up to $\psi = \pi/2$, we may employ the two-dimensional solution with an assumption that ψ has moderately smaller values for almost all values of x extending to r. In other words, ψ becomes large or equal to $\pi/2$, for values of x which are very close to r. So that when necessary $r - x$ may be neglected for $\psi = \pi/2$. On account of the magnitude of x, we have to deal with only one curvature. If R is the radius of curvature at a point (z, x), at which the tangent makes an angle ψ with the x-axis. Then

$$1/R = d\psi/ds = z/a^2$$

Giving

$$(dy/dz)\,\sin y = z/a^2 \quad ...(66)$$

Integrating (66), we get

$$(1/2)(z/a)^2 = C - \cos y \quad ...(67)$$

At the lowest point of meniscus where $\psi = 0$, z/a is very small and hence can be taken as zero. Applying this condition gives $C = 1$

So (67) becomes

$$(1/2)(z/a)^2 = 1 - \cos\psi = 2\sin^2(\psi/2) \text{ giving us}$$

$$(z/a) = 2\sin(\psi/2) \quad ...(68)$$

From (68) $dz/d\psi = a\cos(\psi/2)$ and also we know that $dz/dx = \tan\psi$. These two conditions enable us to express x in terms of ψ as follows

$$\begin{aligned} dx &= dz/\tan\psi = a\,\{\cos(\psi/2)d\psi\}/\tan\psi. \\ &= a\{\cos(\psi/2)d\psi\}\,\{1 - \tan^2(\psi/2)\}/2\tan(\psi/2) \\ &= a\{\cos(\psi/2)\,d\psi\}\{\cos^2(\psi/2) - (\sin^2(\psi/2)\}/2\sin(\psi/2)\cos(\psi/2) \\ &= a[\{1 - 2\sin^2(\psi/2)\}/\sin(\psi/2)]d\,(\psi/2) \\ &= a\{1/\sin(\psi/2) - 2\sin(\psi/2)\}d\,(\psi/2) \end{aligned} \quad ...(69)$$

Integrating (69), we get

$$x/a = \log\{\tan(\psi/4)\} + 2\cos(\psi/2) + C_1 \quad ...(70)$$

The constant of integration is calculated on the basis that at $x = r$, $\psi = \pi/2$, giving us

$$r/a = \log\{\tan(\pi/8)\} + 2\cos(\pi/4) + C_1 \quad ...(71)$$

Subtracting (70) from (71), we get

$$(r - x)/a = \log\{\tan(\pi/8)\} - \log\{\tan(\psi/4)\} + \sqrt{2} - 2\cos(\psi/2) \quad ...(72)$$

In general, for all values of x further from the wall, take ψ small so that $\tan\psi/4 = \psi/4$ and $\cos(\psi/2) = 1$, hence equation (72) becomes

$$(r - x)/a = \log\{\tan(\pi/8)\} - \log\psi + 2\log 2 + \sqrt{2} - 2 \quad ...(73)$$

To eliminate ψ from equations (73) and (65), take logarithm of both sides of (65), which can be written as follows:

$$\log(\psi) = \log(h_0/a) + \log\exp(x/a) - (1/2)\log(2\pi x/a)$$

or

$$0 = \log(h_0/a) + x/a - \log(\psi) - (1/2)\log(2\pi x/a) \quad ...(74)$$

Subtracting (74) from (73), we get

$$(r - x)/a = \log(\sqrt{2} + 1) + 2\log 2 + \sqrt{2} - 2 - \log(h_0/a) - x/a + (1/2)\{\log 2\pi + \log x/a\}$$

It may be noted that $\log\{\tan(\pi/8)\} = \log(\sqrt{2} + 1)$

With sufficient approximation, when h_0 is small enough, we may here substitute r for x and thus

$$r/a - (1/2)\log(r/a) = \log(\sqrt{2} + 1) + \sqrt{2} - 2 + 2\log 2 + (1/2)\log(2\pi) + \log(h_0/a)$$

$$r/a - (1/2)\log(r/a) = 0.8381 + \log(a/h_0) \quad ...(75)$$

From this equation the values of $\log(a/h_0)$ may be calculated for different values of r/a, provided h remains small and variation of z with respect of x is small. Rayleigh showed [2], that these two conditions remain satisfied for $r/a \geq 6$.

7.7.2 Laplace Formula

Go back to equation (25)

$$x\sin\psi = (1/a^2)\int xz\,dx$$

Differentiating it with respect of x, we get

$$\sin\psi + x\cos\psi\, d\psi/dx = xz/a^2. \quad ...(76)$$

writing

$$d\psi/dx = (d\psi/dz)(dz/dx) = \tan\psi\, d\psi/dz$$

Substituting this value of $d\psi/dx$ in (76) and dividing by x, we get

$$\sin\psi\, d\psi/dz + \sin\psi/x = z/a^2.$$

Here also method of successive approximation is used. To start with, the curvature $\sin\psi/x$ is assumed to be negligible, so we get

$$\sin\psi d\psi = zdz/a^2. \quad ...(77)$$

Integrating we get

$$C - \cos\psi = z^2/2a^2. \quad ...(78)$$

Apply the condition that at $\psi = 0$, the lowest point of the meniscus, z is negligibly small, which gives $C = 1$ and (78) becomes

$$1 - \cos\psi = z^2/2a^2.$$

$$2\sin^2(\psi/2) = z^2/2a^2.$$

Giving

$$z/2a = \sin(\psi/2) \quad ...(79)$$

But

$$\sin\psi = 2\sin(\psi/2)\cos\psi/2) = z/a\{1 - z^2/4a^2\}^{1/2}$$

We put the approximate value of the second curvature in (76) by substituting the value of $\sin\psi$ and replacing x by r i.e. we are assuming that x is large enough to be put as r, in other words we are considering the points close to the wall of the wider tube.

$$\sin\psi\, d\psi/dz + (z/ar)(1 - z^2/4a^2)^{1/2} = z/a^2$$

$$\sin\psi d\psi = [z/a^2 - (z/ar)(1 - z^2/4a^2)^{1/2}]dz \qquad ...(80)$$

Integrating (80), we get

$$C - \cos\psi = z^2/2a^2 + (4a/3r)(1 - z^2/4a^2)^{3/2} \qquad ...(81)$$

To calculate C, we know at $\psi = 0$, z is negligibly small, so we get

$$C - 1 = 0 + 4a/3r, \text{ hence (81) becomes}$$

$$1 - \cos\psi = z^2/2a^2 + (4a/3r)\cos^3(\psi/2) - 4a/3r$$

$$z^2/2a^2 = 2\sin^2(\psi/2) + (4a/3r)(1 - \cos^3(\psi/2)) \qquad ...(82)$$

In order to express z in linear form, use (79) i.e. replace $z/2a$ by $\sin(\psi/2)$ in equation (82), we get

$$z/a = 2\sin(\psi/2) + (2a/3r)\{(1 - \cos^3(\psi/2)\}/\sin(\psi/2) \qquad ...(83)$$

On differentiating with respect to ψ, we get *a* better representation of the surface of the meniscus

$$(1/a)\, dz/d\psi = \cos(\psi/2) + (a/3r)[\{3\cos^2(\psi/2).\sin^2(\psi/2) - \cos(\psi/2)(1 - \cos^3(\psi/2)\}/\sin^2(\psi/2)$$

Now we are in position to find x in terms of ψ by using the relation

$$x = \int \cot\psi (dz/d\psi) d\psi \qquad ...(84)$$

$$C + x/a = \log\tan(\psi/4) + 2\cos(\psi/2) - (a/3r)\{1/2(1 - \cos(\psi/2)) + 2\sin^2(\psi/2) - (3/2)\log(1 - \cos(\psi/2) - (3/2)\log\sin(\psi/2)\} \qquad ...(85)$$

At $\psi = \pi/2$, $x = r$ put this boundary condition in equation (85) to get the value of C, giving us

$$C + r/a = \log(\sqrt{2} - 1) + \sqrt{2} - (a/3r)\{1/2(1 - 1/\sqrt{2}) + 2/2 - 3/2\log(1 - 1/\sqrt{2}) - 3/2\log(1/\sqrt{2})\}$$

$$= \log(\sqrt{2} - 1) + \sqrt{2} - (a/3r)\{1 + \sqrt{2}/2 + 1 - 3/2\log(\sqrt{2} - 1)\} \qquad ...(86)$$

Subtracting (85) from (86), we get

$$r/a - x/a = \log(\sqrt{2} - 1) + \sqrt{2} - (a/3r)\{2 - \sqrt{2}/2 - (3/2)\log(\sqrt{2} - 1)\} - \log(\tan\psi/4) - 2\cos(\psi/2) + (a/3r)[1/2(1 + \cos(\psi/2)) + 2\sin^2(\psi/2) - (3/2)\log\{(1 - \cos(\psi/2)/\sin(\psi/2)\}] \qquad ...(87)$$

For small values of ψ, replacing $\cos\psi = 1$ and $\sin\psi = \tan\psi = \psi$, equation (87) becomes

$$r/a - x/a = \log(\sqrt{2} - 1) + \sqrt{2} + 2\log 2 - 2 - (a/3r)\{2 - \sqrt{2}/2 - (3/2)\log(\sqrt{2} - 1)\} + (a/3r)\{1/4 + 0 - 3/2\log(\psi) + 3\log 2\} + \log\psi \qquad ...(88)$$

Writing $\log(\sqrt{2} - 1) + \sqrt{2} + 2\log 2 - 2 = -0.0809 = a_1$

and $2 - \sqrt{2}/2 - 3/2\log(\sqrt{2} - 1) - 1/4 - 3\log 2$

$$= -1/3(\sqrt{2}/6 + (1/2)\log(\sqrt{2} - 1) + \log 2 - 7/12) = -0.0952 = a_2$$

Equation (88) becomes

$$(r - x)/a = a_1 - a_2 a/3r - (1 + a/2r)\log\psi \qquad ...(89)$$

The other equation is derived from the flatter part of the air-liquid interface almost the same way as was done by Rayleigh and is given as

$$\psi = dz/dx = (h_0/a)\, I_1(x/a) = (h_0/a)\,\mathrm{Exp}(x/a).(2\pi x/a)^{-1/2}(1 - 3/8x). \qquad ...(90)$$

Taking natural logarithm of both sides of (90) and approximation, as x/a is large, we get

$$x/a = \log\psi + \log a/h_0 + (1/2)\log 2\pi x/a + 3a/8x \qquad ...(91)$$

Eliminating $\log\psi$ in between (89) and (91), we get

$$r/a - \log(a/h_0) = (a_1 - a_2 a/3r)/(1 + a/2r) + (r - x)/(2r + a) + (1/2)\log 2\pi x/a + 3a/8x \quad ...(91a)$$

Writing $\log 2\pi x/r = \log 2\,\pi r/a + \log(1 - (r - x)/r)$

and $\quad 3a/8x = 3a/8r\,(1 + (r - x)/r)$

Now applying the condition that x is nearly equal to r i.e. $(r - x)$ is small, we get

$$\log 2\pi x/a = \log 2\pi r/a + \log(1 - (r - x)/r) = \log(2\pi r/a) - (r - x)/r$$

and $\quad 3a/8x = 3a/8r$ and substituting these values in (91a), we get

$$r/a - \log(a/h_0) = (a_1 - a_2 a/3r)/(1 + a/r) + 3a/8r + (1/2)\log(2r/a) + (1/2)\log 2\pi \quad ...(92)$$

In (91*a*), x is large and is nearly equal to r so $a\,(r - x)/8r^2$ may be neglected. Also in view of the smallness of a_1 and a_2, it is scarcely necessary to retain the denominator $(1 + a/r)$ of the second term on the right hand side of (92), so that we may write (92) as

$$r/a - \log(a/h_0) = 0.8381 + 0.2798\,a/r + 0.5\log(r/a) \quad ...(93)$$

From equations (92) or (93) we can calculate the value of $\log(a/h_0)$ for different values of r/a and compare them with those obtained from Lord Rayleigh equation (75).

The meniscus volume in each case has been calculated from equation (16). The equation has been reproduced below:

$$2\pi T\cos\theta = (\pi rh + V_m/r)\rho g$$

$$2\pi rT\cos\theta - \pi r^2 h\rho g = V_m \rho g$$

$$V_m = (\pi r)\{2a^2\cos\theta - \rho h\}$$

$$V/\pi a^3 = (r/a)\{2\cos\theta - (r/a)(h/a)\} \quad ...(94)$$

7.8 AUTHOR'S APPROACH

7.8.1 Air-liquid Interface is Never Spherical

For any spherical surface, Figure 7.6 radii of curvature are equal at every point on it, so the pressure difference on the two sides of the air-liquid spherical interface at every point will be equal, and it will also be equal to that at the lowest point. Hence there will be no extra pressure differences on the two sides of the air-liquid spherical interface at different points, which can support the increasing height of the meniscus liquid.

Analytically we may approach as follows. If possible, let us assume that air-liquid interface is spherical of radius r_1. Consider the principle section of the spherical interface in the vertical plane passing through the axis of the tube. Let P be the lowest point and Q be any other point having coordinates (x, z). The axis of the tube is taken as z-axis positive downward and diameter of the contact circle as x-axis positive on the right, shown in Figure 7.8.

In this case, radius of curvature at every point of the meniscus is r_1, so pressure at P as well as at Q will be

$$A - 2T/r_1. \quad ...(95)$$

If projection of Q meets the tangent at the lowest point (P) at the point R, then QR is given by

$$QR = r_1 - r_1\cos(\theta) - z, \text{ a finite height.} \quad ...(96)$$

Here 2θ is the angle subtended by the arc of the circle at its centre.

So the pressure difference between the points Q and R

$$= (r_1 - r_1\cos q - z)rg. \quad ...(97)$$

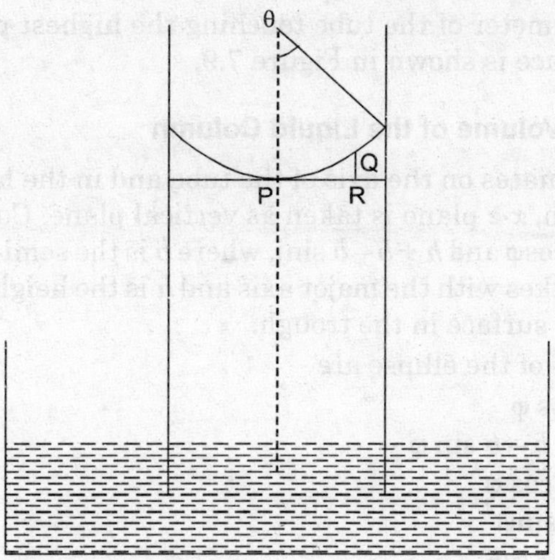

Figure 7.8 Spherical air-liquid interface

But pressure at P and R is equal, as these two points are on the same horizontal plane. In other words pressure difference in between the points Q and R is zero, which contradicts the statement (97). Hence the assumption that air-liquid interface is spherical is not valid.

7.8.2 Air-Liquid Interface is Ellipsoidal

Having established that air-liquid interface cannot be spherical we look for other forms. In order to logically arrive on the form of the air-liquid interface, let us consider the expected requirements from such a surface

Roughly speaking it should have the following qualities.

1. It should be a curved surface.
2. The radius of curvature should vary from point to point.
3. The radius of curvature should decrease as the point moves away from the centre.
4. For circular tubes, surface should be symmetrical about the axis of the tube.
5. The tangent at the lowest point should be horizontal.
6. It should be able to meet the walls of the tube at zero angles or at a very small angle for wetting liquids and at variable angles for other liquids.
7. The surface should become shallower and shallower with the increase in radius of the tube (Day to day observation).

We will see that surface of revolution of an ellipse fulfil the above requirements. Vertical section of such a surface is an ellipse, whose major axis is the diameter of the tube, so that for wetting liquids it meets the tube walls at zero angle and tangent at its lowest point is horizontal. By changing the values of b-the semi-minor axis, the surface may be made as shallower as we wish. The lengths of radii of curvature are different at different points and we can show that these radii decrease as the point moves away from the lowest point.

So following the above arguments, we take air-liquid interface as the surface generated by revolution of quarter of an ellipse whose semi-major axis is the radius of the tube and has a variable semi minor axis b. Axis of the ellipse is the diameter of the contact circle i.e. Major axis coincides with the diameter of the tube touching the highest point of the meniscus. The vertical section of the surface is shown in Figure 7.9.

7.8.3 Equilibrium of the Volume of the Liquid Column

Taking the origin of coordinates on the axis of the tube and in the horizontal plane of the free liquid surface in the trough, x-z plane is taken as vertical plane. Co-ordinates of any point on the ellipse are defined as rcosφ and $h + b - b$ sinj, where b is the semi-minor axis, φ is the angle, which the radius vector makes with the major axis and h is the height of the lowest point of the ellipse from the free liquid surface in the trough.

Parametric equations of the ellipse are

$$x = r\cos\varphi \qquad ...(98)$$

$$z = h + b - b\sin\varphi$$

Giving us $dz/d\varphi = -b\cos\varphi$ ...(99)

$$dx/d\varphi = -r\sin\varphi$$

So $dz/dx = b\cot\varphi/r$ ...(100)

$$\sin\psi = b\cos\varphi/(r^2\sin^2\varphi + b^2\cos^2\varphi)^{1/2},$$

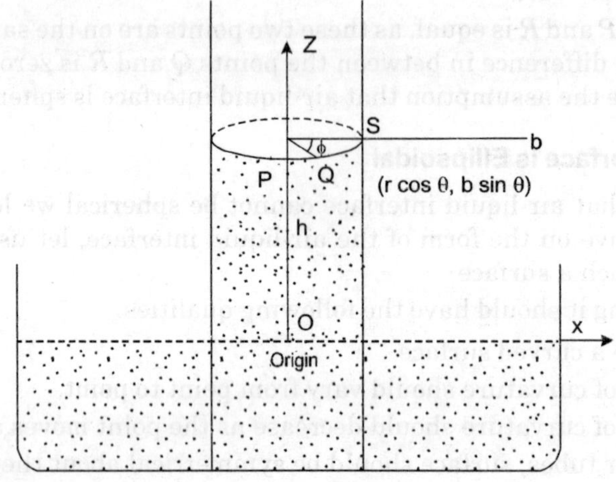

Figure 7.9 Elliptical-interface

Let us consider the pressure equilibrium conditions in this case also at lower most point P and any other point Q (r cos(φ), $h + b - b$ sin(φ)) Figure 7.9. At the point P, each radius of curvature, due to circular symmetry is r^2/b. If r_1 and r_2 are the radii of curvatures at the point Q, then r_1 the radius of curvature in $x - z$ plane is

$$r_1 = [1 + (b^2/r^2)\cot^2\varphi]^{3/2}/(b/r^2\sin^3\varphi)$$

or $$r_1 = [r^2\sin^2\varphi + b^2\cos^2\varphi]^{3/2}/br. = [(r^2 - b^2)\sin^2\varphi + b^2]^{3/2}/br \qquad ...(101)$$

The other radius in $x - y$ plane is

$r_2 = x/\sin\psi$, where ψ is the angle which tangent, at point Q, makes with x-axis and is such that

$$\tan\psi = -b\cos\varphi/r\sin\varphi.$$

Expressing r_2 in terms of ϕ and φ, gives

$$r_2 = (r/b)\,[b^2 + (r^2 - b^2)\sin^2\varphi]^{1/2} \quad ...(102)$$

Here we see that r_1 and r_2 both decrease with decrease in φ, hence the pressure just inside the air-liquid interface decreases as we move away from lowest point of meniscus, satisfying the requirement 3 given above.

Considering the equilibrium of a cylinder of a liquid of radius x and bounded by the ellipsoid on one side and a flat in the horizontal plane of the liquid in the trough on the other. Equation (25), gives us

$$x \sin\psi = r \cos\varphi \sin\psi = 1/a^2 \int zx\, dx \quad ...(103)$$

Limits of x in the integral are from 0 to x

Now $\quad z = h + b - b \sin\varphi \quad ...(104)$

But from (99)

$$dx = -r \sin\varphi\, d\varphi \quad ...(105)$$

For wetting liquid, the angle ψ, which the tangent makes with x-axis, at meeting of the interface with wall is $\pi/2$. Also the angle φ at the wall where the air-liquid interface meets is zero.

Substitution of values of x, z and dx in equation (103) gives

$$r = 1/a^2 \int (h + b - b \sin\varphi)\, r \cos\varphi(-r \sin\varphi) d\varphi \quad ...(106)$$

For entire cross-section, limits of φ in the integral are $\pi/2$ to 0

Putting $\quad \sin\varphi = \zeta$, which makes $\cos\varphi\, d\varphi = d\zeta$

Equation (106) becomes

$$r = -(r^2/a^2) \int (h + b - b\zeta)\zeta d\zeta \quad ...(107)$$

On integrating (107), we get

$$r = -(r^2/a^2)\,[(h + b)\,\zeta^2/2 - b\zeta^3/3]$$

Limits of ζ are from 1 to 0, giving us

$$r = r^2/a^2(h/2 + b/6) \quad ...(108)$$

Substituting the value of h from (18) in (108), we get

$$r = (r^2/a^2)\,(2ba^2/2r^2 + b/6), \text{ giving}$$

$$b = 6r/(6 + r^2/a^2) \quad ...(109)$$

Volume of an ellipsoid obtained by rotating a quarter of the ellipse about its minor axis is $2\pi r^2 b/3$

The volume V_m of the meniscus is the difference in volumes of a cylinder of height b and radius r and the semi-ellipsoid giving us

$$V_m = \pi r^2 b/3 = 2\pi r^3/(6 + r^2/a^2) \quad ...(110)$$

$$V_m/\pi a^3 = 2r^3/a^3/(6 + r^2/a^2) \quad ...(111)$$

$$V/\pi r^3 = 2/(6 + r^2/a^2) \quad ...(112)$$

In equation (111), if r/a is large enough so that 6 is negligible in comparison of $(r/a)^2$, then (111) reduces to

$$V_m/\pi a^3 \cong 2r/a \quad ...(112A)$$

Here $V_m/\pi a^3$ approaches asymptotically to $2r/a$. This is the result what Porter obtained in 1934 [8].

Further from (112) for r/a equal to zero

$$V_m/\pi r^3 = 1/3, \quad \text{...(113)}$$

This is what Lord Rayleigh got while discussing for small-bore tubes for $r = 0$ in equation (60).

Here we have obtained a single relation to find the meniscus volume for all values of r/a. The relation is simple in nature and has been derived on the basis of sound and well-established reasoning. Like Lord Rayleigh, no arbitrary variable has been introduced nor unnecessary approximations from step to step have been resorted to. The aforesaid work has been published in Metrologia(13).

7.8.4 Lord Kelvin's Approach

Lord Rayleigh [2] reported in his paper that Lord Kelvin [6] also considered vertical section of the air liquid interface as an ellipse and gave a relation between a, h and r. The author further developed this relation in the following formulae for meniscus volume.

$$V_m/\pi r^3 = a^2/r^2 \{(1 + r^2/3a^2)^{1/2} - 1\} \quad \text{...(119)}$$

The above relation holds good only for smaller values of r/a.

7.8.5 Discussion of Results

In order to compare the values of $V_m/\pi r^3$ for smaller values of r^2/a^2, the values $V_m/\pi r^3$ for different values of r^2/a^2 using various formulae encountered till now have been indicated in Table 7.1.

- Rayleigh's formula: equation (60) in column 2.
- International critical tables (I.C.T) using Bashforth and Adams Tables in column 3.
- Lord Kelvin's formula equation (119) in column 4.
- The author's formula equation (112) in column 5.

Bashforth and Adams used undisputable differential equation for air-liquid interface and solved it numerically for r^2/a^2 but only up to 10. The values given by Bashforth and Adams have been taken as reference and all values have been compared against these values.

From the data of Table 1 it is evident that values from Lord Kelvin formula agree from those of Bashforth and Adams for r^2/a^2 almost up to 10. While those calculated from Rayleigh's formula agree only up to $r^2/a^2 = 0.7$.

The reasons are as follows:

Rayleigh formula is based upon a spherical air-liquid interface, which is not a valid proposition as has been proved above. For smaller values of r^2/a^2, an ellipse tends to be a circle that is why Rayleigh values agree only for smaller values of r^2/a^2.

Though Kelvin used a correct form of air-liquid interface, but in arriving to the relation in (119) used a relation $h = 2a^2/r$, which is valid for spherical surface only. For smaller values of r/a there is not much difference in the two surfaces. Hence values of meniscus volume agree well for smaller values of r^2/a^2 say up to 10.

Author used the logically correct surface and used a correct relation between h, b, r, and a, the values given in column 5 have been calculated by the universal formula of meniscus volume.

Table 7.1 $V/\pi r^3$ Against r^2/a^2

r^2/a^2	*Rayleigh's formula*	*I.C.T*	*Lord Kelvin*	*Author's Formula*
1	2	3	4	5
0.0	0.3333	0.3333	0.3333	0.3333
0.1	0.3271	0.327	0.3280	0.3279
0.2	0.3213	0.321	0.3229	0.3226
0.3	0.3160	—	0.3182	0.3175
0.4	0.3113	0.311	0.3137	0.3125
0.5	0.30664	—	0.3094	0.3077
0.6	0.30262	0.301	0.3054	0.3030
0.7	0.2990	—	0.3015	0.2985
0.8	0.2959	0.292	0.2978	0.2941
0.9	0.2932	—	0.2943	0.2899
1.0	0.2910	0.284	0.2910	0.2857
1.5	0.2863	0.266	0.2761	0.2667
2.0	0.2926	0.251	0.2638	0.2500
2.5	0.3099	0.238	0.2532	0.2352
3.0	0.3382	0.226	0.2440	0.2222
3.5	0.3775	0.216	0.2359	0.2105
4.0	0.4278	0.206	0.2287	0.2000
4.5	0.4891	0.198	0.2222	0.1905
5.0	0.5614	0.190	0.2163	0.1818
5.5	0.6447	0.184	0.2109	0.1739
6.0	0.7391	0.177	0.2060	0.1667
6.5	0.8444	0.171	0.2014	0.1600
7.0	0.9607	0.1650	0.1972	0.1538
7.5	1.0880	0.1590	0.1933	0.1481
8.0	1.2263	0.1540	0.1896	0.1429
8.5	1.3756	0.1493	0.1861	0.1379
9.0	1.5359	0.1449	0.1829	0.1333
9.5	1.7072	0.1406	0.1798	0.1290
10.0	1.8895	0.1365	0.1769	0.1250

The values given in column 5 agree more closely for r^2/a^2 up to 6. For higher values of r/a, the difference between the values given by the author and those given by Bashforth and Adams is never more than 10% of the meniscus volume.

The values of $V/\pi r^3$ against r^2/a^2 calculated from equation (112) have been plotted in Figure 7.10. The variable r^2/a^2 varies from 0 to 100.

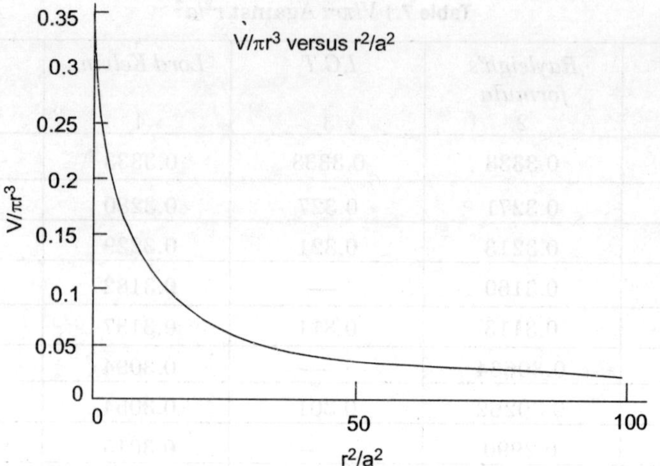

Figure 7.10 $V/\pi r^3$ against r^2/a^2

As expected $V/\pi r^3$ decreases continuously with increase in r^2/a^2. Lower part is almost the branch of a rectangular hyperbola. In fact equation (112) becomes the equation of a rectangular hyperbola if origin is shifted to (–6, 0).

A graph between $V/\pi a^3$ and r^2/a^2 has been drawn and shown in Figure 7.11 for r/a up to 10. This appears to be similar to the branch of a parabola, tapering off slightly. Gradient of the curve is continuously decreasing and finally becomes 2 for very large values of r^2/a^2.

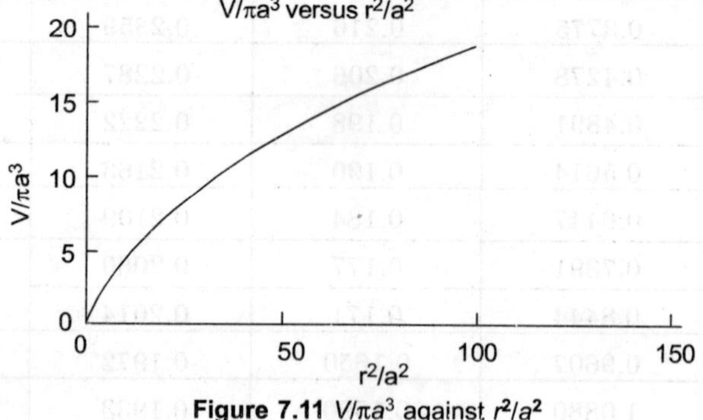

Figure 7.11 $V/\pi a^3$ against r^2/a^2

A graph, between $V/\pi a^3$ versus r/a, has been drawn from author's formula from 0 to 32 for r/a, which has been shown in Figure 7.12.

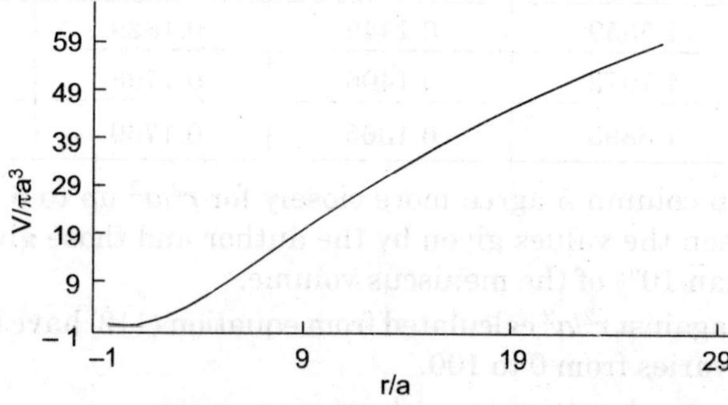

Figure 7.12 $V/\pi a^3$ versus r/a (0 to 32)

One can easily notice a small curvature in the beginning of the curve. For larger values of r/a, $V/\pi a^3$ is almost linear to r/a, ultimately ratio, of $V/\pi a^3$ and r/a, becomes 2.

To show prominently the curvature at the beginning of the curve between $V/\pi a^3$ and r/a, the curve for reduced range of r/a is shown in Figure 7.13. The value of r/a increases from 0 to 6 in steps of 0.1.

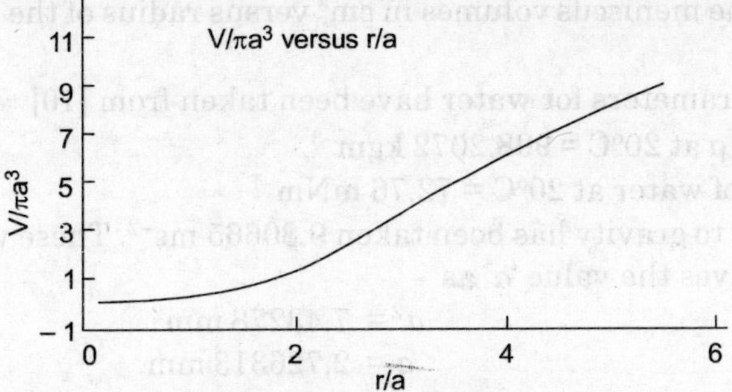

Figure 7.13 $V/\pi a^3$ versus r/a

For wider tubes, the mean of the values of h/a obtained by using the formulae of Lord Rayleigh (75) and Laplace (93) and has been inserted in (94), which gives values of $V_m/\pi a^3$. These values, for r/a equal 5 and onward have been tabulated in the upper row of Table 7.2. For the purpose of comparison the values derived from the Author's formula have been indicated in lower rows of Table 7.2. In general, the values calculated from Authors formula are lesser than those, which were calculated from Laplace and Rayleigh. The values given in the Table 7.2 have been plotted as two separate curves and are shown in Figure 7.14. The difference in the two values goes on decreasing and becomes negligible by the time r/a reaches 32. The maximum difference is at r/a equal to six, which is about 10%.

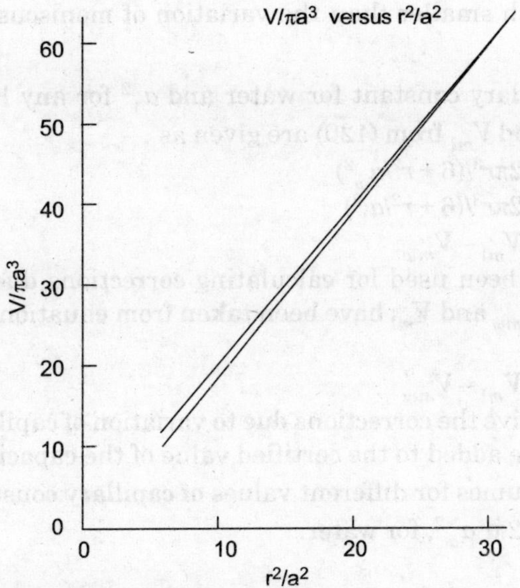

Figure 7.14 Comparative values of $V/\pi a^3$ versus r^2/a^2

Accurate values of $V/\pi a^3$ for r/a from 0 to 6 in steps of 0.1 have been given in Table 7.3.

7.9 VOLUME OF WATER MENISCUS IN RIGHT CIRCULAR TUBES

Equation (111) may be rewritten as

$$V_m = \pi a^3 2\, r^3/a^3/(6 + r^2/a^2)$$

$$= 2\pi r^3/(6 + r^2/a^2) \quad ...(120)$$

Author used (120) to calculate the volume of water meniscus at 20°C for r from 0 to 60 mm in steps of 0.1 mm. The meniscus volumes in cm^3 versus radius of the tube are given in Tables 7.5 to 7.10.

The following parameters for water have been taken from [10]

Density of water ρ at 20°C = 998.2072 kgm^{-3}.

Surface tension of water at 20°C = 72.76 mNm^{-1}

Acceleration due to gravity has been taken 9.80665 ms^{-2}. These values when substituted in the formula of 'a' gives the value 'a' as

$$a^2 = 7.43278 \text{ mm}^2. \quad ...(121)$$

Or

$$a = 2.726313 \text{ mm}.$$

7.10 DEPENDENCE OF MENISCUS VOLUME ON CAPILLARY CONSTANT

From equation (120), it may be seen that volume of meniscus of a wetting liquid depends upon the capillary constant. Water is mostly used, as a media for calibrating all laboratory measures so what is given in the certificate of calibration is the volume of water delivered or contained in the measure. But liquids have much larger variation in capillary constant say from 7.43 mm^2 to almost 2.5 mm^2. So a correction due to variation in capillary constant is necessary.

The capillary constant is a linear function of the ratio of surface tension to the density of the liquid. We know surface tension, as well as, the density of the liquid are temperature dependent quantities. So theoretically meniscus volume will also depend upon temperature, but this variation is much smaller than the variation of meniscus volume due to change in capillary constant.

Let a_w^2 be the capillary constant for water and a_1^2 for any liquid, then corresponding meniscus volumes V_{mw} and V_{ml} from (120) are given as

$$V_{mw} = 2\pi r^3/(6 + r^2/a_w^2) \quad ...(122)$$

$$V_{ml} = 2\pi r^3/(6 + r^2/a_1^2) \quad ...(123)$$

Correction $\quad C = V_{ml} - V_{mw}$. ...(124)

Equation (124) has been used for calculating corrections due to variation in capillary constant. The values of V_{mw} and V_{ml} have been taken from equation (122) and (123) for all the values of r/a.

$$C = V_{ml} - V_{mw}$$

Tables 7.11 to 7.16 give the corrections due to variation of capillary constants. Correction with negative sign is to be added to the certified value of the capacity of the measure.

For wider tubes, volumes for different values of capillary constants is roughly given as

$$V_{mw} = 2\pi r\, a_w^2, \text{ for water.} \quad ...(125)$$

For any other liquid

$$V_{ml} = 2\pi r\, a_1^2. \quad ...(126)$$

So correction to be applied to the stated volume of a measure when liquid of different capillary constant is used for wider tubes is given by

Correction $C = V_{ml} - V_{mw} = 2\pi r\,(a_1^{\,2} - a_w^{\,2})$, ...(127)

Here we see that correction is proportional to the difference in the capillary constants of liquid and water. The fact can be seen easily for tubes having radius more than 40 mm in Table 7.15 and onward.

7.11 FOR LIQUID SYSTEMS HAVING FINITE CONTACT ANGLES

For a particular value of θ, the contact angle, Porter [8,9], used equation (22) to calculate the values of $V/\pi a^3$ from Bashforth and Adams tables for smaller values of r/a and graphed it. For larger values of r/a, he simply extended the curve $V/\pi a^3$ in such a way that the curve becomes asymptotic to the line $y = 2x$. Where x is $(r/a)\cos(\theta)$ and y is $V/\pi a^3$.

$$V/\pi a^3 = 2\cos(q)\,(r/a) \qquad \text{...(128)}$$

The values of $V/\pi a^3$ for different values of r/a and θ are given in Table 7.4

7.11.1 Author's Approach for Liquids having any Contact Angle

Another approach for systems having finite angle of contact is to consider the air-liquid interface as the surface generated by revolving a quarter of an ellipse about the axis of tube having major axis of such a length that the liquid surface meets the walls of the container at the given angle of contact. The major axis is above the diameter of the contact circle. It is just an extension of the case of the liquid system with zero angle of contact.

The vertical section of the interface is an ellipse with α and β as its semi axes. α is greater than r and is related to it by a simple equation

$$\alpha \cos \varphi_o = r.$$

Where φ_o is the angle which radius vector makes with major axis, the angle is positive when measured in clockwise direction.

As in 7.8.3, taking origin at the axis of the tube and in the horizontal plane of air liquid surface in a trough of very large diameter, parametric equations of the ellipse are:

$$x = \alpha\cos(\varphi) \text{ and } z = h + \beta - \beta\sin(\varphi), \text{ giving}$$

$$dx/d\varphi = -\alpha\sin(\varphi) \text{ and } dz/d\varphi = -\beta\cos(\varphi)$$

So dz/dx at walls is given by $\beta\cos(\varphi_o)/\alpha\sin(\varphi_o)$, which is equal to $\cot(\theta)$, where θ is the angle of contact.

or $\cot(\theta) = (\beta/r)\cos^2(\varphi_o)/\sin(\varphi_o)$ ...(129)

Considering again the equilibrium of the complete liquid column rise due to surface tension, as in section 7.8.3, we get the following relation

$$a^2 r\cos(\theta) = r^2.h/2 + r^2\beta\,[1/2 - (1/3)\,\{1 - \sin^3(\varphi_o)\}/\cos^2(\varphi_o)] \qquad \text{...(130)}$$

Now writing $[1/2 - (1/3)\,\{1 - \sin^3(\varphi_o)\}/\cos^2(\varphi_o)] = K$ ...(131)

and multiplying both sides of (120) by 2π, we get

$$2\pi a^2 r\cos(\theta) = \pi r^2 h + 2\pi r^2\beta K \qquad \text{...(132)}$$

Comparing (132) with (15) we get meniscus volume V_m as

$$V_m = 2\pi r^2 \beta K$$

$$V_m/\pi a^3 = 2.(r^3/a^3)(\beta/r)K \quad ...(133)$$

$$V_m/\pi r^3 = 2(\beta/r)K$$

Also $r^2h/2 = \beta a^2$,

Substituting this value in (130), we get

$$a^2 r\cos(\theta) = a^2\beta + r^2\beta K,$$

So we get

$$\beta/r = \cos(\theta)/(1 + Kr^2/a^2) \quad ...(134)$$

Eliminating β/r with help of (134) and (129), we get a cubic equation in $\sin(\varphi_o)$ given as

$$X^3\{r^2/3a^2 - \sin(\theta)\} + X^2\{-\sin(\theta) - 1 - r^2/6a^2\} + X\{\sin(\theta) - 1 - r^2/6a^2)\} + \sin(\theta) = 0 \quad ...(135)$$

Where $X = \sin(\varphi_o)$;

Writing

$$(r^2/3a^2 - \sin(\theta)) = A$$

$$-\sin(\theta) - 1 - r^2/6a^2 = B$$

$$\sin(\theta) - 1 - r^2/6a^2 = C \text{ and} \quad ...(136)$$

$$\sin(\theta) = D$$

In order to reduce (135) to standard form of a cubic equation, put $X = Y + S$, such that the second degree term in (135) becomes zero. Giving us

$$Y^3 + Y(C - B^2/3A)/A + (2B^3/27A^2 - BC/3A + D)/A = 0 \quad ...(137)$$

And $S = -B/3A$

Comparing it with standard cubic equation $Y^3 + QY + R = 0$, ...(138)

we get $Q = (3AC - B^2)/3A^2$

$$R = 2B^3/27A^3 - BC/3A^2 + D/A \quad ...(139)$$

Cube root of (138) is

$$Y = \alpha^{1/3} + \beta^{1/3}$$

Where $\alpha = -R/2 + \sqrt{[R^2/4 + Q^3/27]}$

$$\beta = -R/2 - \sqrt{[R^2/4 + Q^3/27]} \quad ...(140)$$

Giving

$$X = Y - B/3A$$

Once we get the value of φ_o in terms of contact angle θ. Then equation (134) will give us β/r. If the value of β/r is known, equation (133) gives the meniscus volume of a liquid of given angle of contact.

Combining (133) and (134), we get

$$V_m/\pi a^3 = 2.\cos(\theta)(r^3/a^3)/(1/K + r^2/a^2) \quad ...(141)$$

For large values of r/a, such that $1/K$ becomes negligible in comparison to r^2/a^2. Equation (141) gives

$$V_m/\pi a^3 = 2.\cos(\theta)(r/a).$$

This is the result; when h becomes zero in (16) i.e. when r/a is very large.

Table 7.2 Comparative Values of $V/\pi a^3$ Against r/a

r/a	0.0	0.1	0.2	0.3	0.4	0.5	0.6	0.7	0.8	0.9
0.0	— 0.000	— 0.0003	— 0.0026	— 0.0089	— 0.0208	— 0.0400	— 0.0679	— 0.1057	— 0.1542	— 0.2141
1	— 0.2857	— 0.3692	— 0.4645	— 0.5714	— 0.6894	— 0.8182	— 0.9570	— 1.1053	— 1.2623	— 1.4275
2	— 1.6000	— 1.7793	— 1.9646	— 2.1554	— 2.3510	— 2.5510	— 2.7549	— 2.9621	— 3.1723	— 3.3850
3	— 3.6000	— 3.8169	— 4.0355	— 4.2554	— 4.4765	— 4.6986	— 4.9215	— 5.1450	— 5.3691	— 5.5935
4	— 5.8182	— 6.0430	— 6.2680	— 6.4930	— 6.7180	— 6.9429	— 7.1676	— 7.3922	— 7.6165	— 7.8406
5	— 8.0645	9.3487 8.2881	9.5918 8.5114	9.8335 8.7343	10.0736 8.9570	10.3122 9.1793	10.5493 9.4013	10.7848 9.6229	11.0189 9.8442	11.2515 10.0651
6	11.4826 10.2857	11.7123 10.5059	11.9405 10.7258	12.1674 10.9454	12.3930 11.1646	12.6173 11.3834	12.8404 11.6019	13.0623 11.8201	13.2830 12.0380	13.5026 12.2555
7	13.7211 12.4727	13.9386 12.6896	14.1551 12.9062	14.3707 13.1225	14.5854 13.3385	14.7993 13.5542	15.0123 13.7696	15.2246 13.9848	15.4361 14.1996	15.6469 14.4142
8	15.8571 14.6286	16.0666 14.8426	16.2756 15.0565	16.4840 15.2700	16.6919 15.4834	16.8992 15.6965	17.1061 15.9093	17.3126 16.1220	17.5186 16.3344	17.7243 16.5466
9	17.9296 16.7586	18.1345 16.9704	18.3391 17.1820	18.5434 17.3934	18.7474 17.6046	18.9511 17.8156	19.1546 18.0264	19.3579 18.2370	19.5609 18.4475	19.7637 18.6578
10	19.9663 18.8679	20.1688 19.0779	20.3711 19.2877	20.5732 19.4973	20.7751 19.7068	20.9770 19.9161	21.1787 20.1253	21.3802 20.3344	21.5817 20.5433	21.7831 20.7520
11	21.9843 20.9606	22.1855 21.1691	22.3866 21.3775	22.5876 21.5857	22.7885 21.7938	22.9894 22.0018	23.1902 22.2097	23.3909 22.4174	23.5916 22.6251	23.7923 22.8326
12	23.9928 23.0400	24.1934 23.2473	24.3939 23.4545	24.5944 23.6616	24.7948 23.8686	24.9952 24.0755	25.1956 24.2823	25.3959 24.4890	25.5962 24.6956	25.7965 24.9022
13	25.9968 25.1086	26.1971 25.3149	26.3973 25.5212	26.5975 25.7274	26.7977 25.9335	26.9979 26.1395	27.1981 26.3454	27.3982 26.5513	27.5984 26.7570	27.7985 26.9627
14	27.9986 27.1684	28.1987 27.3739	28.3988 27.5794	28.5989 27.7848	28.7990 27.9901	28.9991 28.1954	29.1992 28.4006	29.3992 28.6058	29.5993 28.8108	29.7994 29.0159
15	29.9994 29.2208	30.1995 29.4257	30.3995 29.6306	30.5996 29.8353	30.7996 30.0401	30.9996 30.2447	31.1997 30.4493	31.3997 30.6539	31.5997 30.8584	31.7998 31.0628
16	31.9998 31.2672	32.1998 31.4716	32.3998 31.6759	32.5998 31.8801	32.7999 32.0843	32.9999 32.2885	33.1999 32.4926	33.3999 32.6966	33.5999 32.9006	33.7999 33.1046
17	33.9999 33.3085	34.2000 33.5124	34.4000 33.7163	34.6000 33.9201	34.8000 34.1238	35.0000 34.3275	35.2000 34.5312	35.4000 34.7348	35.6000 34.9384	35.8000 35.1420
18	36.0000 35.3455	36.2000 35.5490	36.4000 35.7525	36.6000 35.9559	36.8000 36.1593	37.0000 36.3626	37.2000 36.5659	37.4000 36.7692	37.6000 36.9724	37.8000 37.1756
19	38.0001 37.3788	38.2001 37.5820	38.4001 37.7851	38.6001 37.9882	38.8001 38.1912	39.0001 38.3943	39.2001 38.5972	39.4001 38.8002	39.6001 39.0032	39.8001 39.2061
20	40.0001 39.4089									

Table 7.3 Values of $V_m/\pi a^3$ Versus r/a

r/a	0.0	0.1	0.2	0.3	0.4	0.5	0.6	0.7	0.8	0.9
0	0.0000	0.0003	0.0026	0.0089	0.0208	0.0400	0.0679	0.1057	0.1542	0.2141
1	0.2857	0.3692	0.4645	0.5714	0.6894	0.8182	0.9570	1.1053	1.2623	1.4275
2	1.6000	1.7793	1.9646	2.1554	2.3510	2.5510	2.7549	2.9621	3.1723	3.3850
3	3.6000	3.8169	4.0355	4.2554	4.4765	4.6986	4.9215	5.1450	5.3691	5.5935
4	5.8182	6.0430	6.2680	6.4930	6.7180	6.9429	7.1676	7.3922	7.6165	7.8406
5	8.0645	8.2881	8.5114	8.7343	8.9570	9.1793	9.4013	9.6229	9.8442	10.0651
6	10.2857	10.5059	10.7258	10.9454	11.1646	11.3834	11.6019	11.8201	12.0380	12.2555
7	12.4727	12.6896	12.9062	13.1225	13.3385	13.5542	13.7696	13.9848	14.1996	14.4142
8	14.6286	14.8426	15.0565	15.2700	15.4834	15.6965	15.9093	16.1220	16.3344	16.5466
9	16.7586	16.9704	17.1820	17.3934	17.6046	17.8156	18.0264	18.2370	18.4475	18.6578
10	18.8679	—	—	—	—	—	—	—	—	—

Table 7.4 $V/\pi a^3$ Versus r/a for Different Angles of Contact

θ°	10°	20°	30°	40°	50°	60°	70°	80°
r/a = $V/\pi a^3$	0.4749 0.0331	0.45513 0.02678	0.42185 0.01882	0.37552 0.01126	0.31706 0.00549	0.2480 0.00198	0.17038 0.00043	0.08674 0.00029
r/a = $V/\pi a^3$	0.87559 0.19126	0.84502 0.1600	0.79168 0.11784	0.71394 0.0739	0.61146 0.03831	0.48495 0.0146	0.3371 0.00331	0.1730 0.00026
r/a = $V/\pi a^3$	2.1567 1.9822	2.1167 1.7378	2.0428 1.4518	1.9262 1.0859	1.7548 0.7163	1.4994 0.3703	1.1598 0.1213	0.65812 0.0120
r/a = $V/\pi a^3$	2.395 2.423	2.355 2.208	2.280 1.874	2.162 1.442	1.987 0.975	1.734 0.532	1.362 0.189	0.801 0.0213
r/a = $V/\pi a^3$	2.593 2.873	2.552 2.599	2.477 2.245	2.359 1.759	2.182 1.220	1.924 0.690	1.540 0.263	0.935 0.033
r/a = $V/\pi a^3$	2.761 3.261	2.720 2.998	2.645 2.582	2.526 2.047	2.349 1.443	2.089 0.843	1.696 0.338	1.059 0.047
r/a = $V/\pi a^3$	2.907 3.613	2.867 3.333	2.792 2.887	2.673 2.309	2.495 1.651	2.233 0.986	1.834 0.413	1.174 0.063
r/a = $V/\pi a^3$	3.152 4.221	3.112 3.912	3.037 3.416	2.918 2.768	2.740 2.021	2.476 1.250	2.071 0.559	1.380 0.098

TABLES 7.5 TO 7.10 VOLUME OF MENISCUS IN cm^3

Table 7.5 Volume of Meniscus in cm^3

r mm	0.0	0.1	0.2	0.3	0.4	0.5	0.6	0.7	0.8	0.9
0	0.0000	0.0000	0.0000	0.0000	0.0001	0.0001	0.0002	0.0004	0.0005	0.0007
1	0.0010	0.0014	0.0018	0.0022	0.0028	0.0034	0.0041	0.0048	0.0057	0.0066
2	0.0077	0.0088	0.0101	0.0114	0.0128	0.0144	0.0160	0.0177	0.0196	0.0215
3	0.0235	0.0257	0.0279	0.0302	0.0327	0.0352	0.0379	0.0406	0.0434	0.0463
4	0.0493	0.0524	0.0556	0.0589	0.0622	0.0656	0.0691	0.0727	0.0764	0.0801
5	0.0839	0.0877	0.0917	0.0957	0.0997	0.1038	0.1080	0.1122	0.1165	0.1208
6	0.1252	0.1296	0.1340	0.1385	0.1431	0.1477	0.1523	0.1570	0.1617	0.1664
7	0.1711	0.1759	0.1808	0.1856	0.1905	0.1954	0.2003	0.2052	0.2102	0.2152
8	0.2202	0.2252	0.2302	0.2353	0.2404	0.2455	0.2506	0.2557	0.2608	0.2659
9	0.2711	0.2762	0.2814	0.2866	0.2917	0.2969	0.3021	0.3073	0.3125	0.3178

Table 7.6 Volume of Meniscus in cm^3

r mm	0.0	0.1	0.2	0.3	0.4	0.5	0.6	0.7	0.8	0.9
10	0.3230	0.3282	0.3334	0.3387	0.3439	0.3491	0.3544	0.3596	0.3649	0.3701
11	0.3754	0.3806	0.3859	0.3911	0.3964	0.4016	0.4069	0.4121	0.4174	0.4226
12	0.4279	0.4332	0.4384	0.4437	0.4489	0.4541	0.4594	0.4646	0.4699	0.4751
13	0.4804	0.4856	0.4908	0.4961	0.5013	0.5065	0.5117	0.5170	0.5222	0.5274
14	0.5326	0.5378	0.5431	0.5483	0.5535	0.5587	0.5639	0.5691	0.5743	0.5795
15	0.5846	0.5898	0.5950	0.6002	0.6054	0.6105	0.6157	0.6209	0.6260	0.6312
16	0.6364	0.6415	0.6467	0.6518	0.6570	0.6621	0.6673	0.6724	0.6775	0.6827
17	0.6878	0.6929	0.6980	0.7032	0.7083	0.7134	0.7185	0.7236	0.7287	0.7338
18	0.7389	0.7440	0.7491	0.7542	0.7593	0.7644	0.7695	0.7745	0.7796	0.7847
19	0.7898	0.7948	0.7999	0.8050	0.8100	0.8151	0.8201	0.8252	0.8302	0.8353

Table 7.7 Volume of Meniscus in cm^3

r mm	0.0	0.1	0.2	0.3	0.4	0.5	0.6	0.7	0.8	0.9
20	0.8403	0.8454	0.8504	0.8555	0.8605	0.8655	0.8706	0.8756	0.8806	0.8856
21	0.8907	0.8957	0.9007	0.9057	0.9107	0.9157	0.9207	0.9257	0.9308	0.9358
22	0.9408	0.9457	0.9507	0.9557	0.9607	0.9657	0.9707	0.9757	0.9807	0.9856
23	0.9906	0.9956	1.0006	1.0055	1.0105	1.0155	1.0204	1.0254	1.0304	1.0353
24	1.0403	1.0452	1.0502	1.0552	1.0601	1.0651	1.0700	1.0750	1.0799	1.0848
25	1.0898	1.0947	1.0997	1.1046	1.1095	1.1145	1.1194	1.1243	1.1292	1.1342
26	1.1391	1.1440	1.1489	1.1539	1.1588	1.1637	1.1686	1.1735	1.1784	1.1833
27	1.1882	1.1932	1.1981	1.2030	1.2079	1.2128	1.2177	1.2226	1.2275	1.2324
28	1.2373	1.2422	1.2470	1.2519	1.2568	1.2617	1.2666	1.2715	1.2764	1.2813
29	1.2861	1.2910	1.2959	1.3008	1.3057	1.3105	1.3154	1.3203	1.3252	1.3300

Table 7.8 Volume of Meniscus in cm^3

r mm	0.0	0.1	0.2	0.3	0.4	0.5	0.6	0.7	0.8	0.9
30	1.3349	1.3398	1.3446	1.3495	1.3544	1.3592	1.3641	1.3690	1.3738	1.3787
31	1.3835	1.3884	1.3933	1.3981	1.4030	1.4078	1.4127	1.4175	1.4224	1.4272
32	1.4321	1.4369	1.4418	1.4466	1.4515	1.4563	1.4612	1.4660	1.4708	1.4757
33	1.4805	1.4854	1.4902	1.4950	1.4999	1.5047	1.5095	1.5144	1.5192	1.5240
34	1.5289	1.5337	1.5385	1.5434	1.5482	1.5530	1.5578	1.5627	1.5675	1.5723
35	1.5771	1.5820	1.5868	1.5916	1.5964	1.6012	1.6061	1.6109	1.6157	1.6205
36	1.6253	1.6301	1.6350	1.6398	1.6446	1.6494	1.6542	1.6590	1.6638	1.6686
37	1.6734	1.6783	1.6831	1.6879	1.6927	1.6975	1.7023	1.7071	1.7119	1.7167
38	1.7215	1.7263	1.7311	1.7359	1.7407	1.7455	1.7503	1.7551	1.7599	1.7647
39	1.7695	1.7743	1.7791	1.7839	1.7887	1.7934	1.7982	1.8030	1.8078	1.8126

Table 7.9 Volume of Meniscus in cm^3

r mm	0.0	0.1	0.2	0.3	0.4	0.5	0.6	0.7	0.8	0.9
40	1.8174	1.8222	1.8270	1.8318	1.8366	1.8413	1.8461	1.8509	1.8557	1.8605
41	1.8653	1.8701	1.8748	1.8796	1.8844	1.8892	1.8940	1.8988	1.9035	1.9083
42	1.9131	1.9179	1.9227	1.9274	1.9322	1.9370	1.9418	1.9465	1.9513	1.9561
43	1.9609	1.9656	1.9704	1.9752	1.9800	1.9847	1.9895	1.9943	1.9991	2.0038
44	2.0086	2.0134	2.0181	2.0229	2.0277	2.0325	2.0372	2.0420	2.0468	2.0515
45	2.0563	2.0611	2.0658	2.0706	2.0754	2.0801	2.0849	2.0896	2.0944	2.0992
46	2.1039	2.1087	2.1135	2.1182	2.1230	2.1277	2.1325	2.1373	2.1420	2.1468
47	2.1515	2.1563	2.1611	2.1658	2.1706	2.1753	2.1801	2.1848	2.1896	2.1944
48	2.1991	2.2039	2.2086	2.2134	2.2181	2.2229	2.2276	2.2324	2.2371	2.2419
49	2.2467	2.2514	2.2562	2.2609	2.2657	2.2704	2.2752	2.2799	2.2847	2.2894

Table 7.10 Volume of Meniscus in cm^3

r mm	0.0	0.1	0.2	0.3	0.4	0.5	0.6	0.7	0.8	0.9
50	2.2942	2.2989	2.3037	2.3084	2.3131	2.3179	2.3226	2.3274	2.3321	2.3369
51	2.3416	2.3464	2.3511	2.3559	2.3606	2.3654	2.3701	2.3748	2.3796	2.3843
52	2.3891	2.3938	2.3986	2.4033	2.4081	2.4128	2.4175	2.4223	2.4270	2.4318
53	2.4365	2.4412	2.4460	2.4507	2.4555	2.4602	2.4649	2.4697	2.4744	2.4792
54	2.4839	2.4886	2.4934	2.4981	2.5029	2.5076	2.5123	2.5171	2.5218	2.5265
55	2.5313	2.5360	2.5407	2.5455	2.5502	2.5550	2.5597	2.5644	2.5692	2.5739
56	2.5786	2.5834	2.5881	2.5928	2.5976	2.6023	2.6070	2.6118	2.6165	2.6212
57	2.6260	2.6307	2.6354	2.6401	2.6449	2.6496	2.6543	2.6591	2.6638	2.6685
58	2.6733	2.6780	2.6827	2.6874	2.6922	2.6969	2.7016	2.7064	2.7111	2.7158
59	2.7205	2.7253	2.7300	2.7347	2.7395	2.7442	2.7489	2.7536	2.7584	2.7631

TABLES 7.11 TO 7.16 CORRECTIONS IN VOLUME cm^3 DUE TO CHANGE IN CAPILLARY CONSTANTS

All CORRECTIONS in Tables 7.11 to 7.16 are negative, hence the number given is to be subtracted from the observed reading.

Table 7.11 Correction in Volume (in cm^3) due to Change in Capillary Constant

Radius	*Capillary constant in mm^2*									
in cm	7.0	6.5	6.0	5.5	5.0	4.5	4.0	3.5	3.0	2.5
0.10	0.0000	0.0000	0.0001	0.0001	0.0001	0.0001	0.0002	0.0002	0.0003	0.0004
0.15	0.0001	0.0002	0.0003	0.0004	0.0005	0.0007	0.0009	0.0011	0.0015	0.0019
0.20	0.0002	0.0004	0.0007	0.0011	0.0015	0.0020	0.0025	0.0033	0.0042	0.0054
0.25	0.0004	0.0010	0.0016	0.0024	0.0032	0.0043	0.0055	0.0070	0.0088	0.0112
0.30	0.0008	0.0018	0.0030	0.0044	0.0059	0.0077	0.0099	0.0124	0.0156	0.0195
0.35	0.0013	0.0030	0.0049	0.0071	0.0096	0.0124	0.0157	0.0196	0.0243	0.0300
0.40	0.0020	0.0045	0.0073	0.0105	0.0140	0.0181	0.0228	0.0282	0.0346	0.0422
0.45	0.0028	0.0063	0.0101	0.0144	0.0192	0.0247	0.0308	0.0379	0.0460	0.0556
0.50	0.0036	0.0082	0.0133	0.0188	0.0250	0.0318	0.0395	0.0482	0.0582	0.0696
0.55	0.0046	0.0103	0.0166	0.0235	0.0310	0.0394	0.0486	0.0589	0.0706	0.0837
0.66	0.0056	0.0126	0.0201	0.0283	0.0372	0.0470	0.0578	0.0697	0.0829	0.0977
0.65	0.0066	0.0148	0.0236	0.0332	0.0435	0.0547	0.0669	0.0803	0.0950	0.1113
0.70	0.0077	0.0171	0.0272	0.0380	0.0496	0.0622	0.0758	0.0906	0.1066	0.1242
0.75	0.0087	0.0193	0.0306	0.0427	0.0556	0.0694	0.0843	0.1004	0.1177	0.1365
0.80	0.0097	0.0215	0.0340	0.0472	0.0613	0.0764	0.0924	0.1096	0.1281	0.1480
0.85	0.0106	0.0235	0.0372	0.0515	0.0668	0.0829	0.1001	0.1184	0.1379	0.1587
0.90	0.0115	0.0255	0.0402	0.0556	0.0719	0.0891	0.1073	0.1266	0.1470	0.1687
0.95	0.0124	0.0274	0.0431	0.0595	0.0768	0.0949	0.1140	0.1342	0.1554	0.1779
1.00	0.0132	0.0292	0.0458	0.0631	0.0813	0.1003	0.1203	0.1412	0.1632	0.1864
1.05	0.0140	0.0308	0.0483	0.0665	0.0856	0.1054	0.1261	0.1478	0.1705	0.1942

Table 7.12 Correction in Volume (in cm^3) due to Change in Capillary Constant

Radius	*Capillary constant in mm^2*									
in cm	7.0	6.5	6.0	5.5	5.0	4.5	4.0	3.5	3.0	2.5
1.10	0.0147	0.0324	0.0507	0.0697	0.0895	0.1101	0.1315	0.1539	0.1772	0.2015
1.15	0.0154	0.0338	0.0529	0.0727	0.0932	0.1144	0.1365	0.1595	0.1833	0.2082
1.20	0.0161	0.0352	0.0550	0.0754	0.0966	0.1185	0.1412	0.1647	0.1890	0.2143
1.25	0.0167	0.0365	0.0569	0.0780	0.0998	0.1222	0.1455	0.1695	0.1943	0.2200
1.30	0.0172	0.0377	0.0587	0.0804	0.1027	0.1257	0.1494	0.1739	0.1992	0.2252
1.35	0.0178	0.0388	0.0604	0.0826	0.1054	0.1289	0.1531	0.1780	0.2037	0.2301
1.40	0.0182	0.0398	0.0620	0.0847	0.1080	0.1319	0.1565	0.1818	0.2078	0.2345
1.45	0.0187	0.0408	0.0634	0.0866	0.1104	0.1347	0.1597	0.1853	0.2117	0.2387
1.50	0.0191	0.0417	0.0648	0.0884	0.1126	0.1373	0.1627	0.1886	0.2152	0.2425
1.55	0.0195	0.0425	0.0660	0.0901	0.1146	0.1397	0.1654	0.1917	0.2185	0.2460
1.60	0.0199	0.0433	0.0672	0.0916	0.1165	0.1420	0.1679	0.1945	0.2216	0.2493
1.65	0.0202	0.0440	0.0683	0.0931	0.1183	0.1440	0.1703	0.1971	0.2245	0.2524
1.70	0.0206	0.0447	0.0694	0.0944	0.1200	0.1460	0.1725	0.1996	0.2271	0.2553
1.75	0.0209	0.0454	0.0703	0.0957	0.1215	0.1478	0.1746	0.2019	0.2296	0.2579
1.80	0.0212	0.0460	0.0712	0.0969	0.1230	0.1495	0.1765	0.2040	0.2319	0.2604
1.85	0.0214	0.0465	0.0721	0.0980	0.1243	0.1511	0.1783	0.2060	0.2341	0.2627
1.90	0.0217	0.0471	0.0729	0.0990	0.1256	0.1526	0.1800	0.2078	0.2361	0.2649
1.95	0.0219	0.0476	0.0736	0.1000	0.1268	0.1540	0.1816	0.2096	0.2380	0.2669
2.00	0.0221	0.0480	0.0743	0.1009	0.1279	0.1553	0.1831	0.2112	0.2398	0.2688
2.05	0.0223	0.0485	0.0750	0.1018	0.1290	0.1565	0.1845	0.2128	0.2415	0.2705

Table 7.13 Correction in Volume (in cm^3) due to Change in Capillary Constant

Radius	*Capillary constant in mm^2*									
in cm	7.0	6.5	6.0	5.5	5.0	4.5	4.0	3.5	3.0	2.5
2.10	0.0225	0.0489	0.0756	0.1026	0.1300	0.1577	0.1858	0.2142	0.2430	0.2722
2.15	0.0227	0.0493	0.0762	0.1034	0.1309	0.1588	0.1870	0.2156	0.2445	0.2738
2.20	0.0229	0.0497	0.0767	0.1041	0.1318	0.1598	0.1882	0.2168	0.2459	0.2753
2.25	0.0231	0.0500	0.0772	0.1048	0.1326	0.1608	0.1893	0.2181	0.2472	0.2766
2.30	0.0232	0.0503	0.0777	0.1054	0.1334	0.1617	0.1903	0.2192	0.2484	0.2780
2.35	0.0234	0.0507	0.0782	0.1060	0.1341	0.1626	0.1913	0.2203	0.2496	0.2792
2.40	0.0235	0.0509	0.0786	0.1066	0.1348	0.1634	0.1922	0.2213	0.2507	0.2804
2.45	0.0237	0.0512	0.0791	0.1072	0.1355	0.1641	0.1931	0.2222	0.2517	0.2815
2.50	0.0238	0.0515	0.0795	0.1077	0.1361	0.1649	0.1939	0.2231	0.2527	0.2825
2.55	0.0239	0.0517	0.0798	0.1082	0.1367	0.1656	0.1947	0.2240	0.2536	0.2835
2.60	0.0240	0.0520	0.0802	0.1086	0.1373	0.1662	0.1954	0.2248	0.2545	0.2844
2.65	0.0241	0.0522	0.0805	0.1091	0.1378	0.1669	0.1961	0.2256	0.2553	0.2853
2.70	0.0242	0.0524	0.0808	0.1095	0.1384	0.1674	0.1968	0.2263	0.2561	0.2862
2.75	0.0243	0.0526	0.0811	0.1099	0.1388	0.1680	0.1974	0.2270	0.2569	0.2870
2.80	0.0244	0.0528	0.0814	0.1103	0.1393	0.1685	0.1980	0.2277	0.2576	0.2877
2.85	0.0245	0.0530	0.0817	0.1106	0.1397	0.1691	0.1986	0.2283	0.2583	0.2885
2.90	0.0246	0.0532	0.0820	0.1110	0.1402	0.1695	0.1991	0.2289	0.2590	0.2892
2.95	0.0247	0.0534	0.0822	0.1113	0.1406	0.1700	0.1997	0.2295	0.2596	0.2898
3.00	0.0248	0.0535	0.0825	0.1116	0.1409	0.1705	0.2002	0.2301	0.2602	0.2905
3.05	0.0248	0.0537	0.0827	0.1119	0.1413	0.1709	0.2006	0.2306	0.2607	0.2911

Table 7.14 Correction in Volume (in cm³) due to Change in Capillary Constant

Radius	*Capillary constant in mm²*									
in cm	7.0	6.5	6.0	5.5	5.0	4.5	4.0	3.5	3.0	2.5
3.10	0.0249	0.0538	0.0829	0.1122	0.1417	0.1713	0.2011	0.2311	0.2613	0.2916
3.15	0.0250	0.0540	0.0831	0.1125	0.1420	0.1717	0.2015	0.2316	0.2618	0.2922
3.20	0.0250	0.0541	0.0833	0.1127	0.1423	0.1720	0.2020	0.2320	0.2623	0.2927
3.25	0.0251	0.0542	0.0835	0.1130	0.1426	0.1724	0.2024	0.2325	0.2628	0.2932
3.30	0.0252	0.0544	0.0837	0.1132	0.1429	0.1727	0.2027	0.2329	0.2632	0.2937
3.35	0.0252	0.0545	0.0839	0.1135	0.1432	0.1731	0.2031	0.2333	0.2636	0.2942
3.40	0.0253	0.0546	0.0841	0.1137	0.1435	0.1734	0.2035	0.2337	0.2641	0.2946
3.45	0.0253	0.0547	0.0842	0.1139	0.1437	0.1737	0.2038	0.2341	0.2645	0.2950
3.50	0.0254	0.0548	0.0844	0.1141	0.1440	0.1740	0.2041	0.2344	0.2648	0.2954
3.55	0.0254	0.0549	0.0845	0.1143	0.1442	0.1742	0.2044	0.2347	0.2652	0.2958
3.60	0.0255	0.0550	0.0847	0.1145	0.1444	0.1745	0.2047	0.2351	0.2656	0.2962
3.65	0.0255	0.0551	0.0848	0.1147	0.1446	0.1748	0.2050	0.2354	0.2659	0.2966
3.70	0.0256	0.0552	0.0850	0.1148	0.1449	0.1750	0.2053	0.2357	0.2662	0.2969
3.75	0.0256	0.0553	0.0851	0.1150	0.1451	0.1752	0.2055	0.2360	0.2665	0.2972
3.80	0.0256	0.0554	0.0852	0.1152	0.1453	0.1755	0.2058	0.2363	0.2668	0.2976
3.85	0.0257	0.0554	0.0853	0.1153	0.1454	0.1757	0.2061	0.2365	0.2671	0.2979
3.90	0.0257	0.0555	0.0854	0.1155	0.1456	0.1759	0.2063	0.2368	0.2674	0.2982
3.95	0.0257	0.0556	0.0855	0.1156	0.1458	0.1761	0.2065	0.2370	0.2677	0.2985
4.00	0.0258	0.0557	0.0857	0.1158	0.1460	0.1763	0.2067	0.2373	0.2680	0.2987
4.05	0.0258	0.0557	0.0858	0.1159	0.1461	0.1765	0.2070	0.2375	0.2682	0.2990

Table 7.15 Correction in Volume (in cm³) due to Change in Capillary Constant

Radius	*Capillary constant in mm²*									
in cm	7.0	6.5	6.0	5.5	5.0	4.5	4.0	3.5	3.0	2.5
4.1	0.0258	0.0558	0.0859	0.1160	0.1463	0.1767	0.2072	0.2377	0.2684	0.2993
4.15	0.0259	0.0559	0.0860	0.1161	0.1464	0.1768	0.2074	0.2380	0.2687	0.2995
4.2	0.0259	0.0559	0.0860	0.1163	0.1466	0.1770	0.2075	0.2382	0.2689	0.2997
4.25	0.0259	0.0560	0.0861	0.1164	0.1467	0.1772	0.2077	0.2384	0.2691	0.3000
4.3	0.0260	0.0560	0.0862	0.1165	0.1469	0.1773	0.2079	0.2386	0.2693	0.3002
4.35	0.0260	0.0561	0.0863	0.1166	0.1470	0.1775	0.2081	0.2388	0.2695	0.3004
4.4	0.0260	0.0562	0.0864	0.1167	0.1471	0.1776	0.2082	0.2389	0.2697	0.3006
4.45	0.0260	0.0562	0.0865	0.1168	0.1473	0.1778	0.2084	0.2391	0.2699	0.3008
4.6	0.0261	0.0563	0.0865	0.1169	0.1474	0.1779	0.2086	0.2393	0.2701	0.3010
4.65	0.0261	0.0563	0.0866	0.1170	0.1475	0.1781	0.2087	0.2395	0.2703	0.3012
4.7	0.0261	0.0564	0.0867	0.1171	0.1476	0.1782	0.2089	0.2396	0.2705	0.3014
4.75	0.0261	0.0564	0.0868	0.1172	0.1477	0.1783	0.2090	0.2398	0.2706	0.3016
4.8	0.0262	.0565	0.0868	0.1173	0.1478	0.1784	0.2091	0.2399	0.2708	0.3018
4.85	0.0262	0.0565	0.0869	0.1174	0.1479	0.1786	0.2093	0.2401	0.2710	0.3019
4.9	0.0262	0.0565	0.0870	0.1175	0.1480	0.1787	0.2094	0.2402	0.2711	0.3021
4.95	0.0262	0.0566	0.0870	0.1175	0.1481	0.1788	0.2095	0.2404	0.2713	0.3022
5.0	0.0262	0.0566	0.0871	0.1176	0.1482	0.1789	0.2097	0.2405	0.2714	0.3024
5.05	0.0263	0.0567	0.0871	0.1177	0.1483	0.1790	0.2098	0.2406	0.2715	0.3025
5.1	0.0263	0.0567	0.0872	0.1178	0.1484	0.1791	0.2099	0.2408	0.2717	0.3027
5.15	0.0263	0.0567	0.0872	0.1178	0.1485	0.1792	0.2100	0.2409	0.2718	0.3028

Table 7.16 Correction in Volume (in cm^3) due to Change in Capillary Constant

Radius	*Capillary constant in* mm^2									
in cm	7.0	6.5	6.0	5.5	5.0	4.5	4.0	3.5	3.0	2.5
5.10	–0.0263	–0.0568	–0.0873	0.1179	0.1486	0.1793	–0.2101	–0.2410	0.2719	0.3030
5.15	0.0263	0.0568	0.0873	0.1180	0.1486	0.1794	0.2102	0.2411	0.2721	0.3031
5.20	0.0263	0.0568	0.0874	0.1180	0.1487	0.1795	0.2103	0.2412	0.2722	0.3032
5.25	0.0264	0.0569	0.0874	0.1181	0.1488	0.1796	0.2104	0.2413	0.2723	0.3033
5.30	0.0264	0.0569	0.0875	0.1182	0.1489	0.1797	0.2105	0.2414	0.2724	0.3035
5.35	0.0264	0.0569	0.0875	0.1182	0.1489	0.1797	0.2106	0.2415	0.2725	0.3036
5.40	0.0264	0.0570	0.0876	0.1183	0.1490	0.1798	0.2107	0.2416	0.2726	0.3037
5.45	0.0264	0.0570	0.0876	0.1183	0.1491	0.1799	0.2108	0.2417	0.2727	0.3038
5.50	0.0264	0.0570	0.0877	0.1184	0.1492	0.1800	0.2109	0.2418	0.2728	0.3039
5.55	0.0264	0.0570	0.0877	0.1184	0.1492	0.1801	0.2110	0.2419	0.2729	0.3040
5.60	0.0265	0.0571	0.0878	0.1185	0.1493	0.1801	0.2110	0.2420	0.2730	0.3041
5.65	0.0265	0.0571	0.0878	0.1185	0.1493	0.1802	0.2111	0.2421	0.2731	0.3042
5.70	0.0265	0.0571	0.0878	0.1186	0.1494	0.1803	0.2112	0.2422	0.2732	0.3043
5.75	0.0265	0.0572	0.0879	0.1186	0.1495	0.1803	0.2113	0.2423	0.2733	0.3044
5.80	0.0265	0.0572	0.0879	0.1187	0.1495	0.1804	0.2114	0.2424	0.2734	0.3045
5.85	0.0265	0.0572	0.0879	0.1187	0.1496	0.1805	0.2114	0.2424	0.2735	0.3046
5.90	0.0265	0.0572	0.0880	0.1188	0.1496	0.1805	0.2115	0.2425	0.2736	0.3047
5.95	0.0265	0.0572	0.0880	0.1188	0.1497	0.1806	0.2116	0.2426	0.2737	0.3048
6.00	0.0265	0.0573	0.0880	0.1189	0.1497	0.1807	0.2116	0.2427	0.2737	0.3049
6.05	0.0266	0.0573	0.0881	0.1189	0.1498	0.1807	0.2117	0.2427	0.2738	0.3050

REFERENCES

[1] Bashforth and Adams; 1883. An Attempt to Test the Theory of Capillary Action, Cambridge University Press.

[2] Lord Rayleigh; 1915. On the Theory of the Capillary Tube; Proc. Roy. Soc (A), **92**, 184–195.

[3] Poisson; 1831. NouvelleTheorie de l'action of capillaire, 112.

[4] Mathieu's ; 1883. Theorie de la Cappillarite, Paris, 46–49.

[5] International Critical Tables of Numerical Data, Physics, Chemistry and Technology, 1926, Edited by E. W. Washburn, Volume **1**, 72–73, McGraw Hill Book Co. Inc. New York

[6] Lord Kelvin; 1886. Popular Lectures and Addresses; Proc. Roy. Institute, I, 40.

[7] Richards and Coombs; 1915. J. Amer. Chemical Society, no. 7.

[8] Porter Alfered W; 1932. On the Volume of the Meniscus at the Surface of a Liquid; Phil. Mag. 14, 694–700.

[9] Porter Alfered W; 1934. On the Volume of the Meniscus at the Surface of a Liquid –Part III; Phil. Mag. 17, 511–517.

[10] Gupta S.V. 2002. Practical Density Measurements, Institute of Physics, U.K.

[11] Ferguson A; 1926. On the Hyperbola Method for Measurement of Surface Tensions; vol. **38**, 193–203.

[12] Sugden Samuel; 1921. The Dtermination of Surface Tension from the Rise in Capillary Tube; J. Chem. Society, **119**, 1483–1492.

[13] Gupta S.V. 2004. Capillary Action in Narrow and Wide Tubes– A Unified Approach, Metrologia, **41**, 361-364.

STORAGE TANKS

8.1 INTRODUCTION

In the next few chapters, I propose to discuss, storage tanks, vehicle tanks, ships and barges including high capacity standard measures. Normally capacity of such tanks lies in between 50 m^3 to 2000 m^3. Aim will be to discuss basics of storage tanks, measurement and calibration, along with observation sheets, and gage tables (volume versus height). The measurement and calibration is vital for petroleum industry itself and Government for variety of reasons. In case of petroleum tanks, all basic measurements are carried out in the field but are collated to prepare gage tables in an office. An incorrect dimensional measurement results in an erroneous gage table, which remain in use for expanded time and thus causes internal accounting problem and dissatisfaction to the concerned parties. Moreover, quite often, such problem does not remain on national level only but spills to International level. In a trade of any kind based on exchange of money by one and quantity of material by the other, the gain of one particular party is a loss of another. Importance of correct measurements in field can be judged by the facts that in most cases, the person who computes the gage tables and who makes actual field measurements are different. The computing man has no direct means of checking such measurements. Further the user of these tables is a different person, so is the user involved in custody transfer of liquid on the basis of the gage table. Therefore the accurate measurements by the field staff are of paramount importance. Error in gage table causes the assessment of tank contents to be inaccurate. The payments, therefore, become disputable. Settlements involving such errors are very difficult and some times impossible, to adjust without loss to one of the parties involved. Hence the procedure of taking measurements and achievement of accuracy in tank calibration is important. One possibility is that all parties interested in subsequent measurement of quantities from tank under calibration may witness all such measurements. However it is not always possible. So calibration of such storage tanks should be carried out by a reliable government agency having well-experienced and educated staff, and having an established traceable system of length and volume measurements. It is hoped that the foregoing discussions will provide an adequate idea of the importance of correct measurements in this particular field.

I propose to discuss in the coming chapters the following:

- Storage tanks
- Vertical storage tanks
- Horizontal storage tanks
- Spheres, spheroids and special purpose containers
- Vehicle tanks,
- Barges and ships

A chapter on large capacity measures used for liquid calibration of tanks is also included.

8.2 DEFINITIONS

There are some specific terms very commonly used in this area, which requires some formal definitions.

1. *Tank strapping*: This is a term used for the overall procedures of measurement to determine dimensions of the storage tank. It includes the following measurements
   - *Depth*: Shell height, oil height, ring height, equalisers line height, and gaging height
   - *Thickness of tank walls*
   - *Circumferences at specified heights*
2. *Deadwood*: Deadwood is any object within the tank, including a floating roof, which displaces liquid and thus reduces the capacity of the tank; also any permanent appurtenances on the outside of the tank, such as cleanouts boxes or manholes, which increases capacity of the tank; deadwood also includes any permanent appurtenances the outside of the tank, such as cleanout boxes or, manholes which increases the capacity of the tank. Dead wood is to be accurately accounted for as to the volume and location. Location should be measured to the nearest millimetre in order to permit:
   - Adequate allowance for the volumes of liquid displaced
   - Or admitted by the various parts and
   - Adequate allocation of the effects at various elevations within the tank.
3. *Sphere*: The stationary tank, which is spherical in shape and is above the ground, it is supported on columns so that the entire tank is above grace.
4. *Spheroid tanks*: A spheroid is a stationary liquid tank having a shell of double curvature. Any horizontal cross-section is circular and a vertical-section is a series of arcs. The height is lesser in comparison of that of the sphere. The bottom rests directly on a prepared grace. The spheroid has a base plating on the grace and projecting beyond the shell. Structural members rest on the base plate. A drip bar is welded to the shell in a horizontal circle just above the structural supports to intercept rainwater. There are two varieties of spheroids used for this purpose, namely
   - A smooth spheroid usually has no inside structural members to support the shell roof.
   - A noded spheroid has abrupt breaks in the vertical curvature called nodes, which are supported by circular grids and structural members inside the tank.
5. *Calibration*: The process of determining the capacity of the tank or the partial capacities corresponding to different heights (levels of liquid) of tank.

6. *Bottom calibration*: The determination of the partial capacities of the lower portion of a tank and the quantity of liquid contained in a tank below the datum point.
7. *Gage table (calibration table)*: Table consisting of volume versus gage height from the datum plate (datum point).
8. *Datum point*: Point used as the base in the preparation of gage tables. It is also known as dip point.
9. *Dip*: Depth of a liquid in a tank. It is also called the innage.
10. *Dip-hatch*: Opening at the top of a tank through which dip rod is inserted and sampling operations are carried out.
11. *Dip plate (datum plate)*: Striking plate positioned below the dip hatch. Bottom or wall movement should not affect its position. The plate whose upper surface is the origin of all measurements either of depth or height.
12. *Dip-rod or dipstick*: A rigid rod of wood or metal usually graduated in units of volume, for measuring the liquid in a tank. A dipstick or dip-rod is associated with a particular tank and is not interchangeable. In a vehicle tank, graduations on one face represent the volume of liquid in a particular compartment only. So a vehicle tanks having four compartments will have a dipstick with four faces each is marked with a number pertaining to a compartment number.
13. *Dip-tape*: Graduated steel tape used for measuring the depth of liquid in a tank either directly by dipping or indirectly by ullage.
14. *Dip-weight*: Weight attached to the steel dip tape of sufficient mass to keep the tape taut and is of such shape as to facilitate the penetration of any sludge that might present on the dip plate (datum plate).
15. *Equivalent of dip*: Depth of liquid in a tank corresponding to a given Ullage.
16. *Floating cover (screen)*: Lightweight cover of either metal or of plastics material designed to float on the surface of the liquid in the tank. The cover rests upon the liquid surface and is used to retard evaporation.
17. *Floating roof tank*: Tank in which the roof floats on the surface of the contents except at a low level when the weight of the roof is taken through its supports by the tank bottom.
18. *Open capacity*: Calculated capacity of a tank or part of it before any dead wood is taken into account.
19. *Types of joints*: Tanks are made usually of circular rings of the height equal to width of the plates available, first these plates are joined to form rings. Then these rings or courses are joined together to form the storage tank of required height. There are four types of joints, namely
    - Lap joint
    - Butt welded
    - Bolted
    - Riveted

    The tanks are named according to the joints used.
20. *Course (ring)*: Tanks are made of sheets or plates of required thickness; these plates are joined together to form rings of required diameter.

21. *Maximum permissible error*: Is the either way deviation allowed of the actual value of the capacity of the tank from its nominal capacity.
22. *Tape positioner*: Guide sliding freely on the strapping tapes and used to pull and hold the tape in the correct position for taking measurements.
23. *Tensioning handles*: Handles fastened to the tape used for pulling it into correct position and applying correct tension.
24. *Ullage*: The capacity of the tank not occupied by the liquid. The distance between the surface of a liquid in a tank and some fixed reference point on the top of dip hatch.
25. *Upper reference point*: Point clearly defined on the dip-hatch and directly above the dip point, from where ullage is measured. As the total distance between this point and the dip point is constant for a given tank. Hence one can find ullage from the reading of the dip-rod.
26. *Water bottom*: Layer of water at the bottom of a tank of such a depth as to cover the bottom completely.
27. *Step-over constant*: Distance between the measuring points of a step over as measured along the arc of the particular course of the tank concerned.
28. *Step-over correction*: Difference between the apparent distance between two points on a tank cell as measured by strapping tape passing over an obstacle and true arc distance as measured by a step-over i.e. the step constant.

8.3 STORAGE TANKS

The storage tanks may be classified according to the:

1. Shape.
2. Position with reference to ground
3. Number of compartments
4. Conditions of maintenance (Influence quantities)
5. Accuracy requirements.

8.3.1 Shape

Storage tanks are available in the following shapes:

1. Vertical cylindrical storage tanks with fixed roof
2. Vertical cylindrical storage tanks with floating roof
3. Horizontal cylindrical storage tanks
4. Spheres and spheroids

Storage tanks with flat, conical, truncated, hemispherical, elliptical, or domed shape bottoms, or having both ends similar may also be found in public use.

8.3.2 Position of the Tank with Respect to Ground

The tanks may be:

1. On the ground
2. Partially underground
3. Completely underground
4. Wholly above the ground

8.3.3 Number of Compartments

The tanks may be of single capacity or of multiple capacities. A vehicle tank carrying petroleum liquids may have several (three to four) compartments each is isolated from other.

8.3.3.1 Single Capacity

There is only one graduation mark, in such a case all the liquid is delivered to one party, or volume delivered/contained in it will always be fixed. So does not have any measuring device. For example a vehicle tank delivers integral number of compartments.

8.3.3.2 Multiple Capacity, having Measuring Devices Like

- A graduated scale with a view window or an external gage tube.
- A graduated rule (dipstick) or a graduated tape with dip-weight or sinker, just like in storage tanks. But here the measurements are manual.
- An automatic level gage (automatic measurement).

8.3.4 Conditions of Maintenance (Influence Quantities)

Main influence quantities, which affect the capacity and calibration of the tank, are pressure and temperature. The pressure including hydrostatic pressure may change the apparent volume by distorting the shell. The hydrostatic pressure depends upon the density of the liquid so density measurement of the liquid is also important. Difference in temperature from its reference temperature will apparently change the volume of the liquid due to expansion or contraction of the liquid and shell. The tanks are classified according to the pressure and temperature maintained in it.

8.3.4.1 In Regard to Pressure, the Tanks may be

- At ambient atmospheric pressure
- Closed at low pressure
- Closed at high pressure

8.3.4.2 In Regard to Temperature Tanks are

- Without heating or cooling (at ambient temperature)
- With heating but without thermal insulation
- With heating and with thermal insulation
- With refrigeration and with thermal insulation

However most common are vertical cylindrical tanks at ambient temperature and pressure with fixed or floating roof.

8.3.5 Accuracy Requirement

From the point of view of accuracy, the storage tanks are classified as

1. Operation control tanks and
2. Custody transfer tank.

8.3.5.1 Operation Control Tanks

Tanks used in the same department of the same plant are called operation control tanks, more correctly as operations control tank. Sometimes these are also called service tanks. These are used for controlling an operation in a plant, viz. mixing different liquids in a set of operations.

In general such tanks do not attract the provisions of Legal Metrology. However for the interest of the user, these need calibration but with a little less accuracy.

8.3.5.2 Custody Transfer Tanks

These tanks are used to transfer liquid from the owner to the user, or are used in inter-department or inter-plant for custody transfer of the liquid on monetary or equivalent basis. So these require calibration with better and known accuracy with all the necessary records and certificates of calibration. Capacity of such tanks is of the order of 2000 m^3.

8.4 CAPACITY OF THE TANKS

Capacity of tanks depends upon it position with respect to ground, shape and pressure or temperature to be maintained. Normally underground tanks are of smaller capacity and vertical cylindrical tanks on the ground have largest capacity. Capacity of storage tanks does vary from 50 m^3 to 2000 m^3 or 50 000 dm^3 to 2000, 000 dm^3.

8.5 MAXIMUM PERMISSIBLE ERRORS OF TANKS OF DIFFERENT SHAPES

The maximum permissible errors, recommended for storage tanks by the International Organization of Legal Metrology (OIML) through OIML-R71 [1] are as follows

± 0.2% for vertical tanks
± 0.3% for horizontal tanks and
± 0.5% for spherical or spheroid tanks

8.6 VERTICAL STORAGE TANK WITH FIXED ROOF

A typical cylindrical storage tank is shown in Figure 8.1. The tank is on the ground having dip-measuring device for volume measurement, likely to attract legal provisions of a country. Basically it is a cylinder with bottom standing upright on the ground.

Cylindrical portion is shell (1) and is made off plates joining together to form a circular ring, which is called as course (ring). Several such rings (courses) are joined together to form the cylindrical shell of the tank. These courses are joined either by riveting each ring with the other or welded together. Welding may be either lap welding or butt- welding i.e. end-to-end welding. Bottom of the tank is marked (2) and is not flat. Roof (3) is fixed. To see the conditions inside one manhole (4) is there. The hole is big enough for man to enter and do the repair, if the tank is empty. Inlet (5) is shown on the left of the reader and outlet (6) is on the right. To drain out the tank for the purpose of cleaning or repairing there is a drain pipe at the lowest point of the bottom and is shown as (7). For gaging the tank there is a gage hatch (8) and guide pipe (9). Lid of the guide pipe is shown as (10) and (11) is a handrail. To access different parts of the shell, there is a ladder with guard-rails. The measurement platform is indicated by (13); next to it is the calibration information plate (17). Dip plate, the reference level, to which all measurements are referred to be indicated as (14). Lower and upper angle irons are indicated by (15) and (16). An opening for inspection is labelled as (18) and vertical axis is (19). Some tanks needs heating system so a heating coil carrying hot liquid is indicated as (20).

PRS is upper reference point and PRI is dipping datum plate. *H* is the reference height, c is ullage, and *h* is the height of the liquid in the tank.

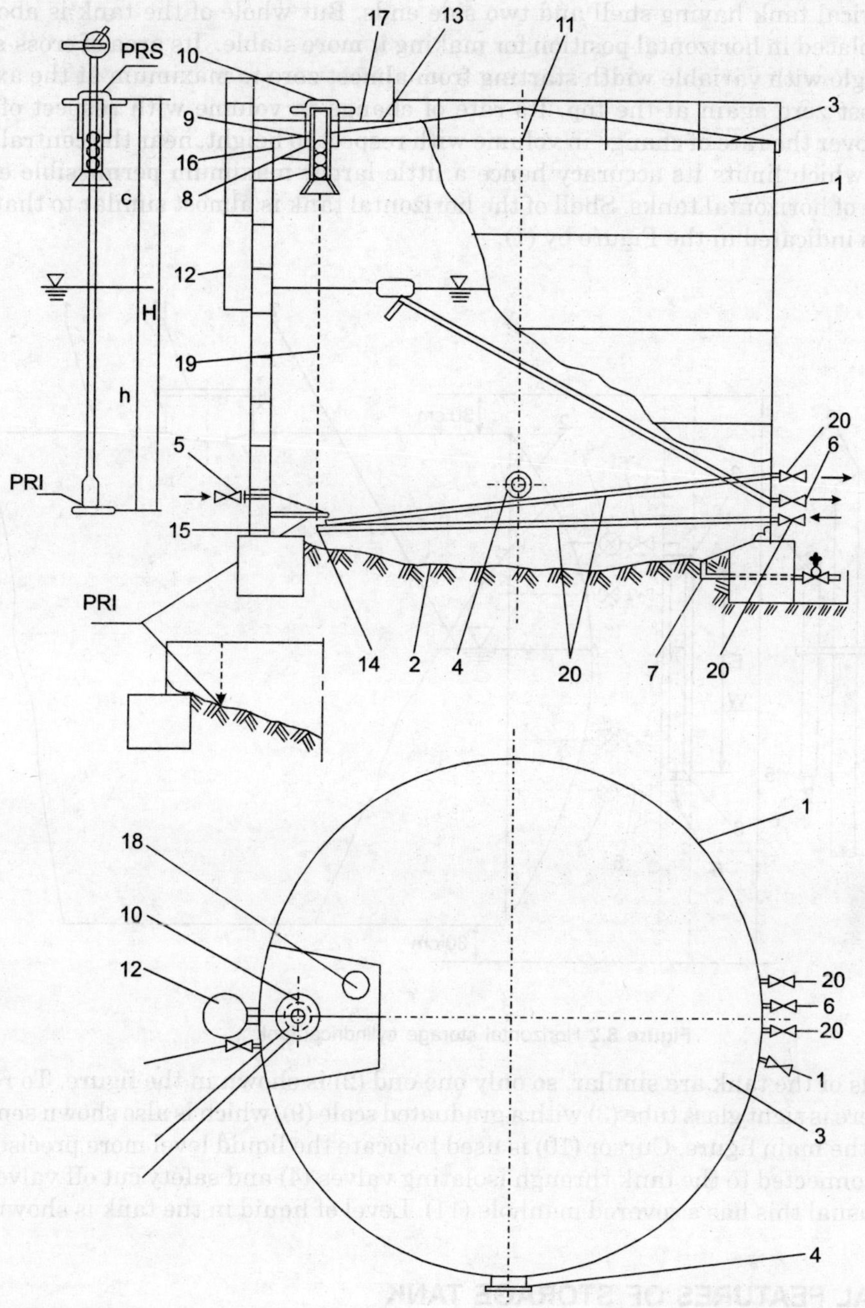

Figure 8.1 Schematic diagram of a vertical storage tank with fixed roof

8.7 HORIZONTAL TANK

The line diagram of a typical horizontal cylindrical tank is given in Figure 8.2. Basically it is again a cylindrical tank having shell and two side ends. But whole of the tank is above the ground and is placed in horizontal position for making it more stable. Its area of cross-section is plane rectangle with variable width starting from almost zero to maximum at the axis and becoming almost zero again at the top. Its rate of change in volume with respect of dip is variable. Moreover the rate of change in volume with respect to height, near the central plane, is quite large, which limits its accuracy hence a little larger maximum permissible error is allowed in case of horizontal tanks. Shell of the horizontal tank is almost similar to that of the vertical tank is indicated in the Figure by (1).

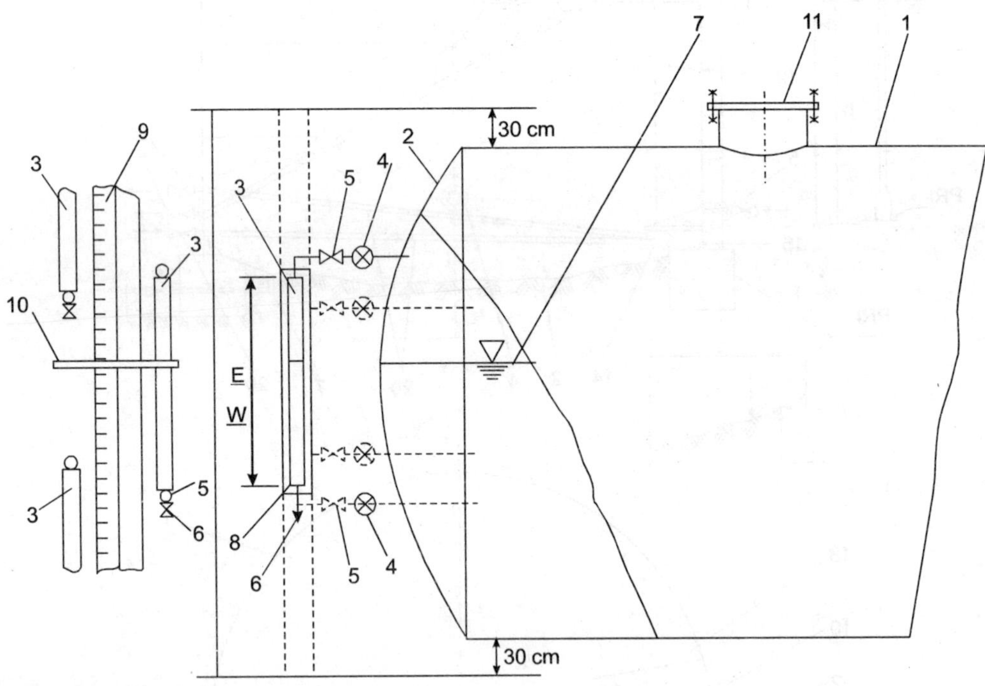

Figure 8.2 Horizontal storage cylindrical tank

Both ends of the tank are similar, so only one end (2) is shown in the figure. To read the liquid level there is sight glass tube (3) with a graduated scale (9), which is also shown separately by the side of the main figure. Cursor (10) is used to locate the liquid level more precisely. The glass tube is connected to the tank through isolating valves (4) and safety cut off valves (5) at each end. As usual this has a covered manhole (11). Level of liquid in the tank is shown as (7).

8.8 GENERAL FEATURES OF STORAGE TANK

1. The tanks are provided with devices to reduce or to prevent loss of liquid due to evaporation.

2. The shape, material, reinforcement, construction and assembly of the tank are such that they can withstand changes in weather conditions especially the changes in pressure and temperature. Pressure changes due to two counts (1) change in atmospheric pressure and (2) due to change in hydrostatic pressure. Such changes should not affect its capacity beyond certain limits.
3. The material of their construction should be such that does not react or is reacted by the liquids likely to be stored in it.
4. The dipping datum plate and the upper reference point should be constructed in such a way that their positions remain practically unchanged irrespective of the state of filling of the tank and other environmental changes.
5. In some cases, especially large tanks of capacity 1000 m^3 and above, it may not be possible to maintain the constancy of datum and upper reference points. Then, the effect, on the reference points with respect of state of filling and changes in temperature and density, should be indicated in the calibration certificate.
6. The shape and interior of the tanks must be so designed that the formation of air pockets while filling and pockets of liquid on draining are prevented.
7. To permit the estimation of capacity of tanks by geometrical means, there should be no deformation, bulges or variation of dimensions affecting the tank capacity and its interpolation.
8. The tanks should be stable on their foundations; this is ensured by anchoring and by adequate methods for stabilisation. The tank is kept full during this period.
9. For vertical cylindrical tanks of capacity larger 2000 m^3, five gage hatches are provided, one of these is as close as possible to centre and others are evenly spaced near the sidewall. The gauge hatch least affected by direct sun light is taken as the reference one.
10. The tanks before calibration and use are pressure tested and should comply with the relevant requirement for leak proofing of the tank.

8.9 METHODS OF CALIBRATION OF STORAGE TANKS

Broadly speaking, there are two methods of calibration of tanks, namely

1. Dimensional method and
2. Volumetric method

However, more often than not the combination of both the methods is used for calibrating a storage tank.

8.9.1 Dimensional Method

The dimensional method is subdivided in the five subgroups.

8.9.1.1 Measuring External Dimensions by Strapping

To determine outer circumferences, tank is strapped at specified positions for each course. Thickness of the plate of different courses is measured at the positions of strapping, the measurements of outer circumference and thickness give the internal circumference and hence the internal diameter of the tank. This gives the area of cross-section at different positions of each course and gives capacity per unit height at these positions. Had it been a purely empty cylinder like a small bucket, we could have found out the capacity straight away. But in tanks of such size there are many other fixed accessories, which change the capacity. The collective name of theses accessories is deadwood. So next step, naturally, is to determine position-wise the volume of this deadwood for all the courses. Having known the capacity per unit height and proportion of volume of the deadwood in that position gives the capacity per unit length. So position wise capacity in small suitable steps is calculated and results are tabulated. We call it as gage (calibration) table.

8.9.1.2 Measuring Internal Dimensions by Strapping

The method is essentially the same as enumerated above, except in this case, position-wise internal diameters are determined. Deadwood is determined likewise and gage table giving position-wise capacity in steps of small height intervals is calculated. For under or partly underground tanks this method is often used.

8.9.1.3 Optical Reference Line

In the two methods enumerated above circumferences/diameters have been measured by physically placing the tapes in that position. For this the observer has to reach those heights by ladder or hanging seat, which is hazardous. To avoid the observer reaching physically there, outer circumference is measured at a convenient height of the bottom course. The diameter of this circumference is taken as reference diameter. All other diameters are measured relative to it. For this purpose, a vertical line parallel to the axis of the tank is created and the distances normal to shell of the tank at every selected place are measured from that reference line. One of the measured distances will be the distance from the position of the reference diameter. If the tank has a same diameter at every point then all measured distances should be equal. So finding differences from the distance at the reference position from every other distance will give the differences of diameters at other positions from the diameter at reference position. The thickness of plates of all courses is measured as before and internal diameters are calculated at the selected places.

The fixed optical line is generated through a right-angled prism fixed in a suitable stand. A scale normal to the surface of the tank is moved along the tank surface with the help of a magnetic trolley. Position of the scale is seen through a short focal length telescope, having an eyepiece with a graduated scale. The trolley carrying the scale is placed at the position at which reference diameter was calculated. One of line of the scale is chosen, eyepiece is so adjusted that the chosen line coincides with the zero of the eyepiece. Position of this line is then measured in eyepiece scale at other selected positions by the moving scale. Thus the differences from the reference point are measured, which will finally give diameters at all other places. Optical system is shown in Figure 8.3.

To average out the irregularities in the cylindrical shell, several reference points are used all along the same manually strapped position. The number of points depends upon the length of circumference. Number of points as per ISO [3] is given below and positions of points are shown in Figure 8.4. No point should be chosen as optical station closer than 30 cm of any vertical seam.

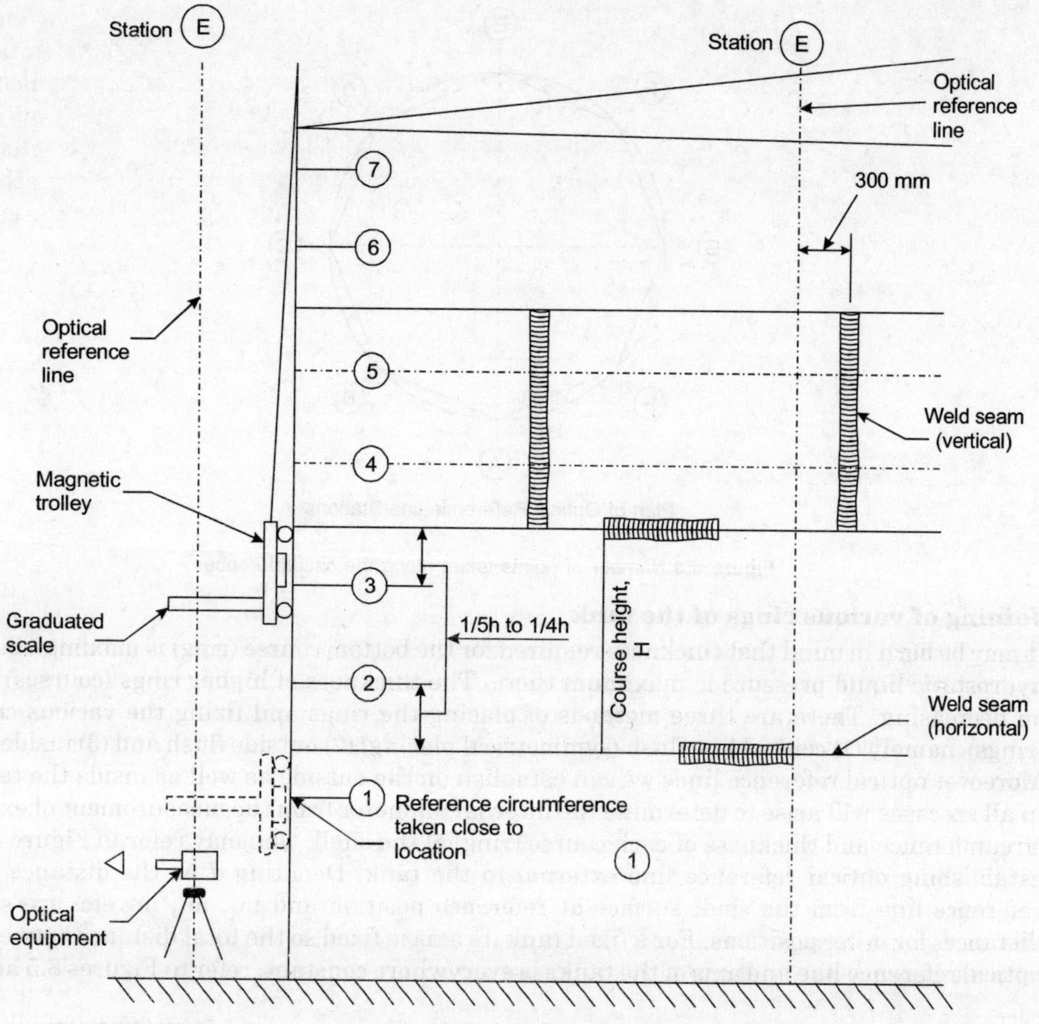

Figure 8.3 Optical arrangement for optical reference line method

Number of points versus circumference length

Circumference in m	*Number of points*
Up to 50	8
Above 50 but up to100	12
Above 100 but up to 150	16
Above 150 but up to 200	20
Above 200 but up to 250	24
Above 250 but up to 300	30
Above 300	36

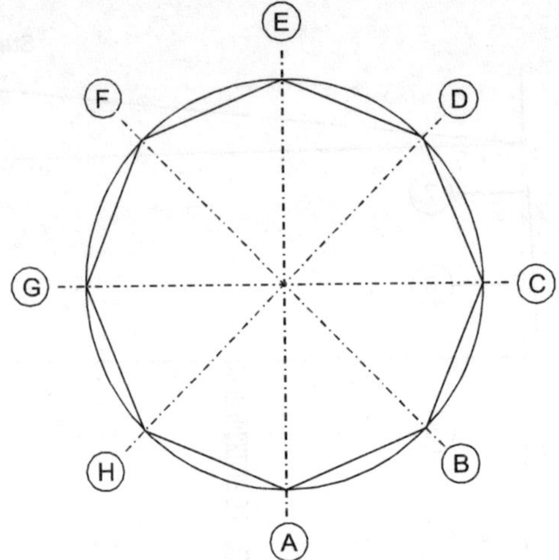

Plan of Optical Reference-line Stations

Figure 8.4 Number of points taken along the circumference

Joining of various rings of the tank

It may be born in mind that thickness required for the bottom course (ring) is maximal, because hydrostatic liquid pressure is maximum there. The thickness of higher rings (courses) keeps on decreasing. There are three methods of placing the rings and fixing the various courses (rings), namely (1) central line flush (symmetrical placing) (2) outside flush and (3) inside flush. Moreover optical reference lines we can establish on the outside as well as inside the tank, so in all six cases will arise to determine the internal diameter from the measurement of external circumference and thickness of each course (ring) of the shell. One may refer to Figure 8.5 for establishing optical reference line external to the tank. Denoting d as the distance of the reference line from the shell surface at reference position and m_1, m_2, m_3 etc. are similar distances for other positions. For a fixed tank its axis is fixed so the total distance between the optical reference line and axis of the tanks is everywhere constant, refer to Figures 8.5 and 8.6.

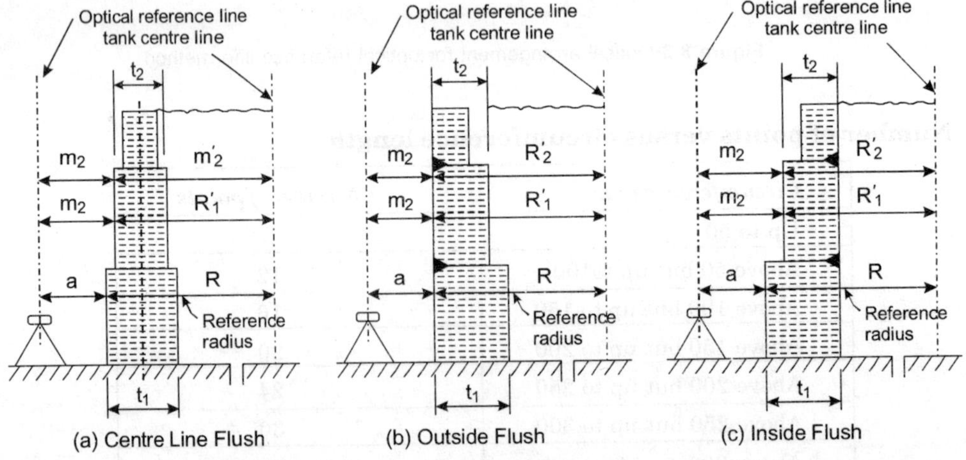

Figure 8.5 Schemes of joining consecutive courses (rings) and reference line is outside the tank

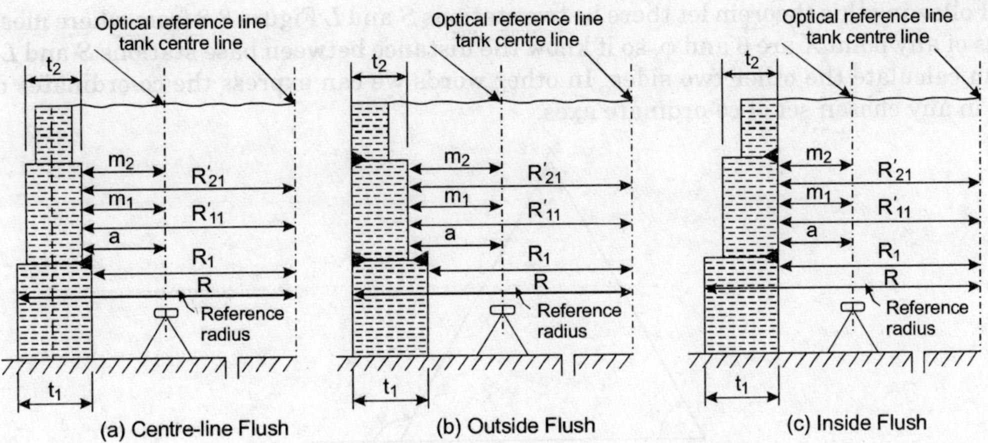

Figure 8.6 Schemes of joining consecutive courses (rings) and reference line is inside the tank

So $\quad R + a + t = R_r' + t_r + m_r$ for all values for r

But $\quad R = (C/2\pi) - t$, giving

$$R_r' = C/2\pi + a - m_r - t_r.$$

The equation is valid for all three arrangement of fixing the consecutive courses (rings).

If the optical reference line is established inside the tank, then

$$R + t - a = R_r' + t_r - m_r$$

Giving $\quad R_r' = C/2\pi - a + m_r - t_r$

Here R is the radius and C is the circumference of the bottom course (ring) at the reference position;

t_r is the thickness of the shell plate at the rth observation;

and m_r is the value of rth observation.

On each course at least two sets of observations at following locations should be taken.

- $1/5^{th}$ to $1/4^{th}$ of the course height above the horizontal lower seam.
- $1/5^{th}$ to $1/4^{th}$ of the course height below the horizontal upper seam.

8.9.1.4 Optical Triangulation Method for Vertical Tanks, Spheres and Spheroids

The principle of this method is based upon a theorem that if in a triangle, two angles and length of one side is known that all its sides are known. Consider a triangle with base 'a' and other two base angles as γ and β (Figure 8.7). Then we have the following relations

$$a/\sin(\pi - \beta - \gamma)) = b/\sin(\beta) = c/\sin(\gamma)$$

$$a/\sin(\gamma + \beta) = b/\sin(\beta) = c/\sin(\gamma)$$

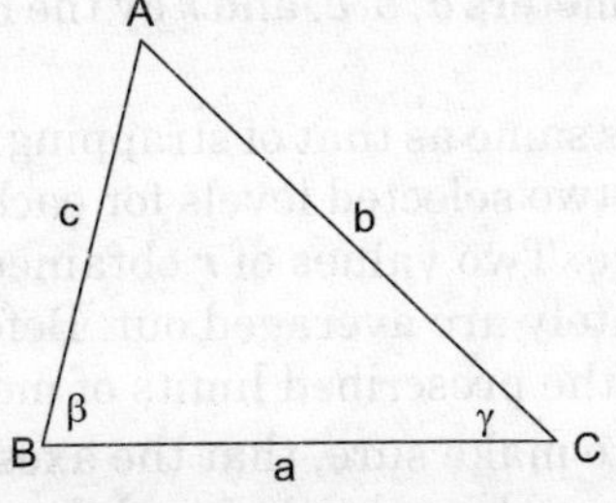

Figure 8.7 Sine law

Following this theorem let there be two stations S and L Figure 8.8 from where measured angles of any point X are θ and φ, so if know the distance between base stations S and L, then we can calculate the other two sides. In other words we can express the co-ordinates of this point in any chosen set of co-ordinate axes.

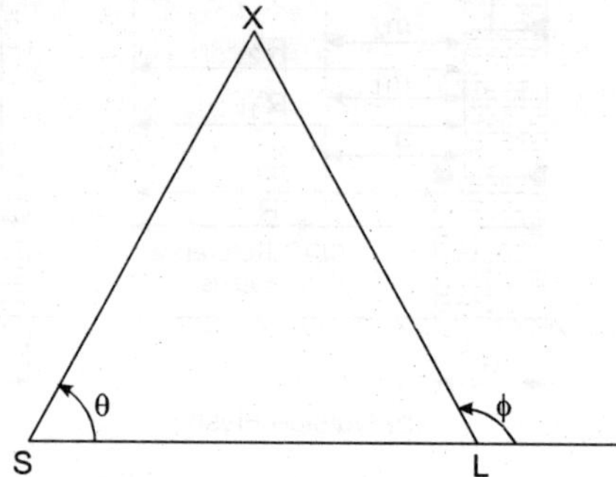

Figure 8.8 Coordinates of a point

This technique is applied for measuring internal diameters of a vertical tank; we use two theodolites, whose optical axes have been made collinear. Angles θ and φ for any point P, are respectively measured from the x-axis in the horizontal plane and with the z-axis which is normal to the horizontal plane. Then the knowledge of the distance between the two measuring stations would give us the coordinates of the observed points. Projection of the points on a cylindrical surface is a circle of say radius r. Let coordinates of the centre of this circle are (a, b). The equations of the circle passing through each observed point having same azimuth angle, should satisfy the following equation for all points i.e. for all values of p.

$$(x_p - a)^2 + (y_p - b)^2 = r^2. \quad ...(1)$$

So we have n equations from which we can always find the best estimates of values of a, b and r by minimising the function $F(a, b, c)$ given by:

$$F(a, b, c) = (x_p - a)^2 + (y_p - b)^2 - r^2, \quad ...(2)$$

a, b, and r as three parameters.

Putting each of $\delta F/\delta a$, $\delta F/\delta b$ and $\delta F/\delta r$ to zero gives us three normal equations. Solution of these equation give the best estimates of parameters a, b and r.

We are interested in the value of r only, twice of which gives us the diameter of the vertical cylindrical tank.

The same technique can be used for measuring diameters of spherical and spheroid tanks. In that case we will have n equations of flowing nature

$$(x_p - r)^2 + (y_p - b)^2 + (z_p - c)^2 - r^2 = 0 \quad ...(3)$$

Where (a, b, c) are co-ordinates of the centre and r is radius of the sphere. So we have to solve n equations for four parameters a, b, c, and r by the method of least squares as enunciated above.

The rest of methodology is same as that of strapping. In this case, target point are marked all along the circumference at two selected levels for each course (ring) and solve equation (1) for points at one level at a time. Two values of r obtained by treating upper and lower target point for the same ring separately are averaged out. Before taking the average we may see if the two values of r are within the prescribed limits of measurement.

Important precaution is to make sure, that the axes of two theodolites are collinear. The two theodolites are directed towards each other and the position of one is so adjusted that the illuminated cross wire of the opposite theodolite falls exactly upon the cross wire of the observing theodolite. The same procedure is used for the other theodolite. The process is repeated till the cross wire of one lies on the other, seen from either of the two theodolites.

8.9.1.5 Electro-optical Method for Vertical Storage Tanks

Instead of using two theodolites, in this method, we use only one theodolite and range finding laser device. The laser ranging device measures distance and theodolite gives azimuth and horizontal angles simultaneously. We measure optical distances of various target points Figure 8.9 with their angular positions, i.e. their azimuth angle and angle from some arbitrary line in the horizontal plane. Then taking the arbitrary line and a line perpendicular to it in the same horizontal plane as x and y axes respectively and the line passing through the intersection of these two lines and normal to the horizontal plane as z-axis. We can express the coordinates of any target point in terms of distance and two angles. Coordinates of any point X_p are given as

$$X_p = D_p \cos(\theta)\sin(\phi),\ Y_p = D_p \sin(\theta)\sin(\phi)$$

$$Z_p = D_p \cos(\phi).$$

Where D_p is the distance of pth point as measured by laser ranging device.

Projection of a set of points, at one level i.e. with same azimuth angle ϕ, on *x-y* plane will be a circle. So co-ordinates X_p and Y_p will lie on a circle, coordinates of whose centre is say (*a*, *b*) and its radius *r*, so

$$(X_p - a)^2 + (Y_p - b)^2 - r^2 = 0$$

We can use the technique of least square for finding the best estimates of parameters *a*, *b*, and *r*.

It may be mentioned that $d\sin(\phi)$ for target points at one level of observation will always be the same provided the origin is on the axis of the cylinder and it coincides with *z*-axis.

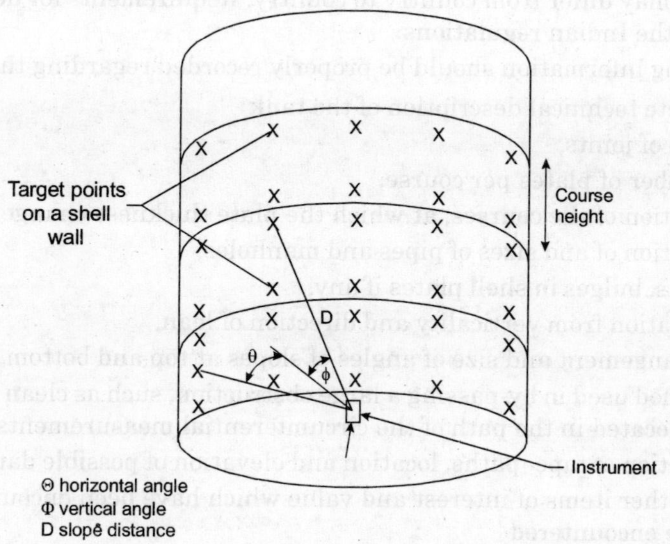

Figure 8.9 Electro-optical method

Measuring instrument used

It consists of essentially two devices, namely

(1) Optico-electro distance measuring device and

(2) Azimuthal and horizontal angles measuring device.

The first device consists of a laser ranging system to measure the distance of target points. The second device is a simple optical theodolite. Both devices are merged together so that both have common axis. The combined system is known as electro-optical distance ranging system.

In sections 8.9.1.4 and 8.9.1.5, all angular measurements are carried out with a readability of 0.000 2 gon, repeatability of 0.000 5 gon with over all uncertainty of 0.001 gon, where gon is a new measure of the angle such that 100 gon = one right angle i.e. gon is 100th of a right angle [4].

8.9.2 Volumetric Method

For comparatively smaller tanks, or for bottoms of irregular shape or the ring, which has too much deadwood of a complex geometry, volumetric or liquid calibration method is employed. For this we can use portable tanks, positive displacement meter, fixed service tanks or weighing of the input/output of liquid. The details of each are given in section 8.13.

8.10 DESCRIPTIVE DATA

We have seen above that there are two methods for calibration of tanks. One is by dimensional measurements of the tank and other is volumetric or liquid calibration. In fact both methods are often used for one tank. Whichever method we adopt certain descriptions about the tank, site, ownership etc. are necessary to write in the calibration certificate. Prescribed format and details asked for may differ from country to country. Requirements for descriptive data given below are as per the Indian regulations.

The following information should be properly recorded regarding the tank:

1. Complete technical description of the tank
   - Type of joints,
   - Number of plates per course,
   - Location of the courses, at which the plate thickness changes,
   - Location of and sizes of pipes and manholes,
   - Dents, bulges in shell plates if any,
   - Deviation from verticality and direction of lean,
   - Arrangement and size of angles of slopes at top and bottom of the shell,
   - Method used in by-passing a large obstruction, such as clean out box or insulation box located in the path of the circumferential measurements
   - Location of tape- paths, location and elevation of possible datum plate and
   - All other items of interest and value which have been encountered with or likely to be encountered
2. Ownership
3. Plant or name of the property
4. Location

5. Manufacturer of tank
6. Erecting Company
7. Description of tank
8. Height (Shell height)
9. Gauge of the plates used
10. Type of roof
11. Weight of floating roof if any
12. Tank contents (Name of liquid)
13. Average liquid temperature
14. Gauge in cm/mm
15. Innage to self
16. Floor or outage
17. Hydrometer reading............at........ °C
18. Sample temperature
19. To prepare.........copies.........increments in ....
20. Gauging reference point to top of top angle.../mm...........normal
21. Service

8.11 STRAPPING METHOD

8.11.1 Precautions

1. Due allowance is given for the expansion of walls of the tank due to hydrostatic pressure when it is full with the liquid. So to get error free measurements by strapping, the tank is filled at least once in the present location with the liquid, which it is likely to contain to the expected height or with water or any other liquid to its equivalent height. At least 24 hours are allowed for settling.
2. All data and methods, whereby measurements are obtained, necessary for the preparation of gage table, should follow sound engineering principles.
3. When drawings of the tank are available all measurements taken should be compared with the given dimensions. Observations showing differences more than the specified maximum permissible errors should be repeated and re-verified.
4. All linear measurements are taken after the tank is subjected to the test given in 1 above.
5. The calibration process depends by and large on the cleanliness inside the tank. The interior of the upright cylindrical surface, and roof supporting members, such as columns and braces in the tank, should be clean and free from any foreign matter including but not limited to the residue of commodities adhering to the side, rust, dirt, emulsion and paraffin. If found dirty on inspection, the tank should be cleaned. Internal incrustation or adhesion has same effect on the capacity of the tank as deadwood and so should be accounted for in same manner.
6. If ladders are used, all rungs should be inspected and tested at ground level. Ladders should not be outstretched for convenience beyond their safe working range. It should be understood that inherent danger in using the ladders increases with soggy footings, relatively smooth upper bearing surface, strong gust of winds, or sudden slack or pull in circumference tape etc.

7. All measurements and descriptive data taken at tank site should be checked and properly recorded. It is preferable to assign this job to a single person.
8. Please take time to do a good job. It should be checked that all descriptive data taken about the tank is properly and legibly recorded.
9. Tanks with a nominal capacity of 2000 dm^3 or less may be strapped in any condition of fill, provided they have been filled at least once at their present location. Small movement of oil into or out of such tanks may be allowed even while strapping is in progress.
10. Bolted tanks with nominal capacity of more than 100 m^3 must have been filled at least once at the present location and must remain at least 2/3rd full when strapping is carried out. Small movement of oil into or out of such tanks may be allowed even while strapping is in progress.
11. Riveted/welded tanks with nominal capacity of more than 100 m^3 may be strapped in any condition of fill, provided they have been filled at least once at their present location. No movement of oil into or out of such tanks may be allowed while strapping is in progress.
12. Complete description of the tank should be recorded and should form the part of the strapping data.
13. A tank, which has been re-strapped, should be identifiable either by re-numbering or some other adequate method.
14. If the calibration of the tank is required to be interrupted, it may be resumed with minimum delay, without repetition of previous observations provided that:
    - There is no major change in equipment and, as far as possible, in personnel.
    - All records of previous observations are complete and legible.
    - Same hydrostatic head is maintained in the tank.

8.11.2 Equipment used in Strapping

8.11.2.1 Steel Tapes

Steel tapes of 30 m, 50 m and 100 m in length are normally used for strapping. All the steel tapes should be pre-calibrated from a laboratory, which has a record of traceable measurements. The certificates should indicate the errors at the prescribed points with expanded uncertainty at 95% confidence level (2σ level).

8.11.2.2 Spring Balance

Calibrated at a laboratory having traceable measurements and compiling with the relevant standard specification.

8.11.2.3 Dynamometer

To apply required tension to the tape, a dynamometer is used. It should be pre-calibrated by a laboratory having traceable measurement. It should also conform to the national or international standard specifications.

8.11.2.4 Step-over(s)

It is a frame holding rigidly two scribing points with adjustable distance between the scribing points. A typical step over is shown Figure 8.10. The frame is constructed of wood or steel; if required the step-over may be painted. In some cases a step-over with fixed distance between the scribing points is used. This is used to correct, deviation of the tape from the normal circular part.

When the tape crosses an obstruction, such as projection deformity, fitting or lapped joints, and its path deviates from a true circle and causes error in circumference measurement. To overcome this error a step-over is used, which gives correction to be applied to arrive at the true length of the circumference.

Construction

A step over is a frame consisting of two right-angled L shape arms. To hold the two arms rigidly, bolts and nuts may also be used. The two arms are joined together rigidly holding two arms at right angles to the connecting arms. One arm is made slide fit to another. At the end of each of the arm there is a scribing point. The connecting rod and two arms are of such lengths that the points may be applied to the tape well clear of the obstruction and of its effect on the tape path, while the frame itself does not touch either the obstruction or the tank shell. Rigidity of construction is essential. A suitable step-over is shown in Figure 8.10. The step-over should have sufficient distance between the two arms, so that each void between the tape and the surface of the shell can be measured. The arms are of sufficient length to prevent contact between the interconnecting members and tank plate or obstruction.

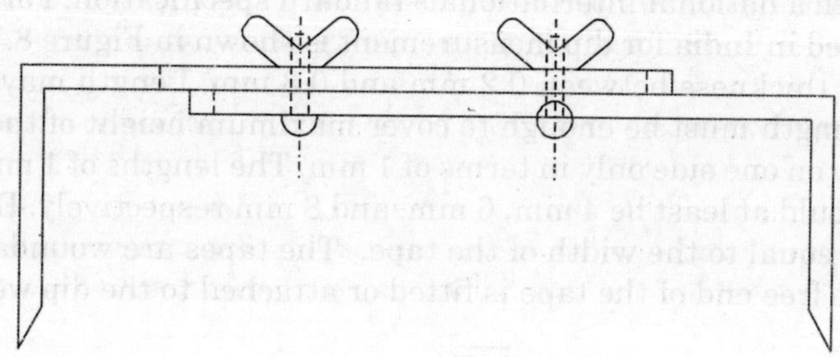

Step-over

Figure 8.10 Step-over

Use of step-over

1. For obstructions, the strapping tape is stretched as is used in measurement of circumference of the tank, which is being calibrated, but not within 30 cm of any horizontal seam. The scribing points are then applied to the tape near the middle of a plate where the tape is fully in contact. The length between the points, as measured on the curved tape is then read off as closely as possible, fractions of one mm is estimated. The observations are repeated on minimum of two and maximum of four plates equally spaced around the circumference. The average of all observations is taken. Let the mean be X mm. The step-over will vary with the tank diameter and the course concerned since they are made on surface differently curved.
2. With the help of the tape still in the same position and under tension used in strapping, the step over is applied to the tape on either side of the obstruction on the tape path and length between the points of the scribers is read on the tape. Let it be Y mm. Then $Y - X$ is the correction to be applied for this particular obstruction. Similarly the corrections for other obstacles will be $Y_r - X$.
3. We have assumed in the above paragraph that the step over is always placed on the tape in the horizontal plane. To acquire this situation unmistakeably, a spirit level is

attached on one side of the connected rods and scriber is so placed that the air-bubble is always in the centre.

4. When the butt-strap of lap joints or the tank shell include rivets or other features which exert uneven effects on the void so produced between the tape and tank surface from joint to joint. In such cases, the step over is employed for each joint separately. The span of the instrument is measured prior to its use in accordance with the step 1 above.
5. Stretch the tape over the joints and place the step over in position at each location of void between tape and shell surface completely spanning the void so that scribing points contact the shell at an edge of the tape. The length of the tape encompassed by the subscribing points, with the tape in proper position and having required tension, is estimated nearest to 0.5 mm. At each step-over location, the difference between the length encompassed by the scribing points and the known span of the instrument is the effect of the void, at that point of the circumference as measured. The sum of such difference in any given path subtracted from the measured circumference will give the corrected circumference at that level.

8.11.2.5 Dip-tape and Dip Weight

It should comply with a national/international standard specification. For example the type of tape measure adopted in India for dip measurement is shown in Figure 8.11. The tape is 13 or 16 mm wide having thickness between 0.2 mm and 0.3 mm. Length may be according to the use and one-piece length must be enough to cover maximum height of the tank. It is marked legibly and indelibly on one side only in terms of 1 mm. The lengths of 1 mm, 5 mm and 10 mm graduations line should at least be 4 mm, 6 mm, and 8 mm respectively. Decimetre and metre graduations should equal to the width of the tape. The tapes are woundable on a reel with a protecting case. The free end of the tape is fitted or attached to the dip weight.

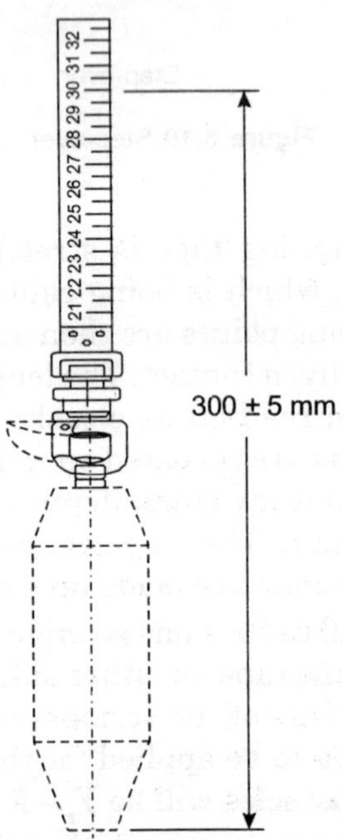

Figure 8.11 Dip-tape with swivel hook

Dip Weight

Dip weights are of two types, light and heavy and are cylindrical torpedo in shape, shown in Figure 8.12, heavier weight is attached by swivel hook as shown in Figure 8.10 and should weigh 1500 ± 50 g. Lighter one should weigh 700 ± 50 g. The effective length of the weight including attaching arrangement is 150 mm. The dip weight is graduated in a manner similar to tape. The graduations on the dip weight begins from its bottom face and is carried over in such a way than when weight is attached to the tape, the graduations are continuous from the weight to the tape.

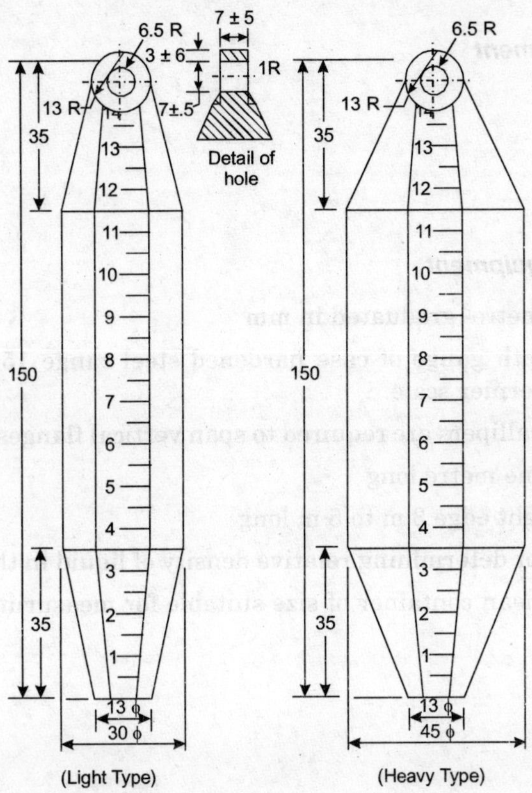

Figure 8.12 Two types of dip weights

Maximum permissible error for tapes

The error in length supported on a horizontal surface with a tension of 50 newtons should not exceed

- Between any two consecutive mm and cm marks Not more than ±0.2 mm
- Between any two consecutive decimetre marks Not more than ±0.4 mm
- From zero to
  - One metre mark ±0.4 mm
  - Two metre mark ±0.6 mm
  - Five metre mark ±1.0 mm

- Any one metre mark ±1.0 mm for the first five metres beyond the first five metres plus 0.5 mm for each additional five metres or part thereof. However maximum error should not increase by 2.0 mm.

8.11.2.6 Loops and Cords

One or more metal loops, which can slide freely on the tape, are also required. Two chords are attached to it; each chord is of sufficient length so as to reach from the top of the tanks to the ground. The tape is positioned and its tension is evenly distributed by passing these loops around the tank.

8.11.2.7 Accessory Equipment

- Ropes
- Seat Hooks
- Safety belts
- Ladders

8.11.2.8 Miscellaneous Equipment

- Steel ruler one metre- graduated in mm
- Depth gage: Depth gauge of case hardened steel range 15 cm and readability of 0.1 mm on the vernier scale.
- Calliper: 15 cm callipers are required to span vertical flanges and bolt heads.
- Straight edges one metre long
- Engineer's straight edge 3 m to 5 m long
- Hydrometers (For determining relative density of liquid in the tank)
- Sample can- A clean container of size suitable for measuring relative density with hydrometers-2
- Spirit level
- Tape positioners
- Awl and Making
- Marking crayon
- Record paper
- Plumb line
- Dumpy level
- Positive displacement bulk meter
- Special clamps: These are required for spanning vertical obstructions in making circumference measurements.

8.11.3 Strapping Procedure

The tank is strapped by the method given below. A tension of 45 ± 5 N, which in terms of weight is 4.5 ± 0.5 kg, is applied to the tape. If necessary, freely sliding loops are used to transmit the tension uniformly through out its length, the loops being passed around the tank by operators with the aid of light chain cords. The tape path should always be parallel to the circumferential seams of the tank.

1. If the tape to be used is not long enough to encircle the tank completely, then after the level of the tape path is chosen, fine line are scribed perpendicular to this path so that the circumference is measured in sections.
2. If the tape to be used can encircle the tank completely, then after the level of the tape has been done, the tape is passed around the circumference and held so that the first graduated centimetre lies within the middle circumferential third of any plate. The other end of the tape is brought alongside. The tension is then applied through the spring balance and transmitted throughout the length of the tape.
3. After one set of circumference measurement is complete, the tape is shifted a little around the tank, is brought to the same level as before, tension is applied and another observation is taken. The final reading is the mean of the two observations. Measurements are taken in terms of mm.

8.11.4 Maximum Permissible Errors in Circumference Measurement

When observations are repeated, mean should lie within specified maximum permissible errors (MPE). In India, we have the following MPE

Measured length	*MPE*
Up to 30 metres	± 2 mm
Over 30 and up to 50 metre	± 4 mm
Over 50 and up to 70 metres	± 6 mm
Over 70 but up to 90 metres	± 8 mm
Over 90 metres	± 10 mm

All the tapes used in the strapping process are calibrated and should especially conform to the requirements of the national or international standard specification in respect of length. The tapes should be calibrated with over all uncertainty better than one third of the figures given above.

8.12 CORRECTIONS APPLICABLE TO MEASURED VALUES

The corrections are to be applied for,

- Over coming obstacles (step- over),
- Sagging of tape under its own weight,
- Plate thickness and
- For temperature differences.

8.12.1 Step Over Correction

1. Subtract distance between the two legs of the step over from the observed distance on the tape as it passes over an obstacle. This error is subtracted from the length recorded.
2. Step over correction is also applicable if it passes over the vertical seams provided that the tape path is clear from the rivet heads. Average step over correction due to

seam is determined for a given course. To obtain total correction for the measured circumference multiply it by number of seams and subtract it from the measured circumference.

3. Applicable correction is ignored for a single obstruction if the error is less than 2 mm.
4. The use of step-over corrects for error encounter due to external projections, but could not account for internal projections or depressions. These are taken as deadwood and are indicated location wise in the deadwood column of the gage table.

By choosing tape path in such a way that appurtenances are avoided, use of step over could be minimised to a great extent.

8.12.2 Temperature Correction

Temperature correction is applicable due to two counts, (1) due to difference in reference temperatures of tape and tank. Usually tapes are calibrated at 20 °C while tank is calibrated at 15 °C. (2) The second correction is due to the fact that tank measurements are taken at a temperature other than its reference temperature. The coefficients of linear expansion of tape and the tank material are required to apply these corrections. Instead of additive correction a factor to multiply the observed length is calculated.

The factor to be used is $[1 + (\gamma_t - \gamma_m)(t - 20)]$. This is to bring the dimensions of the tank to 20 °C, but the tank is to be calibrated at 15 °C, for which another factor $[1 - \gamma_m(20 - 15)]$ is required.

Here γ_m is coefficient of linear expansion for material of the tank and γ_t is for the tape.

8.12.3 Correction Due to Sag

Assuming that the tape will take a shape of a catenary

The correction Z due to the sag is given as

$$Z = W^2S^3/24T^2 \text{ in m}$$

Where

S is span of the tape in m,

T is tension applied in kg force

W is the mass of the tape in kg/m.

If we put tape related variables together as K, then K is given as

$$K = W^2/24T^2.$$

For a tape of 10 mm wide and 0.25 mm thick, made of steel having a density of 7850 kg/m^3, the values of K for different values of tension applied to it are

T	K
4.4 kg	8.29×10^{-5} per m^2
4.5 kg	7.92×10^{-5} per m^2
4.6 kg	7.58×10^{-5} per m^2

For a length of 40 m the sag at 4.5 kg tension will be 5.0688 mm. This is the correction in diameter measurement. This correction is to be subtracted from the observed reading. However, no correction in the measurement of outer circumference due to sagging is required as the tape in this case is everywhere in contact of the tank surface and its horizontality is monitored.

Subtract Z for sag and add the length of the dynamometer to average observed diameter of each course (ring). The length of the dynamometer is taken when it is registering a pull of 4.5 kg force.

Correction due to stretching is not required because the tension applied is same at which the tape was calibrated.

The temperature correction for a temperature difference of 7°C for a length of 40 m is $11.10^{-6} \times 7 \times 40 = 3$ mm, it should be added to the observed reading.

For circumference of the same tank the correction will become roughly 9 mm.

8.13 VOLUMETRIC METHOD (LIQUID CALIBRATION)

Volumetric or Liquid Calibration is a method of determining incremental volumes and capacities of tanks by transfer of known quantities of a liquid to or from the tank under test. The procedure is suitable for preparing a gage table for the tank under-test of any shape and design except for a meter prover.

Liquid calibration is a very general term used for calibrating a given tank at its different levels, against a standard tank of known capacity though a liquid, or, using a calibrated positive displacement meter.

The procedure is selected keeping in mind the accuracy requirement and available equipment at the site. The procedure should be such which can be completed in the shortest time and so that a better accuracy is maintained during that time.

8.13.1 Portable Tank

A portable volumetric tank can be used in calibrating comparatively smaller capacities tanks of say from 10 m^3 (10 000 dm^3) to 100 m^3 (100 000 dm^3). The procedure generally gives highest degree of accuracy but it is rather time consuming. Portable tanks are of much smaller diameter and capacity so a better accuracy is ensured in them. These tanks are calibrated by gravimetric method using water as standard of density. The method is beneficial as it gives better accuracy for the calibrating tank.

8.13.2 Positive Displacement Meter

Positive displacement meter of 0.1% accuracy are used in calibrating a tank, especially those portions of it, which are not in regular geometric shape. A portable meter of the said accuracy class may be used for larger capacity of a tank than necessary. Of course here the inaccuracy will always be greater than that of the calibrating meter.

8.13.3 Fixed Service Tank

At some installations fixed service tanks are available. These are calibrated by strapping with greatest possible accuracy without worrying about time involved. The tanks are, then, used for calibrating the other storage tanks. The diameters of service tank should be smaller than that of the tank under calibration to give better precision in calibration. In case diameter of the fixed tank is bigger than that of the tank under-test, the fixed tank may be calibrated by using prover tank or a master meter.

8.13.4 Weighing Liquid

If the tank under calibration is meant for storing liquid, which is viscous and has a tendency to adhere to the walls of the tank, it is preferable to use the liquid weighing procedure. For calibrating such tanks weighed liquid is delivered in to the tank. As in this case, density of the liquid used is taken from the literature, liquid should be free from any sediment and water.

8.14 LIQUID CALIBRATION PROCESS

8.14.1 Priming

Before actual calibration, the tanks should be filled at least once to its maximum capacity with the liquid, which it intends to store or by water of proportional height. The purpose is to apply hydrostatic pressure to the walls of the tank, as it is likely to experience in actual use.

8.14.2 Material Required

1. Liquid in sufficient quantity, liquid should be non-volatile type and its density should be nearly equal to that of the liquid, which the tank is intended to store.
2. Gagging equipment like dip tapes, thermometers and other measuring instruments used in calibration should conform to their respective national or international standard specifications.
3. Suitable formats for recording, calculating capacity and presentation of the records should be pre-decided and sufficient number of such forms must be available at the calibration site.
4. If uncertainty is required to be calculated and to be reported than the procedure set in ISO/OIML guide [16] should be used.
5. When using the portable tank method, one or more such tanks with proper identifications along with a calibration rig should be available. Number of filling required should be pre-calculated taking the capacities of the portable and under calibration tank in to consideration.
6. Prior to start of the calibration procedure it should be ensured that the proper national authority has calibrated the portable tanks and their calibration certificate is available for inspection and applying corrections to the portable tanks.
7. When using a positive displacement meter, it should be properly selected taking into consideration the capacity of the tank and especially that of the bottom. The meter should be of non-temperature compensated, and they should be equipped with continuous correction type calibrator. Generally instruments like pressure regulators, gauges and meter proving tank should be available so that the meter under use may be calibrated during use. On-line thermometers with recorder, air eliminators, strainers pumps, quick acting valves and related pipe fittings should also be made available at the site. Standard meter should conform to the relevant national or international standard specifications.

8.14.3 Considerations to be Kept in Mind

1. The size of the incremental step in preparing the gage table is determined by the deadwood and its volume distribution with respect of height. Tank shape and or a particular zone to be calibrated demands due consideration.

2. Due attention should be paid for all hose and pipe connections properly tightened. While installing, proper steps should be taken for elimination of air or vapour locking. All piping should be filled fully with calibrating liquid before the test is commenced and should remain full all the time.
3. If a pump is used to transfer the liquid, caution should be observed to ensure that the liquid level in the delivering tank is never lowered to allow suction of air in the system. The suction pump and its piping should be of appropriate size to avoid pulling a vacuum on the system.
4. If a meter is used, care is to be taken to avoid pulling a vacuum on the system. The meter will register the volume of vapour or air as if liquid has passed through it.
5. The job of calibration should be completed in one go without any interruption. If due to certain emergencies the calibration process is interrupted and liquid level in either tank is changed due to temperature, the calibration process may be resumed after applying proper correction to the volume of each tank, before continuing the job further.
6. Better results are obtained when the ambient and liquid temperatures are almost equal.
7. Accurate temperature measurement is a prerequisite for volumetric measurement. So some guidelines, keeping in view the volume, depth and shape of the liquid, are framed for the locations at which temperature is measured. These are given below:
   - The temperature to be used for the purpose of preparing gage table should be the mean of all measured temperatures.
   - Normally one temperature measurement in the middle of the liquid having volume less than 200 dm^3 is sufficient. For volumes in between 200 dm^3 and 1000 dm^3, two temperature measurements, one at the middle of the upper half and second at the middle of lower, should be taken. If the height of 1000 dm^3 is less than 0.5 m, then one temperature measurement at the middle is sufficient. If height of the liquid is more than 0.5 m but less than 1m, then two temperatures as mentioned above should be taken. For liquids having depth more than one metre, temperature at three equally spaced points, should be measured.
   - Calibrated thermometers of smallest graduations of not more than 0.2 °C should be used. Efforts are made to estimate and record up to half the width of graduations.
   - Temperature corrections to the metered or gauged volumes should be applied so as to bring the volume of the liquid transferred to the measured temperatures of the liquid in the tank. For the use of pure distilled water, correction factors F due to temperature difference are given in Tables 8.1 and 8.2. For other liquids, coefficient of cubical expansion must be known to prepare Gage tables.
   - Environmental temperature and wind or rain conditions should be recorded at the time of test.
8. Irrespective of the method used for calibrating the tank, one should stop for taking the readings of (1) hand-line gauge of liquid receiving tank, (2) liquid temperature in

receiving tank, (3) meter or gage reading of delivery tank, and (4) temperature of delivery tank (5) Automatic float gage, which has been set when it first starts floating:

- When liquid first hits hand-gauging point. (Where the tank has downward cone bottom, two gauging or striking points may be used.
- When tank bottom is completely covered.
- At lower and upper limits of all deadwood.
- When float gauge floats freely, adjustment should be made with hand-line gage.
- At bottom edge and fully floating position of the floating roof and at sufficient number of points to establish incremental values desired.
- Every 5 to 10 cm cylindrical portion of the tank.
- At top of each course (ring).
- Hand-line gage should be read in terms of 1 mm and temperature to 0.1°C and meter to 0.05 dm^3.

8.15 TEMPERATURE CORRECTION IN LIQUID TRANSFER METHOD

Quite often the liquid in the tank under test is taken out and metered somewhere else, or vice versa. The temperature of the liquid in the tank and at the place of metering may not be the same. If the temperature of the liquid in the tank is higher than gage table would indicate some volume V_t, but the volume V_s shown by the meter at lower temperature will be less.

To establish the relation between V_t and V_s, we use the fact that the mass of the liquid involved will remain unchanged. If d_t and d_s are respectively the density of liquid in the tank and where metered, then we have

$$V_t d_t = V_s d_s, \text{ giving}$$

$$V_s = (d_t/d_s)\, V_t \text{ or } V_s = F\, V_t.$$

To find out the error in the gage table, volume indicated by the gage table is multiplied by the factor F. F will normally be less than unity if tank is at a higher temperature than the meter and will be more than unity if tank is at lower temperature than metering device. To calculate this factor F, we need the knowledge of density of liquid at different temperatures, which is usually not known. However density of pure distilled water is known with great accuracy at all temperatures of interest. Hence water is used for such purpose. The values of the factor F are given, for all temperature normally encountered in the field with 5°C above or below the temperature in the tank, in Tables 8.1 and 8.2.

The factors to multiply the gauge volume are given in Table 8.1, if the temperature of liquid in the tank is less than that when liquid is metered.

The multiplying factors given in Table 8.2 are used when temperature of liquid in the tank was more than when the liquid was metered.

Table 8.1 Values of Factor *F* when the Temperature of Liquid in the Tank was Less than when it is Metered (*T* Sands for Tank Temperature)

T	0.5	1.0	1.5	2.0	2.5	3.0	3.5	4.0	4.5	5
5	1.000006	1.000008	1.000006	1.000000	0.999990	0.999976	0.999958	0.999935	0.999908	0.999876
6	1.000014	1.000024	1.000030	1.000032	1.000030	1.000024	1.000014	1.000000	0.999981	0.999958
7	1.000021	1.000039	1.000052	1.000062	1.000068	1.000070	1.000068	1.000062	1.000052	1.000038
8	1.000028	1.000053	1.000074	1.000092	1.000105	1.000115	1.000121	1.000123	1.000122	1.000116
9	1.000036	1.000067	1.000096	1.000121	1.000142	1.000159	1.000173	1.000183	1.000189	1.000191
10	1.000042	1.000081	1.000117	1.000149	1.000177	1.000202	1.000223	1.000240	1.000254	1.000264
11	1.000049	1.000095	1.000137	1.000176	1.000211	1.000243	1.000272	1.000296	1.000317	1.000335
12	1.000055	1.000108	1.000157	1.000202	1.000245	1.000283	1.000319	1.000351	1.000379	1.000404
13	1.000062	1.000120	1.000176	1.000228	1.000277	1.000323	1.000365	1.000404	1.000439	1.000471
14	1.000068	1.000133	1.000195	1.000253	1.000309	1.000361	1.000410	1.000456	1.000498	1.000537
15	1.000074	1.000145	1.000213	1.000278	1.000340	1.000398	1.000454	1.000506	1.000555	1.000601
16	1.000080	1.000157	1.000231	1.000302	1.000370	1.000435	1.000497	1.000555	1.000611	1.000663
17	1.000086	1.000168	1.000248	1.000325	1.000399	1.000470	1.000538	1.000603	1.000665	1.000724
18	1.000091	1.000180	1.000265	1.000348	1.000428	1.000505	1.000579	1.000650	1.000718	1.000783
19	1.000097	1.000191	1.000282	1.000370	1.000456	1.000539	1.000619	1.000696	1.000770	1.000841
20	1.000102	1.000201	1.000298	1.000392	1.000484	1.000572	1.000658	1.000740	1.000820	1.000897
21	1.000107	1.000212	1.000314	1.000414	1.000510	1.000605	1.000696	1.000784	1.000870	1.000953
22	1.000113	1.000222	1.000330	1.000435	1.000537	1.000636	1.000733	1.000827	1.000918	1.001007
23	1.000118	1.000233	1.000345	1.000455	1.000563	1.000667	1.000769	1.000869	1.000966	1.001060
24	1.000123	1.000243	1.000360	1.000475	1.000588	1.000698	1.000805	1.000910	1.001012	1.001112
25	1.000127	1.000252	1.000375	1.000495	1.000613	1.000728	1.000841	1.000951	1.001058	1.001163
26	1.000132	1.000262	1.000389	1.000515	1.000637	1.000757	1.000875	1.000990	1.001103	1.001213
27	1.000137	1.000272	1.000404	1.000534	1.000661	1.000786	1.000909	1.001029	1.001147	1.001262
28	1.000142	1.000281	1.000418	1.000552	1.000685	1.000815	1.000942	1.001067	1.001190	1.001310
29	1.000146	1.000290	1.000432	1.000571	1.000708	1.000843	1.000975	1.001105	1.001232	1.001358
30	1.000151	1.000299	1.000445	1.000589	1.000731	1.000870	1.001007	1.001142	1.001274	1.001404
31	1.000155	1.000308	1.000458	1.000607	1.000753	1.000897	1.001039	1.001178	1.001315	1.001450
32	1.000159	1.000317	1.000472	1.000624	1.000775	1.000924	1.001070	1.001214	1.001356	1.001495
33	1.000164	1.000325	1.000484	1.000642	1.000797	1.000950	1.001101	1.001249	1.001395	1.001539
34	1.000168	1.000334	1.000497	1.000659	1.000818	1.000975	1.001131	1.001284	1.001434	1.001583
35	1.000172	1.000342	1.000510	1.000676	1.000839	1.001001	1.001160	1.001318	1.001473	1.001626
36	1.000176	1.000350	1.000522	1.000692	1.000860	1.001026	1.001190	1.001351	1.001511	1.001668
37	1.000180	1.000358	1.000534	1.000708	1.000880	1.001050	1.001218	1.001384	1.001548	1.001710
38	1.000184	1.000366	1.000546	1.000724	1.000900	1.001075	1.001247	1.001417	1.001585	1.001751
39	1.000188	1.000374	1.000558	1.000740	1.000920	1.001098	1.001275	1.001449	1.001621	1.001791
40	1.000192	1.000382	1.000569	1.000756	1.000940	1.001122	1.001302	1.001480	1.001657	1.001831

Table 8.2 Values of Factor *F* when the Temperature of Liquid was More than when it is Metered (*T* Stands for Tank Temperature)

T	0.5	1.0	1.5	2.0	2.5	3.0	3.5	4.0	4.5	5.0
5	0.999990	0.999976	0.999959	0.999938	0.999913	0.999885	0.999853	0.999817	0.999779	0.999736
6	0.999983	0.999961	0.999937	0.999908	0.999876	0.999841	0.999802	0.999760	0.999714	0.999665
7	0.999975	0.999947	0.999915	0.999879	0.999841	0.999798	0.999753	0.999704	0.999652	0.999596
8	0.999968	0.999933	0.999894	0.999851	0.999806	0.999757	0.999705	0.999649	0.999591	0.999529
9	0.999961	0.999919	0.999873	0.999824	0.999772	0.999717	0.999658	0.999596	0.999531	0.999463
10	0.999954	0.999905	0.999853	0.999798	0.999739	0.999677	0.999613	0.999545	0.999474	0.999400
11	0.999948	0.999892	0.999834	0.999772	0.999707	0.999639	0.999568	0.999494	0.999417	0.999338
12	0.999941	0.999880	0.999815	0.999747	0.999676	0.999602	0.999525	0.999445	0.999362	0.999277
13	0.999935	0.999867	0.999796	0.999722	0.999645	0.999566	0.999483	0.999397	0.999309	0.999218
14	0.999929	0.999855	0.999778	0.999698	0.999616	0.999530	0.999442	0.999351	0.999257	0.999160
15	0.999923	0.999843	0.999761	0.999675	0.999587	0.999495	0.999401	0.999305	0.999205	0.999103
16	0.999917	0.999832	0.999743	0.999652	0.999558	0.999461	0.999362	0.999260	0.999155	0.999048
17	0.999912	0.999820	0.999726	0.999630	0.999530	0.999428	0.999324	0.999216	0.999107	0.998994
18	0.999906	0.999809	0.999710	0.999608	0.999503	0.999396	0.999286	0.999174	0.999059	0.998941
19	0.999901	0.999799	0.999694	0.999587	0.999477	0.999364	0.999249	0.999132	0.999012	0.998889
20	0.999895	0.999788	0.999678	0.999565	0.999451	0.999333	0.999213	0.999091	0.998966	0.998838
21	0.999890	0.999778	0.999663	0.999545	0.999425	0.999303	0.999178	0.999050	0.998921	0.998789
22	0.999885	0.999767	0.999647	0.999525	0.999400	0.999273	0.999143	0.999011	0.998876	0.998740
23	0.999880	0.999757	0.999632	0.999505	0.999375	0.999243	0.999109	0.998972	0.998833	0.998691
24	0.999875	0.999748	0.999618	0.999486	0.999351	0.999214	0.999075	0.998934	0.998790	0.998644
25	0.999870	0.999738	0.999603	0.999467	0.999327	0.999186	0.999042	0.998896	0.998748	0.998598
26	0.999865	0.999729	0.999589	0.999448	0.999304	0.999158	0.999010	0.998859	0.998707	0.998552
27	0.999861	0.999719	0.999575	0.999429	0.999281	0.999131	0.998978	0.998823	0.998666	0.998507
28	0.999856	0.999710	0.999562	0.999411	0.999259	0.999104	0.998947	0.998788	0.998626	0.998463
29	0.999852	0.999701	0.999548	0.999393	0.999236	0.999077	0.998916	0.998753	0.998587	0.998420
30	0.999847	0.999692	0.999535	0.999376	0.999215	0.999051	0.998886	0.998718	0.998548	0.998377
31	0.999843	0.999684	0.999522	0.999359	0.999193	0.999025	0.998856	0.998684	0.998510	0.998335
32	0.999839	0.999675	0.999509	0.999342	0.999172	0.999000	0.998826	0.998651	0.998473	0.998293
33	0.999834	0.999667	0.999497	0.999325	0.999151	0.998975	0.998797	0.998618	0.998436	0.998252
34	0.999830	0.999658	0.999484	0.999308	0.999131	0.998951	0.998769	0.998585	0.998399	0.998212
35	0.999826	0.999650	0.999472	0.999292	0.999110	0.998927	0.998741	0.998553	0.998364	0.998172
36	0.999822	0.999642	0.999460	0.999276	0.999090	0.998903	0.998713	0.998522	0.998329	0.998133
37	0.999818	0.999634	0.999448	0.999260	0.999071	0.998879	0.998686	0.998491	0.998294	0.998095
38	0.999814	0.999626	0.999437	0.999245	0.999052	0.998856	0.998659	0.998460	0.998260	0.998057
39	0.999810	0.999619	0.999425	0.999230	0.999033	0.998834	0.998633	0.998430	0.998226	0.998020
40	0.999806	0.999611	0.999414	0.999215	0.999014	0.998811	0.998607	0.998401	0.998193	0.997984

REFERENCES

[1] API 2550. Measurement and Calibration of Upright Cylindrical Tanks.

[2] ISO 7507-1, 1993. Calibration of Vertical Storage Tanks (Strapping method).

[3] ISO 7507-2, 1993. Calibration of Vertical Storage Tanks (Optical reference method).

[4] ISO 7507-3, 1993. Calibration of Vertical Storage Tanks (Optical triangulation method).

[5] Hoa G Nguyen and Micheal R Blackburn; 1995. A Simple Method for Range Finding via Laser Triangulation, Technical doc. 2734, US Navy.

[6] ISO 7507-4, 1995. Calibration of Vertical Storage Tanks (internal electro optical ranging method).

[7] ISO 7507-6, 1997. Calibration of Vertical Storage Tanks (monitoring).

[8] Manual of Weights and Measures, 1975. Directorate of Weights and Measures, Ministry of Industry and Civil Supplies, Govt. of India.

[9] Gupta S V. A Treatise on Standards of Weights *and Measures*, 2003. Commercial Law Publishers (India) Private Limited.

[10] API standard 2555, 1965. Method of Liquid Calibration of Tanks.

Specifications for measuring instruments, which are used in calibration of tanks:

[11] API 2543 or ASTM D 1086. Measuring the Temperature of Petroleum and Petroleum Products.

[12] API 2544 or ASTM D 287. Test for API Gravity of Crude Petroleum and Petroleum Products.

[13] API 2546 D1085. Gaging Petroleum and Petroleum Products.

[14] API 3546 or ASTM D 270. Sampling Petroleum and Petroleum Products.

[15] API 1101. Measurement of Petroleum Liquids Hydrocarbons by Positive Displacement Meter.

[16] ISO, OIML and IEC, 1992. Guide to the Expression of Uncertainty of Measurement.

CALIBRATION OF VERTICAL STORAGE TANK

9.1 MEASUREMENT OF CIRCUMFERENCE

For the purpose of understanding the plan of strapping levels, let us consider structure of the shell of the tank. You may notice that the vertical shell (wall) of the tank is comprised of several rings, known as courses and are made from thick plates in the form of arcs of a circle, these are joined in different modes, namely

- Riveting
- Welding

Welding itself may be

1. Lap welding
2. Butt-welding (End to end welding)

Strapping levels are decided upon the method used to join the shell plates of the tank.

9.1.1 Strapping Levels (Locations) for Vertical Storage Tanks

9.1.1.1 Strapping Levels Riveted Tanks

1. Circumference is measured at 7 to 10% [1,2] of the height of exposed portions of each course (ring) above the level of the top of the bottom angle iron of the tank and above the upper edge of each horizontal overlap between courses (rings) see arrows A in Figure 9.1 and Figure 9.2. Numerical sub-scripts indicate the order of the course starting from bottom.
2. Circumference is measured at 7 to 10% of exposed portion of each course (ring) below the lower edge of each horizontal overlap between courses (ring) and below the level of the lowest part of the top angle on the tank (see arrows B in Figure 9.1 and 9.2).

For each course (ring) measurements are taken at two levels, Level A and level B.

9.1.1.2 *Strapping Levels (Locations) for Welded Tanks*

Circumference is measured at two levels for each course see arrows A and B in Figure 9.2 at the top and bottom of each course and at 20% of the height of the exposed portion of the respective course away from the angle irons or seam.

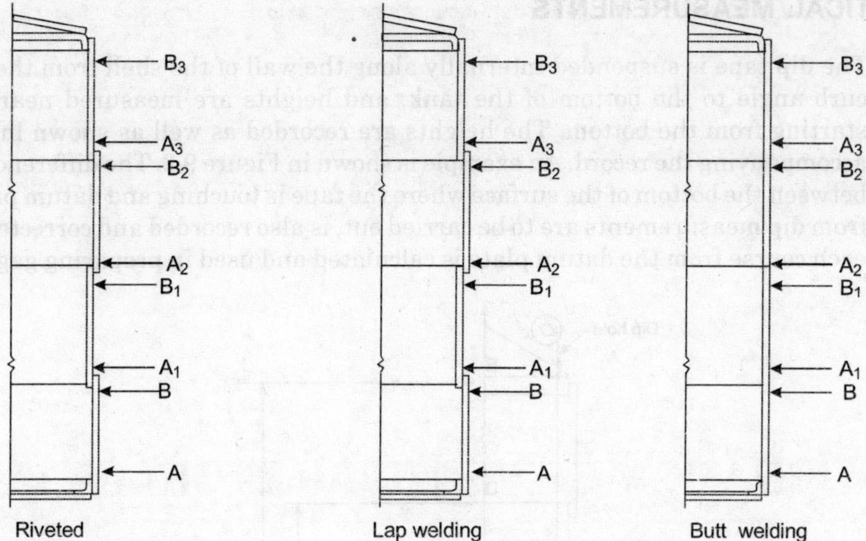

Figure 9.1 Riveted joints **Fig. 9.2** Lap welding and Butt welding

9.1.1.3 General Precautions

- Circumferential tape paths, having been located at elevations as prescribed in 9.1.1.1 and 9.1.1.2 are examined for obstructions and type of vertical joints. Projections due to dirt and scale are removed along each path.
- Occasionally, some features of construction such as manholes or insulation box make it impractical to make measurements at the prescribed level. If the obstruction cannot be conveniently spanned by a step-over, then a substitute path located as near as to the prescribed one is chosen. However the strapping records will show the new path and reasons for change.
- The type and characteristics of vertical joints shall be determined by close examination in order to establish the method of measurement and equipment required. If the tape is not in close contact with the surface of the tank throughout its whole length owing to vertical joints, a step-over is used so that correction is applied to adjust the gross difference for this effect.

9.2 MEASUREMENT OF THICKNESS OF THE SHELL PLATE

1. Where the type of construction of the shell is such that leaves the plate edges exposed, a minimum of four thickness measurements are made on each course. The points should be equi-space spread all over the circumference. The average of the thickness measurements for the course is recorded. Further all thickness locations must be properly labelled and a record is maintained. The thickness is not to be measured where the plate edges have been distorted by caulking.
2. Where plates are concealed due to the type of joints used for example butt joints, it should be clearly stated in the records and alternative step as described below is taken.
3. In absence of direct measurement, the thickness reported by the fabricator in the drawings may be used.

9.3 VERTICAL MEASUREMENTS

1. The dip tape is suspended internally along the wall of the shell from the top of the curb angle to the bottom of the tank, and heights are measured nearest to mm starting from the bottom. The heights are recorded as well as shown in the figure accompanying the record. An example is shown in Figure 9.3. The difference in height between the bottom of the surface where the tape is touching and datum plate, where from dip measurements are to be carried out, is also recorded and corrected height of each course from the datum plate is calculated and used in preparing gage table.

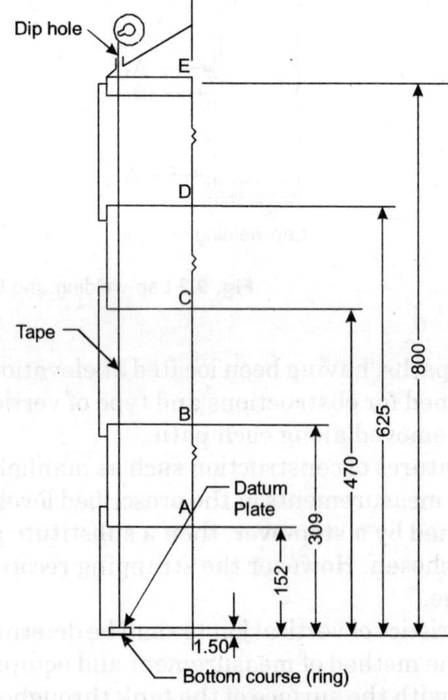

Figure 9.3 Vertical measurements

2. As shown in Figure 9.3, the height of datum plate from the bottom of the surface where tape was touching is 1.5 cm. So to obtain correct height from the datum plate of each course 1.5 cm is subtracted from the indicated height of each course.

Example

So correct heights of each course from datum plate are:

$$A = 152 - 1.5 = 150.5 \text{ cm}$$
$$B = 309 - 1.5 = 307.5 \text{ cm}$$
$$C = 470 - 1.5 = 468.5 \text{ cm}$$
$$D = 625 - 1.5 = 623.5 \text{ cm}$$
$$E = 800 - 1.5 = 798.5 \text{ cm}$$

3. When it is not convenient to measure the course heights internally, then these are measured from outside the tank and due allowance is made for the effect of horizontal seam overlaps. The heights thus obtained will be vertical distances of the successive edges of the courses, as exposed externally so to obtain the correct heights, in case of

lap joints, the width of the lap in each course is to be measured and necessary corrections together with that due to datum are applied to obtain internal course heights.

4. If necessary, external height of each course is measured at several points equally distributed along the circumference and average value is calculated.

9.4 DEADWOOD

Deadwood is measured, if possible, within the tank itself. Dimensions shown in the drawings supplied by the fabricator may be accepted if actual measurements are not practicable.

1. Measurement of deadwood should also show the lowest and highest levels measured from the datum plate adjacent to the shell, at which deadwood affects the capacity of the tank. Measurements should be in increments, which permits allowance for its varying effect on the tank capacity at various elevations.
2. Large deadwood of irregular shape is measured in suitably chosen separate sections.
3. Work sheet on which details of deadwood are sketched should also contain location dimensions. The sketch should be clearly identifiable and be part of the strapping records.
4. For variable deadwood, such as nozzles and manholes especially encountered in the first two courses from the bottom, average deadwood correction is worked out and applied.

9.5 BOTTOM OF TANK

The different tanks have, in general, different types of bottoms. The bottom of the tank may be flat, conical, hemi-spherical or semi-ellipsoidal or a combination of these.

9.5.1 Flat Bottom

1. Tank bottom, which is flat and stable under varying liquid loads, will have no effect on tank capacity. If necessary, its depression due to varying load may be calculated by known geometric principles. So in either case, there will be no threat in making of the correct gage table.
2. Tank bottom, which has irregular slope and is unstable such that its correct capacity cannot be determined conveniently from linear measurements alone, will require either liquid calibration or a floor survey.

9.5.1.1 Liquid Calibration

The procedure in carrying out the liquid calibration is to fill into the tank, quantities of known volume of water or other non-volatile liquid until the datum point is just covered and the total volume of liquid is recorded. Additional known volume of water or liquid is added till the highest point of the bottom is just covered. This may be done in one or several stages depending upon the irregularity in slop of the bottom. Dip reading and volume added is recorded at each stage. The dip step of about 5 cm seems to be all right. For liquid calibration, a calibrated positive displacement meter may also be used as described in Chapter 8. Volumes for the tank calibration gage-table above this point are computed from linear measurements.

9.5.1.2 Floor survey

The floor survey consists of recording levels by means of a dumpy level with the help of spirit level the cross-sections and longitudinal sections of the entire floor are computed. The levels when plotted will define the profile and the geometric pattern of the bottom of the tank. Thus the capacity of the tank is finally calculated.

During the tank bottom calibration the difference in height between datum plate and the bottom of the bottom course, wherever is possible, are recorded.

9.5.2 Bottom with Conical, Hemispherical, Semi-ellipsoidal or having Spherical Segment

Volume of the tank bottoms conforming to geometrical shapes may either be computed from

(1) Linear dimensions,
(2) Measurement of liquid volumes by filling in small steps or
(3) By floor survey.

Any appreciable differences in shape affecting the volume such as knuckles, etc. are measured and recorded in sufficient details to permit computation of the true volume. In either of these methods good number of measurements should be taken at different points of the bottom.

9.6 MEASUREMENT OF TILT OF THE TANK

Normally the storage tank should be vertical, however due to several reasons it may not be truly vertical. So the measurements are taken to find out the tilt, if it exists. This can be conveniently done by suspending a plumb line from the top and measuring the offset at the bottom as shown in Figure 9.4. The angle of tilt will then be the offset divided by the tank height. Offset is the distance from the bottom of the point where plumb line is supposed to touch the levelled ground. The distance is measured along the radius of the circular shell. Alternatively, if the tank is calibrated by floor survey with the help of dumpy level, the tilt can be estimated by taking readings along the periphery of the tank bottom. In any of these methods, a sufficient number of measurements are taken at the different points of the circumference to determine the correct offset.

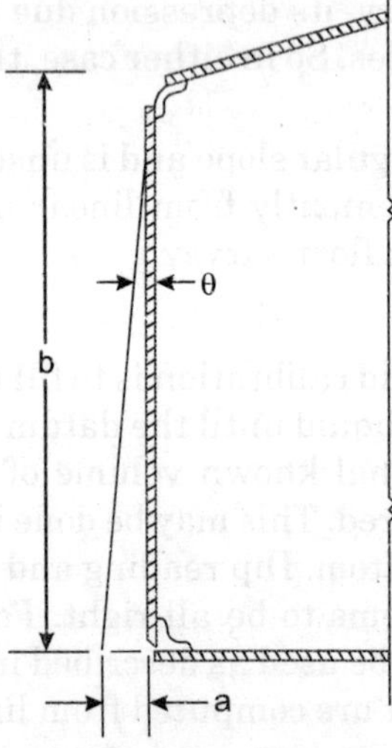

Fig. 9.4 Tilt of the tank.

9.7 FLOATING ROOF TANKS

9.7.1 Liquid Calibration for Displacement by the Floating-roof

Corrections for displaced volume because of the weight of the roof and deadwood associated with it are accounted for while preparing gage table (calibration table).

If the weight of the floating-roof is accurately known then from the density of liquid at the temperature of measurement, one can find out the volume of displaced liquid.

Alternatively, displacement due to the floating-roof and deadwood may be determined by admitting liquid till the dip reading is just below the lowest point of the roof. Accurately known volume of liquid is then admitted to the tank and corresponding dip reading are taken and recorded at a number of suitable intervals till the roof becomes fully liquid borne. Record the density and temperature of the liquid used.

1. It is advisable to use the liquid of same density the tank is supposed to store. If it is not practical, water may be used and suitable corrections are applied. Use of water has an advantage that its density versus temperature relation is very well known.
2. During liquid calibration any space under the roof that may trap air or gas should be vented to the atmosphere.
3. Before liquid calibration the height of the lowest joint of the roof with reference to datum is recorded, wherever possible.
4. To asses the point at which roof becomes fully liquid/water borne the following procedure may be adopted:

With roof resting fully on its support, paint four short horizontal white lines about 3 cm wide on the tank sides in such a position that can be viewed from some definite point, their lower edges are just above four similar lines marked on the roof edges or shoes. Then slowly pump liquid into the tank; when all the point markings are seen to have moved upwards, at this position, the roof becomes liquid borne. Take the dip at this point and record the reading. Alternatively, from some chosen view- point on the dipping platform, note the position of the roof against rivet heads on the vertical seam or other markings on the tank shell instead of paint marks. In both cases extend the points of reference round the greater part of the tank wall and see movement relative to all points.

9.7.1.1 Weight of Floating Roof

The floating of the entire roof will include weight of roof plus half the weight of the rolling ladder and other hinged and flexibly supported accessories that are carried up and down in the tank with roof.

9.7.1.2 Fixed Deadwood of Roof

Fixed deadwood of roof is measured as described in section 9.4 on deadwood calibration. The drain lines and other accessories attached to the underside of the roof are treated as fixed deadwood in position they occupy when the roof is at rest on its supports.

When all or part of the weight of the roof is resting on its supports, the roof is deadwood itself and as the liquid level rises around the roof, its geometric shape will determine how it should be deducted. The geometric shape may be taken from the fabricators drawings or measured in the field with the aid of an engineer's level, while the roof is resting on its support.

9.7.2 Variable Volume Roofs

Roofs with flexible membrane such as lifter, breather or balloon, require special deadwood measurements for roof parts that are sometimes submerged. When these parts such as columns are fixed relative to the tank shell, they should be measured as deadwood in the usual way. When these parts move with the roof and hang down into the liquid, these are taken as fixed deadwood, with the roof in the lowest position. Details may be secured from the fabricators drawings or measured in the field.

Some variable volume roofs have flexible members, which may float on the surface when the membrane is deflated and the liquid level is high. The floating weight of the membrane displaces a small volume of liquid. Data on the floating weight should be secured from the fabricators drawing and supplemented, if necessary by measurements in the field.

Some variable volume roofs have liquid seal troughs or other appurtenances, which make the upper outside part of the shell inaccessible for outside circumference measurements. Liquid calibration of this portion of the shell may be made or (1) Dimensions may be taken from the fabricator's drawings or (2) the highest measurable circumference may be used as a basis for the portion of the tank that cannot be measured. The method used, should be indicated on the gage table (calibration table).

9.8 CALIBRATION BY INTERNAL MEASUREMENTS

9.8.1 Outline of the Method

The method is based on the measurement of internal diameters.

1. Diameters are measured only after the tank has been filled at least once in the present location with the liquid to its working capacity or with water to its equivalent height. A time of 24 hours are allowed for setting.
2. The stipulated number of internal diameters is obtained in the following way:
   - The measurement is taken between diametrically opposite points at the following levels on each course. The minimum number of diameters is two at each level at right angles to each other.

For riveted tanks

- At 10% of the height of the exposed portion of each course above the level of the top of the bottom angle iron of the tank and above the upper edge of each horizontal over-lap between courses and below the level of the lower part of the top angle iron of the tank.
- At 10% of the height of the exposed portion of each course, below the level of the lower edge of each horizontal over-lap between courses and below the level of the lowest part of the top angle iron of the tank.

For welded tanks

- Two levels, are selected, each is at 20% of the height of the exposed portion of the respective course away from the angle irons or seams.

For all tanks

- No measurement should be taken nearer than the 30 cm to any vertical seam.

3. Where practical, outer circumference is also measured at approximately same height at which internal diameters was measured. Measurement of thickness of plates gives internal diameter. The values when compared will serve a good method of estimating accuracy of measurement and compatibility of instruments used.
4. It may be necessary in practice to refer all tanks dips to a datum point other than the datum point used for tank calibration (gage table). If so, the difference between the two datum points is also determined either by normal survey method or by other suitable means.
5. The overall height is measured using dip-tape with dip-weight from the dipping datum point mentioned in 4 above to the reference point (the dipping reference point) on the dip hatch. This overall height is recorded and also marked on the tank at the dip hatch.

9.8.2 Equipment

Equipment needed for internal measurements is practically the same as given in section 8.9.2 of previous chapter. The tape should be greased well before use and grease is evened out before use.

9.8.2.1 Diameter Measurements

1. All diameter measurements in this case also are done with a tape under a tension of (45 ± 5) N i.e. (4.5 ±0.5) kg. The tension in the tape is applied and indicated by dynamometer. The tension is necessary as all steel tapes are calibrated with the aforesaid tension.
2. All measurements are recorded as read, if one reads up to 1 mm then all readings must be recorded within 1 mm, even if 1mm is only an estimated value. Also do not exclude the length of the dynamometer.
3. The dynamometer length will be measured when it is showing a tension of 45 N or 4.5 kg before it is put into use. Its length is also checked during its use in diameter measurement.
4. The internal diameter will be measured in locations as given in point 2 of 9.8.1.
5. If for any reason it is impracticable to take measurements at the prescribed position, then the diameters are measured as close to the prescribed position as possible. Select every location at least 30 cm away from the seam.
6. If measurements have been taken at non-prescribed levels, the position of level should be recorded together with reasons for leaving the prescribed level.

9.8.2.2 Procedure to Carryout Measurements

1. Measurements are taken with the zero end of the steel tape attached to the dynamometer, one operator placing the dynamometer on the predetermined point and second operator placing the other ruler end on the point diametrically opposite. Zero of the ruler coincides with the shell. The tape is then pulled along the ruler until the requisite tension is applied, which sometimes is indicated by sounding of a buzzer. The graduated side of the tape is kept facing upward. The relative position of the tape and ruler is maintained by a firm grip until the reading of the tape and ruler are read. The ruler is then removed. Total measurement is sum of the readings on the tape and on the ruler. The operation is repeated at various positions at which

measurements are required throughout the tank. The measurements are recorded clearly in white chalk on the steel plates to indicate that measurements are taken there. Here one can notice that in each case, readings shown on the tape will be less by the length of the dynamometer. So length of the dynamometer indicating 45 N or 4.5 kg is to be finally added to the mean values of the diameter shown by the tape.

2. Each measurement of diameter is recorded to the nearest mm.
3. All other measurements are carried out in the same manner as are carried out in external measurement procedure.

9.9 COMPUTATION OF CAPACITY OF A TANK AND PREPARING GAUGE TABLE FOR VERTICAL STORAGE TANK

The major portion of the tank is cylindrical. So relation between circumference and area of cross-section of the tank may be written in terms of circumference as follows

Circumference $C = 2\pi r$ or $C^2 = 4\pi^2 r^2$

Also Area of cross-section $S = \pi r^2$.

Dividing we get $S = C^2/4\pi$

This will become

Volume of cylinder per unit length = $C^2/4\pi$ in m^3, if C and unit length are taken in metres.

Or = $1000.C^2/4\pi$ dm^3

This will become $10\ C^2/4\pi$ dm^3 if the height (unit length) is taken as one cm.

OR Volume per cm = $10\ C^2/4\pi\ dm^3/cm = 0.795778\ C^2\ dm^3/cm$

Here it should be remembered that C is still in metres

9.9.1 Principle of Preparing Gauge Table (Calibration Table)

1. The intervals of dip at which the tables are made should not be too great otherwise there will be inaccuracies in interpolating the value of volume at a particular dip not listed in the table. Normally 5 cm interval is sufficient, along with a proportional table, calculated on the basis of average difference for the chosen interval. Interval of the proportional table should be in mm. Such table will be able to give volumes in dm^3 (litres). However for lap joints, the proportional parts table will be based on the average difference for each course separately. Levels affected by bottom irregularities and deadwood is not included in calculating the average difference in volume per unit depth used in preparing the proportional table. This table is not applicable for interpolations of these levels.
2. The tables may be set out more fully if greater speed in calculation is desired. But it should be remembered that the table set out in one page is quicker in use than the one occupying several pages.
3. It should be kept in mind that no liquid measurement requires better relative accuracy of one part in ten thousand. Commercial table never requires a fraction of litre; any table, which is able to calculate within one litre, is more than sufficient.
4. Keeping all these points in view, 5cm interval with difference table has been found to be acceptable.

A typical blank gauge table is given in Table 9.1.

Table 9.1A Gauge Table (Volume versus Dip)

Proportional table			*Main table*							
S.N.	*mm*	*dm³*	*cm*	*dm³*	*cm*	*dm³*	*cm*	*dm³*	*cm*	*dm³*
1	0		00		200		400		600	
2	1		05		05		05		05	
3	2		10		10		10		10	
4	3		15		15		15		15	
5	4		20		20		20		20	
6	5		25		25		25		25	
7	6		30		30		30		30	
8	7		35		35		35		35	
9	8		40		40		40		40	
10	9		45		45		45		45	
11	10		50		250		450		550	
12	11		55		55		55		55	
13	12		60		60		60		60	
14	13		65		65		65		65	
15	14		70		70		70		70	
16	15		75		75		75		75	
17	16		80		80		80		80	
18	17		85		85		85		85	
19	18		90		90		90		90	
20	19		95		95		95		95	
20	20		100		300		500		700	
21	20		05		05		05		05	
22	21		10		10		10		10	
23	22		15		15		15		15	
24	23		20		20		20		20	
25	24		25		25		25		25	
26	25		30		30		30		30	
27	26		35		35		35		35	
28	27		40		40		40		40	
29	28		45		45		45		45	
30	29		150		350		550		750	
31	30		55		55		55		55	
32	31		60		60		60		60	
33	32		65		65		65		65	
34	33		70		70		70		70	
35	34		75		75		75		75	
36	35		80		80		80		80	
37	36		85		85		85		85	
38	37		90		90		90		90	
39	38		95		95		95		95	

(Contd.)

40	39		200		400		600		800	
41	40									
42	41									
43	42									
44	43									
45	44									
46	45									
47	46									
48	47									
49	48									
50	49									
51	50									

Table 9.1B Gauge Table (Volume versus Dip)

Proportional table			*Main table*							
S.N.	*mm*	dm^3	*cm*	dm^3	*cm*	dm^3	*cm*	dm^3	*cm*	dm^3
1	0		800		1000		1200		1400	
2	1		05		05		05		05	
3	2		10		10		10		10	
4	3		15		15		15		15	
5	4		20		20		20		20	
6	5		25		25		25		25	
7	6		30		30		30		30	
8	7		35		35		35		35	
9	8		40		40		40		40	
10	9		45		45		45		45	
11	10		850		1050		1250		1450	
12	11		55		55		55		55	
13	12		60		60		60		60	
14	13		65		65		65		65	
15	14		70		70		70		70	
16	15		75		75		75		75	
17	16		80		80		80		80	
18	17		85		85		85		85	
19	18		90		90		90		90	
20	19		95		95		95		95	
20	20		900		1100		1300		1500	
21	20		05		05		05		05	
22	21		10		10		10		10	
23	22		15		15		15		15	
24	23		20		20		20		20	
25	24		25		25		25		25	

(Contd.)

26	25		30		30		30		30	
27	26		35		35		35		35	
28	27		40		40		40		40	
29	28		45		45		45		45	
30	29		950		1150		1350		1550	
31	30		55		55		55		55	
32	31		60		60		60		60	
33	32		65		65		65		65	
34	33		70		70		70		70	
35	34		75		75		75		75	
36	35		80		80		80		80	
37	36		85		85		85		85	
38	37		90		90		90		90	
39	38		95		95		95		95	
40	39		1000		1200		1400		1600	
41	40									
42	41									
43	42									
44	43									
45	44									
46	45									
47	46									
48	47									
49	48									
50	49									
51	50									

It may be noted that second and third columns represent proportional table.

9.10 CALCULATIONS

The mean diameter is the average of the separate tape measurements corrected for sag plus the length of dynamometer.

The procedure is:

Average of tape readings is obtained for each course by dividing the sum of all the readings by the number of measurements taken. Round off the average to the nearest 0.1 mm.

Correct the mean for sag; correction in this case is negative.

Add to the result the length of the dynamometer.

Apply the temperature correction as indicated in section 8.10 of previous chapter. If the reference temperature is 20ºC and we wish to prepare the calibration table at 15ºC, then a correction factor is (1-0.00009) is necessary, so multiply the corrected reading by 0.99991.

Calculate the open capacity of each course without giving any consideration to deadwood. In these calculations we assume that the course is a true cylinder of internal diameter equal to the mean of the measured values of internal diameters.

The open capacity of each course per cm is

$(\pi d^2/4)$ cm^3/cm, d the diameter is expressed in cm, or
$(\pi d^2/4)/1000$ dm^3 per cm
$= 0.000\ 7854\ d^2$dm^3 per cm, where d is still in cm

When the level or levels from which oil depths will be measured differ from the datum level from which the tank table is first prepared, correction for difference is made in the final calibration table.

9.11 DEADWOOD

1. The open capacity of each course is adjusted for any deadwood it contains.
2. The total volume of each piece of deadwood is calculated to the nearest dm^3. In this context, the term 'piece of deadwood' includes such items as the rivet heads in one line around the tank, taken collectively as a single 'piece' of deadwood.
3. The effect of small pieces of deadwood may be neglected provided
   (*i*) That it affects the gage table not exceeding 0.005% of the capacity of whole course.
   (*ii*) Any deadwood so neglected is distributed evenly or substantially so over the whole height of the course. In calculating the table however it will be permissible to include the effect of any deadwood, howsoever small it may be.

9.12 TANK BOTTOM

1. When the tank bottom is substantially horizontal, for example when the tank is build on a level concrete raft or steel structure, then bottom irregularities can be neglected.
2. When the tank bottom has been calibrated by measuring of suitable known volumes of liquid, the gage table for these levels is prepared from these measurements. The highest level and capacity shown in this part of the table so prepared will then be the datum level and capacity. From this level onward the rest of the table will be prepared by calculation as described above.

9.13 FLOATING ROOF TANKS

The gaUge table for the floating roof tanks will also be made as described above except the following modifications:

1. Allowance for deadwood is made as described in section 9.12.
2. The drain pipes and other accessories attached to the underside of the roof will be included as fixed deadwood in the positions they occupy when the roof is at rest on its supports. The position of these accessories should be specified in the gage table.
3. Two levels are defined both by an exact number of centimetres above the datum point from which dip readings are taken. The first level designated as A, will not be less than 4 cm and not more than 6 cm below the lowest point of the roof plates when the roof is at rest. The second level designated B will be not less than 4 cm and more than 6 cm above the free liquid surface, when the roof is at its lowest liquid borne position.

4. The floating weight of the entire roof includes weight of the roof plus half the weight of the rolling ladder and other hinged and flexibly supported accessories, which are carried up and down in the tank with the roof. The volume in dm^3 displaced by the roof weight can be calculated from.

Roof weight in kg/[density of stored liquid in kg per dm^3 at tank temperature].

This displacement, minus the volume of the deadwood already accounted for in 2 above, will be considered as an item of deadwood applicable to all levels above the point B. It will be either entered as such on a supplementary table or taken into account in the preparation of the final table as a deduction for deadwood at all levels above B. For level between A and B, the proportional of roof displacement is to be taken into account as deadwood. This may be calculated from the dimensions of the floating roof. These partial displacements may either be entered as such in the supplementary table as applicable for the levels in between A and B or taken in account in preparation of the final gage table. Alternatively, where measured quantities of liquid have been admitted to the tank and corresponding levels of free liquid surface determined by the dipping, the necessary adjustment to the tank capacity within the range of the levels A and B is computed from the data. The part of the table between A and B is marked "not accurate".

5. It is not feasible to allow in the tank table, for the effect of extraneous matter retained by the roof, varying friction of the roof shoes and varying immersion of roof supports.

9.14 COMPUTATION OF GAUGE TABLES IN CASE OF TANKS INCLINED WITH THE VERTICAL

9.14.1 Correction for Tilt

If the tank is inclined to angle θ, then effectively the vertical height H is given by:

$H = L\cos(\theta)$, where L is the length along the tank, which will be vertical height if tank is vertical.

$$H = L\cos(\theta), \text{ giving}$$
$$L = H\sec(\theta)$$

Hence volume in dm^3 per cm (unit length along the walls of the tank) becomes as

Volume per cm along the tank = $C^2\sec(\theta)/4\pi$

Or $0.795778C^2 \sec(\theta)$ dm^3 per cm.

Where C is the internal circumference in meters.

$$\text{Similarly the volume (in dm}^3\text{) per cm} = 0.0007854\,D^2\sec(\theta)$$
$$= 0.000\,7854\,D^2/(1+\theta^2/2)$$
$$= 0.000\,7854\,D^2\,(1+\theta^2/2)$$
$$\text{So error due to tilt} = 0.000\,7854\,D^2\,(1+\theta^2/2) - 0.000\,7854\,D^2$$
$$= 0.000\,7854\,D^2\,\theta^2/2$$
$$\text{So relative (fractional) error} = 0.000\,7854\,D^2\,\theta^2/2/[0.000\,7854\,D^2] = \theta^2/2$$

Or fraction error is $\theta^2/2$, giving percentage error as

Percentage error $=50 \times \theta^2$

It may be noted that

$$\theta = \text{Horizontal offset/ height of the tank.}$$

Here we see that if tilt in the tank is less than one part in fifty (horizontal offset/height of the tank) the error due to tilt will be less than 0.02%, which can be ignored. Similar correction is applied when the circumference is calculated by measuring circumference.

The value of π has been taken, as 3.1415962 and this value will be used in further calculations.

9.14.2 Example of Strapping Method

For the purpose of designating the different courses (rings) of the tank, courses have been numbered from the bottom course. Writing

C for external circumference;

Sc for step over correction; and

T for plate thickness, the data is tabulated below.

9.14.2.1 *Data Obtained by Measurements*

Course No.	*C*	*Sc*	*T*	*Internal heights of the courses*	
	m	*m*	*m*	*Individual cm*	*Cumulative cm*
8 Top	115.080	0.002	0.007	187.0	1475.0
8 Middle	115.075	0.002	0.007		
8 Bottom	115.086	0.002	0.007		
7 Top	115.125	0.002	0.007	179.0	1288.0
7 Middle	115.125	0.002	0.007		
7 Bottom	115.130	0.002	0.007		
6 Top	115.095	0.003	0.010	190.0	1109.0
6 Middle	115.091	0.003	0.010		
6 Bottom	115.092	0.003	0.010		
5 Top	115.145	0.004	0.013	179.0	919.0
5 Middle	115.160	0.004	0.013		
5 Bottom	115.162	0.004	0.013		
4 Top	115.085	0.010	0.013	191.0	740.0
4 Middle	115.088	0.010	0.013		
4 Bottom	115.094	0.010	0.013		
3 Top	115.175	0.010	0.016	178.0	549.0
3 Middle	115.176	0.010	0.016		
3 Bottom	115.172	0.010	0.016		
2 Top	115.077	0.013	0.018	191.0	371.0
2 Middle	115.085	0.013	0.018		
2 Bottom	115.071	0.013	0.018		
1Top	115.188	0.015	0.020	180	180.0
1 Middle	115.188	0.015	0.020		
1 Bottom	115.175	0.015	0.020		

9.14.2.2 Deadwood Data

Course No	Applicable height cm		Deadwood		Total Deadwood dm^3
	From	to	dm^3 (litres)	dm^3 per cm	
8	1466	1475	–350	–38.889	
8	1415	1466	–508	–9.961	
8	1350	1415	–2336	–35.938	–3194
8	1288	1350	Nil	Nil	
7	1109	1288	Nil	Nil	
6	919	1109	Nil	Nil	
5	740	919	Nil	Nil	
4	549	740	–195	–1.021	–195
3	371	549	–259	–1.455	–259
2	180	371	–309	–1.618	–309
1	107	180	–145	–1.986	
1	51	107	+59	+1.054	
1	46	51	–36	–7.200	
1	0	46	Nil	Nil	–122

Calculation to obtain corrected circumference for course No. 8 Top		
Measured external circumference at 20°C		115.080 m
Correction due to difference in reference temperatures		– 0.010 2 m
Calculated circumference at 15°C		115.069 8 m
Step-over correction		–0.002 0 m
Correction due to plate thickness	$7 \times 2\pi = 6.28316 \times 0.007$	–0.044 0 m
Corrected internal circumference C_i		115.023 8 m

Similar calculations are carried out for other circumferences.

9.14.2.3 Calculation of Open Capacity

Course No.	C_i *m*	C_{imean} *m*	*Open capacity of course*	
			Litres per cm	*In dm³ (litres)*
8 Top	115.023 8			
8 Middle	115.018 8	115. 024 1	10528.566	1968842
8 Bottom	115.029 8			
7 Top	115.068 8			
7 Middle	115.068 8	115.070 4	10537.053	1886132
7 Bottom	115.073 8			
6 Top	115.069 0			
6 Middle	115.065 0	115.066 7	10536.376	2001911
6 Bottom	115.066 0			
5 Top	115.049 1			
5 Middle	115.064 1	115.059 8	10535.112	1885785
5 Bottom	115.066 1			
4 Top	114.983 1			
4 Middle	114.986 1	114.987 1	10521.803	2009664
4 Bottom	114.992 1			
3 Top	115.050 3			
3 Middle	115 055 3	115. 052 3	10533.739	1875006
3 Bottom	115.051 3			
2 Top	114.940 7			
2 Middle	114.948 7	114.941 4	10513.441	2008067
2 Bottom	114.934 7			
1 Top	115.037 1			
1 Middle	115.037 1	115.032 8	10530.168	1895430
1 Bottom	115.024 1			

Here C_{imean} represents the mean value of corrected internal circumference for that course. Combining the data from the deadwood table, we can construct a gauge- table giving the heights, at which rate of volume changes, is given below. From this table one can calculate the gauge table in steps of 5 cm height.

9.14.2.4 Gauge Table at which Rate of Volume Changes with Respect of Height

S.No.	*Liquid dip cm*		*Interval*	*Deadwood dm^3/cm*	*Capacity dm^3/cm*	*Net capacity dm^3/cm*	*Capacity dm^3 (L)*
1	0	46	46	Nil	10530.02	10530.02	484381
1	46	51	5	−7.20	10530.02	10522.82	52614
1	51	107	56	+1.05	10530.02	10531.07	589740
1	107	180	73	−1.99	10530.02	10528.03	768547
2	180	371	191	−1.62	10513.441	10511.821	2007758
3	371	549	178	−1.46	10533.739	10532.279	1874746
4	549	740	191	−1.02	10521.803	10520.783	2009469
5	740	919	179	Nil	10535.112	10535.112	1885785
6	919	1109	190	Nil	10536.376	10536.376	2001911
7	1109	1288	179	Nil	10537.053	10537.053	1886132
8	1288	1350	62	Nil	10528.566	1528.566	94771
8	1350	1415	65	−35.94	10528.566	10492.626	682021
8	1415	1466	51	−9.96	10528.566	10518.606	536449
8	1466	1475	9	−38.89	10528.566	10489.676	94407

9.15 EXAMPLE OF INTERNAL MEASUREMENT METHOD

9.15.1 Data Obtained by Internal Measurement

To consider the computation of a gauge table from internal measurements, let us take the same tank for which gauge table has been prepared in section 9.14. That is positions and volumes of deadwood are same. Measure several diameters in the specified locations, and take the mean. Apply corrections for each mean value in three locations for difference in reference temperatures of tape and tank and step-over corrections. Take the mean of the corrected mean diameters to give the mean diameter for the course. Further apply sag correction and add the length of the dynamometer.

From these diameters calculate the open capacity of the course per cm (Cap/cm) and total open capacity (Cap) of the course. The results together with mean corrected internal diameter

D_{imean} are shown in table below.

Course no.	*Course height in cm*	D_{imean} *cm*	*Cap / cm* dm^3 */ cm*	*Cap* dm^3
8	187	3661.35	10528.60	1968848.2
7	179	3662.82	10537.06	1886133.7
6	190	3662.70	10536.36	2002041.4
5	179	3662.48	10535.09	1886008.4
4	191	3660.17	10521.8	2009663.8
3	178	3662.24	10533.74	1975082.2
2	191	3658.71	10513.42	2008063.2
1	180	3661.60	10530.04	1895407.2

9.15.2 Gauge Table Volume Versus Height

S.No.	*Liquid dip cm*		*Interval*	*Deadwood* dm^3/cm	*Capacity* dm^3/cm	*Net capacity* dm^3/cm	*Capacity* dm^3
1	0	46	46	Nil	10530.04	10530.04	484382
1	46	51	5	–7.20	10530.04	10522.84	52614
1	51	107	56	+1.05	10530.04	10531.09	589741
1	107	180	73	–1.99	10530.04	10528.05	768548
2	180	371	191	–1.62	10513.42	10511.80	2007754
3	371	549	178	–1.46	10533.74	10532.28	1874746
4	549	740	191	–1.02	10521.80	10520.78	2009469
5	740	919	179	Nil	10535.09	10535.09	1885781
6	919	1109	190	Nil	10536.36	10536.36	2001908
7	1109	1288	179	Nil	10537.06	10537.06	1886133
8	1288	1350	62	Nil	10528.60	10528.60	652773
8	1350	1415	65	–35.94	10528.60	10492.66	682023
8	1415	1466	51	–9.96	10528.60	10518.64	536451
8	1466	1475	9	–38.89	10528.60	10489.71	94407

9.16 DEFORMATION OF TANKS

When vertical tank is full then hydrostatic pressure on the lower courses will be more than on the upper one. The hydrostatic pressure will increase the tank diameter thereby reducing the height. The reduction in height of the courses will cause lowering of the upper part of the shell. Referring to Figure 9.5, the relative reduction of tank height is calculated by the formula given below with the following notations [3,4]:

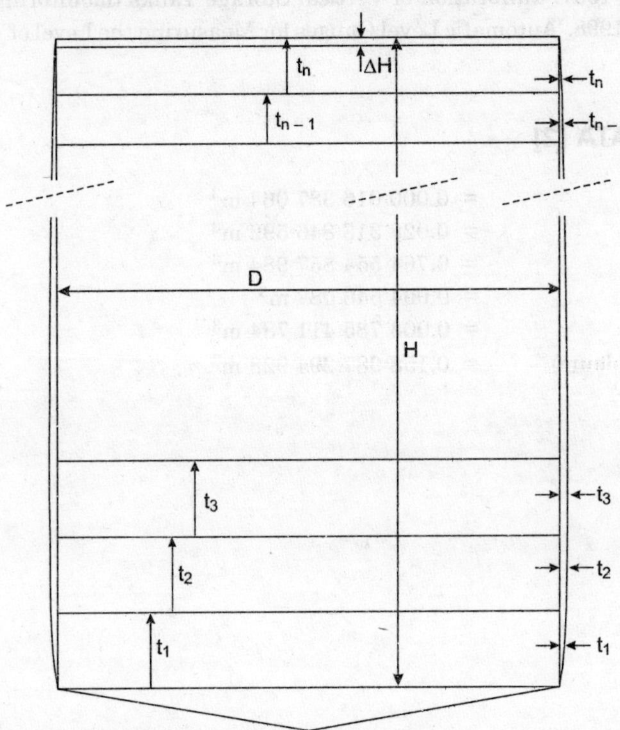

Figure 9.5

ρ – Density of the liquid expressed in kg/m^3

D – Diameter of the tank in m

E – Modulus of elasticity in N/m^2

μ – Poisson's ratio

h_n is height and t_n is the height of the nth course (ring) counted from bottom.

H is height of the tank in m

Then $\Delta H/H$ relative reduction in height is expressed as

$$\Delta H/H = (D\rho g/4\mu E)\ [H/t_1 + \{(H - h_1)^2/H\}\ (1/t_2 - 1/t_1) + \{(H - h_1 - h_2)^2/H\}(1/t_3 - 1/t_2) + \ldots\{\{H - (h_1 + h_2 + \ldots\ldots h_{n-1})\}^2/H\}\ (1/t_n - 1/t_{n-1})]$$

REFERENCES

[1] Manual of Weights and Measures, 1975. Directorate of Weights and Measures, Ministry of Industry and Civil Supplies, Govt. of India.
[2] Gupta S.V. A Treatise on Standards of Weights and Measures, 2003. Commercial Law Publishers (India) Private Limited.
[3] ISO 7507-6, 1997. Calibration of Vertical Storage Tanks (monitoring).
[4] OIML R 85-1998. Automatic Level Gauges for Measuring the Level of Liquid in Fixed Storage Tanks.

SOME USEFUL DATA [2]

1 cubic inch	= 0.000 016 387 064 m^3
1 cubic foot	= 0.028 316 846 592 m^3
1 cubic yard	= 0.764 554 857 984 m^3
1 gallon (UK)	= 0.004 546 087 m^3
1 gallon (USA)	= 0.003 785 411 784 m^3
1 barrel (for petroleum)	= 0.158 987 294 928 m^3.

10

CHAPTER

HORIZONTAL STORAGE TANKS

10.1 INTRODUCTION

Shape wise there is no difference in between the vertical and a horizontal storage tanks. In a vertical cylindrical tank, vertical section is a simple rectangle of fixed dimensions and horizontal section is a circle, which remains practically of same size, so having measured the internal diameter, we can straight away find out the volume for unit rise in the level of liquid. However, in case of a horizontal tank, horizontal cross-section of liquid in it is a rectangle of variable width. The width is zero for zero depth, becomes equal to diameter for a depth equal to the radius of the tank and will thereafter decrease with the increase in depth and will eventually be zero at a depth equal to the diameter of the tank. The vertical section, in this case, is a segment of circle, whose height is the depth of liquid present.

For total capacity of the tank, the method basically remains same, i.e. measure the diameter at the selected places by strapping and length of the tank. But in this case, as the horizontal cross-sectional area of liquid in the tank does not remain same, so the gauge table is not straight away in terms of volume versus height of the liquid in the tank. In case of horizontal tanks, instead of height in units of length it is the ratio of the height of the liquid in the tank to the diameter of the tank and volume is again in terms of ratio of the volume of the liquid present to the capacity of the tank are normally considered.

10.2 EQUIPMENT REQUIRED

The equipment required is same as described in section 8.9.2 of Chapter 8 on storage tanks.

10.3 STRAPPING LOCATIONS FOR HORIZONTAL TANKS

The tanks are made of a number of rings, which are joined together. To join these rings there are three options:

One end of one ring is joined to that of the other. For which it is necessary that faces of the rings are machined square to the axis and are reasonably flat. This is known as

- Butt-welding

Alternate rings are of such diameters that a ring just fit into it on each end. That is outer diameter of one is just equal to the inner diameter of the other. The portions of rings to be overlapped are machined properly so that these can fit into each other. To fix these rings, there are two choices, one is welding and another is riveting. Riveting will again require better-finished and polished ends of each ring. So we have

- Lapped welding and
- Riveting the overlapping parts of the rings.

10.3.1 Butt-welded Tank

A typical Butt-welded tank is shown in Figure 10.1. The points marked as *X* indicate the beginning and end of each course (ring). The points marked as *A* are the locations at which circumference is measured. The points *A* are at 20% and 80% of the length of each ring. *B* is the length of each ring between two consecutive points marked as *X*. *C* and *D* indicate the length of the straight flange and bulge of each head respectively.

So total length of the cylindrical portion of the tank is the sum of the length of the main cylinder and twice the length of each flange.

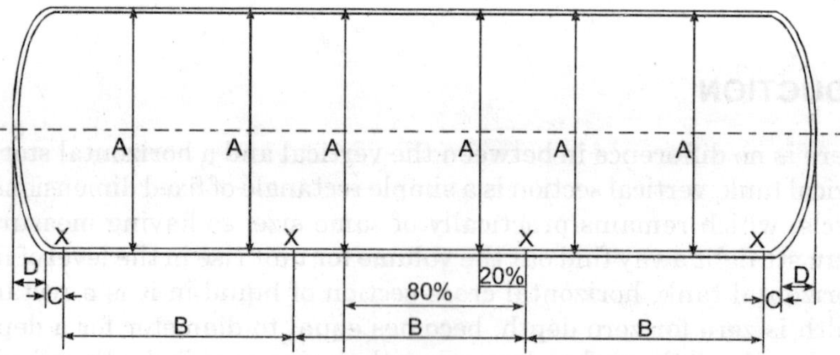

Figure 10.1 Locations for a butt-welded tank

10.3.2 Lap-welded Tank

Figure 10.2 gives a sketch of a lap-welded tank. A course (ring) of smaller diameter is fitted with courses (rings) of larger diameter on each side and then welded. The exposed portion of every smaller diameter course will be lesser than that of larger diameter. Points marked *A* indicate the locations for circumference measurements. These points are at 20% and 80% of the length of the course (ring). *D* is the length of overlap. *B* and *C* are exposed lengths of the rings. *E* and *F* are distances from tangent point and projection of head from joint respectively.

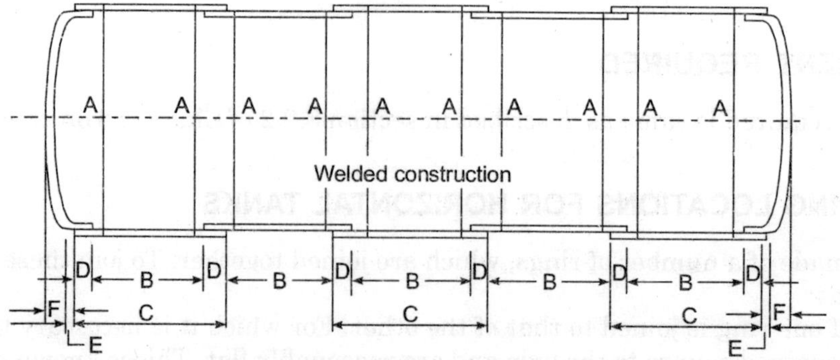

Figure 10.2 Locations for lap-welded tanks

10.3.3 Riveted Over Lap Tank

A lap riveted tank is indicated in the Figure 10.3. All notations are similar as in the lap-welded tank. Locations of circumference measurements are the same as in Figure 10.2 of each course (ring). *B* is un-lapped length of each course (ring). *C* is the exposed length of each course (ring). *D* is the width of each overlap and *E* distance from joints to tangent point and *F* is projection of each head from joint.

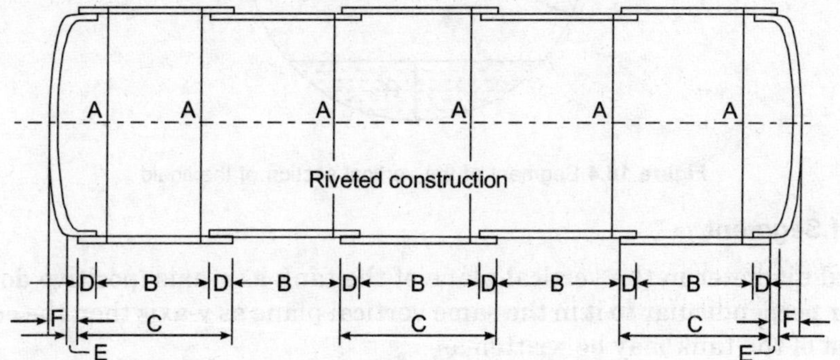

Figure 10.3 Locations for lap-welded riveted tanks

We can see that in case of lap-welded or riveted tanks

$$C = B + 2D$$

10.3.4 Locations

If the tank is formed of complete rings, circumferences are measured at two locations of each course (ring.), namely one at 20% and the other at 80% of the length of the course (ring). These positions, marked as *A*, are respectively shown in Figure 10.1 for butt-welded, in Figure 10.2 for lap-welded and in Figure 10.3 for riveted cylindrical tanks.

If the tank is composed of a bottom sheet and two longitudinal sheets to make the complete tank or alternatively has a bottom sheet and several partial rings, then circumference are measured at almost equally spaced four points. Measurements, for example, are carried out at 15%, 30%, 50%, and 85 % of the length. Countries using non-metric system take measurements at 1/8th, 3/8th, 5/8th and 7/8th of the whole length of the tank.

10.3.5 Precautions

If the measurements taken on successive rings indicate unusual variations or distortions in the rings, then additional measurements are taken at all the locations to satisfy the observers of his measurement capability.

10.4 PARTIAL VOLUME IN MAIN CYLINDRICAL TANK

We know that in case of horizontal cylindrical tank, rectangular horizontal cross-section is not of constant dimensions. So we consider the area of the vertical section of liquid inside the tank which is a segment of a circle of radius equal to the inside radius of the tank. The depth of liquid is *H* then height of the segment is also *H*. Please refer to Figure 10.4.

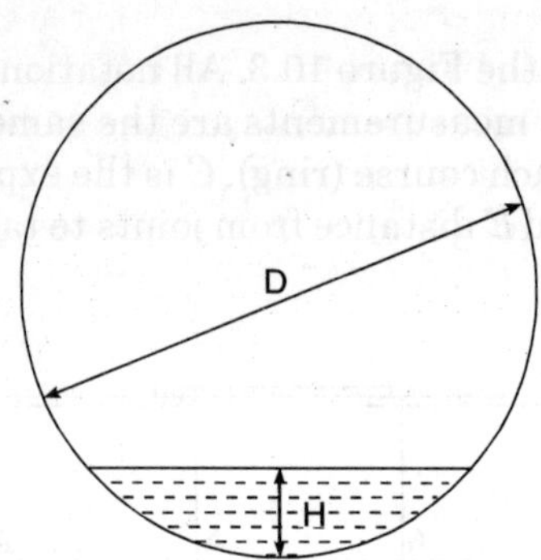

Figure 10.4 Segment of the vertical section of the liquid

10.4.1 Area of Segment

Taking vertical diameter in the vertical plane of the tank as x-axis (positive downward) and other diameter perpendicular to it in the same vertical plane as y-axis then the equation of the vertical section of the tank may be written as

$$x^2 + y^2 = a^2 \quad ...(1)$$

Where 'a' is the average radius of the tank at the course concerned.

So Area of the section containing liquid up to the depth H is given as

$$\text{Area} = 2\pi \int y dx$$

Limits of integration are from $a - H$ to a

$$= 2\pi \int (a^2 - x^2)^{1/2} dx$$

$$= \pi[a^2 \sin^{-1}(x/a) + x(a^2 - x^2)^{1/2}]$$

Here Lower limit is a – H and upper limit is a, substituting the limits, we get

Area of vertical section

$$= a^2[(\pi/2) - \sin^{-1}(1 - H/a) - (1 - H/a)(2H/a - H^2/a^2)^{1/2}] \quad ...(2)$$

Assuming the ends of the tank as flat and distance between the two ends as L.

Volume of the liquid present

$$= L a^2[\pi/2 - \sin^{-1}(1 - H/a) - \{(1 - H/a)(2H/a - H^2/a^2)^{1/2}\}]$$

One method is to find the radius a, which is deduced by measurements of circumference by strapping and thickness of the shell. Then apply the above formula and calculate the volume of liquid versus gage height H, which is a gauge table without deadwood. But in this way, every body has to do calculations for is own tank, which is not very convenient.

Let us find a relation applicable for every tank. The capacity of the shell of length L with no deadwood is $\pi r^2 L$.

Now we define K- as the ratio of volume of liquid present to the capacity of the shell of the tank, so K is given by:

$$K = [\pi/2 - \sin^{-1}(1 - H/a) - \{(1 - H/a)(2H/a - H^2/a^2)^{1/2}\}]/\pi$$

$$K = [\pi/2 - \sin^{-1}(1 - 2H/D) - \{(1 - 2H/D)(4H/D - 4H^2/D^2)^{1/2}\}]/\pi \quad ...(3)$$

Here D is diameter of the shell.

Expression for K is independent of the particular sets of units used in measurement or dimensions of an individual tank. Equation (3) is of universal nature and is applicable to any tank to give ratio of the volume of liquid contained in it to its capacity provided we know H/D.

The actual gage volume in the cylindrical portion of a particular tank, for given value of *H/D*, is the product of K and the capacity of the shell.

So *K* values for *H/D* from 0 to 0.5 have been calculated in steps of 0.001 with differences between successive values of *H/D* and have been given in Table 10.1 to 10.5. To economise on space, a set of 25 *K* values with consecutive differences corresponding to each value of *H/D* have been given in one column, and four such columns have been accommodated in one table. Differences given in the third sub-column of each column are used for interpolating the value of *K* if *H/D* is taken up to 4th decimal place.

10.5 PARTIAL VOLUMES IN THE TWO HEADS

However our gage table problem is not yet over, because the ends of the horizontal cylindrical tanks are never flat. In general there are three types of heads.

(1) Heads (Ends) are combination of spherical surface at each end (flange) and flat surface in the centre, (Knuckle head) Figure 10.5.
(2) Heads (Ends) are part of an ellipsoidal or spherical surface, Figure 10.6.
(3) Bumped heads, Figure 10.7.

To solve this problem, in this case also *K* the ratio of liquid volume contained to a certain gauge height '*H*' and the capacity of the head is theoretically calculated. Capacity of the head is the volume of liquid, which will fully fill the head between the internal surface of the head and an imaginary plane normal to the axis at the end of the cylindrical shell. *K* values for *H/D* from 0 to 0.5 in steps of 0.001 have been calculated for both types of heads, namely Ellipsoidal [2] and bumped heads [3]. The values are tabulated in Tables 10.6-10.10 and 10.11 to 10.15.

10.5.1 Partial Volumes for Knuckle Heads

The end is a surface generated by full revolution of the combination of a straight line and a circular arc about *GK* as axis. *E* is the centre of the circle with radius $r = EB$ and *EC*. The vertical section of the one head is shown in Figure 10.5.

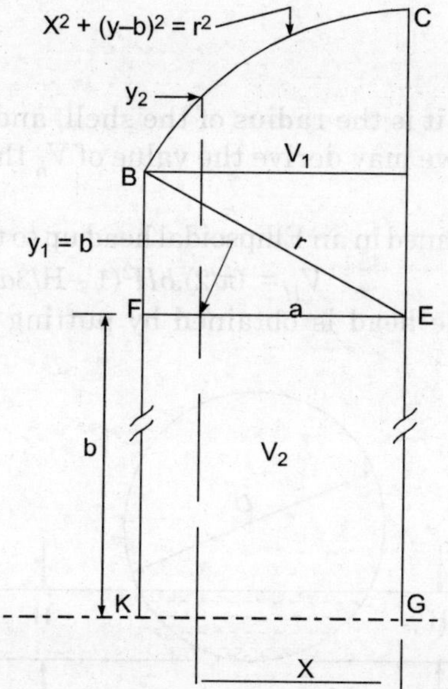

Figure 10.5 Vertical section – a knuckle head

Taking GC and GK as axes of coordinates, the equation of the circular arc BC with E as its centre and radius r is

$$x^2 + (y - b)^2 = r^2.$$

Where $b = GE$ is the ordinate of the centre.

The y_2 the ordinate of any point on the arc BC is given by

$$y_2 = b + (r^2 - x^2)^{1/2}.$$

Height of the straight portion is $y_1 = b$,

Volume V_1 of the portion obtained by the revolution of the arc BC about the x-axis is given by

$$V_1 = \pi\int(y_2{}^2 - y_1{}^2)dx,$$

The limits of x are from $x = 0$ to $x = GK = a$

$$= \pi\int[\{b + (r^2 - x^2)\}^2 - b^2]dx$$

$$= \pi\int[b^2 + 2b(r^2 - x^2)^{1/2} + r^2 - x^2 - b^2]dx$$

$$= \pi\int[r^2 - x^2 + 2b(r^2 - x^2)^{1/2}]dx$$

$$= \pi[r^2x - x^3/3 + 2b\{(x/2)(r^2 - x^2)^{1/2} + (r^2/2)\sin^{-1}(x/r)\}]$$

Limits of x are from 0 to a, substituting the values of limits we get

$$V_1 = \pi[r^2a - a^3/3 + ab(r^2 - a^2)^{1/2} + br^2\sin^{-1}(a/r)] \quad ...(5)$$

V_2 the volume of revolution, obtained by revolving the rectangle of length b and width a about GK -the x-axis, is given by

$$V_2 = \pi ab^2. \quad ...(6)$$

Total capacity of knuckle head is $V_{Head} = V_1 + V_2$

$$V_{Head} = \pi[ab^2 + r^2a - a^3/3 + ab(r^2 - a^2)^{1/2} + (br^2)\sin^{-1}(a/r)] \quad ...(7)$$

Remember there are two heads so total capacity of the tank

$$= \pi a^2L + 2\,V_{Head} \quad ...(8)$$

10.5.2 Ellipsoidal or Spherical Heads

Let us first consider an ellipsoidal head as from there we can straight away drive for the case of a spherical head.

10.5.2.1 Ellipsoidal Head

Let the semi-major axis be a, as it is the radius of the shell, and b is the minor axis. Then by simple mathematical relations, we may derive the value of V_h the volume of the liquid having the height H in one head.

Volume V_h of the liquid contained in an Ellipsoidal head up to the height H may be written as:

$$V_H = (\pi/2).bH^2(1 - H/3a) \quad ...(9)$$

Volume of the liquid in one head is obtained by putting $H = 2a = D$, where D is the diameter of the shell as

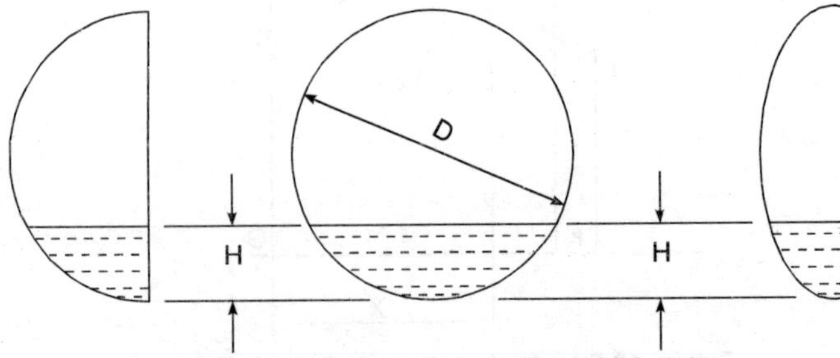

Figure 10.6 Spherical or ellipsoidal head

$$V_{2a} = (\pi/2)\, 4a^2b(1-2/3) = (\pi/2)\, (D^2)b(1/3) = (\pi/6)D^2b \qquad ...(10)$$

So Factor K is given as

$$K = (\pi/2).bH^2(1-H/3a)/(\pi/6)D^2b$$

$$(H/D)^2(3-2.H/D) \qquad ...(11)$$

It is again independent of the actual dimensions of the tank, so as before has been used to generate Tables 10.6 to 10.10.

10.5.2.2 Spherical Head

We observe from above that if we make $b = a$, it becomes the case of a spherical head. So V_h of liquid of height H in a spherical head is given by

$$V_h = (\pi/2).\, H^2a(1-H/3a)$$

So V, the volume of each head is obtained by putting $H = 2a$, giving us

$$V_{2a} = (\pi/2).4a^2a(1-2a/3a) = (\pi/6).D^3, \qquad ..(12)$$

Dividing V_h/V_{2a} we get

$$K = (H/D)^2(3-2H/D). \qquad ...(13)$$

We see that K factor for Ellipsoidal and spherical heads is same. So the Tables 10.6 to 10.10 will be used for spherical heads also.

10.5.3 Bumped (Dished Heads)

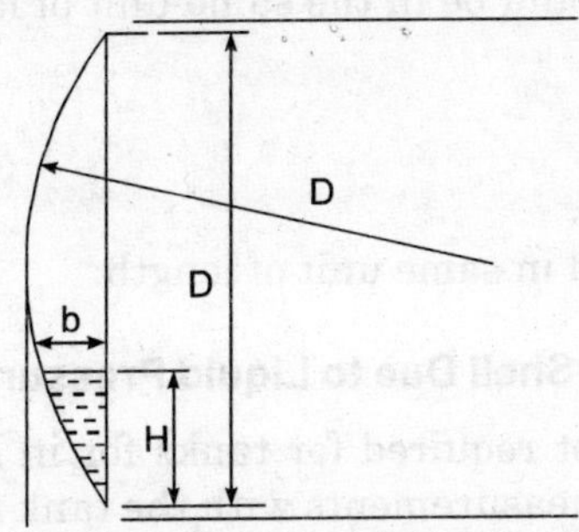

Figure 10.7 Bumped heads

Following a similar procedure as above, the values of K for bumped head versus H/D have been tabulated in Tables 10.11 to 10.15. H/D changes from 0.000 to 0.500 in steps of 0.001.

10.5.4 Volume in the Tank

It may be mentioned that in each case there are two heads (ends), hence the volume of liquid in a particular horizontal tank is given by

$$V = K_1.V_C + 2K_2.V_{Head}, \qquad ...(14)$$

Here V_C, K_1 and V_H, K_2 are respectively the volume and K factor of the cylindrical shell and one head.

10.5.5 Values of *K* for *H/D* > 0.5

For tanks which are more than half full, H/D will be greater than 0.5. K factor may be determined by the following formula.

$$K(H/D) = 1 - K(1 - H/D)$$

For example

For $H/D = 0.6$

$$K(0.6) = 1 - K(1 - 0.6) = 1 - K(0.4)$$

10.6 APPLICABLE CORRECTIONS

10.6.1 Tape Rise Corrections

For larger obstacles, which normally are less in number, step-over method is convenient to use, however for smaller obstacles like rivets in a riveted tank or but straps on the path of tape, step-over method is not convenient. For such obstacles the following formula may be applied.

10.6.1.1 For Butt Straps

Correction = $2N.T.W/D + 8(N.T/3)(T/D)1/2$,

Where

N is number of obstacles (butt straps)

W is width of a projection (strap)

T is rise of projection (strap)

D = nominal diameter of the tank shell

All linear measurements should be in the same unit of length, be it mm, cm, or even in inches or feet.

10.6.1.2 For Lap Joints

Correction = $(4\text{N}.T/3)(T/D)^{1/2}$,

All dimensions are expressed in same unit of length.

10.6.2 Expansion/Contraction of Shell Due to Liquid Pressure

Normally such corrections are not required for tanks for in house process control. Simpler method is to take circumference measurements with the tank full as well as when it is empty, Take the mean of two circumferences and use it in gage-table preparation.

For more accurate work, graphical solution given in [2] may be used.

10.6.3 Flat Heads Due to Liquid Pressure

To determine the increase in volume, the head is considered as a dished head with the radius of the dish from the measured bulge.

10.6.4 Effects of Internal Temperature on Tank Volume

The effects of internal temperature on tanks, which are kept under ambient conditions, this correction is not necessary. However for temperature-controlled tanks, the method of calculation is fully explained in API 2541[3].

10.6.5 Effects on Volume of Off Level Tanks

Let the axis of the horizontal tank is not horizontal but is inclined by an amount E/D, where E is elevation of one point over the other end of the tank axis. For E/D smaller than 0.01, the correction is negligible, for higher tilts, graphical method as enunciated in API 2551 [2] is used.

Table 10.1 (For Main Section of the Shell)

K* values for different values of *H/D

H/D	*K value*	*dif*	*H/D*	*K value*	*dif*	*H/D*	*K value*	*dif*	*H/D*	*K value*	*dif*
.001	0.000063	63	.026	0.007062	402	.051	0.019251	558	.076	0.034746	672
.002	0.000153	89	.027	0.007471	409	.052	0.019814	562	.077	0.035423	676
.003	0.000280	126	.028	0.007887	416	.053	0.020381	567	.078	0.036104	680
.004	0.000430	150	.029	0.008311	423	.054	0.020954	573	.079	0.036789	684
.005	0.000600	169	.030	0.008742	430	.055	0.021532	578	.080	0.037478	688
.006	0.000788	188	.031	0.009179	437	.056	0.022116	583	.081	0.038171	692
.007	0.000993	204	.032	0.009624	445	.057	0.022703	587	.082	0.038868	696
.008	0.001212	219	.033	0.010076	451	.058	0.023296	592	.083	0.039568	700
.009	0.001446	233	.034	0.010534	458	.059	0.023894	597	.084	0.040273	704
.010	0.001693	246	.035	0.010999	464	.060	0.024496	602	.085	0.040981	708
.011	0.001952	259	.036	0.011470	471	.061	0.025103	607	.086	0.041693	712
.012	0.002224	271	.037	0.011947	477	.062	0.025715	611	.087	0.042409	715
.013	0.002507	282	.038	0.012431	483	.063	0.026332	616	.088	0.043128	719
.014	0.002801	293	.039	0.012921	489	.064	0.026953	621	.089	0.043852	723
.015	0.003105	304	.040	0.013417	496	.065	0.027578	625	.090	0.044578	726
.016	0.003419	314	.041	0.013919	501	.066	0.028208	629	.091	0.045309	730
.017	0.003744	324	.042	0.014427	507	.067	0.028843	634	.092	0.046043	734
.018	0.004078	333	.043	0.014941	513	.068	0.029481	638	.093	0.046781	737
.019	0.004421	343	.044	0.015460	519	.069	0.030125	643	.094	0.047522	741
.020	0.004773	351	.045	0.015985	525	.070	0.030772	647	.095	0.048267	744
.021	0.005134	360	.046	0.016516	530	.071	0.031424	651	.096	0.049016	748
.022	0.005503	369	.047	0.017052	536	.072	0.032080	656	.097	0.049768	751
.023	0.005881	377	.048	0.017594	541	.073	0.032741	660	.098	0.050523	755
.024	0.006266	385	.049	0.018141	546	.074	0.033405	664	.099	0.051282	758
.025	0.006660	393	.050	0.018693	552	.075	0.034074	668	.100	0.052044	762

Table 10.2 (For Main Section of the Shell)

K* values for different values of *H/D

H/D	*K value*	*dif*	*H/D*	*K value*	*dif*	*H/D*	*K value*	*dif*	*H/D*	*K value*	*dif*
.101	0.052810	766	.126	0.072990	843	.151	0.094971	911	.176	0.118506	969
.102	0.053579	769	.127	0.073837	846	.152	0.095884	913	.177	0.119477	970
.103	0.054351	772	.128	0.074686	849	.153	0.096799	915	.178	0.120450	973
.104	0.055127	775	.129	0.075538	852	.154	0.097717	917	.179	0.121425	975
.105	0.055906	779	.130	0.076393	854	.155	0.098638	920	.180	0.122402	977
.106	0.056688	782	.131	0.077251	857	.156	0.099560	922	.181	0.123382	979
.107	0.057473	785	.132	0.078112	860	.157	0.100485	925	.182	0.124363	981
.108	0.058262	788	.133	0.078975	863	.158	0.101413	927	.183	0.125347	983
.109	0.059054	791	.134	0.079841	866	.159	0.102343	929	.184	0.126332	985
.110	0.059849	795	.135	0.080710	868	.160	0.103276	932	.185	0.127320	987
.111	0.060648	798	.136	0.081582	871	.161	0.104210	934	.186	0.128310	989
.112	0.061449	801	.137	0.082456	874	.162	0.105147	937	.187	0.129302	991
.113	0.062254	804	.138	0.083333	876	.163	0.106087	939	.188	0.130296	993
.114	0.063062	807	.139	0.084212	879	.164	0.107028	941	.189	0.131292	995
.115	0.063873	810	.140	0.085095	882	.165	0.107973	944	.190	0.132290	998
.116	0.064686	813	.141	0.085980	884	.166	0.108919	946	.191	0.133290	999
.117	0.065503	816	.142	0.086867	887	.167	0.109868	948	.192	0.134292	1001
.118	0.066323	820	.143	0.087757	890	.168	0.110818	950	.193	0.135296	1003
.119	0.067146	823	.144	0.088650	892	.169	0.111772	953	.194	0.136302	1005
.120	0.067972	826	.145	0.089545	895	.170	0.112727	955	.195	0.137310	1007
.121	0.068802	829	.146	0.090443	897	.171	0.113685	957	.196	0.138320	1009
.122	0.069633	831	.147	0.091344	900	.172	0.114645	959	.197	0.139331	1011
.123	0.070468	834	.148	0.092247	902	.173	0.115607	962	.198	0.140345	1013
.124	0.071306	837	.149	0.093152	905	.174	0.116571	964	.199	0.141361	1015
.125	0.072147	840	.150	0.094060	907	.175	0.117537	966	.200	0.142379	1017

Table 10.3 (For Main Section of the Shell)

K* values for different values of *H/D

H/D	*K value*	*dif*	*H/D*	*K value*	*dif*	*H/D*	*K value*	*dif*	*H/D*	*K value*	*dif*
.201	0.143398	1019	.226	0.169479	1064	.251	0.196605	1104	.276	0.224645	1138
.202	0.144419	1021	.227	0.170545	1065	.252	0.197709	1104	.277	0.225784	1138
.203	0.145443	1023	.228	0.171613	1067	.253	0.198816	1106	.278	0.226924	1140
.204	0.146468	1025	.229	0.172682	1069	.254	0.199923	1107	.279	0.228065	1141
.205	0.147495	1027	.230	0.173753	1070	.255	0.201033	1109	.280	0.229208	1142
.206	0.148524	1028	.231	0.174825	1072	.256	0.202143	1110	.281	0.230352	1143
.207	0.149555	1030	.232	0.175899	1074	.257	0.203255	1111	.282	0.231497	1145
.208	0.150587	1032	.233	0.176975	1075	.258	0.204369	1113	.283	0.232644	1146
.209	0.151622	1034	.234	0.178052	1077	.259	0.205484	1114	.284	0.233791	1147
.210	0.152658	1036	.235	0.179131	1078	.260	0.206600	1116	.285	0.234940	1148
.211	0.153696	1038	.236	0.180212	1080	.261	0.207717	1117	.286	0.236090	1150
.212	0.154736	1039	.237	0.181294	1082	.262	0.208836	1119	.287	0.237242	1151
.213	0.155778	1041	.238	0.182377	1083	.263	0.209957	1120	.288	0.238394	1152
.214	0.156822	1043	.239	0.183463	1085	.264	0.211079	1121	.289	0.239548	1153
.215	0.157867	1045	.240	0.184549	1086	.265	0.212202	1123	.290	0.240703	1154
.216	0.158914	1047	.241	0.185638	1088	.266	0.213326	1124	.291	0.241859	1156
.217	0.159963	1048	.242	0.186728	1089	.267	0.214452	1125	.292	0.243016	1157
.218	0.161013	1050	.243	0.187819	1091	.268	0.215579	1127	.293	0.244175	1158
.219	0.162065	1052	.244	0.188912	1092	.269	0.216708	1128	.294	0.245334	1159
.220	0.163119	1053	.245	0.190006	1094	.270	0.217838	1129	.295	0.246495	1160
.221	0.164175	1055	.246	0.191102	1095	.271	0.218969	1131	.296	0.247657	1161
.222	0.165233	1057	.247	0.192200	1097	.272	0.220101	1132	.297	0.248820	1162
.223	0.166292	1059	.248	0.193299	1098	.273	0.221235	1133	.298	0.249984	1164
.224	0.167353	1060	.249	0.194399	1100	.274	0.222370	1135	.299	0.251149	1165
.225	0.168415	1062	.250	0.195501	1101	.275	0.223507	1136	.300	0.252315	1166

Table 10.4 (For Main Section of the Shell)

K* values for different values of *H/D

H/D	*K value*	*dif*	*H/D*	*K value*	*dif*	*H/D*	*K value*	*dif*	*H/D*	*K value*	*dif*
.301	0.253483	1168	.326	0.283013	1194	.351	0.313134	1216	.376	0.343752	1234
.302	0.254652	1168	.327	0.284207	1194	.352	0.314350	1215	.377	0.344986	1233
.303	0.255822	1169	.328	0.285402	1195	.353	0.315566	1216	.378	0.346220	1234
.304	0.256992	1170	.329	0.286598	1195	.354	0.316783	1217	.379	0.347455	1235
.305	0.258164	1171	.330	0.287795	1196	.355	0.318002	1218	.380	0.348691	1235
.306	0.259337	1172	.331	0.288993	1197	.356	0.319220	1218	.381	0.349927	1236
.307	0.260511	1174	.332	0.290192	1198	.357	0.320440	1219	.382	0.351164	1236
.308	0.261686	1175	.333	0.291391	1199	.358	0.321661	1220	.383	0.352402	1237
.309	0.262862	1176	.334	0.292592	1200	.359	0.322882	1221	.384	0.353640	1238
.310	0.264040	1177	.335	0.293793	1201	.360	0.324104	1221	.385	0.354879	1238
.311	0.265218	1178	.336	0.294996	1202	.361	0.325326	1222	.386	0.356118	1239
.312	0.266397	1179	.337	0.296199	1203	.362	0.326550	1223	.387	0.357358	1239
.313	0.267577	1180	.338	0.297403	1204	.363	0.327774	1224	.388	0.358599	1240
.314	0.268759	1181	.339	0.298608	1204	.364	0.328999	1224	.389	0.359840	1241
.315	0.269941	1182	.340	0.299814	1205	.365	0.330224	1225	.390	0.361082	1241
.316	0.271124	1183	.341	0.301020	1206	.366	0.331451	1226	.391	0.362324	1242
.317	0.272309	1184	.342	0.302228	1207	.367	0.332678	1226	.392	0.363567	1242
.318	0.273494	1185	.343	0.303436	1208	.368	0.333905	1227	.393	0.364810	1243
.319	0.274681	1186	.344	0.304646	1209	.369	0.335134	1228	.394	0.366054	1244
.320	0.275868	1187	.345	0.305856	1210	.370	0.336363	1229	.395	0.367299	1244
.321	0.277056	1188	.346	0.307067	1210	.371	0.337593	1229	.396	0.368544	1245
.322	0.278246	1189	.347	0.308278	1211	.372	0.338823	1230	.397	0.369790	1245
.323	0.279436	1190	.348	0.309491	1212	.373	0.340054	1231	.398	0.371036	1246
.324	0.280627	1191	.349	0.310704	1213	.374	0.341286	1231	.399	0.372282	1246
.325	0.281819	1192	.350	0.311918	1214	.375	0.342518	1232	.400	0.373530	1247

Table 10.5 (For Main Section of the Shell)

K* values for different values of *H/D

H/D	*K value*	*dif*	*H/D*	*K value*	*dif*	*H/D*	*K value*	*dif*	*H/D*	*K value*	*dif*
.401	0.374778	1248	.426	0.406125	1259	.451	0.437711	1267	.476	0.469454	1272
.402	0.376026	1248	.427	0.407385	1259	.452	0.438978	1267	.477	0.470726	1271
.403	0.377275	1248	.428	0.408645	1259	.453	0.440246	1267	.478	0.471998	1271
.404	0.378524	1249	.429	0.409905	1260	.454	0.441514	1267	.479	0.473270	1272
.405	0.379774	1249	.430	0.411165	1260	.455	0.442782	1267	.480	0.474542	1272
.406	0.381024	1250	.431	0.412426	1260	.456	0.444050	1268	.481	0.475814	1272
.407	0.382275	1250	.432	0.413687	1261	.457	0.445318	1268	.482	0.477087	1272
.408	0.383526	1251	.433	0.414949	1261	.458	0.446587	1268	.483	0.478359	1272
.409	0.384778	1251	.434	0.416211	1261	.459	0.447856	1268	.484	0.479631	1272
.410	0.386030	1252	.435	0.417473	1262	.460	0.449125	1269	.485	0.480904	1272
.411	0.387283	1252	.436	0.418736	1262	.461	0.450394	1269	.486	0.482177	1272
.412	0.388536	1253	.437	0.419998	1262	.462	0.451663	1269	.487	0.483450	1272
.413	0.389789	1253	.438	0.421262	1263	.463	0.452933	1269	.488	0.484722	1272
.414	0.391044	1254	.439	0.422525	1263	.464	0.454203	1269	.489	0.485995	1272
.415	0.392298	1254	.440	0.423789	1263	.465	0.455473	1269	.490	0.487268	1272
.416	0.393553	1254	.441	0.425053	1264	.466	0.456743	1270	.491	0.488541	1272
.417	0.394808	1255	.442	0.426318	1264	.467	0.458013	1270	.492	0.489814	1273
.418	0.396064	1255	.443	0.427583	1264	.468	0.459284	1270	.493	0.491087	1273
.419	0.397320	1256	.444	0.428848	1265	.469	0.460555	1270	.494	0.492360	1273
.420	0.398577	1256	.445	0.430113	1265	.470	0.461825	1270	.495	0.493634	1273
.421	0.399834	1257	.446	0.431379	1265	.471	0.463096	1271	.496	0.494907	1273
.422	0.401091	1257	.447	0.432645	1265	.472	0.464368	1271	.497	0.496180	1273
.423	0.402349	1257	.448	0.433911	1266	.473	0.465639	1271	.498	0.497453	1273
.424	0.403607	1258	.449	0.435177	1266	.474	0.466910	1271	.499	0.498726	1273
.425	0.404866	1258	.450	0.436444	1266	.475	0.468182	1271	.500	0.500000	1273

Table 10.6 (Ellipsoidal and Spherical Heads)

K* values for different values of *H/D

H/D	*K*	*diff*	*H/D*	*K*	*diff*	*H/D*	*K*	*diff*	*H/D*	*K*	*diff*
.000	0.00000	3	.026	.0001993	155	.051	0.007538	293	.076	0.016450	424
.001	0.000003	9	.027	0.002148	160	.052	0.007831	298	.077	0.016874	429
.002	0.000012	15	.028	0.002308	166	.053	0.008129	304	.078	0.017303	434
.003	0.000027	21	.029	0.002474	172	.054	0.008433	309	.079	0.027737	439
.004	0.000048	27	.030	0.002646	177	.055	0.008742	315	.080	0.018176	444
.005	0.000075	33	.031	0.002823	183	.056	0.009057	320	.081	0.018262	449
.006	0.000108	38	.032	0.002006	189	.057	0.009377	325	.082	0.019069	454
.007	0.000146	45	.033	0.03195	194	.058	0.009702	330	.083	0.019523	460
.008	0.000191	51	.034	0.003389	200	.059	0.010032	336	.084	0.019983	464
.009	0.000242	56	.035	0.003589	206	.060	0.010368	341	.085	0.020447	469
.010	0.000298	62	.036	0.003795	211	.061	0.010709	346	.086	0.020916	474
.011	0.000360	69	.037	0.001006	216	.062	0.011055	352	.087	0.021390	479
.012	0.000429	74	.038	0.004222	222	.063	0.011407	357	.088	0.021689	484
.013	0.000503	80	.039	0.004414	228	.064	0.011764	362	.089	0.022353	489
.014	0.000583	85	.040	0.004672	233	.065	0.012126	367	.090	0.022842	494
.015	0.000668	92	.041	0.004905	239	.066	0.012493	372	.091	0.023336	499
.016	0.000760	97	.042	0.005244	244	.067	0.012865	378	.092	0.023835	503
.017	0.000857	130	.043	0.005388	250	.068	0.013243	383	.093	0.024338	509
.018	0.000960	109	.044	0.005638	255	.069	0.013626	388	.094	0.024847	513
.019	0.001069	115	.045	0.005893	260	.070	0.016014	393	.095	0.025360	519
.020	0.001084	120	.046	0.006153	266	.071	0.014407	399	.096	0.025870	523
.021	0.001304	127	.047	0.006419	272	.072	0.014806	403	.097	0.026402	528
.022	0.001531	132	.048	0.006691	277	.073	0.015209	409	.098	0.026930	532
.023	0.001563	137	.049	0.006968	282	.074	0.015618	413	.099	0.027462	538
.024	0.001700	144	.050	0.007250	288	.075	0.016031	419	.100	0.028000	542
.025	0.001844	149									

Table 10.7 (Ellipsoidal and Spherical Heads)

K* values for different values of *H/D

H/D	*K*	*diff*	*H/D*	*K*	*diff*	*H/D*	*K*	*diff*	*H/D*	*K*	*diff*
.101	0.028542	548	.126	0.043627	663	.151	0.061517	771	.176	0.082024	873
.102	0.029090	552	.127	0.044290	668	.152	0.062288	776	.177	0.082897	875
.103	0.029642	556	.128	0.044958	672	.153	0.063064	779	.178	0.083772	880
.104	0.030198	562	.129	0.045630	676	.154	0.063843	784	.179	0.084652	884
.105	0.030760	566	.130	0.046306	681	.155	0.064627	788	.180	0.085536	888
.106	0.031326	571	.131	0.046987	685	.156	0.065415	792	.181	0.086424	891
.107	0.031897	576	.132	0.047672	690	.157	0.066207	796	.182	0.087315	895
.108	0.032473	580	.133	0.048362	694	.158	0.067003	801	.183	0.088210	899
.109	0.033053	585	.134	0.049056	698	.159	0.067804	804	.184	0.089109	903
.110	0.033638	590	.135	0.049754	703	.160	0.068608	808	.185	0.090012	906
.111	0.034228	594	.136	0.050457	707	.161	0.069416	813	.186	0.090918	910
.112	0.034822	599	.137	0.051164	712	.162	0.070299	817	.187	0.091828	915
.113	0.035421	604	.138	0.051876	716	.163	0.071046	820	.188	0.092743	917
.114	0.036025	608	.139	0.052592	720	.164	0.071866	825	.189	0.093660	922
.115	0.036633	613	.140	0.053312	725	.165	0.072691	829	.190	0.094582	925
.116	0.037246	618	.141	0.054037	728	.166	0.073520	832	.191	0.095507	929
.117	0.037846	622	.142	0.054765	734	.167	0.074352	837	.192	0.096436	933
.118	0.038486	627	.143	0.055499	737	.168	0.075189	840	.193	0.097369	936
.119	0.039113	631	.144	0.056236	742	.169	0.076029	845	.194	0.098305	940
.120	0.039744	636	.145	0.056978	746	.170	0.076874	849	.195	0.099245	944
.121	0.040380	640	.146	0.057724	750	.171	0.077723	852	.196	0.100189	947
.122	0.041020	645	.147	0.058474	754	.172	0.078575	857	.197	0.101136	951
.123	0.041665	650	.148	0.059228	754	.173	0.079432	860	.198	0.102087	955
.124	0.042315	654	.149	0.059987	759	.174	0.080292	864	.199	0.103042	958
.125	0.042969	658	.150	0.060750	763	.175	0.081156	868	.200	0.104000	962

Table 10.8 (Ellipsoidal and Spherical Heads)

K* values for different values of *H/D

H/D	*K*	*diff*	*H/D*	*K*	*diff*	*H/D*	*K*	*diff*	*H/D*	*K*	*diff*
.201	0.104962	965	.226	0.130142	1051	.251	0.157376	1130	.276	0.186479	1200
.202	0.105927	969	.227	0.131193	1054	.252	0.158506	1132	.277	0.187679	1203
.203	0.106896	973	.228	0.132247	1058	.253	0.159638	1136	.278	0.188882	1206
.204	0.107869	976	.229	0.133305	1061	.254	0.160774	1138	.279	0.190088	1208
.205	0.108845	979	.230	0.134366	1064	.255	0.161912	1142	.280	0.191296	1211
.206	0.109824	984	.231	0.135430	1068	.256	0.163054	1144	.281	0.192507	1213
.207	0.110808	986	.232	0.136498	1070	.257	0.164198	1147	.282	0.193720	1217
.208	0.111794	990	.233	0.137568	1074	.258	0.135345	1150	.283	0.194937	1218
.209	0.112784	994	.234	0.138642	1077	.259	0.166495	1153	.284	0.196155	1222
.210	0.113778	997	.235	0.139719	1080	.260	0.167648	1156	.285	0.197377	1224
.211	0.114775	1001	.236	0.140799	1088	.261	0.168804	1159	.286	0.198601	1226
.212	0.115776	1004	.237	0.141882	1087	.262	0.169963	1161	.287	0.199827	1229
.213	0.116780	1007	.238	0.142969	1090	.263	0.171124	1165	.288	0.201056	1232
.214	0.117787	1011	.239	0.144059	1093	.264	0.172289	1167	.289	0.202288	1234
.215	0.118798	1015	.240	0.145152	1096	.265	0.173456	1170	.290	0.203522	1237
.216	0.119813	1017	.241	0.146248	1099	.266	0.174626	1173	.291	0.204759	1239
.217	0.120830	1022	.242	0.147347	1102	.267	0.175799	1175	.292	0.205998	1241
.218	0.121852	1024	.243	0.148449	1105	.268	0.176974	1179	.293	0.207239	1245
.219	0.122876	1028	.244	0.149554	1109	.269	0.178153	1181	.294	0.208484	1246
.220	0.123904	1031	.245	0.150663	1111	.270	0.179334	1184	.295	0.209790	1249
.221	0.124935	1035	.246	0.151774	1114	.271	0.180518	1187	.296	0.210979	1252
.222	0.125970	1038	.247	0.152888	1118	.272	0.181705	1189	.297	0.212231	1254
.223	0.127008	1041	.248	0.154006	1121	.273	0.182894	1192	.298	0.213485	1256
.224	0.128049	1045	.249	0.155127	1123	.274	0.184086	1195	.299	0.214741	1259
.225	0.129094	1048	.250	0.156250	1126	.275	0.185281	1198	300	0.216000	1261

Table 10.9 (Ellipsoidal and Spherical Heads)

K* values for different values of *H/D

H/D	*K*	*diff*	*H/D*	*K*	*diff*	*H/D*	*K*	*diff*	*H/D*	*K*	*diff*
.301	0.217261	1264	.326	0.249536	1319	.351	0.283116	1368	.376	0.317813	1409
.302	0.218525	1266	.327	0.250855	1322	.352	0.284484	1369	.377	0.319222	1410
.303	0.219791	1268	.328	0.252177	1323	.353	0.285853	1371	.378	0.320632	1411
.304	0.221059	1271	.329	0.253500	1326	.354	0.287224	1373	.379	0.322043	1413
.305	0.222330	1273	.330	0.254826	1328	.355	0.288597	1375	.380	0.323456	1414
.306	0.223608	1275	.331	0.256154	1329	.356	0.289972	1376	.381	0.324870	1416
.307	0.224878	1278	.332	0.257483	1332	.337	0.291348	1378	.382	0.326286	1417
.308	0.226156	1280	.333	0.258815	1334	.358	0.292726	1380	.383	0.327703	1419
.309	0.227436	1282	.334	0.260149	1335	.359	0.294106	1382	.384	0.329122	1420
.310	0.228718	1285	.335	0.261484	1338	.360	0.295488	1383	.385	0.330542	1421
.311	0.230003	1286	.336	0.262822	1339	.361	0.296871	1385	.386	0.331963	1423
.312	0.231289	1289	.337	0.264161	1342	.362	0.298256	1387	.387	0.333386	1424
.313	0.232578	1292	.338	0.265503	1344	.363	0.299643	1388	.388	0.324810	1425
.314	0.233870	1293	.339	0.266847	1345	.364	0.301031	1390	.389	0.336235	1427
.315	0.235163	1296	.340	0.268192	1347	.365	0.302421	1391	.390	0.337662	1428
.316	0.236459	1298	.341	0.269539	1350	.366	0.303812	1393	.391	0.339090	1429
.317	0.237757	1300	.342	0.270889	1351	.367	0.305205	1395	.392	0.340519	1431
.318	0.239057	1302	.343	0.272240	1353	.368	0.306600	1396	.393	0.341950	1432
.319	0.240359	1305	.344	0.273592	1355	.369	0.307996	1398	.394	0.343382	1433
.320	0.241664	1307	.345	0.274948	1357	.370	0.309394	1399	.395	0.344815	1435
.321	0.243971	1309	.346	0.276305	1358	.371	0.310793	1401	.396	0.346250	1435
.322	0.244280	1311	.347	0.277663	1361	.372	0.312194	1403	.397	0.347685	1437
.323	0.245591	1313	.348	0.279024	1362	.373	0.313597	1404	.398	0.349122	1439
.324	0.246904	1315	.349	0.280386	1364	.374	0.315001	1405	.399	0.350561	1439
.325	0.248219	1317	.350	0.281750	1366	.375	0.316406	1407	.400	0.352000	1441

Table 10.10 (Ellipsoidal and Spherical Heads)

K* values for different values of *H/D

H/D	*K*	*diff*	*H/D*	*K*	*diff*	*H/D*	*K*	*diff*	*H/D*	*K*	*diff*
.401	0.353441	1441	.426	0.389811	1468	.451	0.426735	1486	.476	0.464028	1496
.402	0.354882	1443	.427	0.391279	1468	.452	0.428221	1487	.477	0.465524	1497
.403	0.356325	1444	.428	0.392747	1469	.453	0.429708	1487	.478	0.467021	1498
.404	0.357769	1446	.429	0.394216	1470	.454	0.431195	1487	.479	0.468519	1497
.405	0.359215	1447	.430	0.395686	1471	.455	0.432682	1488	.480	0.470016	1498
.406	0.360662	1447	.431	0.397157	1472	.456	0.434170	1489	.481	0.471514	1498
.407	0.362109	1448	.432	0.398629	1473	.437	0.435659	1489	.482	0.473012	1498
.408	0.363557	1450	.433	0.400102	1473	.458	0.437148	1490	.483	0.474510	1498
.409	0.365007	1451	.434	0.401575	1474	.459	0.438638	1490	.484	0.476008	1499
.410	0.366458	1452	.435	0.403049	1475	.460	0.440128	1491	.485	0.477507	1498
.411	0.367910	1453	.436	0.404524	1476	.461	0.414619	1491	.486	0.479005	1499
.412	0.369363	1454	.437	0.406000	1477	.462	0.443110	1491	.487	0.480504	1499
.413	0.370817	1455	.438	0.407477	1477	.463	0.444601	1492	.488	0.482003	1500
.414	0.372272	1456	.439	0.408954	1478	.464	0.446093	1493	.489	0.483503	1499
.415	0.373728	1457	.440	0.410432	1479	.465	0.447586	1493	.490	0.485002	1499
.416	0.375185	1459	.441	0.411911	1479	.466	0.449079	1493	.491	0.486501	1500
.417	0.376644	1459	.442	0.413390	1480	.447	0.450572	1494	.492	0.488001	1500
.418	0.378103	1460	.443	0.414870	1481	.468	0.452066	1494	.493	0.489501	1499
.419	0.379563	1461	.444	0.416351	1482	.469	0.453560	1494	.494	0.491000	1500
.420	0.381024	1462	.445	0.417833	1482	.470	0.455054	1495	.495	0.492500	1500
.421	0.382486	1463	.446	0.419315	1483	.471	0.456549	1495	.496	0.494000	1500
.422	0.383949	1464	.447	0.420798	1483	.472	0.458044	1495	.497	0.495500	1500
.423	0.385413	1465	.448	0.422281	1484	.473	0.459539	1496	.498	0.497000	1500
.424	0.386878	1466	.449	0.423765	1485	.474	0.461035	1496	.499	0.498500	1500
.425	0.388344	1467	.450	0.425250	1485	.475	0.462531	1497	500	0.500000	1500

Table 10.11 (Bumped Head)

K* values for different values of *H/D

H/D	*K*	*diff*	*H/D*	*K*	*diff*	*H/D*	*K*	*diff*	*H/D*	*K*	*diff*
.000	0.00000		.026	0.00061	6	.051	0.00315	15	.076	0.00827	26
.001	0.00000	0	.027	0.00067	5	.052	0.00330	16	.077	0.00853	27
.002	0.00000	0	.028	0.00072	7	.053	0.00346	16	.078	0.00880	27
.003	0.00000	1	.029	0.00079	7	.054	0.00362	17	.079	0.00907	28
.004	0.00001	0	.030	0.00086	7	.055	0.00379	17	.080	0.00935	28
.005	0.00001	10	.031	0.00093	8	.056	0.00396	17	.081	0.00963	29
.006	0.00002	1	.032	0.00101	8	.057	0.00413	18	.082	0.00992	29
.007	0.00002	1	.033	0.00109	8	.058	0.00431	18	.083	0.01021	30
.008	0.00003	2	.034	0.00117	9	.059	0.00449	19	.084	0.01051	30
.009	0.00004	1	.035	0.00126	9	.060	0.00468	19	.085	0.01081	31
.010	0.00006	2	.036	0.00135	9	.061	0.00487	19	.086	0.01112	31
.011	0.00007	2	.037	0.00144	10	.062	0.00506	20	.087	0.01143	31
.012	0.00009	2	.038	0.00154	10	.063	0.00526	21	.088	0.01174	32
.013	0.00011	3	.039	0.00164	10	.064	0.00547	20	.089	0.01206	33
.014	0.00013	2	.040	0.00174	11	.065	0.00567	22	.090	0.01239	33
.015	0.00016	3	.041	0.00185	11	.066	0.00589	21	.091	0.01272	34
.016	0.00018	3	.042	0.00196	12	.067	0.00610	23	.092	0.01306	34
.017	0.00021	4	.043	0.00208	12	.068	0.00633	22	.093	0.01340	34
.018	0.00024	4	.044	0.00220	12	.069	0.00655	23	.094	0.01374	35
.019	0.00028	4	.045	0.00232	13	.070	0.00678	24	.095	0.01409	36
.020	0.00032	4	.046	0.00245	13	.071	0.00702	24	.096	0.01445	35
.021	0.00036	4	.047	0.00258	14	.072	0.00726	25	.097	0.01480	37
.022	0.00040	5	.048	0.00272	14	.073	0.00751	25	.098	0.01517	37
.023	0.00045	5	.049	0.00286	14	.074	0.00776	25	.099	0.01554	37
.024	0.00050	5	.050	0.00300	15	.075	0.00801	26	.100	0.01591	38
.025	0.00055	6									

Table 10.12 (Bumped Head)

K* values for different values of *H/D

H/D	*K*	*diff*	*H/D*	*K*	*diff*	*H/D*	*K*	*diff*	*H/D*	*K*	*diff*
.101	0.01629	39	.126	0.02741	51	.151	0.04170	63	.176	0.05915	76
.102	0.01668	39	.127	0.02792	51	.152	0.04233	65	.177	0.05991	77
.103	0.01707	39	.128	0.02843	53	.153	0.04298	64	.178	0.06068	78
.104	0.01746	40	.129	0.02896	52	.154	0.04362	66	.179	0.06146	77
.105	0.01786	40	.130	0.02948	53	.155	0.04428	66	.180	0.06223	79
.106	0.01826	41	.131	0.03001	54	.156	0.04494	66	.181	0.06302	78
.107	0.01867	42	.132	0.03055	55	.157	0.04560	67	.182	0.06380	80
.108	0.01909	42	.133	0.03109	54	.158	0.04627	67	.183	0.06460	80
.109	0.01951	42	.134	0.03168	55	.159	0.04694	68	.184	0.06540	80
.110	0.01993	43	.135	0.03218	56	.160	0.04762	68	.185	0.06620	81
.111	0.02036	44	.136	0.03274	56	.161	0.04830	69	.186	0.06701	81
.112	0.02080	44	.137	0.03330	57	.162	0.04889	69	.187	0.06782	82
.113	0.02124	44	.138	0.03387	57	.163	0.04968	70	.188	0.06864	82
.114	0.02168	45	.139	0.03444	58	.164	0.05038	70	.189	0.06946	83
.115	0.02213	45	.140	0.03502	58	.165	0.05108	71	.190	0.07029	83
.116	0.02258	46	.141	0.03560	59	.166	0.05179	71	.191	0.07112	84
.117	0.02304	47	.142	0.03619	59	.167	0.05250	72	.192	0.07196	84
.118	0.02351	47	.143	0.03678	59	.168	0.05322	73	.193	0.07280	85
.119	0.02398	47	.144	0.03737	61	.169	0.05395	72	.194	0.07365	85
.120	0.02445	48	.145	0.03798	60	.170	0.05467	74	.195	0.07450	86
.121	0.02493	49	.146	0.03858	61	.171	0.05541	74	.196	0.07536	86
.122	0.02542	49	.147	0.03919	62	.172	0.05615	74	.197	0.07622	87
.123	0.02591	49	.148	0.03981	63	.173	0.05689	75	.198	0.07709	87
.124	0.02640	50	.149	0.04044	62	.174	0.05764	75	.199	0.07796	88
.125	0.02690	51	.150	0.04106	64	.175	0.05839	76	.200	0.07884	88

Table 10.13 (Bumped Head)

K* values for different values of *H/D

H/D	*K*	*Diff*	*H/D*	*K*	*Diff*	*H/D*	*K*	*Diff*	*H/D*	*K*	*Diff*
.201	0.07972	88	.226	0.10329	100	.251	0.12972	111	.276	0.15882	122
.202	0.08060	90	.227	0.10429	101	.252	0.13083	112	.277	0.16004	122
.203	0.08150	89	.228	0.10530	101	.253	0.13195	112	.278	0.16126	123
.204	0.08239	90	.229	0.10631	102	.254	0.13307	113	.279	0.16249	123
.205	0.08329	91	.230	0.10733	102	.255	0.13420	113	.280	0.16372	123
.206	0.08420	90	.231	0.10835	103	.256	0.13533	113	.281	0.16495	124
.207	0.08510	92	.232	0.10938	103	.257	0.13646	114	.282	0.16619	124
.208	0.08602	92	.233	0.11041	103	.258	0.13760	115	.283	0.16743	124
.209	0.08694	92	.234	0.11144	104	.259	0.13875	115	.284	0.16867	125
.210	0.08786	93	.235	0.11248	105	.260	0.13990	115	.285	0.16992	125
.211	0.08879	94	.236	0.11353	104	.261	0.14105	115	.286	0.17117	126
.212	0.08973	94	.237	0.11457	106	.262	0.14220	116	.287	0.17243	126
.213	0.09067	94	.238	0.11563	105	.263	0.14336	117	.288	0.17369	126
.214	0.09161	95	.239	0.11668	106	.264	0.14453	117	.289	0.17495	127
.215	0.09256	95	.240	0.11774	107	.265	0.14570	117	.290	0.17622	127
.216	0.09351	96	.241	0.11881	107	.266	0.14687	118	.291	0.17740	128
.217	0.09447	96	.242	0.11988	108	.267	0.14805	118	.292	0.17877	127
.218	0.09543	96	.243	0.12096	108	.268	0.14923	119	.293	0.18004	129
.219	0.09639	97	.244	0.12204	108	.269	0.15042	118	.294	0.18133	128
.220	0.09736	98	.245	0.12312	109	.270	0.15160	120	.295	0.18261	129
.221	0.09834	98	.246	0.12421	109	.271	0.15280	119	.296	0.18390	130
.222	0.09932	98	.247	0.12530	110	.272	0.15399	121	.297	0.18520	129
.223	0.10030	99	.248	0.12640	110	.273	0.15520	120	.298	0.18649	130
.224	0.10129	100	.249	0.12750	111	.274	0.15640	121	.299	0.18779	131
.225	0.10229	100	.250	0.12861	111	.275	0.15761	121	.300	0.18910	131

Table 10.14 (Bumped Head)

K* values for different values of *H/D

H/D	*K*	*Diff*	*H/D*	*K*	*Diff*	*H/D*	*K*	*Diff*	*H/D*	*K*	*Diff*
.301	0.19041	131	.326	0.22424	139	.351	0.26007	147	.376	0.29764	154
.302	0.19172	131	.327	0.22563	140	.352	0.26154	147	.377	0.29918	154
.303	0.19303	132	.328	0.22703	141	.353	0.26301	148	.378	0.30072	154
.304	0.19435	132	.329	0.22844	140	.354	0.26449	148	.379	0.30226	154
.305	0.19567	133	.330	0.22984	141	.355	0.26597	148	.380	0.30380	154
.306	0.19700	133	.331	0.23125	141	.356	0.26745	149	.381	0.30534	155
.307	0.19833	133	.332	0.23266	142	.357	0.26894	149	.382	0.30689	155
.308	0.19966	134	.333	0.23408	142	.358	0.27043	149	.383	0.30844	155
.309	0.20100	134	.334	0.23550	142	.359	0.27192	149	.384	0.30999	155
.310	0.20235	134	.335	0.23692	142	.360	0.27341	149	.385	0.31154	156
.311	0.20368	135	.336	0.23834	143	.361	0.27490	150	.386	0.31310	156
.312	0.20503	135	.337	0.23977	143	.362	0.27640	150	.387	0.31466	156
.313	0.20638	135	.338	0.24120	144	.363	0.27790	151	.388	0.31622	156
.314	0.20773	136	.339	0.24264	143	.364	0.27941	150	.389	0.31778	156
.315	0.20909	136	.340	0.24407	144	.365	0.28091	151	.390	0.31934	157
.316	0.21045	136	.341	0.24551	144	.366	0.28242	151	.391	0.32091	157
.317	0.21181	137	.342	0.24695	145	.367	0.28393	152	.392	0.32248	157
.318	0.21318	137	.343	0.24840	145	.368	0.28545	151	.393	0.32405	157
.319	0.21455	138	.344	0.24985	145	.369	0.28696	152	.394	0.32562	157
.320	0.21593	137	.345	0.25130	145	.370	0.28848	152	.395	0.32719	158
.321	0.21730	138	.346	0.25275	146	.371	0.29000	153	.396	0.32877	158
.322	0.21868	139	.347	0.25421	146	.372	0.29153	152	.397	0.33035	158
.323	0.22007	138	.348	0.25567	146	.373	0.29305	153	.398	0.33193	158
.324	0.22145	139	.349	0.25713	147	.374	0.29458	153	.399	0.33351	159
.325	0.22284	140	.350	0.25860	147	.375	0.29611	153	.400	0.33510	158

Table 10.15 (Bumped Head)
K* values for different values of *H/D

H/D	*K*	*Diff*	*H/D*	*K*	*Diff*	*H/D*	*K*	*Diff*	*H/D*	*K*	*Diff*
.401	0.33668	159	.426	0.37690	163	.451	0.41802	165	.476	0.45972	168
.402	0.33827	159	.427	0.37853	163	.452	0.41967	166	.477	0.46140	167
.403	0.33986	159	.428	0.38016	164	.453	0.42133	167	.478	0.46307	168
.404	0.34145	160	.429	0.38180	163	.454	0.42300	166	.479	0.46475	168
.405	0.34305	159	.430	0.38343	163	.455	0.42466	166	.480	0.46643	168
.406	0.34464	160	.431	0.38506	164	.456	0.42632	166	.481	0.46811	167
.407	0.34624	160	.432	0.38670	164	.457	0.42798	167	.482	0.46978	168
.408	0.34784	160	.433	0.38834	164	.458	0.42965	166	.483	0.47146	168
.409	0.34944	160	.434	0.38998	164	.459	0.43131	167	.484	0.47314	168
.410	0.35104	160	.435	0.39162	164	.460	0.43298	166	.485	0.47482	168
.411	0.35262	161	.436	0.39326	164	.461	0.43464	167	.486	0.47650	168
.412	0.35425	161	.437	0.39490	164	.462	0.43631	167	.487	0.47818	168
.413	0.35586	161	.438	0.39654	165	.463	0.43798	167	.488	0.47986	168
.414	0.35747	161	.439	0.39819	165	.464	0.43965	167	.489	0.48154	168
.415	0.35908	161	.440	0.39984	164	.465	0.44132	167	.490	0.48832	168
.416	0.36069	162	.441	0.40148	165	.466	0.44209	167	.491	0.48890	168
.417	0.36231	161	.442	0.40313	165	.467	0.44466	167	.492	0.48658	168
.418	0.36392	162	.443	0.40478	165	.468	0.44633	167	.493	0.48827	168
.419	0.36554	162	.444	0.40643	165	.469	0.44800	167	.494	0.48995	168
.420	0.36716	162	.445	0.40808	165	.470	0.44967	168	.495	0.49163	168
.421	0.36878	162	.446	0.40974	165	.471	0.45135	167	.496	0.49331	168
.422	0.37040	162	.447	0.41139	165	.472	0.45302	168	.497	0.49499	168
.423	0.37202	163	.448	0.41304	165	.473	0.45470	167	.498	0.49667	168
.424	0.37365	163	.449	0.41470	166	.474	0.45637	167	.499	0.49835	165
.425	0.37528	162	.450	0.41636	166	.475	0.45804	168	.500	0.50000	165

REFERENCES

[1] OIML R 71-1985. Fixed storage tanks, OIML, Paris.
[2] API standard 2551:1965. Measurement and calibration of horizontal tanks.
[3] API standard 2541: 1950. ASTM tables for positive displacement meter prover tanks.

11
CHAPTER

CALIBRATION OF SPHERES, SPHEROIDS AND CASKS

11.1 SPHERICAL TANK

A tank whose shell is a complete sphere is known as spherical tank. Capacity of such tanks is small in comparison to tanks in cylindrical form. Such tanks in general are stationary in nature and are used for storing liquids only. The tank is supported in such a way that whole shell of the tank is above the ground. Unlike vertical upright or horizontal tanks, there is no internal structural member. So no problem of deadwood and its volume distribution in this case. A diagram of typical spherical tank is shown in Figure 11.1.

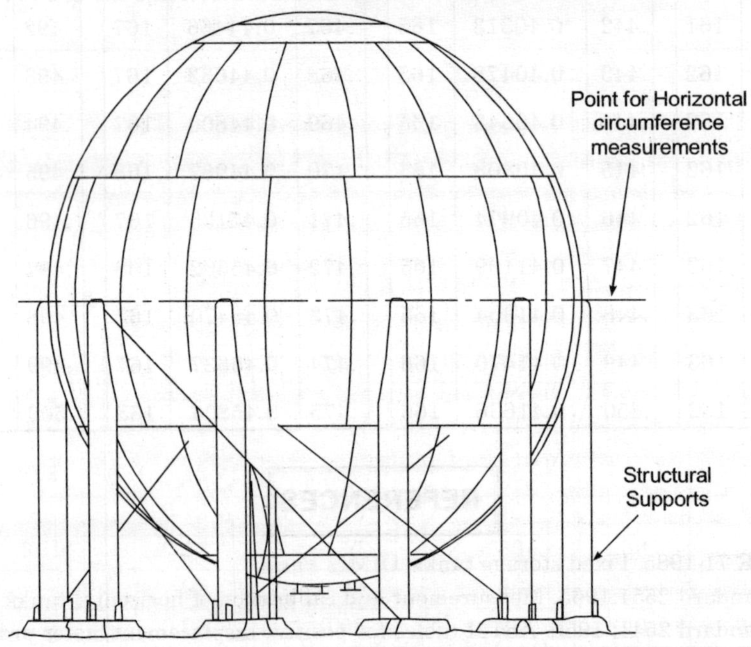

Figure 11.1 Diagram of a typical spherical tank

11.2 CALIBRATION

11.2.1 Strapping Method

11.2.1.1 Equipment

The capacity of such tanks is also determined by strapping procedure. Equipment used for strapping a spherical tank is same as described in section 8.9.2 of chapter 8. Every instrument mentioned there need not be used but the specification and other requirements like the use of only calibrated measuring tapes for circumference or depth measurement are the same. General precautions are also the same. That is, tank should be completely filled at least once before strapping and during calibration, should remain full with the liquid it intends to store or equivalent amount of water head.

11.2.1.2 Locations

There are practical problems in locating the great circles. Moreover it is rather difficult to place the measuring tapes flat all along the great circle and avoid its slipping while measuring the circumference. So only three great circles are chosen for strapping. One is the equatorial circumference and other two are vertical great circles passing through its poles. Here one may see that the terminology used is that of earth. To locate the largest horizontal circumference, equatorial great circle, the builder is supposed to tack-weld short rods normal to the shell at distances of not more than 3 m. Moreover the upper face of each short rod should lie in the same horizontal plane.

11.2.2.3 Field Measurements

The circumference measurement of equator may present difficulty as generally supporting pillars come in the way. To circumvent it, suitable step-over(s) is used and necessary corrections are applied. Sometimes if it is not possible to take measurement at the equatorial position, the measurement is carried out at a height of H normal to the equator and corrected value of circumference at the equator is calculated by the formula

$$C_e = \{C_h^2 + (2\pi H)^2\}^{1/2} \qquad \text{...(1)}$$

Besides measurement of the equatorial circumference, the circumferences of any two mutually perpendicular vertical great circles i.e. circles passing through the poles are measured. Let C_1, C_2 and C_e be respectively the circumferences of two vertical great circles and equator. The inspection of the three values will give a fairly good idea if the shell is truly spherical. If the values of all circumferences are within the prescribed tolerance then the shell may be assumed as spherical and volume V of the tank is given by

$$V = (C_1 C_2 C_e)/6\pi^2 \qquad \text{...(2)}$$

The total inside height D is measured along the central axis of the shell. Usually there is no manhole or other fittings along this line. If it is not feasible to take measurement along this line, one may measure at a convenient distance say m units from the central line. Then radius and hence diameter may be calculated by assuming that D_m the measured vertical height is a chord of a great circle passing through the poles, so D the diameter-total depth along the central line is given by

$$(D/2)^2 = (D_m/2)^2 + m^2$$

giving

$$D = (D_m^2 + 4m^2)^{1/2} \qquad \text{...(3)}$$

11.2.2 Liquid Calibration

In case of sphere and spheroid tanks, liquid calibration method is profitable by way of relatively better accuracy. In strapping method there are only few locations for sphere tanks and still fewer for spheroid tanks. Liquid calibration may be carried out taking either calibration tanks or positive displacement meters as standard.

11.2.2.1 Calibrating Tank as Standard

The tank should be filled with water to the top capacity. The water is discharged into calibrated tank where it is accurately measured. The capacity of the calibration tank should be such that water delivered, for each incremental decrease in height, is measured conveniently. The capacity of the calibration tank should not be smaller than the largest volume of one half of increment value for the tank. The capacity should not be greater than the largest volume of the one increment value of the tank. Calibration should be obtained for each 2.5 cm of the upper 25% and the lower 25% of the height between the bottom and top capacity lines and every 5 cm for the intervening height. The incremental discharged should be measured by means of tape and bob or gage glass readings, or any other level measuring device.

Equipment used is same as discussed in Chapter 8 under liquid calibration method.

11.2.2.2 P. D. Meter as Standard

Meter readings should be taken at every 2.5 cm interval for upper and lower 25% of the height. For intervening heights, the interval is doubled.

11.3 COMPUTATIONS

11.3.1 Direct from Formula and Tables

Volume of the segment of a sphere of radius r and height H is given by

$$V_h = (\pi/3)(H)^2[3r - H] \quad ...(4)$$

Volume of the sphere $V = (4\pi/3)r^3$, giving us

$$V_h/V = (1/4)(H/r)^2\,[3 - (H/r)]$$

Expressing V_h/V as a factor K, then

$$K = (1/4)(H/r)^2\,[3 - (H/r)] \quad ...(5)$$

If D is diameter then $D = 2r$ and K factor in terms of H/D is given as

$$K = (H/D)^2\,[3 - 2H/D] \quad ...(6)$$

11.3.2 Alternative Method (Reduction Formula)

There is another approach to establish gauge-table calculations. We take a fixed value of increment say G and represent internal height along the axis of the sphere as $2r$. Let on the scale from its centre, $m + 1^{st}$ be the point, which coincides with the datum line, i.e. volume at this point is zero. If we denote V_{m+1}, as volume at the $m + 1^{st}$ point, then

$$V_{m+1} = 0$$

Taking G as the incremental height and r is the one-half of the vertical inside height i.e. radius of the sphere.

So volume of the liquid V_m at the first increment i.e. at the mth point from centre of the scale using (4) is given by

$$V_m = (V/4)(G/r)^2\{3 - (G/r)\} \quad ...(7)$$

$$= (V/4)(G/r)\{3\,G/r - (G/r)^2 + 3 - 3\}$$
$$= (V/4)(G/r)\{3 - (G/r)^2\} + (V/4)\{3(G/r)^2 - 3(G/r)\}$$
$$= (V/4)(G/r)\{3 - (G/r)^2\} + (3V/4)(G/r)^3\{r/G - (r/G)^2\}$$

Writing $r/G = m$, we get

$$V_m = (V/4)(G/r)\{3 - (G/r)^2\} + (3V/4)(G/r)^3\{m - m^2\}/2 \quad ...(8)$$

Write $K_1 = (V/4)(G/r)\{3 - (G/r)^2\}$ and ...(9)

$K_2 = (3V/2)(G/r)^3$, we get a relation

$$V_m = K_1 - (m^2 - m)K_2/2 \quad ...(10)$$

So V_m is the first volume increment and V_1 is the increment in volume on either side of the centre of the sphere for height G.

Extending use of (10) for next lower point i.e. $m + 1^{st}$, we can express V_{m+1} as

$$V_{m+1} = K_1 - \{(m+1)^2 - (m+1)\}K_2/2 \quad ...(11)$$

Subtracting (11) from (10), we get

$$V_m - V_{m+1} = \{(m+1)^2 - (m+1) - m^2 + m\}K_2/2$$
$$V_m = V_{m+1} + mK_2. \quad ...(12)$$

This is a reduction formula between two consecutive points on the scale. The volume of each increment above the bottom increment is mK_2.

Giving m all positive integral values we get the following set of equations

$$V_m = V_{m+1} + mK_2$$
$$V_{m-1} = V_m + (m-1)K_2$$
$$V_{m-2} = V_{m-1} + (m-2)K_2$$
$$V_{m-3} = V_{m-2} + (m-3)K_2$$

....

....(13)

....

$$V_3 = V_4 + 3K_2$$
$$V_2 = V_3 + 2K_2$$
$$V_1 = V_2 + 1K_2$$

Adding all the equations in set (13), we get

$$V_1 = V_{m+1} + \Sigma(m\,K_2)$$

Or Simply

$$V_1 = [m(m+1)/2]K_2.$$

Mind V_1 is the value of volume increment at the centre of the sphere but one scale division lower.

Hence total volume till the midpoint (centre of the sphere) will be

$$\Sigma V_r = (V_1 + V_2 + V_3 + \ldots\ldots\ldots V_m) = (1/2)\,\Sigma r(r+1)\,K_2 = [m(m+1)(m+2)/6]K_2$$

Height H can be expressed as n times the increment G, then V_h up to the height H will be given by the sum of all increment from m to n (n is less than m) i.e.

$$V_h = \Sigma V_r \quad (14)$$

r takes integral values from m to n.

The values of V_h/V can be calculated from either of the expressions namely (5) or (12). The values V_h/V have been given in Tables 11.3 to 11.7 for H/D from 0 to 0.5 in steps of 0.001. It is the same increment as has been used in the tables for horizontal storage tanks in Chapter 10.

11.3.3 Example of Calculation for Sphere

The field data about a tank is as follows:

Horizontal circumference was measured not in the equatorial plane but in a horizontal plane at height of 250 mm and it is measured as 40004 mm.

The other two vertical circumferences measured respectively are 40036 mm and 40032 mm. The internal height measured along the chord at a horizontal distance of 250 mm is 12730 mm. Average plate thickness is 18.5 mm

Equatorial circumference $C_e = \{40004^2 + (2\pi\, 250)^2\}^{1/2} = 40035$ mm

Subtract $2\pi t$ from each circumference, where t is thickness of the shell and value of t in the present case is 1.85 cm, so inner circumferences are:

39 92.2 cm, 3992.3 cm and 39 91.9 cm

So volume V of the sphere $= C_e\, C_1\, C_2/6\pi^2$.

$= 39.922 \times 39.923 \times 39.919/59.2177601 = 1074.393\ \text{m}^3$ or $= 1074\ 393\ \text{dm}^3$

Now the diameter along the central line of the sphere $= \{12730^2 + 4 \times 250^2\}$

$= 12740$ mm

Thus radius $r = 6370$ mm

Let the increment is 25 mm

Then $G/r = 3.924646782 \times 10^{-3}$

$(G/r)^2 = 1.540285236 \times 10^{-5}$

$(G/r)^3 = 6.045124374 \times 10^{-8}$

$K_1 = (V/4)(G/r)\{3 - (G/r)^2\} = 1074393 \times 0.003924678 \times 2.999984563/4$

$= 3162.4\ \text{dm}^3$

$m = r/G = 254.8$

$H/r = G/r$

$K_2 = 1.5 \times 1074393 \times 6.045124374 \times 10^{-8} = 0.09742\ \text{dm}^3$

$(r/H)K_2 = 248.226\ \text{dm}^3$

$V_m = K_1 - (m^2 - m)K_2/2$

$= 3162 - (254.8 \times 254.8 - 254.8) \times 0.09742/2 = 3162 - 3149.99$

$12.41\ \text{dm}^3$

$V_{m-1} = V_m + (m-1)K_2 = 12.41 + 24.73 = 37.14$

Partial gauge table by strapping method

H mm	*r*	V_r	rK_2	V_{r-1}	*Partial volume H*	*Partial volume From (4)*	*Partial volume from tables*
25	254.8	—	—	12.41	12.41	12.49	12.5
50	253.8	12.41	24.73	37.14	49.55	49.89	49.65
75	252.8	37.14	24.63	61.77	111.21	112.12	111.7
100	251.8	61.77	24.53	86.20	197.41	199.07	197.96
125	250.8	87.4	24.43	110.63	308.04	310.64	308.64
150	249.8	110.33	24.34	134.97	443.01	446.73	445.8

One may see that using equation (12) for partial volumes involve too many calculations involving very large or very small numbers, which affects the accuracy of final result. Equation (4) does not involve much calculation or very large or small numbers. Even use of the tables involves multiplication of *K* factor by the volume of the tank, which is in 7 significant figures if rounded in dm^3 so difference of one dm^3 is obvious. So it is safer to use equation (4) for calculation of partial volumes, though universal use of tables may be better from consistency point of view.

Considering the example again, in which

H/*D* for 25 mm is equal to 0.00196 giving $K = 0.00001184$ and volume = 12.5 dm^3

H/*D* for 50 mm is equal to 0.00392 giving $K = 0.00004622$ and volume = 49.6 dm^3

H/*D* for 75 mm is equal to 0.00589 giving $K = 0.00009398$ and volume = 101.0 dm^3

H/*D* for 100 mm is equal to 0.00785 giving $K = 0.00018430$ and volume = 197.96 dm^3

H/*D* for 125 mm is equal to 0.00981 giving $K = 0.00028727$ and volume = 308.6 dm^3

H/*D* for 150 mm is equal to 0.00118 giving $K = 0.00041486$ and volume = 445.75 dm^3

11.4 SPHEROID

A spheroid is a stationary liquid storage tank having a shell of double curvature. Any horizontal cross-section is a circle and vertical cross-section is an arc of some other circle for smooth spheroid and series of circular arcs for nodded spheroid. The height of the tank is lesser compared to that of a sphere. The bottom of the tank rests directly on a prepared ground. The spheroid has a base plate resting on the ground and projecting beyond the shell. Structural members rest on the base plate and support the overhanging part of the shell for a short distance above the base plate. A drip bar is welded to the shell in a horizontal circle just above the structural supports to intercept rainwater.

A smooth spheroid shown in Figure 11.2 usually has no inside structural members to support the shell roof.

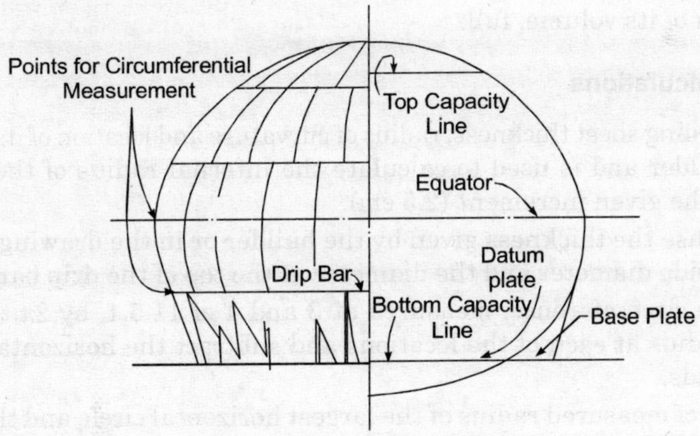

Figure 11.2 Smooth spheroid tank

A noded spheroid is shown in Figure 11.3. It has abrupt breaks in the vertical curvature called nodes, which are supported by a circular girder and structural members inside the tank.

11.5 CALIBRATION

11.5.1 Strapping

Due to structural problems it is not practical to strap at more than two locations at the upper edge of the drip bar, and at the position where horizontal circumference is largest.

The following measurements are taken:

1. Elevation of datum plate relative to bottom is measured.
2. The elevation of the top of the drip bar, relative to the bottom capacity line, is measured at equally spaced four points around the spheroid.

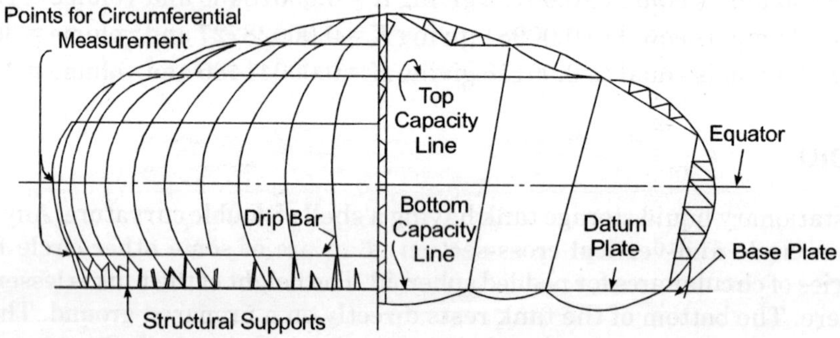

Figure 11.3 Nodded spheroid tank

3. Outside circumference is measured at the level where the tangents to the spheroid are vertical. This will give maximum circumference of the shell.
4. Another out side circumference of the spheroid is measured on the upper edge of the drip bar.

During the measurements of the circumferences at 3 and 4, the spheroid should remain, at least three fourth of its volume, full.

11.5.2 Step wise Calculations

1. Data regarding sheet thickness, radius of curvature and location of its centre is supplied by the builder and is used to calculate the internal radius of the shell at the mid height of the given increment (2.5 cm).
2. Similarly use the thickness given by the builder or in the drawings, to calculate the largest inside diameter and the diameter at the top of the drip bar.
3. Divide the circumference, measured at 3 and 4 of 11.5.1, by 2π to get the average outside radius at each of the locations and subtract the horizontal thickness to get inside radius.
4. Find ratio of measured radius of the largest horizontal circle and that of given by the builder. Adjust all horizontal radii in the upper portion of the shell by multiplying each by this ratio.
5. A similar ratio of measured radius at the top of the drip bar to the blueprint radius is calculated to adjust the radii for lower portion of the shell.
6. Correct for any dead wood.

7. Complete the gage table by totalling the net incremental volume, starting with zero at the bottom capacity line. The gage table may be prepared by any desired increment using graphs or mathematical relation to establish a smooth curve.
8. Record on the gauge table the elevation of the datum plate from the bottom capacity line.
9. A clear indication whether the capacity table was made from the data obtained by liquid calibration or by the strapping method should be made.

11.5.3 Example for Partial Volumes of a Spheroid

11.5.3.1 Measurements

Measurement data

Datum plate is set at the elevation of bottom capacity line. So measurement mentioned at 1 of 11.5.1 is zero.

Height of top of drip bar to bottom capacity line (2 of 11.5.1) at 4 equally spaced points 145.8 mm, 146.0 mm, 146.0 mm, 146.3 mm

Average height of top of drip bar	146.0 mm
Maximum circumference of the shell (3 of 11.5.1)	39526 mm
Outside circumference at drip bar (4 of 11.5.1)	36067 mm

Data from the blue print

1. Outside radius of the vertical curvature R	4450 mm
2. Height of the centre of the vertical curvature from the bottom capacity line a	4025 mm
3. Horizontal distance from drip bar (axis of the tank) to the vertical from centre of curvature *L*	1927 mm
Radius of vertical curvature	3797 mm

Plate thickness at drip bar = 10 mm

Inside radius of the circumference at the top of drip bar = 5725

158 987 928

11.5.3.2 Computations

Inside radius at maximum circumference = 39526/2π – 10 = 6280.75 = 6281 mm

Inside radius at maximum circumference (builder/blueprint) = 6280.95 = 6281 mm

Multiplying factor for adjusting radii in upper portion of tank = 6280.75/6280.95 = 0.999968

Inside radius of the circumference at the top of drip bar = 36067/2π – 10 = 5730 mm

Multiplying factor *M* for adjusting radii in the lower portion of tank = 5730/5725 = 1.000873

11.5.3.3 Elementary Volume

The surface of the shell is the surface of the revolution of arcs of circles of different radii with centres at certain distances from the vertical axis of the spheroid. Using radii and their locations of their centres of curvature, we can determine the horizontal distances of the shell at mid point of a small vertical increment. For a small increment along the axis each portion may be taken as a cylinder of radius equal to the distance calculated and height equal to the increment. Vertical distances are measured from the bottom capacity line. Sum of volumes of these cylinders gives the volume of the shell.

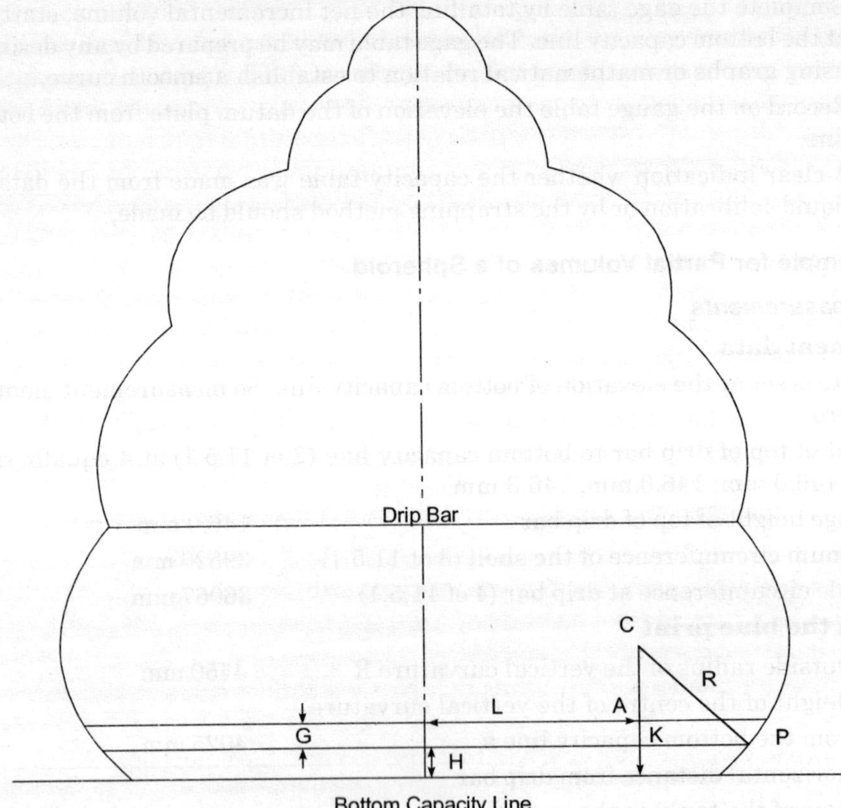

Figure 11.4 Radius of elementary cylinder

Let there be point P on the surface of spheroid and C be the centre of curvature of the surface at P, then CP is radius of curvature R. If vertical distance of C from the bottom capacity line is A, then PK is given by

$$PK = \sqrt{R^2 - (A - H)^2}$$

If L is the horizontal distance of the point C from the axis of the spheroid, then $L + PK$ may be taken as the radius of a cylinder of very small height G. Giving its volume $= \pi G(L + PK)^2 = \pi G(L + B)^2$

Example of effective radius of the elementary cylinder

Horizontal thickness of plate (4449/3797)10 = 11.7 mm

Height	*H*	*A*	*R*	$B = \sqrt{R^2 - A^2}$	*L = 1927* *Radius = L + B*	*Effective* *Radius = M × Radius*
25 mm	12.5	4012.5	4450	1924.1	3851.1	3854.5
50 mm	37.5	3987.5	4450	1975.4	3902.4	3905.8
75 mm	62.5	3962.5	4450	2025.1	3952.1	3955.6

Partial volume with deadwood

Height	*Incremental volume dm^3*	*Volume every 2 mm in dm^3*	*Deadwood every 2 mm*	*Volume every 2 mm in dm^3*	*Net incremental volume*	*Partial volume dm^3*
25	1166.9	93.352	0.365	92.987	1162	1162
50	1198.1	95.848	0.3656	95.483	1194	2356
75	1228.9	98.312	0.3656	97.947	1224	3580

11.6 TEMPERATURE CORRECTION

The effect of expansion or contraction of tanks containing liquid at normal temperature is disregarded. Corrections are not necessary unless the measurements at very high accuracy are needed. For temperature corrections, it is necessary to estimate service temperature, of the contents and compute volume correction by using the formula

Fractional volume correction = 3 γ (T – reference temperature).

The values of γ coefficient of linear expansion for most commonly used materials like steel and aluminium is given in Tabies 11.1 and 11.2 for different temperature ranges.

For non-insulated metal tanks, the temperature of the shell may be taken as the mean of adjacent liquid and ambient air temperatures i.e. mean of the temperatures on the inside and outside of the shell at the same location. To apply the temperature correction to spheres and spheroids, only the horizontal dimensions are functions of tank calibration corrections. The liquids height dimension is a function of gauging the liquid level. Hence thermal effect corrections are separately considered for innage and outage gauge readings.

11.6.1 Coefficients of Volume Expansion for Steel and Aluminium

Generally all tanks are made of steel or Aluminium their coefficients of linear expansion are given respectively in Tables 11.1 and 11.2

11.7 STORAGE TANKS FOR SPECIAL PURPOSES

There is a class of vessels used for brewing, maturing, storage and delivery of alcoholic liquids. Casks, Barrels and Vats are some specific examples. Vats available are from few thousand dm^3 to one hundred thousand dm^3. Vats are simple stationary storage tanks used for storing and maturing an alcoholic liquid. So they need calibration as any other petroleum vessel. In these cases also calibration tables (Volume versus dip height) are made.

Casks and barrels are portable vessels of much smaller capacity used for transporting liquors. So Vats and casks or barrels fall under the purview of Excise Departments and hence are to be calibrated by a Government agency.

11.7.1 Casks and Barrels

Casks are essentially containers, which can be rolled and are used for the transport and delivery of liquids when completely full i.e. liquid is delivered in terms of one full unit of cask or barrel.

Hereafter the term Cask(s) will include barrel(s) so for brevity only the term cask(s) will be used.

11.7.1.1 Capacity

All casks are content measures and their capacity is defined when completely full at 20°C. Minimum capacity of a cask is 2 dm^3. Casks may be of any capacity but in multiples of 5 dm^3 if the capacity is less than or equal to 100 dm^3. For larger capacity these may be in multiples of 50 dm^3.

In general casks are available in two accuracy classes namely class *A* and class *B*.

11.7.1.2 Material Requirements

Casks are normally made of wooden planks and strips or of metallic sheets or any other suitable material. The coefficients of expansion of the material must be such that their capacity, for a temperature changes from 10°C to 30°C, do not change by more than

0.25% for class *A* and

0.5% for class *B* casks.

Similarly their capacity should not change by the above said quantity, when an excess pressure of 10^5 Pa (almost equal to atmospheric pressure) is maintained for 72 hours.

Further the elasticity of the material must be such that after subjecting the cask for excess pressure for 72 hours, and returning to normal pressure for another 72 hours the capacity should not differ from its initial capacity by more than 0.025% for class *A* and 0.05% for class *B*.

11.7.1.3 Shape

The casks when made of solid wood, with butted staves held together by metal hoops are curved body with the greatest perimeter being at the mid-point of the body, and two flat or slightly curved ends. Ideally a cask may be considered as either a combination of two frusta of a cone or a surface of revolution of a part of parabola joined base to base or ellipse about an axis parallel to their axes.

Casks made of any other material may also be cylindrical or spherical in shape.

The position of bunghole is such that allows for complete filling of the cask and is of such form so that no air pockets are formed. In addition to the bunghole, the cask may have one or more orifices.

11.7.1.4 Maximum Permissible Error for Casks [9]

A. At the time of verification

± 0.5 percent but not less than 0.1 dm^3 for class *A* casks

± 1 percent but not less than 0.15 dm^3 for class *B* casks

B. At the time of inspection when a cask is in service

B.1 For class *A* casks

± 1 percent but not less than 0.2 dm^3

B.2 For class *B* casks

± 4 percent for casks up to 5 dm^3 capacity

± 0.3 dm^3 for casks over 5 dm^3 to 15 dm^3

± 1 dm^3 for casks over 15 dm^3 to 60 dm^3

± 1.5 dm^3 for casks over 60 dm^3 to 75 dm^3

± 2 percent for casks over 75 dm^3

11.8 GEOMETRIC SHAPES AND VOLUMES OF CASKS

11.8.1 Cask Composed of two Frusta of Cone

Referring to Figure 11.5, let r_1 and r_2 are radii of top and base of the frustum with height h.

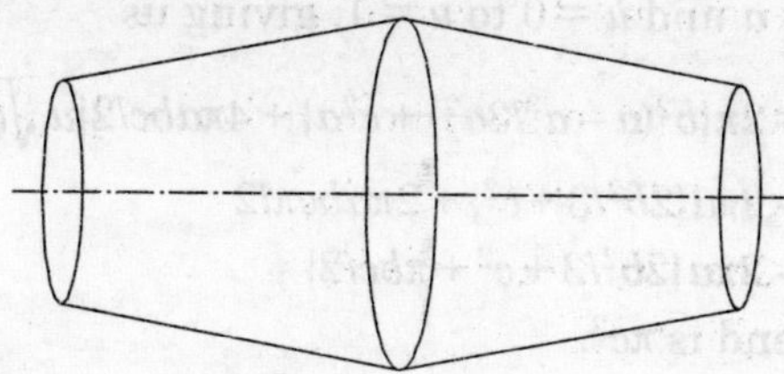

Figure. 11.5 Combination of frusta of a cone

Volume of half of the cask is equal to that of a frustum of a cone having two circular ends of radii r_1 and r_2 whose volume is given by

$$\pi h(r_1{}^2 + r_2{}^2 + r_1 r_2)/3, \text{ giving us}$$

$$\text{Volume of cask } = 2\pi h\ (r_1{}^2 + r_2{}^2 + r_1 r_2)/3 \qquad ...(15)$$

Above expression can be written in terms of the semi-vertical angle α of the cone and its end radius. Radius r_2 may be written as

$$r_2 = r_1 + h\tan(\alpha), \text{ giving us}$$

$$\text{Volume of cask} = 2\pi h\{3r_1{}^2 + 3r_1 h\tan(\alpha) + h^2\tan^2(\alpha)\}/3 \qquad ..(16)$$

11.8.2 Cask-volume of Revolution of an Ellipse

Let there be an ellipse with semi minor and major axes as b and a respectively, referring Figure 11.6.

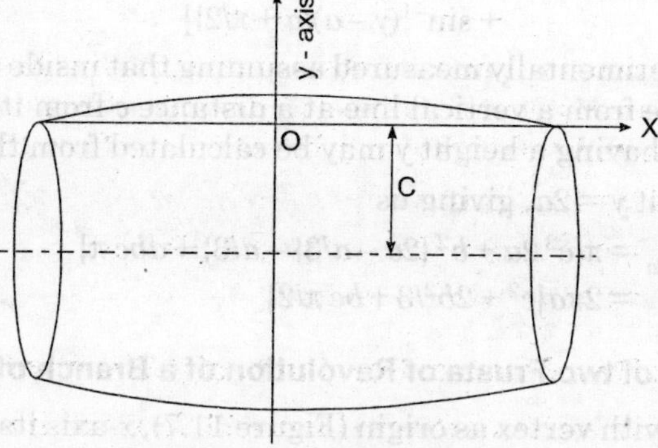

Figure 11.6 Ellipsoidal cask

If major axis is horizontal then equation of the ellipse taking its centre as origin and its axes as coordinate axes, its equation is given as

$$x^2/a^2 + y^2/b^2 = 1$$

The semi-ellipse is rotated about a vertical line at a distance of c from its axis then V the volume of revolution is given as

$$V = 2\pi\int (y + c)^2 dx$$

Limits of integration are $x = 0$ to $x = a$.

So volume of cask V is given as

$$V = 2\pi\int(y^2 + c^2 + 2cy)dx$$

$$V = 2\pi\int(1 - x^2/a^2)b^2 + c^2\}dx + 4\pi bc\int\sqrt{(1 - x^2/a^2)}dx$$

$$V = 2\,\pi[b^2(x - x^3/3a^2) + c^2x)] + 4\pi abc\int\sqrt{1-u^2)}\ du, \text{ where } u = x/a$$

Limits are $x = 0$ to $x = a$ and $u = 0$ to $u = 1$, giving us

$$V = 2\pi\{b^2(a - a^3/3a^2) + c^2a\} + 4\pi abc/2[u\sqrt{(1-u^2)} + \sin^{-1}(u)]$$

$$= 2\pi a\{2b^2/3 + c^2\} + 2\pi abc\pi/2$$

$$= 2\pi a\{2b^2/3 + c^2 + \pi bc/2\} \qquad ...(17)$$

Area of each circular end is πc^2.

Transferring the origin to A the end of its major axis and turning the axes by 90° (axis of the cask vertical), the equation of ellipse becomes

$$x^2/b^2 + (y - a)^2/a^2 = 1$$

Volume of liquid of height y V_y is given by

$$V_y = \pi\int(x + c)^2dy$$

Limits of y are $y = 0$ to $y = y$

$$V_y = \pi\int x^2 + c^2 + 2\,cx)dy$$

$$= \pi\int[c^2 + b^2(1 - (y - a)^2/a^2) + 2\,cb\,\sqrt{(1 - (y - a)^2/a^2)}]dy$$

$$= \pi[c^2y + b^2\{y - (y - a)^3/3a^2\} + abc\{(y - a)/a\,\sqrt{(1 - (y - a)^2/a^2)}$$

$$+ \sin^{-1}\{(y - a)/a\}] \qquad ...(18)$$

The limits of y are from $y = 0$ to $y = y$ giving us the volume V_y of liquid up to the height y as

$$V_y = \pi[c^2y + b^2\{y - (y - a)^3/3a^2 - a/3\} + abc\ [\{(y - a)/a\,\sqrt{(1 - (y - a)^2/a^2)}$$

$$+ \sin^{-1}(y - a)/a + \pi/2\}] \qquad ...(19)$$

If a, b and c are experimentally measured assuming that inside surface of cask is a surface of revolution of an ellipse from a vertical line at a distance c from its major axis, then volume of the liquid in the cask having a height y may be calculated from the above equation.

For total volume put $y = 2a$, giving us

$$V_{2a} = \pi[c^2\,2a + b^2\,\{2a - a/3\} - a/3\} + abc\ \pi]$$

$$= 2\pi a[c^2 + 2b^2/3 + bc\ \pi/2] \qquad ...(20)$$

11.8.3 Cask Composed of two Frusta of Revolution of a Branch of a Parabola

Let there be a parabola with vertex as origin (Figure 11.7), x-axis its axis and latus-rectum $4a$, then

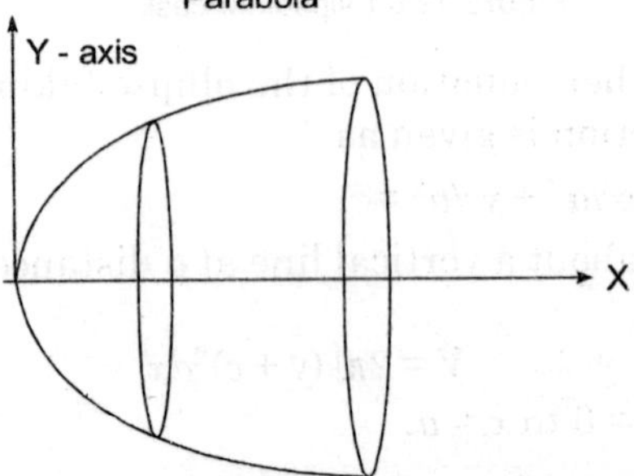

Figure 11.7 Cask with surface of revolution of a parabola

$$V_y = \pi\left[c^2 y + b^2\left\{y - \frac{(y-a)^2}{3a^2}\right\} + abc\left\{\frac{(y-a)}{a}\sqrt{1-(y-a^2)a^2} + \sin^{-1}(y-a)/a + \frac{\pi}{2}\right\}\right]$$

The equation of the parabola is

$$y^2 = 4ax$$

Consider two points A and B with coordinates (x_1, y_1) and (x_2, y_2) on it. Let this branch AB is rotated about its axis then V the volume of the revolution is given by

$$V = \pi\int y^2dx = 4a\pi\int xdx$$

Limits of integration are from $x = x_1$ to $x = x_2$ giving us

$$V = 4a\pi[x^2/2] = 2a\pi(x_2{}^2 - x_1{}^2)$$
$$= 2a\pi\{(y_2{}^2/4a)^2 - (y_1{}^2/4a)^2\}$$
$$= (\pi/8a)\{(y_2)^4 - (y_1)^4\}.$$

So volume of the cask having two such frusta joined together base to base (Figure11.7) is given by

$$(\pi/4a)\{(y_2)^4 - (y_1)^4\} \quad ...(21)$$

If r_1 and r_2 are the radii of one end and middle of the cask having surface of revolution due to part of parabola (Figure 11.7), then volume of the cask is

$$(\pi/4a)\{(r_2)^4 - (r_1)^4\}.$$

11.9 CALIBRATION/ VERIFICATION OF CASKS

11.9.1 Reporting/Marking the Values Rounded Upto

According to OIML R-45 [9], casks made of metal and having capacity up to 100 dm³, their capacity must be marked before submitting for calibration/verification. Casks of capacity greater than 100 dm³ have an option of marking the capacity.

In case of casks on which its capacity is not marked, calibration means finding its capacity at 20 °C and marking the capacity nearest to values given in the table below:

Table rounding of the values of capacity of the casks

Range of Capacity in dm³	*Class A casks Rounded to dm³*	*Class B casks Rounded to dm³*
Up to 5	0.05	0.05
Over 5 to 15	0.1	0.1
Over 15 to 60	0.1	0.5
Over 60 t0 150	0.2	1
Over 150 to 300	0.5	1
Over 300 to 600	1	1
Over 600 to 1500	2	2
Over 1500	5	5

11.9.2 Uncertainty in Measurement

In case of casks marked with capacity, calibration and verification means determination of their capacity at 20°C and verifying that the marked capacity is within the prescribed limits of maximum permissible errors. While determining the capacity, the uncertainty in measurement should not be more than:

± 0.1 dm^3 for casks of capacity less than and up to 100 dm^3

± 0.3 percent for capacity greater than 100 dm^3

11.9.3 Calibration Procedures

We can use either of the two methods of calibration of a cask. The two methods as described in chapter 3 and 4 respectively are gravimetric and volumetric.

11.9.3.1 Calibrating a Cask (Volumetric Method)

For calibrating a cask, a standard measure of delivery type is used. There are two possibilities (1) Capacity of the cask is nominally equal to that of the standard measure and (2) capacity of the cask is some integral multiple of that of standard measure.

Case (1) The cask under test is cleaned and dried. Standard measured is filled with water under gravity. Temperature of water is recorded. Water is filled from the standard measure till the cask is completely full, record the reading of the standard if capacity of cask under-test is less than that of standard. If capacity of the cask-under test is more than of standard add water till the cask is completely full through a calibrated measuring cylinder. The capacity of cask is then calculated at 20°C, which is the reference temperature for cask. We need here coefficients of expansion of the materials of standard and cask and also density of water at the temperature of measurement. Details have been explained in 4.3.1 of Chapter 4.

Case (2) The method is same as before except a better temperature stability is required in this case. Temperature of water delivered should not change by 0.5°C in the entire process. The cask is filled with water and the number of times the standard measure is used is noted, say it is n. If cask requires more water but less than the capacity of the standard measure then fill it completely with a calibrated measuring cylinder. The volume of water transferred is then n times the capacity of the standard measure plus volume of water transferred by the measuring cylinder. This value of volume is used to calculate the capacity of cask at 20°C.

11.9.3.2 Calibration of Casks (Gravimetric Method)

In addition of marked capacity, the following two weights are marked on the cask.

1. Dry tare weight: It is the weight of an empty dry cask, including its plugs, bungs etc used to close orifices and is measured prior to wetting.
2. Wet tare weight: It is the weight of an empty cask including its plugs, bungs etc used to close the orifices and is measured after wetting of the interior and draining it for 30 seconds.

These two weights are useful to calculate the volume of the liquor contained in the cask by simply weighing it and knowing the density of the liquor. It is assumed that same liquor is filled every time. So once density is measured, it will work for quite sometime.

Casks especially of smaller capacities can easily be calibrated by weighing the water contained in it. In using this method dry tare is determined first. Clean and dry the cask and weigh it. Apply necessary correction for air buoyancy etc. to get the dry tare. Fill the measure completely with distilled water and weigh it. From the difference of apparent masses with and without water, we get apparent mass of water. Then apply corrections as explained in Chapter 3 and get the capacity of the measure at 20°C. Empty the water and allow 30 seconds of drainage time and again weigh it to get wet tare weight of the cask.

11.10 VATS

The vats are used for storing and maturing alcohol-based liquids. In most of the countries, excise duty is levied on this liquid, so the capacity of such vats is determined by a government agency. In fact what the departments of excise duty wish to know is the volume of liquid contained in the vat or drawn out from it. For this purpose they use dipstick to measure heights of the liquid before and after each delivery. So the dipstick is calibrated to indicate volume of liquid contained versus its height indicated by the dipstick.

11.10.1 Shape

Most of the vats are cylindrical or part of a frustum of a cone with vertical axis.

11.10.2 Material

Quite a variety of materials are used in fabrication of vats. Steel or copper sheets are used for vats. Sometimes vats are made of wooden planks/strips bounded by metal strips. To keep the flavour of the liquor intact vats are lined with suitable materials. Material coating also helps to protect the shell material. To measure the wall thickness of vats is always a problem. So internal strapping will be advantageous as it is difficult to measure wall thickness of vats.

11.10.3 Calibration

11.10.3.1 Strapping

Measures the internal diameter at several places say 5 to 6 places. See the trend by plotting the length of the diameters on x-axis and height at y-axis if the ends of diameters are in a straight lines, then shell of the vat is a part of the frustum of cone. From the graph we can find the equation of the straight line, which will give diameter at any height. Hence volume of given increment at any height can be calculated and gauge table can be prepared.

If the diameters vary randomly, and variation is small within experimental error then mean of the diameters is taken, as the diameter of the cylindrical shell and gauge table may be prepared.

If variation is large then diameters are to be taken at a larger number of positions as specified in the relevant national or international standards and gauge table is prepared as in the case of vertical storage tank.

11.10.3.2 Liquid Calibration

Alternatively liquid calibration method may be employed to produce calibration tables. We may use large capacity delivery measures or positive displacement metres, which are calibrated immediately before calibrating a vat. Positive displacement metres with 0.1% accuracy are easily available which are quite suitable for liquid calibration. Known amount of water/liquid is withdrawn and dipstick reading is recorded. The process is repeated till the entire dipstick is covered. For verification, the indication of volume on the dipstick should never differ from the measured volume of liquid by the maximum permissible error. For calibration dipstick is successively marked with the volume drawn. The entire dipstick is covered. To indicate the zero of the dipstick an inverted arrow is marked. The head of the arrow coincides with the flat end of the dipstick. At the time of verification, this mark is examined to ensure that dipstick has not been tampered with.

11.11 RE-CALIBRATION OF ANY STORAGE TANK WHEN DUE

A storage tank of any form becomes due for calibration after a fixed period of time. Normally the department of legal metrology of a country fixes the time interval between any two consecutive calibrations/verifications.

Recalibration of a tank also becomes due under any of the following conditions:

1. When any deadwood is changed, such as concrete is installed inside the tank to reinforce it.
2. When the tank is repaired or changed in any manner, which may affect the total or incremental volume.
3. When tank is moved from one place to another.
4. When any deformation becomes noticeable. After extended service, the tank, sometimes deforms at the saddle or at other supports.
5. When it is evident that previous circumferential measurements were taken at points other than prescribed by the competent authority.
6. When measurements taken for the purpose of checking the accuracy of the existing records, found to differ by more than the prescribed limits. Any such measurement should be taken at the previous positions only.
7. When any tank is restored to service after remaining disconnected or abandoned.

Table 11.1 Coefficient of Linear Expansion of Steel in Different Ranges of Temperature

Temperature range °C	*Steel* $(°C)^{-1}$
– 57 to – 30	0.0000108
– 30 to 2	0.0000110
2 to 25	0.0000112
25 to 53	0.0000113
53 to 81	0.0000115
81 to 109	0.0000117
109 to 136	0.0000119
136 to 163	0.0000121
163 to 191	0.0000122
191 to 218	0.0000124

Table 11.2 Coefficient of Linear Expansion of Aluminium in Different Ranges of Temperature

Temperature range °C	*Steel* $(°C)^{-1}$
– 57 to – 24	0.0000220
– 24 to 4	0.0000223
4 to 43	0.0000227
43 to 76	0.0000230
76 to 109	0.0000234
109 to 143	0.0000238
143 to 177	0.0000241
177 to 209	0.0000245

Table of V_h / V versus H/D for spheres

Table 11.3

H/D	V_h/V	H/D	V_h/V	H/D	V_h/V	H/D	V_h/V
0.001	0.0000030	0.026	0.0019928	0.051	0.0075377	0.076	0.0164500
0.002	0.0000120	0.027	0.0021476	0.052	0.0078308	0.077	0.0168739
0.003	0.0000269	0.028	0.0023081	0.053	0.0081292	0.078	0.0173029
0.004	0.0000479	0.029	0.0024742	0.054	0.0084331	0.079	0.0177369
0.005	0.0000747	0.030	0.0026460	0.055	0.0087423	0.080	0.0181760
0.006	0.0001076	0.031	0.0028234	0.056	0.0090568	0.081	0.0186201
0.007	0.0001463	0.032	0.0030065	0.057	0.0093766	0.082	0.0190693
0.008	0.0001910	0.033	0.0031951	0.058	0.0097018	0.083	0.0195234
0.009	0.0002415	0.034	0.0033894	0.059	0.0100322	0.084	0.0199826
0.010	0.0002980	0.035	0.0035893	0.060	0.0103680	0.085	0.0204467
0.011	0.0003603	0.036	0.0037947	0.061	0.0107090	0.086	0.0209159
0.012	0.0004285	0.037	0.0040057	0.062	0.0110553	0.087	0.0213900
0.013	0.0005026	0.038	0.0042223	0.063	0.0114069	0.088	0.0218691
0.014	0.0005825	0.039	0.0044444	0.064	0.0117637	0.089	0.0223531
0.015	0.0006682	0.040	0.0046720	0.065	0.0121258	0.090	0.0228420
0.016	0.0007598	0.041	0.0049052	0.066	0.0124930	0.091	0.0233359
0.017	0.0008572	0.042	0.0051438	0.067	0.0128655	0.092	0.0238346
0.018	0.0009603	0.043	0.0053880	0.068	0.0132431	0.093	0.0243383
0.019	0.0010693	0.044	0.0056376	0.069	0.0136260	0.094	0.0248468
0.020	0.0011840	0.045	0.0058928	0.070	0.0140140	0.095	0.0253603
0.021	0.0013045	0.046	0.0061533	0.071	0.0144072	0.096	0.0258785
0.022	0.0014307	0.047	0.0064194	0.072	0.0148055	0.097	0.0264017
0.023	0.0015627	0.048	0.0066908	0.073	0.0152090	0.098	0.0269296
0.024	0.0017004	0.049	0.0069677	0.074	0.0156176	0.099	0.0274624
0.025	0.0018437	0.050	0.0072500	0.075	0.0160313	0.100	0.0280000

Table 11.4

H/D	V_h/V	H/D	V_h/V	H/D	V_h/V	H/D	V_h/V
0.101	0.0285424	0.126	0.0436273	0.151	0.0615171	0.176	0.0820245
0.102	0.0290896	0.127	0.0442902	0.152	0.0622884	0.177	0.0828966
0.103	0.0296415	0.128	0.0449577	0.153	0.0630638	0.178	0.0837725
0.104	0.0301983	0.129	0.0456296	0.154	0.0638435	0.179	0.0846523
0.105	0.0307597	0.130	0.0463060	0.155	0.0646273	0.180	0.0855360
0.106	0.0313260	0.131	0.0469868	0.156	0.0654152	0.181	0.0864235
0.107	0.0318969	0.132	0.0476721	0.157	0.0662072	0.182	0.0873149
0.108	0.0324726	0.133	0.0483617	0.158	0.0670034	0.183	0.0882100
0.109	0.0330529	0.134	0.0490558	0.159	0.0678037	0.184	0.0891090
0.110	0.0336380	0.135	0.0497543	0.160	0.0686080	0.185	0.0900118
0.111	0.0342277	0.136	0.0504571	0.161	0.0694164	0.186	0.0909183
0.112	0.0348221	0.137	0.0511643	0.162	0.0702290	0.187	0.0918286
0.113	0.0354212	0.138	0.0518759	0.163	0.0710455	0.188	0.0927427
0.114	0.0360249	0.139	0.0525918	0.164	0.0718661	0.189	0.0936605
0.115	0.0366333	0.140	0.0533120	0.165	0.0726908	0.190	0.0945820
0.116	0.0372462	0.141	0.0540366	0.166	0.0735194	0.191	0.0955073
0.117	0.0378638	0.142	0.0547654	0.167	0.0743521	0.192	0.0964362
0.118	0.0384859	0.143	0.0554986	0.168	0.0751888	0.193	0.0973689
0.119	0.0391127	0.144	0.0562360	0.169	0.0760294	0.194	0.0983052
0.120	0.0397440	0.145	0.0569778	0.170	0.0768740	0.195	0.0992453
0.121	0.0403799	0.146	0.0577237	0.171	0.0777226	0.196	0.1001889
0.122	0.0410203	0.147	0.0584740	0.172	0.0785751	0.197	0.1011363
0.123	0.0416653	0.148	0.0592284	0.173	0.0794316	0.198	0.1020872
0.124	0.0423148	0.149	0.0599871	0.174	0.0802920	0.199	0.1030418
0.125	0.0429688	0.150	0.0607500	0.175	0.0811563	0.200	0.1040000

Table 11.5

H/D	V_h/V	*H/D*	V_h/V	*H/D*	VV_h/V	*H/D*	V_h/V
0.201	0.1049618	0.226	0.1301417	0.251	0.1573765	0.276	0.1864789
0.202	0.1059272	0.227	0.1311929	0.252	0.1585060	0.277	0.1876792
0.203	0.1068962	0.228	0.1322473	0.253	0.1596385	0.278	0.1888821
0.204	0.1078687	0.229	0.1333051	0.254	0.1607739	0.279	0.1900877
0.205	0.1088448	0.230	0.1343660	0.255	0.1619123	0.280	0.1912960
0.206	0.1098244	0.231	0.1354302	0.256	0.1630536	0.281	0.1925069
0.207	0.1108075	0.232	0.1364977	0.257	0.1641979	0.282	0.1937205
0.208	0.1117942	0.233	0.1375684	0.258	0.1653450	0.283	0.1949366
0.209	0.1127844	0.234	0.1386422	0.259	0.1664951	0.284	0.1961554
0.210	0.1137780	0.235	0.1397193	0.260	0.1676481	0.285	0.1973768
0.211	0.1147752	0.236	0.1407995	0.261	0.1688039	0.286	0.1986007
0.212	0.1157758	0.237	0.1418829	0.262	0.1699626	0.287	0.1998272
0.213	0.1167798	0.238	0.1429695	0.263	0.1711242	0.288	0.2010563
0.214	0.1177873	0.239	0.1440592	0.264	0.1722886	0.289	0.2022879
0.215	0.1187983	0.240	0.1451520	0.265	0.1734558	0.290	0.2035220
0.216	0.1198126	0.241	0.1462480	0.266	0.1746259	0.291	0.2047587
0.217	0.1208304	0.242	0.1473470	0.267	0.1757987	0.292	0.2059978
0.218	0.1218516	0.243	0.1484492	0.268	0.1769744	0.293	0.2072395
0.219	0.1228761	0.244	0.1495545	0.269	0.1781528	0.294	0.2084837
0.220	0.1239040	0.245	0.1506628	0.270	0.1793340	0.295	0.2097303
0.221	0.1249353	0.246	0.1517742	0.271	0.1805180	0.296	0.2109793
0.222	0.1259699	0.247	0.1528886	0.272	0.1817047	0.297	0.2122309
0.223	0.1270079	0.248	0.1540060	0.273	0.1828942	0.298	0.2134848
0.224	0.1280492	0.249	0.1551265	0.274	0.1840864	0.299	0.2147412
0.225	0.1290938	0.250	0.1562500	0.275	0.1852813	0.300	0.2160000

Table 11.6

H/D	V_h/V	*H/D*	V_h/V	*H/D*	V_h/V	*H/D*	V_h/V
0.301	0.2172612	0.326	0.2495361	0.351	0.2831159	0.376	0.3178133
0.302	0.2185248	0.327	0.2508554	0.352	0.2844836	0.377	0.3192218
0.303	0.2197908	0.328	0.2521769	0.353	0.2858531	0.378	0.3206318
0.304	0.2210591	0.329	0.2535004	0.354	0.2872244	0.379	0.3220432
0.305	0.2223298	0.330	0.2548260	0.355	0.2885973	0.380	0.3234561
0.306	0.2236028	0.331	0.2561536	0.356	0.2899721	0.381	0.3248704
0.307	0.2248781	0.332	0.2574833	0.357	0.2913485	0.382	0.3262862
0.308	0.2261558	0.333	0.2588149	0.358	0.2927267	0.383	0.3277033
0.309	0.2274358	0.334	0.2601486	0.359	0.2941065	0.384	0.3291219
0.310	0.2287180	0.335	0.2614842	0.360	0.2954881	0.385	0.3305418
0.311	0.2300026	0.336	0.2628219	0.361	0.2968713	0.386	0.3319632
0.312	0.2312894	0.337	0.2641615	0.362	0.2982562	0.387	0.3333859
0.313	0.2325784	0.338	0.2655030	0.363	0.2996428	0.388	0.3348100
0.314	0.2338697	0.339	0.2668466	0.364	0.3010310	0.389	0.3362354
0.315	0.2351633	0.340	0.2681920	0.365	0.3024208	0.390	0.3376621
0.316	0.2364590	0.341	0.2695394	0.366	0.3038123	0.391	0.3390902
0.317	0.2377570	0.342	0.2708886	0.367	0.3052054	0.392	0.3405195
0.318	0.2390572	0.343	0.2722398	0.368	0.3066000	0.393	0.3419502
0.319	0.2403595	0.344	0.2735928	0.369	0.3079963	0.394	0.3433821
0.320	0.2416640	0.345	0.2749478	0.370	0.3093941	0.395	0.3448154
0.321	0.2429707	0.346	0.2763045	0.371	0.3107935	0.396	0.3462498
0.322	0.2442795	0.347	0.2776632	0.372	0.3121944	0.397	0.3476856
0.323	0.2455905	0.348	0.2790236	0.373	0.3135968	0.398	0.3491225
0.324	0.2469036	0.349	0.2803859	0.374	0.3150008	0.399	0.3505607
0.325	0.2482188	0.350	0.2817501	0.375	0.3164063	0.400	0.3520001

Table 11.7

H/D	V_h/V	H/D	V_h/V	H/D	V_h/V	H/D	V_h/V
0.401	0.3534406	0.426	0.3898105	0.451	0.4267353	0.476	0.4640277
0.402	0.3548825	0.427	0.3912781	0.452	0.4282212	0.477	0.4655243
0.403	0.3563254	0.428	0.3927465	0.453	0.4297076	0.478	0.4670213
0.404	0.3577696	0.429	0.3942159	0.454	0.4311947	0.479	0.4685185
0.405	0.3592148	0.430	0.3956860	0.455	0.4326822	0.480	0.4700160
0.406	0.3606613	0.431	0.3971570	0.456	0.4341704	0.481	0.4715137
0.407	0.3621088	0.432	0.3986289	0.457	0.4356590	0.482	0.4730117
0.408	0.3635575	0.433	0.4001015	0.458	0.4371482	0.483	0.4745098
0.409	0.3650073	0.434	0.4015750	0.459	0.4386379	0.484	0.4760082
0.410	0.3664581	0.435	0.4030493	0.460	0.4401280	0.485	0.4775068
0.411	0.3679101	0.436	0.4045243	0.461	0.4416187	0.486	0.4790055
0.412	0.3693631	0.437	0.4060001	0.462	0.4431098	0.487	0.4805044
0.413	0.3708171	0.438	0.4074767	0.463	0.4446013	0.488	0.4820035
0.414	0.3722722	0.439	0.4089540	0.464	0.4460933	0.489	0.4835027
0.415	0.3737284	0.440	0.4104320	0.465	0.4475858	0.490	0.4850020
0.416	0.3751855	0.441	0.4119108	0.466	0.4490786	0.491	0.4865014
0.417	0.3766437	0.442	0.4133902	0.467	0.4505719	0.492	0.4880010
0.418	0.3781028	0.443	0.4148704	0.468	0.4520656	0.493	0.4895006
0.419	0.3795630	0.444	0.4163513	0.469	0.4535596	0.494	0.4910004
0.420	0.3810241	0.445	0.4178328	0.470	0.4550540	0.495	0.4925002
0.421	0.3824862	0.446	0.4193150	0.471	0.4565488	0.496	0.4940001
0.422	0.3839492	0.447	0.4207978	0.472	0.4580439	0.497	0.4955000
0.423	0.3854131	0.448	0.4222812	0.473	0.4595394	0.498	0.4970000
0.424	0.3868780	0.449	0.4237653	0.474	0.4610351	0.499	0.4985000
0.425	0.3883438	0.450	0.4252500	0.475	0.4625312	0.500	0.5000000

REFERENCES

[1] API 2555. Liquid Calibration of Tanks.
[2] API 2552. Measurement an Calibration Spheres and Spheroid.
[3] ISO 9091–1 1991. Calibration of Spherical Tanks in Ships (Stereo Photogrammetry).
[4] ISO 9091–2 1991. Calibration of Spherical Tanks in Ships (Refrigerated Light Hydro-carbon).
[5] API 2552. Measurement an Calibration Spheres and Spheroid.
[6] API 2551. Measurement and Calibration of Horizontal Tanks.
[7] ISO 9091–1 1991. Calibration of Spherical Tanks in Ships (Stereo Photogrammetry).
[8] ISO 9091–2 1991. Calibration of Spherical Tanks in Ships (Refrigerated Light Hydro-carbon).
[9] OIML R–45 1976. Casks and Barrels.

LARGE CAPACITY MEASURES

12.1 INTRODUCTION

There is no hard and fast demarcation between the capacities of small and large capacity measures. Normally any measure having capacity 50 dm^3 or more is termed as large capacity measure. Though there is no regulatory provision about their capacities, most of these are with nominal values of 50, 100, 200, 500, 1000 and 2000, 5000, 10000 dm^3 etc. Measures of capacity 250 dm^3 and 2500 dm^3 are also used in special cases. In this chapter, we will discuss material, form /design, dimensions, maximum permissible errors of a few of them.

12.2 ESSENTIAL PARTS OF A MEASURE

A measure basically consists of three parts.

1. A measuring neck with a window and a graduated scale
2. Body of the measure. The body constitutes the major capacity of the measure.
3. A delivery pipe with proper valves for a delivery measure.

12.2.1 Graduated Scale of the Measure

It is located at the top of a content measure and is below the main body for a delivery measure. Its nominal volume is 1% to 2% of the nominal capacity of the measure.

The distance between smallest graduations should not be less than 1 mm. The volume value between two consecutive marks is normally 1/10th of the maximum permissible error (MPE) of the measure. Volume of the graduated portion of the neck should not be less than the MPE of the measure to be verified against it. Taking in to consideration of maximum permissible errors of standard measures used at various levels, we may drive the values of the diameter and length of the neck of a measure Figure 12.1. The rectangle with solid lines is the scale ***S*** on a transparent glass. Outer rectangle *W* with dashed lines is the jacket in which the scale is snugly fit with no leakage of liquid.

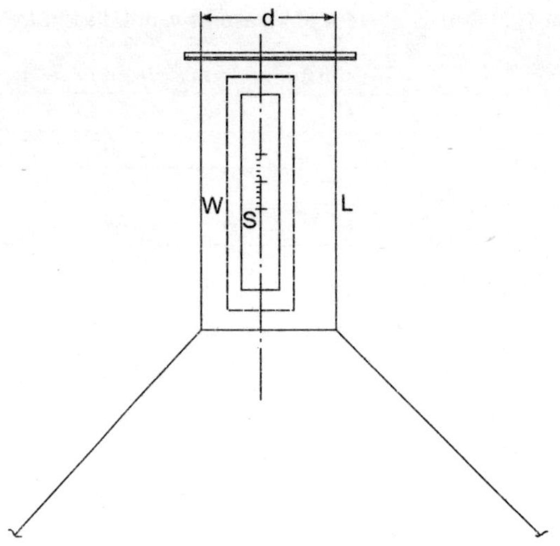

Figure 12.1 Design of the neck of a measure

12.2.1.1 General Approach

Criteria

MPE	0.01%	0.03%	0.1%
Distance between two successive graduation lines	1 mm	2 mm	3 mm

For 0.01% MPE and having a distance of 1 mm between two consecutive graduation marks, radius r of the neck must be such that

$$\pi r^2 \times 0.1 = V/100000 \; V \text{ is in cm}^3 \text{ and } r \text{ in cm.}$$

$$= V/100 \text{ If } V \text{ is in dm}^3 \text{ and } r \text{ in cm.}$$

$$\pi r^2 = V/10$$

$$r = \sqrt{(V/10\pi)} \text{ cm}$$

$$r = 0.178\,412\,31\sqrt{V} \text{ cm} \qquad ...(1)$$

For 0.03% MPE, but having a distance of 2 mm in between two consecutive graduation marks

$$\pi r^2 \times 0.2 = 3V/100 \text{ where } V \text{ is in dm}^3 \text{ and } r \text{ in cm.}$$

$$r = \sqrt{(3V/20\,\pi)} = 0.218\,509\,562\,\sqrt{V} \text{ cm} \qquad ...(2)$$

For 0.1% MPE, and having a distance of 3 mm between two consecutive graduation marks

$$0.3\pi r^2 = V/10000, \; r \text{ in cm } V \text{ in cm}^3$$

$$= V/10, \text{ where } V \text{ is in dm}^3 \text{ and } r \text{ in cm.}$$

$$r = \sqrt{(V/3\pi)} = 0.325\,74\,824\,\sqrt{V} \text{ cm} \qquad ...(3)$$

Using above formulae the neck radius for measures of 10 000 dm^3 (litres) to 100 dm^3 (litres) are given in Table12.1:

Table 12.1 Radius of Neck of Measures with Different MPE in cm

Capacity in dm³	*Maximum permissible error in percent of capacity*		
	0.01%	0.03%	0.1%
10 000	17.84	21.85	32.57
5000	12.62	15.45	23.03
2000	7.98	9.77	14.68
1000	5.64	6.91	10.30
500	3.98	4.89	7.23
200	2.52	3.09	4.61
100	1.78	2.18	3.25

12.2.1.2 Specific Set of Criteria

1. Ten smallest graduations are equal to the MPE of the standard measure.
2. MPE of the standard measure under consideration is 0.1% of its capacity.
3. Height of the smallest interval is 2 mm.
4. The capacity of the graduated scale should be at least equal to the MPE of the measure, which could be tested against it.
5. MPE of the measure under test should at least be 3 times the MPE of the standard measure.
6. The capacity of the neck is 2% of the nominal capacity of the measure.

Combining 1, and 2, gives the radius of the neck. Satisfying the criteria 1 to 4 gives the length of the graduated scale and number of scale intervals on either side of the graduation indicating its nominal capacity. Criteria 2 and 6 give the length of the neck.

$$\pi r^2.2.10^{-3} = 10^{-4}.V,$$

$$r^2 = V/20\ \pi, \text{ where } r \text{ is in m and } V \text{ in m}^3. \qquad ...(4)$$

By giving V the values of nominal capacity, we get the values of r and hence $2r$, but calculated value of r may have many decimal places, so we have to round of this value, which I have called as rounded off value and has also been indicated in the Table 12.2. Please refer to Figure 12.1.

Once the rounded off values of diameter of the neck is obtained then length of neck in metres can be given by the following formula

$$\pi r^2 L = 2V/100 \text{ or } V/50$$

$$L = V/50\pi r^2 = 0.006366199\ V/r^2 \qquad ...(5)$$

Table 12.2 Diameter and Length of the Neck of Measures with Specific Criteria

Capacity of the measure dm³	50	100	200	500	1000	2000	5000	10000
Calculated value of $2r$ in mm	60	80	112	180	252	356	564	796
Rounded off value of $2r$ in mm	60	80	120	80	250	360	560	800
Maximum permissible error cm³	50	100	200	500	1000	2000	5000	10000
Value of minor graduation cm³	5	10	20	50	100	200	500	1000
Length of neck cm	350	397.8	353.4	392.9	407.4	392.9	406.0	398.0
Rounded off value in mm	350	400	400	400	400	400	400	400
Volume of neck in cm³	990	2011	4524	10179	19635	40365	98520	1999334
Diff from 2% of V	–10	+11	+524	+179	–365	+365	–1480	–666

12.3 DESIGN CONSIDERATIONS FOR MAIN BODY

12.3.1 Measure Inscribed within a Sphere

The main body for the given volume should have the minimum surface area [1]. Minimum surface area not only reduces the cost of the material but also reduces errors due to change in relative humidity and temperature. For given volume, the surface area of a spherical cell is least. But the fabrication difficulties and flow of liquid along the surface of the measure do not allow us to have a spherical shape. Most common shape of a measure consists of a cylindrical body with two surmounted cones on either side of it. Let it be inscribed in a sphere of diameter D and radius R, Figure 12.2. Let volume of each surmounted cones be V_1 and that of cylindrical body be V_2.

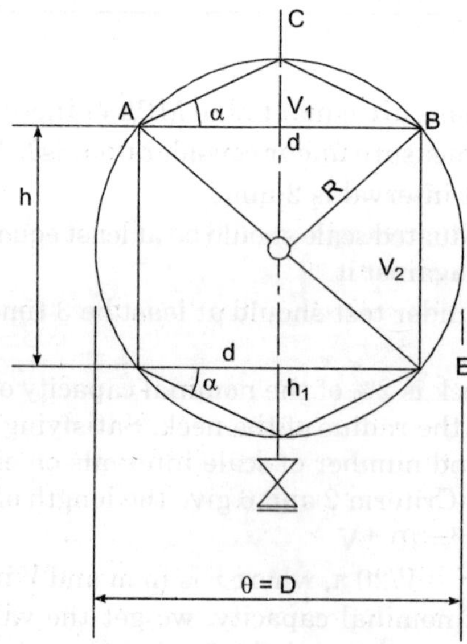

Figure 12.2 Cylinder surmounted by two cones inscribed in a sphere

V-the volume of the body of the measure is given as

$$V = V_2 + 2V_1 \qquad \text{...(6)}$$

Let h_1 be the height of each cone and h the height of the cylinder, then

$$h + 2h_1 = 2R \text{ Here } R \text{ is the radius of the sphere} \qquad \text{...(7)}$$

If the length of the chord of the circle, bounding the segment of height h_1, is 'd' then Volume of cone V_1 will be given as

$$V_1 = \pi d^2.h_1/12,$$

Similarly V_2 the volume of the cylindrical portion is

$V_2 = \pi\, d^2.h/4$, giving V-the total volume of the main body

$$V = 2V_1 + V_2, \text{ giving}$$

$$= (\pi\, d^2/6)h_1 + (\pi\, d^2/4)h$$

Substituting the value of h_1 in terms of h and R

$$V = (\pi\, d^2/4)(2h_1/3 + h) = (\pi\, d^2/4)\ \{(2R - h\,)/3 + h)\}$$

$$= (\pi\, d^2/4)[2R - h + 3h)/3.$$

$$V = (\pi\, d^2/6)(R + h) \qquad \text{...(8)}$$

Also from triangle ABE, Figure 12.2

$$d^2 = 4R^2 - h^2, \text{ giving } V \text{ as}$$

$$V = (\pi/6)(R+h)(4R^2 - h^2)$$

Differentiating V with respect of h we get

$$dV/dh = (\pi/6)[1(4R^2 - h^2) + (R+h)(-2h)] = (\pi/6)[-3h^2 - 2hR + 4R^2],$$

Putting dV/dh equal to zero gives us α relation between R and h for which volume is maximum. So from above, we get

$$3h^2 + 2hR - 4R^2 = 0, \text{ or}$$

$$h = R(-1+\sqrt{13})/3 \text{ and} \qquad ...(9)$$

$$d = (R/3)\sqrt{(22+2\sqrt{13})} \qquad ...(10)$$

But from the triangle ABC Figure 12.2, if α is the incline of the surface of the cone with horizontal, then

$$\tan(\alpha) = (2R - h)/d,$$

Substituting the value of h and d in terms of R, we get

$$\tan(\alpha) = [6R - R(-1+\sqrt{13})]/[R\sqrt{\{(22+2\sqrt{13})\}}]$$

$$\tan(\alpha) = (7-\sqrt{13})/[\sqrt{(22+2\sqrt{13})}], \text{ giving}$$

$$\alpha = 32^\circ\, 8' \qquad ...(11)$$

Substituting the value of $(R+h)$ from (9) and (10) in (8), we get V as

$$V = (\pi d^2/6)[(-1+\sqrt{13})/3 + 1]R$$

$$= (\pi d^2/6)[(2+\sqrt{13})]R/3, \text{ substituting the value } R \text{ in terms of } d$$

$$V = (\pi d^3/6)[2+\sqrt{13})]/\sqrt{(22+2\sqrt{13})}$$

$$d^3 = 1.841438852\, V \qquad ...(12)$$

$$h/d = (\sqrt{13}-1)/\sqrt{(22+2\sqrt{13})}, \text{ giving}$$

$$h = 0.482087.d \qquad ...(13)$$

If h_1 be the height of each cone, then $h_1 = (d/2)\tan(\alpha)$, giving

$$h_1 = 0.3140\,258.d \qquad ...(14)$$

From (12), the values of d of capacity measures from 50 to 10 000 dm^3 are calculated for $\alpha = 32^\circ\, 8'$, corresponding value of h_1 and h have also been calculated from (13) and (14) and are given in the Table 12.3. Refer Figure 12.2.

Table 12.3 Dimensions of Capacity Measures for α=32° 8′

Capacity	*d in mm*	*h in mm*	h_1 *in mm*
10000	2641	1273	829
5000	2096	1010	658
2000	1544	744	484
1000	1225	591	384
500	973	469	305
200	716	345	224
100	569	274	179
50	452	218	142

Above dimensions are only approximate, as we have not taken in to account the volume of the neck and volume of two non-existing cones by virtue of fitting the necks to the body having base diameter equal to that of the neck and height equal to (d^2) tan(α). We have also not considered the effect of fixing the necessary valves at the bottom on the capacity of the measure. Besides that there will be quite a few small ports for holding the temperature measuring devices.

12.3.2 General Case

Let us consider a very general case of a measure having a cylindrical body of diameter d, height h and angle α, which the cone is making at its base, refer to Figure 12.3. The volume V can be expressed as:

$$V = \pi h d^2/4 + \pi d^3 \tan(\alpha)/12 \qquad ...(14)$$

If S is the surface area then S is given as

$$S = \pi h d + (\pi/2) d^2 \sec(\alpha) \qquad ...(15)$$

Eliminating h from the above equations, we get

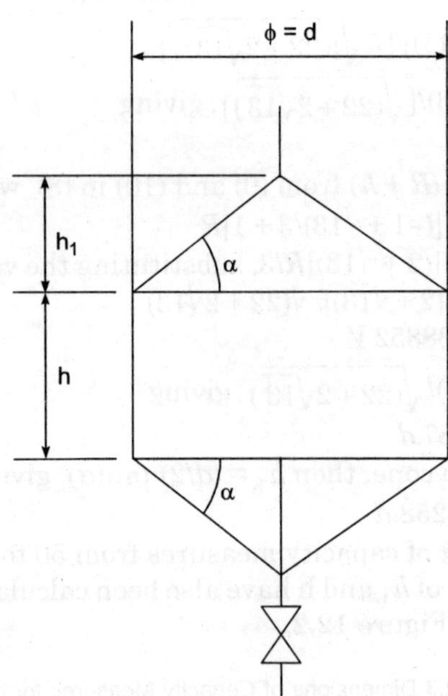

Figure 12.3 General form of cylinder with cones

$$S = 4V/d - \pi d^2 \tan(\alpha)/3 + \pi \sec(\alpha)(d^2/2), \qquad ...(15A)$$

There are two independent variables d and (α), so to make S minimum, partially differentiate (15A) with respect to each giving us

$$\delta S/\delta\alpha = 0 - (\pi/3)\, d^2 \sec^2(\alpha) + (\pi/2)\, d^2 \sec(\alpha)\tan(\alpha) \qquad ...(16)$$

$$\delta S/\delta d = -4V/d^2 - (2\pi/3) d \tan(\alpha) + (\pi 2.\, d/2)(\sec(\alpha)), \qquad ...(17)$$

For S to be minimum each of the partial derivative must be separately zero, giving us

$$\delta S/\delta\alpha = 0 = -\pi d^2 \sec^2(\alpha)[1/3 - \sin(\alpha)/2], \text{ giving}$$

$$\sin(\alpha) = 2/3, \text{ or}$$

$$\tan(\alpha) = 2/\sqrt{5} \text{ and} \qquad ...(18)$$

$$\cos(\alpha) = \sqrt{5}/3$$

Equating $\delta S/\delta d$ equal to zero gives

$$\delta S/\delta d = 0 = -4V/d^2 - 2\pi d[\tan(\alpha)/3 - 1/(2\cos(\alpha)] \quad ...(19)$$

Writing the values of $\tan(\alpha)$ and $\cos(\alpha)$ from (18), we get

$$-4V/d^2 - 2\pi d[2/3 - 3/2]/(\sqrt{5}) = -4V/d^2 + \pi d.(5/3\sqrt{5}) = 0, \text{ giving us}$$

$$d^3 = 12\sqrt{5}\, V/(5\pi) \quad ...(20)$$

from (14) we may write h as

$$h = 4V/\pi d^2 - (d/3)\tan(\alpha) \quad ...(21)$$

$$= 4V/\pi d^2 - 2d/(3\sqrt{5}) \quad ...(22)$$

If h_1 be the height of each cone then

$$h_1 = (d/2)\tan(\alpha) \quad ...(23)$$

$$= d/\sqrt{5}, \quad ...(24)$$

From (18)

$$\sin(\alpha) = 2/3 \text{ gives}$$

$$\alpha = 41^\circ 50'.$$

Using (20), (22) and (24), we can express the values of d, h and h_1 for $\alpha = 41^\circ 50'$ as follows:

$$d^3 = 1.70823046\, V \quad ...(20A)$$

$$h = 4V/\pi d^2 - 2d/(3\sqrt{5}) = 1.27323981\, V/d^2 - 0.298142397\, d \quad ...(22A)$$

$$h_1 = 0.447231595\, d \quad ...(23A)$$

The values of d, h, and h_1 obtained for different capacity measures are given in Table 12.4

Table 12.4 Dimensions of Capacity Measures for 41°, 50′ in mm

Capacity	*d*	*H*	h_1
10000 dm³	2575.4	1151.8	1151.8
5000 dm³	2044.1	914.2	914.2
2000 dm³	1506.1	673.6	673.6
1000 dm³	1195.4	534.6	534.6
500 dm³	948.8	424.3	424.3
200 dm³	699.1	312.6	312.6
100 dm³	554.9	248.1	248.1
50 dm³	440.4	196.9	196.9

However from the point of view of fabrication, to measure and thus maintain $\alpha = 41^\circ 50'$ is difficult, so instead of this value if we take $\alpha = 45^\circ$

then $h = d$

Use of (19), (21) and (23) give

$$d^3 = 12V/\pi(3\sqrt{2} - 2) = 1.703\,2197\, V \quad ...(25)$$

$$h = 4V/\pi d^2 - d/3 \quad ...(26)$$

$$h_1 = d/2 \quad ...(27)$$

Calculating from (25) the values of d for different capacity of measures and substituting the values of d in expressions (26) for h and in (27) for h_1, we get the values of these parameters.

So dimensions of these parameters for measures of all capacities are calculated and are given in Table 12.5 (Reference Figure 12.3).

Table 12.5 Dimensions of Capacity Measures for α = 45°

Capacity	*d mm*	*h mm*	h_1 *mm*
10000	2572.9	1065.7	1286.5
5000	2042.1	845.9	1021.1
2000	1504.6	623.2	752.3
1000	1194.2	496.7	597.1
500	947.9	392.6	473.9
200	698.4	289.3	349.2
100	554.3	229.6	277.2
50	440.0	182.2	220.0

12.4 DELIVERY PIPE

The delivery pipe may be straight and cylindrical, or a slant conical pipe fitted with proper valves. Straight pipes though are simpler to design and construct, pose a problem of fitting the inlet and outlet valves. The valves are available only with certain dimensions, which may not suit to the diameter obtained by calculations. Alternative is, to use reducers of proper sizes to suit the size of the valve. But this will obstruct the flow as well as retain a variable and unknown volume of the liquid. So instead of straight vertical cylinder, conical pipes inclined at different angles to horizontal may be used. Conical pipes have an advantage of having end section suitable to the size of the valve. Further not only flow in conical pipes is smooth but also retention of volume is minimal and constant if any.

12.4.1 Slant Cone at the Bottom

We may choose a slant cone for lower portion. The radius of the cone is equal to the base of the main body (cylindrical portion) and one side is vertical while the other is slant. The slant side of the cone will help in delivering the liquid fast and with better reproducibility, refer to Figure 12.4.

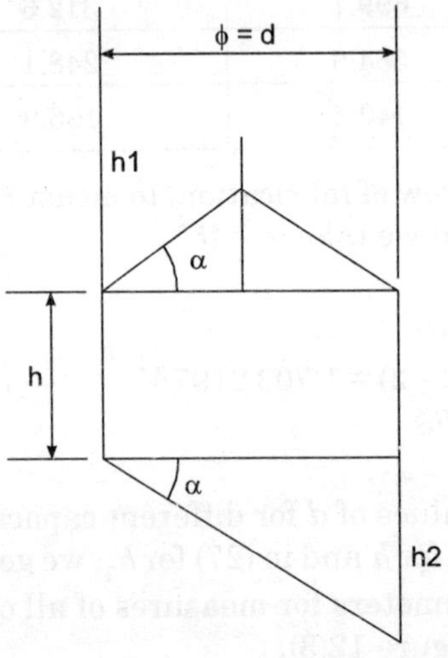

Figure 12.4 Measure with slant cone as delivery pipe

Using similar notations as before the Volume V and surface S of the measure may be expressed as

$$V = \pi d^3 \tan(\alpha)/24 + \pi d^2 h/4 + \pi d^3 \tan(\alpha)/12$$
$$= (\pi/8)\, d^3 \tan(\alpha) + (\pi/4) d^2 h$$
$$S = (\pi/4)\, d^2 \sec(\alpha) + \pi d h + (\pi/4)\, d^2 \sec(\alpha) \sqrt{\{1 + 3\sin^2(\alpha)\}}$$
$$= (\pi/4)\,(d^2 \sec(\alpha)\{1 + \sqrt{\{1+3\sin^2(\alpha)\}} + 4V/d - (\pi/2)\, d^2 \tan(\alpha)$$

Taking a fixed arbitrary value of α and minimizing the surface S by differentiating with respect to d and putting it to zero, we get

$$d^3 = (8V \sec(\alpha)/\pi)/[\{1 - 2\sin(\alpha)\} + \sqrt{\{(1+3\sin^2(\alpha)\}}]$$

Other parameters, like h height of the cylindrical portion and h_1 the height of the cone at the top and h_2 in terms of d are given as

$$h = 4V/\pi d^2 - (d/2)\tan(\alpha)$$
$$h_1 = d \tan(\alpha)/2 \text{ and height } h_2 \text{ of the lower portion as}$$
$$h_2 = d \tan(\alpha)$$

The values of d, h, h_1, and h_2 have been calculated for $\alpha = 30°$(refer to Figure 12.4) for capacity measures of various capacities and are given in the Table 12.6.

Table 12.6 Dimensions of Capacity Measures for $\alpha = 30°$

Capacity dm³	*d mm*	*h mm*	h_1 *mm*	h_2 *mm*
10000	2555	1213	738	1476
5000	2028	963	585	1170
2000	1494	710	431	862
1000	1186	563	342	684
500	941	447	272	544
200	693	330	200	400
100	550	265	158	316

Again referring to Figure 12.4, taking $\alpha = 45°$ the corresponding values of d, h, h_1 and h_2 are given in Table 12.7.

Table 12.7 Dimensions of Capacity Measures for $\alpha = 45°$

Capacity dm³	*d mm*	*h mm*	h_1 *mm*	h_2 *mm*
10000	2489.6	809.5	1248.8	2489.6
5000	1976.0	642.5	988.0	19760
2000	1455.9	642.5	728.0	1455.9
1000	1155.6	473.4	577.8	1155.6
500	917.2	298.2	458.6	917.2
200	675.8	219.7	337.9	675.8
100	536.4	174.4	268.2	536.4
50	425.7	138.4	2212.9	425.7

12.4.2 Measures with Cylindrical Delivery Pipe

The measures have neck and delivery pipe of same dimensions.

12.5 SMALL ARITHMETICAL CALCULATION ERRORS

While establishing a relation between *d*-the diameters of the cylindrical portion of the main body with *V* the volume, we have taken *V* as the nominal capacity of the measure. In fact we should reduce *V* by the volume of the neck. In case, neck and delivery pipe are of same dimensions and shape, we should reduce *V* by two times the neck volume. Further in calculating the volume of the body we have taken the full volume of the surmounting cones, but in fact body volume will be less by the volume of the cone having base diameter equal to that of neck and semi-vertical angle that of the upper cone of the body. We have also not considered a device, which can adjust for the error, which may occur in fabricating the measure in a workshop. So while finally calculating the dimensions of the body we take into account for the neck volume and volume of the cone. I may emphasise that volume of neck is to be subtracted from *V* while volume of cone, which does not exists, is to be added to *V*. I also intend to give a simple device to adjust the capacity of the measure, which can be used to fix a device for temperature measurement also.

12.5.1 Adjusting Device

Instead of adjusting any parameter of the main body a separate adjusting device may be provided. The capability of the adjusting device may be up to the tune of 1% of the nominal capacity of the measure. But the adjusting device is feasible for smaller measures. It is a close hollow threaded cylinder *C* of appropriate dimension fitted at right angles to any slant surface of the cone Figure 12.5. The cylinder moves on a nut W welded to the measure. It is moved out for increasing the capacity and drawn in for decreasing the capacity of the measure. To fix the position of cylinder a locking nut L is provided. The nut L moves on the cylinder and is brought in contact to the welded nut through a quarter pin it is fixed with the welded nut.

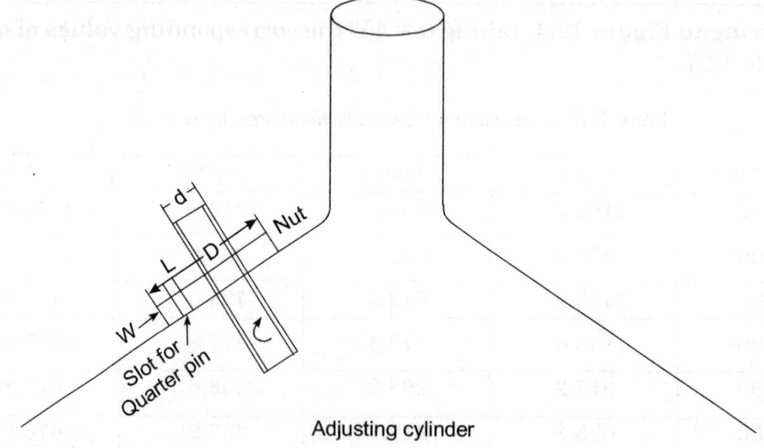

Figure 12.5 Adjusting device

12.6 DESIGNING OF CAPACITY MEASURES

We have discussed the design essentially for two types of main body of the measures. Namely (1) with cylindrical neck and delivery pipe, and body comprising of a cylinder surmounted by two cones, and (2) with cylindrical neck with a body comprising of cylinder surmounted by a cone on top and a slant cone at the bottom. Design (1) is perfectly symmetrical but design (2) is asymmetrical.

So for complete design of the measure, design the neck first, length and diameter of the neck rounded off in terms of mm for measures of capacities 50 to 1000 dm^3 and rounded off to 5 mm for larger measures. Recalculate volume of the neck with final values of its diameter and length. Calculate the volume of non-existing cone having base diameter equal to that of the neck and height given by the value of angle α assumed for surmounted cone of the main body. Subtract the volume of the neck and add volume of the non-existing cone to the nominal capacity of the measure. The value of this volume is used to calculate d diameter and h height of the cylindrical body and the height h_1 of the surmounting cone or cones.

Use the rounded off values of d–the diameter of the body and adjust the height of the cylindrical body to give the desired volume. Instead of height of the cone give the rounded off values of the slant height of the cone. For a workshop worker it is much easier to work out the sheet for a frustum of a cone with its slant height rather than vertical height. In rounding off process adjust the angle α rather than any other dimension.

12.6.1 Symmetrical Content Measures

The diagram of one such measure is given in Figure 12.6. Neck is graduated and body is made of cylinder surmounted with one cone. Base of the cylinder is fitted with an out let valve directly nearest to the bottom. A rim of sufficient strength is attached at the bottom end, so that bottom does not touch the ground or the surface at which the measure is placed.

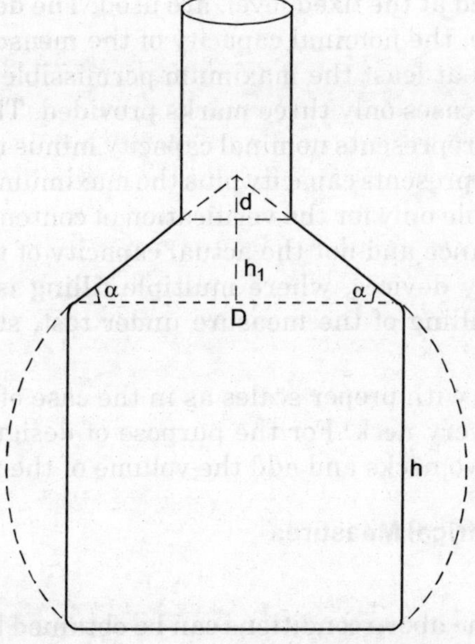

Figure 12.6 Symmetrical content measure

Table 12.8 Dimensions of Content Measures Data Table 12.5 (Figure 12.6)

Cone angle ALPHA = 45°

Capacity dm^3	*Diameter of body mm*	*Volume of cone dm^3*	*Volume of body dm^3*	*Volume of neck dm^3*	*h_1 mm*	*h mm*
10000	2570	2154.9	7644.0	201.1	1286	1473.4
5000	2045	1095.9	3803.8	100.3	1021	1158.0
2000	1505	440.1	1519.1	40.7	752	853.9
1000	1195	221.3	759.0	19.6	597	677
500	950	111.5	378.4	10.2	474	534
200	700	44.7	151.3	3.9	349	393
100	550	21.7	76.3	2.0	277	321
50	440	11.1	37.89	0.990	220	249

12.6.2 Asymmetrical Content Measure (with a Conical Outlet)

Process of the calculations will be practically the same, except instead of final adjustment of α adjust the height of the cylindrical body.

12.6.3 Measures with Cylindrical Delivery Pipe

It is convenient to have the delivery pipe as the measuring device. The rim of the upper neck is made bevelled and flat. The capacity of the measure is defined till the rim of the upper neck. For ensuring we may use a glass flat plate and filling the measure and sliding the glass plate on the rim and looking for any air bubble. Alternatively some sort of devices, by which measure is filled up to a certain fixed level only for example some level detecting device or simply over flowing through a hole provided at the fixed level, are used. The delivery neck is graduated in full. On the centre of the scale, the nominal capacity of the measure is marked prominently. The either side of scale covers at least the maximum permissible error of the measure to be calibrated against it. In some cases only three marks provided. The central mark represents nominal capacity, upper mark represents nominal capacity minus maximum permissible error in deficiency and lower mark represents capacity plus the maximum permissible error in excess. This type of measures is suitable only for the verification of content measures. In verification, we have to see for the compliance and not the actual capacity of the measure. But in case of calibrations of larger capacity devices, where multiple filling is required, the neck has a measuring device. For final filling of the measure under test, standard measures of much smaller capacity are used.

Design the delivery neck with proper scales as in the case of neck in section 12.5.1; use same dimensions for the delivery neck. For the purpose of designing the parameters of the body subtract the volume of two necks and add the volume of the two cones.

12.6.4 Dimensions of Symmetrical Measures

12.6.4.1 Content Measures

A content measure fulfilling the above conditions can be obtained by removing the lower neck as shown in Figure 12.7. The outline of the measure is shown by solid line.

12.6.4.2 Delivery Measures

Calculated dimensions of measures with cylinder as delivery pipe whose body is inscribed inside a circle (α = 32°, 8′) are given in Table 12.9. Dimensions of those measures whose body surface area has been minimised giving α = 41°, 50′ and those of with α = 45° (arbitrarily chosen) are given in Tables 12.10 and 12.11 respectively. Refer Figure 12.7.

Dimensions of neck and delivery pipe are same.

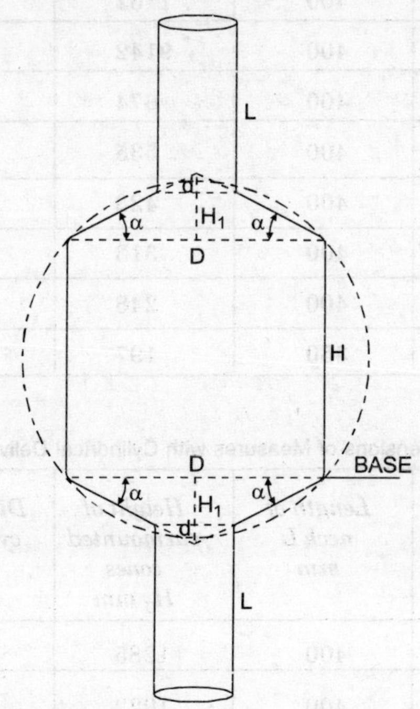

Figure 12.7 Symmetrical delivery measure

Table 12.9 (α = 32°,8) Dimensions of Measures with Cylinder as Delivery Pipe (Figure 12.7)

Capacity dm³	*Diameter of neck d mm*	*Length of neck L mm*	*Height of surmounted cones H_1 mm*	*Diameter of cylindrical body D mm*	*Height of cylindrical body H mm*
10000	800	400	829	2640	1088
5000	565	400	658	2095	1259
32000	360	400	484	1545	919
1000	250	400	384	1225	738
500	180	400	305	973	584
200	112	400	224	716	234
100	80	400	179	570	342
50	60	350	138	440	266

Table 12.10 (41°, 50) Dimensions of Measures with Cylindrical Delivery Pipe (Figure 12.7)

Capacity dm^3	*Diameter of neck d mm*	*Length of neck L mm*	*Height of surmounted cones* H_1 *mm*	*Diameter of cylindrical body D mm*	*Height of cylindrical body H mm*
10000	800	400	1152	2570	1182
5000	565	400	9142	2040	925
2000	360	400	674	1505	681
1000	250	400	535	1195	538
500	180	400	423	949	426
200	112	400	313	699	314
100	80	400	248	555	257
50	60	350	197	440	198

Table 12.11 (α = 45°) Dimensions of Measures with Cylindrical Delivery Pipe (Figure 12.7)

Capacity dm^3	*Diameter of neck d mm*	*Length of neck L mm*	*Height of surmounted cones* H_1 *mm*	*Diameter of cylindrical body D mm*	*Height of cylindrical body H mm*
10000	800	400	1285	2570	1097
5000	565	400	1022	2045	855
2000	360	400	752	1505	629
1000	250	400	597	1194	497
500	180	400	474	948	393
200	112	400	349	698	289
100	80	400	277	555	238
50	60	350	220	440	182

12.6.5 Delivery Measures with Slant Cone as Delivery Pipe

Designing procedure is same as in section 12.6.2. Taking above points in to considerations, the dimensions for two types of measures are given below:

Design data for the neck would remain same for both sets of measures. In the first set the slant cone makes angle of 30° with horizontal, other parameters of the measures are given in Table12.6, and second set for which the axis of cone makes angle of 45° and its other data is given in Table 12.7 above. Refer to Figure 12.8.

12.6.5.1 Dimensions of Measures with Slant Cone as Delivery Pipe

Dimensions of measures including those of neck are given in Table 12.12. The semi-vertical angle of surmounted cone is 60°.

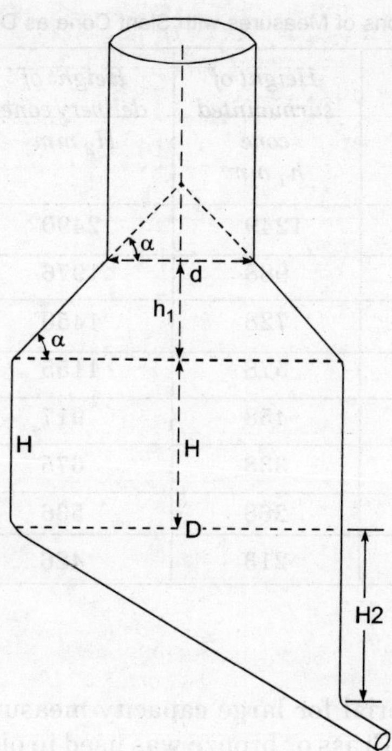

Figure 12.8 Delivery measures with slant cone as delivery pipe

Table 12.12 (α = 30°) Dimensions of Measures with Slant Cone as Delivery Pipe (Figure 12.8)

Capacity dm^3	*Diameter of neck d mm*	*Length of neck L mm*	*Height of surmounted cones* h_1 *mm*	*Height of delivery cone* H_2 *mm*	*Diameter of cylindrical body D mm*	*Height of cylindrical body H mm*
10000	800	400	738	1480	2560	1211
5000	565	400	585	1170	2030	970
2000	360	400	431	862	1500	701
1000	250	400	342	684	1185	566
500	180	400	272	544	941	448
200	112	400	200	400	691	334
100	80	400	158	316	550	262
50	60	350	127	254	440	202

12.6.5.2 Dimensions of Measures with Slant Cone as Delivery Pipe

Dimensions of measures including those of neck are given in Table 12.13. The semi-vertical angle of surmounted cone is 45°.

Table 12.13 (α = 45°) Dimensions of Measures with Slant Cone as Delivery Pipe (Figure 12.8)

Capacity dm^3	*Diameter of neck mm*	*Height of neck mm*	*Height of surmounted cone h_1 mm*	*Height of delivery cone H_2 mm*	*Diameter of D of cylindrical body mm*	*Height of cylindrical body H mm*
10000	800	400	1249	2490	2490	676
5000	565	400	988	1976	1975	537
2000	360	400	728	1450	1450	385
1000	250	400	578	1155	1155	316
500	180	400	458	917	917	250
200	112	400	338	675	675	188
100	80	400	268	536	536	146
50	60	350	213	426	426	109

12.7 MATERIAL

The most common and best material for large capacity measures is stainless steel. Next in order of merit is galvanised steel. Brass or bronze was used in old days especially for standard measures established by a Law of the Country. The aforesaid materials, in form of sheet only, are used in fabrication of measures, especially large capacity measures. The necks and delivery portion are made of mild steel pipes or plates. In case of mild steel it is advisable to have a corrosion resistance coating all over the inner and outer surface. Inside surface is made smooth. Special care should be taken in smoothing out all the joints. All extra welding material should be carefully removed.

12.7.1 Thickness of Sheet used

After careful consideration of rigidity of material and hydrostatic pressure which the walls of the measures are supposed to experience. Preferable thickness of the sheet of mild steel to be used in capacity measure is given in Table 12.14.

Table 12.14 Thickness of Mild Steel Sheet Used

Capacity of measure in dm^3	*Minimum thickness in mm*
Upto 500 dm^3	1.8
1000 and 1500 dm^3	2.5
1500 to 5000 dm^3	4.0
Above 5000 dm^3	6.0

12.8 CONSTRUCTION OF MEASURES

12.8.1 Steps for Construction

Steps-wise construction

1. Neck and delivery pipe is made according to the design. Care is to be taken for a good welding. Use a hand grinder to make the welding smooth and free from burs.
2. Capacity of neck and delivery pipe is checked, by using two plane glass discs as base and top.
3. Make the frustum of the upper cone and see for nice welding, weld it to neck. Make the welding smooth and free from burs.
4. Make the cylindrical body and smooth out the welding. Join it with the delivery pipe with frustum of lower cone if any. Make sure about the good welding. At this stage, rough estimate of the capacity is again made to see if some changes in to the delivery pipe are required. We may increase the height and diameter to certain extent. See that inside surface is smooth and is without dents.
5. Join the neck to the main body and fill it with water through a calibrated measure by volume transfer method. Adjust the position of the scale, in such a way that central line of the scale indicates the nominal capacity.

12.8.2 Requirements of Construction

1. The measure should be fully welded construction, all welds being external and continuous with good penetrations with no scales. All joints should be smooth and free from projections. Resort to grinding if necessary.
2. Measures should be free from surface defects and indentations. External surface may be painted and inside surface may be coated with good quality epoxy resins. If the surface is water repellent it will work better.
3. Content measures with no delivery pipe should be provided with filling pipes reaching to almost bottom of the measure. This ensures that water is not poured but starts moving up slowly without dissolving air.
4. A baffle plate may be provided to minimise the turbulence and vortex formation in delivery measures. Baffles should be so designed that they do not trap air or liquid during filling and emptying.
5. The outlet valve should be so constructed and fitted to the delivery device, so that measure is completely emptied.
6. A manhole or hand hole may be provided to enable cleaning and inspection.

12.8.3 Stationary Measure

A stationary measure should be so installed that its axis is vertical on permanent supports secured to the ground. A typical arrangement is shown in Figure 12.9.

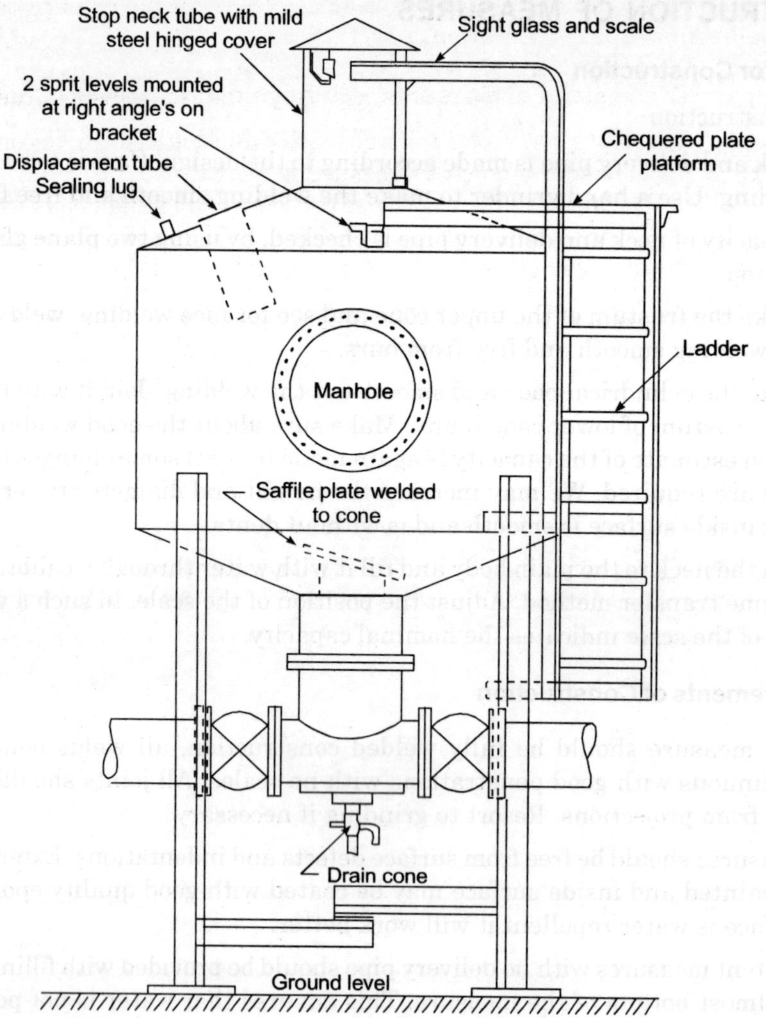

Figure 12.9 A permanently installed measure

12.8.4 Portable Measure

A portable measure supported by legs should be provided with sufficient jacks attached permanently to base of the cradle to enable it to be leveled in two planes. The measure should be provided with two spirit levels fixed at right angles to each other. Some measures may be vehicle mounted.

12.9 DIMENSIONS OF MEASURES OF SPECIFIC DESIGNS

There are quite a few modifications to the designs of measures used by different national metrology laboratories. Normally design of neck is universal. The difference is only in the delivery system. Some have used arbitrary round off values in terms of centimetres for smaller measures and in terms of decimetres for larger measures. They have not restricted themselves to have minimum surface area condition.

12.9.1 Design and Dimensions of Measures with Asymmetric Delivery Cone

The base cone is obtained by revolving a triangle with different base angles once about the axis of the measure. A basic design of such a measure is shown in 12.10. SIM, France has been using such measures. The author procured these drawings and dimensions while on study tour in the year 1967. The author wishes to thank SIM for providing the literature.

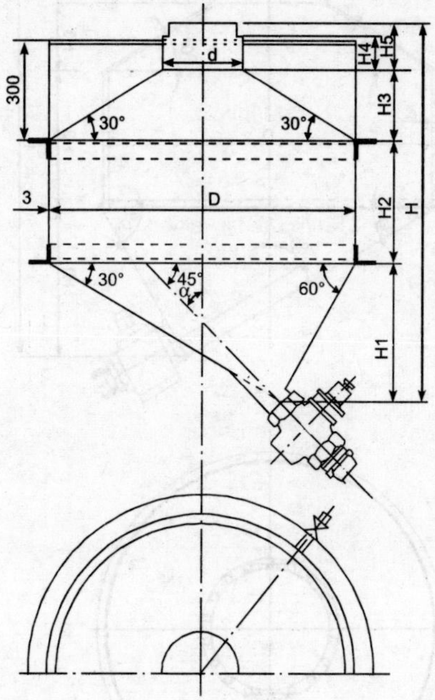

Figure 12.10 Measure with asymmetric base

12.9.1.1 Recommended Dimensions

The dimensions given in Table 12.15 (Figure 12.10) belong to the measures with shorter neck without any measuring scale.

Table 12.15 Measures with Asymmetric Base and Small Neck

Capacity	*D*	*d*	*H*1	*H*2	*H*3	*H*4	*H*5	α°	*H*
1000	1200	350	520	590	245	110	145	45	1500
500	1000	250	435	390	216	110	159	45	1200
200	800	201	350	198	180	64	120	60	848
100	550	139	240	183	120	72	120	60	763
50	550	94	240	75	133	55	120	60	568

Irregularities in some dimensions like H4, H3 are due to final adjustment of the nominal capacity of the measure.

The dimensions given in Table 12.16, Figure 12.11 belongs to the measures with longer neck having a measuring scale. *Ls* and *Ln* are respectively the length of scale and the neck.

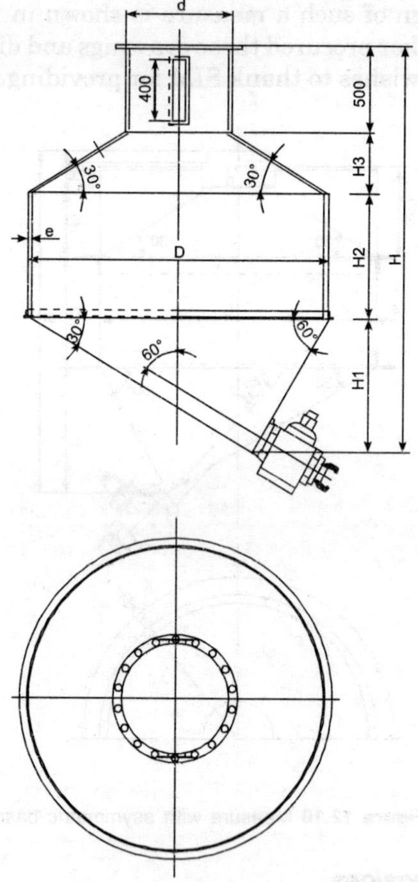

Figure 12.11 Standard measures with longer necks

Table 12.16 Measures with Asymmetric Base with Neck having a Measuring Scale

Capacity	*D*	*d*	*H*1	*H*2	*H*3	*Ls*	*Ln*	*H*
3000	1800	615	780	720	348	400	500	2348
2000	1500	500	650	748	288	400	500	2186
200	800	201	350	192	180	240	310	1032
100	550	139	240	177	120	240	310	947
50	550	94	240	72	133	240	310	755

12.9.2 Measures Designed at NPL, India

We designed our own capacity measures at National Physical Laboratory, New Delhi, India.

The measures are symmetrical with neck and delivery pipe of same design. The neck or the delivery pipe both have measuring scales so that a single measure may be used for calibrating both content as well as delivery measures. A typical measure, with graduated scale, is shown in Figure 12.12.

The recommended dimensions are given in Table 12.17. *L* in the figure represent all around *L* shaped strip welded to the measure for supporting it on a tripod with a circular ring.

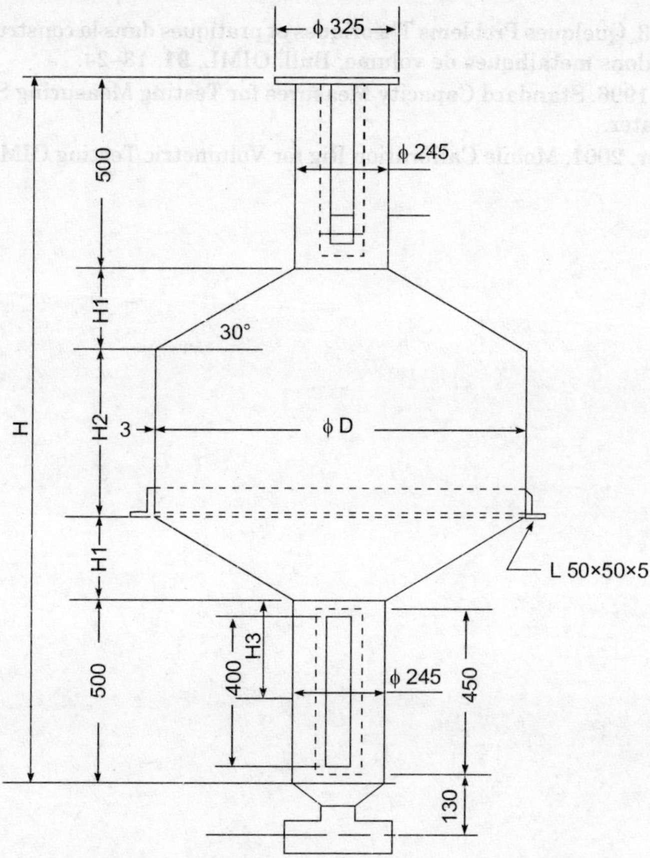

Figure 12.12 NPL designed measures with windows

Table 12.17 Dimensions of Measures Designed at NPL India

Capacity	*D*	*H1*	*H2*	*H3*	*d*
2000	1500	288	800	240	501
1000	1200	245	617	240	352
500	1000	220	415	240	245
200	700	160	356	240	150
100	540	124	312	240	110
50	420	100	265	240	75

$$H = 1000 + 2H1 + H2$$

Here efforts are made to have rounded off values for diameters of the cylindrical body and to see that the set of measures if placed in a room give an aesthetic look. Not that a bigger measure has a smaller dimension than the smaller measure.

REFERENCES

[1] Nadolo, A.1983. Quelques Problems Theoriques et pratiques dans la construction et etalonnage des jauges etalons metalliques de volume, Bull. OIML, **91**, 13–24.

[2] OIML R-120, 1996. Standard Capacity Measures for Testing Measuring Systems for Liquids other than Water.

[3] Kleppan Roger, 2001. Mobile Calibration Rig for Volumetric Testing OIML Bulletin, Volume **XLII**, 5-8.

VEHICLE TANKS AND RAIL TANKERS

13.1 INTRODUCTION

Different names are given to tanks mounted on a vehicle. We, in India, call them vehicle tanks [1], in USA and Canada these are called tank cars [2], while Europeans call them as road tankers [3]. These vehicle tanks are used for transporting milk, petroleum and its products. These are not only used for transport of a petroleum product or milk but to vend it. Vending is carried out either in units of one full tank at a time or one compartment of it or sometimes in a continuous quantity also. In such cases, vehicle is provided with meter as output measuring device. So these come in the purview of legal metrology commonly known as Weights and Measures. When vehicle tank delivers all its content in one step than tank is taken as a capacity measure with no partition, otherwise each compartment is taken as a separate capacity measure. If the tank can vend partial volumes in continuous form, then it is taken as a measuring instrument.

13.1.1 Definitions

13.1.1.1 Vehicle Tank

An assembly used for transport and delivery of liquids. It comprises of a tank, which may or may not be divided in to compartments, and a vehicle. If the driving vehicle is motor driven, then the system is vehicle tank, if it is a train, then it is called rail tanker. Basically both are same in use and purpose. The only basic difference is the capacity of the tank. Vehicle tanks have a capacity range of 0.5 m^3 to 50 m^3 while that of rail tanks is from 10 m^3 to 120 m^3. The tank may be permanently mounted on the chassis of a vehicle or on a detachable temporary mount on the vehicle. The tanks are either attached to a trailer or it is mounted on the chassis of truck, in this case one may call it as self-propelled.

13.1.1.2 Shell

Shell is cylindrical portion of the tank.

13.1.1.3 Heads

Heads are the closing ends of the shell.

13.1.1.4 Nominal Capacity

Nominal capacity of a rail or road tanker is the volume of the liquid at the reference temperature, which the tank contains under operating conditions.

13.1.1.5 Total Contents

The maximum volume at reference temperature of the liquid, which the tank can contain to the stage of overflowing, under rated operating conditions.

13.1.1.6 Expansion Volume

The difference in the total content and the nominal capacity is expansion volume.

13.1.1.7 Calibration

The calibration consists of all operations necessary to determine the capacity of the tank at one or several levels. The levels may be marked on a scale (dipstick) or are realisable in some other way.

13.1.1.8 Dipstick

Dipstick is a square or a rectangular metal bar of brass or any other hard material suitable to be used in the liquid, which the vehicle tank intends to transport and deliver.

13.1.1.9 Ullage Stick

It is a T-shaped metal bar of brass or other suitable material used to determine the depth of the level of the liquid from proof level.

13.1.1.10 Proof Level

Proof level is the reference level to which all depth measurements are related.

13.1.1.11 Dip Pipe

A pipe rigidly attached to the top of the tank extending vertically downward up to approximately 15 cm from the bottom of the tank. The pipe has perforations at the top above the maximum level.

13.1.1.12 Vertical Measurement Axis

The vertical line along which levels of liquid are gauged is the vertical measurement axis.

13.1.1.13 Reference Point

A point on the vertical measurement axis with reference to which ullage height is measured.

13.1.1.14 Reference Height H

The distance, measured along the measurement axis, between the reference point and foot of the vertical measurement axis, on the inner surface of the tank or on datum plate.

13.1.1.15 Ullage Height (C)

The distance between the free surface of the liquid and the reference point (proof level), measured along the vertical measurement axis.

13.1.1.16 Sensitivity of a Tank

Sensitivity of the tank in the vicinity of the filling level is the change in level Δh, divided b, the corresponding relative change in volume- $\Delta V/V$ corresponding to the level h. So Sensitivity S is given by

$$S = V\Delta h/\Delta V$$

13.1.1.17 Calibration Table (Gauge Table)

Gage table is the expression, in the form of a table, or of the mathematical function $V(h)$ representing the relation between height h as independent variable and the volume $V(h)$ as dependent variable.

13.1.2 Basic Construction

Essentially it consists of a cylindrical horizontal tank mounted on a vehicle. Horizontal tank is divided in three to four parts; each is called as compartment. Compartments if exist are totally isolated from each and are connected to an outlet valve in such a way that at a time only one compartment is connected to it. The tank is mounted with at least at 2° degrees of inclination with the horizontal on the vehicle so that it drains out completely. Thickness of the sheet used in construction is about 2.5 mm to 3 mm for shell and 3 to 4 mm for ends. However thickness may be more for rail tankers, which are of much larger capacity.

The discharge device is comprised of a discharge pipe with stop valve at its end. Sometimes positive displacement meter is provided for discharge measurement. A foot valve may be provided to stop flow of liquid between tank and the discharge pipe. Some tanks may incorporate devices fitted at the lowest point for water separation.

Vehicle tanks are provided with a ladder giving access to the dome and a platform for the operator affecting the measurements or checking the tank.

Liquids of one or more than one compartments may be vended at a time. Some times especially milk tanks are fitted with a calibrated meter, so that any desired volume may be taken out at a time.

13.1.3 Pumping and Metering

Some tanks are provided with

- Pumping station: It consists of a filter and very short pipes with no valves or branch connections. The installation is such that it can be drained completely each time the tank is emptied under gravity only without the need of any special means.
- A flow measuring assembly including a flow meter with or without a pump. Normally these meters are positive displacement type with measuring accuracy of 0.1 percent

The connections between the stop valves of the tank and these installations are by means of short, easy to install and detachable coupling.

13.1.4 Other Devices

Some times for special liquids and situations, level warning devices and level indicators are also fitted along with the tank.

13.2 CLASSIFICATION OF VEHICLE TANKS

The vehicle tanks are classified according to method of mounting, ancillary instruments, influence factors, like temperature and pressure, in use, and capacity.

1. Tank may be mounted permanently on the chassis of a vehicle, trailer or be self-propelled.
2. Detachable tank mounted temporarily on the vehicle by means suitable fasteners, which ensure that the position of the tank remains unchanged during transit.

3. Tank fitted with metering device for measuring partial volume continuously.
4. Tank is fitted with no metering device so that it delivers only in terms of compartments if it has any or of whole tank.
5. Tank in which liquid is maintained at a particular temperature by cooling or heating.
6. Tank in which a product is maintained at specific pressure and works at atmospheric pressure.
7. Coated tanks with a suitable material for a specific product.
8. Capacity of tanks may vary from 500 dm^3 to 50, 000 dm^3. However tanks of capacity 8000 dm^3 to 15000 dm^3 are quite common.

13.2.1 Pressure Tanks

In item 6, it is mentioned that tanks may work under partial vacuum or excessive pressure. From that angle vehicle tanks may be further classified as follows:

13.2.1.1 Atmospheric Tanks

Those tanks, which store or transport material under atmospheric pressure and the material are flown out at atmospheric pressure. The material is filled under atmospheric pressure.

13.2.1.2 Pressure Discharge Tanks

Those tanks, which are filled at atmospheric pressure but discharge their material at a pressure greater than atmospheric.

13.2.1.3 Vacuum Filling/ Pressure Discharge Tanks

Those tanks, which are filled under reduced pressure but discharge their material under pressure greater than atmospheric.

13.2.2 Pressure Testing

13.2.2.1 Pressure Testing for Atmospheric Tanks

Tanks, which are supposed to work under atmospheric pressure, are tested as follows [4]:

Fill the tank up to the maximum capacity (up to the brim) with cold clean water;
Close all the valves and dome. Apply hydraulic pressure.
Increase the pressure slowly and slowly till it is about 14 kPa above atmospheric.
Maintain this pressure for at least 30 minutes.
See for any leakage through the valves or elsewhere and also for any distortion.
Empty the tank and dry it thoroughly.

13.2.2.2 Pressure Testing for Pressure Discharge Tanks

Tanks which are supposed to discharge with pressure in excess of atmospheric are tested by almost the same procedure as was used for atmospheric tanks except, the excess of pressure to be maintained for 30 minutes, in this case, is 140 kPa above atmospheric.

13.2.2.3 Pressure Relief Devices

Every tank should have a proper pressure relief device. Any device of minimum effective area of 300 cm^2 and capable of allowing at least 84.5 m^3 of air to pass in 1.3 s is good enough to counteract the vacuum arising from the rapid change during discharge.

13.2.3 Temperature Controlled Tanks

The tanks carrying and delivering liquids at a particular temperature are known as temperature-controlled tanks.

For effective temperature control, the insulated materials are of such a thickness that temperature of liquid does not change by more than 0.5 °C [4], when the difference in outside and inside temperatures of 5 °C is maintained for 24 hours. The ambient temperature is around 24 °C. The material used for this purpose should be

Non-hygroscopic,

An effective vapour barrier,

Non-setting type,

Fire-resistant and

Have low free chloride ion.

Usually Polystyrene is used for such purpose.

The outlet pipe is also insulated up to the outlet valve using a closed cell, non-hygroscopic plastic material, with solid flexible outer and inner skins of thickness 25 mm.

13.2.3.1 Cladding

Cladding of the insulated tank is done with stainless steel or glass-fibre reinforced plastics sheets, with moulded or formed caps.

The cladding is supported and secured by means of stainless steel foundation rings welded to the tank shell. Timber should not be used for this purpose. The cladding is sealed around man-ways using a collar. Any joint in the cladding is sealed to render them waterproof.

13.3 REQUIREMENTS

13.3.1 National Requirements

As most of the tank cars come under the control of the Departments of Legal Metrology, shapes, material for construction of the shell, reinforcing elements, safety devices etc. have to abide the national rules and regulation of Legal Metrology. If the liquids being transported are inflammable then such tanks should also abide the rules and regulations of the departments concerned with fire hazards and safety.

13.3.2 Material Requirements

For potable liquids like milk, the structural characteristics of the tank like shape and material should have no adverse effect on the quality of the liquid transported and advice of the health authorities are binding to such tanks.

13.3.2.1 Special Material Requirements

For milk and milk products, stainless steel of specific grades is recommended [4]. For example British Standard [4] prescribe either X5CrNi18-10 1 4301 or grade X5CrNi Mo17-12-2 1 4401 as the materials for tank shell or any part of it which comes in contact with the milk stored or transported.

The tanks storing or transporting milk and its products in liquid form, require some special welding. Welding for such tanks should be either metal-inert-gas (MIG) or the tungsten inert-gas (TIG) process.

13.3.3 Change in Reference Height

The reference height H of any tank or its compartment, during filling, should not vary by more than 2 mm or 0.1% of the reference height which ever is larger.

13.3.4 Change in Capacity

For a tank having compartments, capacity of the compartment should not change by more than 0.1% of its capacity when the adjoining compartments are filled or emptied.

13.3.5 Air Trapping

Every tank or its compartment should be such that no air is trapped while filling and no liquid is retained on emptying in all normal positions of use.

Spouts, mouldings or vent pipes valves may be used in order to comply with the above requirements.

13.3.6 For Better Emptying

To ensure complete drainage, the cylindrical shell should have a slop of 2° with the vehicle on the level ground.

The volume of liquid remained sticking in the tank and outlet pipe may be at the most $1/5^{th}$ of the maximum permissible error allowed on the tank.

Anti-wave devices and reinforcing elements that may be fitted in the tank should be of shape such that filling, draining and checking emptiness of the tank in not impeded. Appropriate orifices should be provided for this purpose.

13.3.7 Deadwood Positioning

The deadwood placed inside the tank for the purpose of adjusting the capacity to a given value or any other body, should be permanently fixed to ensure that capacity is not modified by an inadvertent removal or displacement of such deadwood.

The placing of deadwood inside the tank for the purpose of adjusting the capacity to a given value or any other body which when removed or changed could modify the capacity of the tanks, is prohibited.

13.3.8 Dome and Level Gauging Device

The dome, on the top of the shell may be fitted, to serve as a manhole and an expansion chamber.

The dome, when fitted, shall be on the upper part of the shell body and should be welded to it. In general, the level-measuring device shall be inside the dome.

13.3.8.1 Shape of Dome

The dome may be a cylindrical or parallelepiped in form (Figure 13.1) and welded to the shell. If it is in parallelepiped form, it should be all along the full length of the shell and permanently fixed by welding. If the sidewalls of the dome are mounted so that they penetrate the shell, providing cut out or orifices at the level of the internal generator avoid the formation of air pockets in the upper part of the shell.

13.3.8.2 Size of the Dome

The transverse section of the shell and dome should be such so as to allow inspection of the interior of tank. A diameter of 500 mm is supposed to be all right for such purpose.

13.3.8.3 Accessories

The following accessories are provided in the dom

(i) A filling aperture, fitted with leak proof cover;
(ii) An orifice for the observation of filling;
(iii) A venting device or double acting valve;
(iv) The level index is in the dome or in the upper part of the shell.

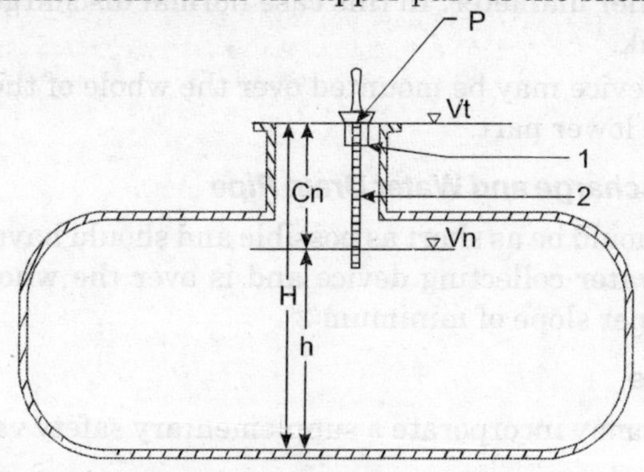

Figure 13.1 Dome

13.3.8.4 Gauging Device

The level-measuring device should ensure a safe, easy and unambiguous readout almost independent of tank tilt under rated operating conditions.

13.3.9 Shape of the Shell

The shape of the shell of the tank should be such that in the zone where the level of the contained liquid is gauged, the sensitivity as defined in 13.1.1.16 is such that for Δh equal to 2 mm $\Delta V/V$ is 1/1000.

13.3.10 Maximum Filling Level for Vehicle Tanks

Normally the national or international regulations prescribe the maximum level of filling for inflammable liquids. However for international transport of dangerous goods by road, there appears to be an agreement with in European countries, that maximum level of filling is F times the reference height H. F is the ratio of density values of liquid at 50°C and 15°C.

$$F = d_{50}/d_{15}$$

For petrol of density = 0.700 kg/dm^3, the calculated volume expansion is about 3% for variation of 35°C [3].

For potable liquids (like milk), wine an expansion of volume of 0.5% is considered reasonable for temperate climates [3].

13.4 DISCHARGE DEVICE

The discharge device will ensure complete and rapid discharge of the liquid contained in the tank; for this purpose the device is connected to the lowest part of the tank shell.

13.4.1 Single Drain Pipe and Stop Valve

Each tank should have a single drain orifice and a single stop valve.

13.4.1.1 Location of Discharge and Drain Pipe

In case of special vehicle tanks for airports, the fitting of a device, to collect water and impurities precipitated by the liquid contained in the tank, is permitted. This device has a separate drainpipe, but should be of smaller diameter. In this case normal discharge pipe is not connected to the lowest part of the tank.

The collecting device may be mounted over the whole of the lower part of the tank or on a reduced area of the lower part.

13.4.1.2 Slope of Discharge and Water Drain Pipe

The discharge pipe should be as short as possible and should have a slope of minimum 2°. If the tank is fitted with water collecting device and is over the whole length of the tank then it should also have proper slope of minimum 2°.

13.4.1.3 Safety Valve

The discharge device may incorporate a supplementary safety valve generally operable by the foot.

13.4.1.4 Separate Discharge Pipe

If a vehicle tank has number of compartments then each compartment should have a device to discharge it independently.

For vehicle tanks intended to deliver larger volumes at one place and carrying only one product, may have multiple discharges. However, it should be clearly displayed.

13.4.1.5 Stop Valves

Stop valves should be readily accessible.

13.5 MAXIMUM PERMISSIBLE ERRORS

(*i*) The maximum permissible error of calibration as per OIML [3] recommendation is ± 0.2%.

(*ii*) But if a vehicle tank is checked while operating, the maximum permissible error may be ± 0.5%.

(*iii*) The volume of the tank is normally determined, with the stop valve closed.

(*iv*) When the tank is fitted with a vapours collecting device, the volume of this device is included in the volume of the tank.

13.6 LEVEL MEASURING DEVICES

13.6.1 Dipstick

(*i*) A dipstick, which is normally square in cross-section and of sufficient length to hold while resting on the datum plate of the tank. For tanks having compartments, one face of the dipstick will indicate the volume for one particular compartment. A dipstick is associated with a particular tank, so should have proper indication to tell to which tank it belongs to.

(ii) A dipstick may be with single mark to indicate the capacity of a particular compartment or tank. In such cases the tank will deliver liquid in terms of one compartment at a time. The mark on one face will represent the capacity of one compartment. It will be clearly indicated as to which compartment it refers to.

(iii) Some times a dipstick may have continuous scale marks to indicate volume of liquid contained in the compartment/ tank up to that mark. In that case the tank will be taken as a measuring instrument.

13.6.2 Level Measuring Device

13.6.2.1 Mechanical Level Indicating Device

Normally such a device indicates the position of the level of a stationary liquid in reference to a fixed reference point. Reference point is on the top reachable from the dom. Greater value of the level indication means less volume of the liquid contained in the compartment or tank.

Principle of working

A precise grooved drum carries a fine wire at the end of which is attached a flat disc. The magnetic coupling of the drum is such that the drum stops moving as soon as the tension in the string becomes less than a predetermined value. The tension in the string is due to gravitational force of whole mass of the flat disc, until the flat comes in contact of the liquid. The density of the flat is much larger than that of liquid. The flat sinks and displaces liquid so the resultant tension in the string is reduced due to upward thrust on the flat due to liquid. By adjusting magnetic coupling and the volume of flat, it is made possible that as soon as the flat enters in to the liquid up to a certain depth, resultant tension in the string reduces beyond the predetermined value, so that drum stops. The length of the string from reference point is measured and gives the position of the liquid level.

In equilibrium condition

$$T = W - Vdg,$$

Where T is tension, W is gravitation pull due to the mass of the flat, V is volume of liquid displaced, d is its density and g is acceleration due to gravity.

As the liquid recedes down, the tension in the string increases so the drum moves till flat again comes to rest in the next lower position of the liquid. The length of the string from the reference point gives new level of the liquid.

Continuous monitoring of the drum movement can give the vertical velocity of the liquid level. If the movement of the drum or flat is indicated automatically, then the device becomes automatic level indicating device.

Similar principles are used to measure the velocity of the moving liquid level moving vertically downward.

13.6.2.2 Electronic Level Indicating Device

Such devices indicate liquid level, for stationary as well as for moving liquid. If the devices use electronics for indication of the level, these are known as electronic level indicating devices.

13.6.2.3 Accuracy Classes

OIML recommends [5] that the level indicating devices may be of two types namely class 3 and class 2. Class 3 is applicable for tanks containing refrigerated (hydro-carbons) fluids. Class 2 is applicable to all tanks carrying any other liquid. A typical level indicating arrangement is shown in Figure 13.2.

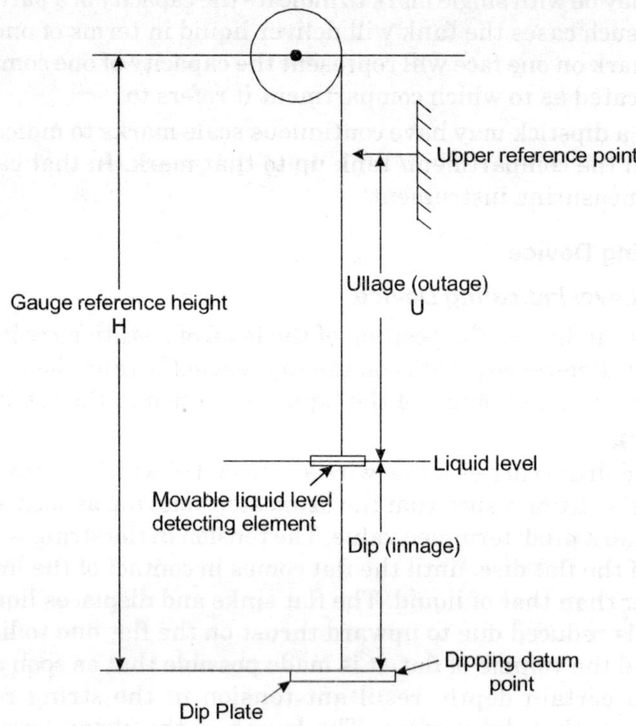

Figyre 13.2 Level indicating arrangements

13.6.2.4 Maximum Permissible Errors

Maximum permissible errors are given as percentage of depth of the liquid level as well as in absolute terms of distance. First and third rows of Table 13.1 give maximum permissible errors at the time of verification while those in second and fourth are at time of inspection or when the device is in use.

Table 13.1

Verification or inspection	*Maximum permissible errors*	
	Class 2	*Class 3*
At the time of verification	0.02%	0.03%
At the time of inspection	0.04%	0.06%
At the time of verification	2 mm	3 mm
At the time of inspection	3 mm	4 mm

13.7 VOLUME/CAPACITY DETERMINATION

The volume of liquid or capacity of the vehicle tanks can be determined by either of the following method:

1. By finding the volume of water required to be filled through a set of calibrated tanks.

We call it a volume transfer method and the place with equipment, where such method is used, is called as water gage plant.

2. By finding out mass of water filled by weighing it. In this case whole vehicle tank is weighed before and after filling the tank. The procedure is called water-weighing method.
3. By measuring the dimensions of the tank and computing its volume from the knowledge of internal diameter and shape of the two heads (ends). This is strapping method. This method has been discussed in detail in Chapter 11 for horizontal tank.

Out of these three methods, volume transfer method (water gage plant method) is supposed to be most accurate and reliable.

13.7.1 Water Gauge Plant

The gauge plant, Figure 13.3, essentially consists of a number of calibrated measures with water filling arrangement and mounted well above the height of a vehicle tank. Each measure is able to deliver water in to the vehicle tank on command.

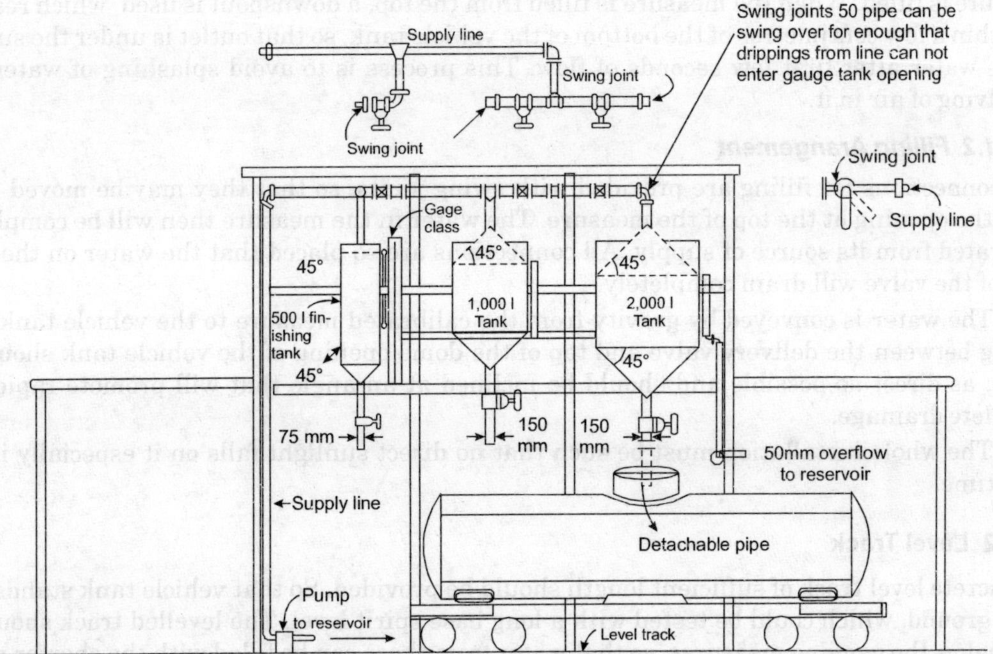

Figure 13.3 Gauge plant

13.7.1.1 Calibrated Measures or Gauge Tanks

Calibrated capacity measures shown in Gauge plant rig (Figure 13.3) are some times also called as gauge tanks. Number and capacity of calibrated capacity measures depend upon, the type of vehicle tanks to be verified against it. Normally two to three measures of varying capacity will do. Measures of 2000 dm^3, 1000 dm^3 and 500 dm^3 are often used. All measures are calibrated for delivery with water. The measures may be with cylindrical body surmounted with cones on either side. For better drainage, semi-vertical angle of the cone is not greater than 45°. Finishing tank of 500 dm^3 capacity should be having an external gage glass tube through out its length and be so connected that water level in the glass tube and in the tank stands at the same level. This ensures that gage glass tube indicates the correct volume delivered. The gage tube on smaller tank is shown on the left of the tank. It facilitates to measure continuously the volume of water delivered in to the vehicle tank. If by regulations, vehicle tanks and its compartments are to have capacity only in terms of say 500 dm^3, then finishing tank of 500-dm^3 capacities with no gage glass tube will do.

The opening of the calibrated measure may be circular or elliptical but should have true and absolutely level edges. The opening may be the *V* notches so that when fitted to overflowing, the measures necessarily contain a fixed volume of water.

Each measure is provided with a draw off valve at the bottom and suitable connections at the top for filling.

When desired, the measures may be provided with suitable connections for filling from the bottom. In this case the valves to supply line leading to the measure should be doubled and the nipples between the valves are equipped with drain cock or bleeder. To ascertain that no water is leaking into or out of the supply line, the drain cock or bleeder is opened after the measure is filled. When the measure is filled from the top, a downspout is used, which reaches to within a few centimetres of the bottom of the vehicle tank, so that outlet is under the surface of the water after first few seconds of flow. This process is to avoid splashing of water and dissolving of air in it.

13.7.1.2 Filling Arangement

The connections for filling are provided with swing joints, so that they may be moved away from the opening at the top of the measure. The water in the measure then will be completely separated from its source of supply. All connections are so placed that the water on the tank side of the valve will drain completely.

The water is conveyed by gravity from the calibrated measure to the vehicle tank. The piping between the delivery valve and top of the dome opening of the vehicle tank should be short, as direct as possible and should be inclined at an angle that will promote rapid and complete drainage.

The whole installation must be such that no direct sunlight falls on it especially in the noontime.

13.7.2 Level Track

A concrete level track of sufficient length should be provided. So that vehicle tank stands on a level ground, which could be tested with a long base spirit level. The levelled track should be just below the gagging measures, so that water from these can be filled with the shorter pipes.

13.8 CALIBRATING A SINGLE COMPARTMENT VEHICLE TANK

13.8.1 General Precautions

1. In hot weather, before calibration, cool down the tank by spraying water on its outside.
2. Check the tyres of the four wheels that these were equally worn out. Check the air pressure of each tyre, which should be the same in all the wheels. This ensures that axis of the tank on the vehicle becomes horizontal when the vehicle tank is made to stand on an especially made level track.

3. The inside of the tank should be checked for proper fittings and cleanliness. Get it cleaned for all foreign substances such as dirt or rust etc.
4. The isolation of chambers from each other should be ensured.
5. Special attention should be given that no extra means have been fixed, which may defraud the suppliers or the receiver.
6. The tank may be rinsed with water followed by proper draining just before it is filled with water from the calibrated measures.
7. It should be ensured that no direct sunlight is falling on the vehicle tank and the calibration measures.
8. Thorough checking of the calibration rig should be carried out. See that the scale behind the gagging tube on the finishing measure is tight and in proper position. The glass gage tube will also help in checking that water being filled in vehicle tank is without any sedimentation.
9. The calibrated measure is filled with clean water.
10. After each delivery one should inspect the measures for sedimentation. If any sedimentation is present, the water is stirred through thoroughly to carry it away when emptied.
11. The operator then should examine all valves of the measure to see that none is leaking. If any one valve is leaking or not fitting well, it should be removed or repaired as the need be.

13.8.2 Filling of the Vehicle Tank

The vehicle is driven on the level track specially prepared and year-marked for it. It is ensured that dome is directly below the outlet tube of the calibrated measure. One measure is emptied in to the vehicle tank at a time. Five minutes drainage time is allowed. During the waiting time, the outlet valve of the vehicle tank is examined. If the leakage is more that 20 drops per minute, water is drained out of the vehicle tank and its outlet valve is repaired first and ensured that leakage is not beyond the permissible limit of 20 drops per minute. If necessary another one full measure may be drained into the vehicle tank. When only a fraction of the capacity of the measure of water is to be filled, then we can switch on to the next measure of lower capacity or the finishing measure depending upon the volume of water required to be filled in the vehicle tank. Before commencing the next measure, the number of full measures has been emptied in the vehicle tank is recorded. This will happen with vehicle tanks of larger capacity with no compartments. The finishing measure is always filled up to its capacity and not to any intermediate mark on the gage glass tube before emptying it into the vehicle tank.

Now two situations arise. Either we have to perform (1) Calibration or (2) Verification.

13.8.3 Calibration of a Vehicle Tank

The vehicle tank is to be calibrated that it delivers the marked volume on it. Entire surface of tank is rinsed with water and drainage time of 5 minutes is allowed. Then fill the vehicle tank with required volume of water with the help of the finishing measure. Wait for 5 minutes for water to settle down. Put the dipstick on the datum plate in vertical position and mark the level of water on it. Graduate this mark by etching or engraving and filling with indelible ink. The mark is to be as fine as possible but at the same time convenient to be used. The method of proper marking is given below. By this procedure we ensure that vehicle tank contains the quantity when filled up to the graduation mark is such that it is able to deliver the marked quantity within permissible limits, as the vehicle tank was rinsed with water prior to actual filling. This process is known as calibration. In reality we have calibrated the dipstick.

13.8.4 Verification of the Vehicle Tank

The dipstick is already marked but we wish to see, if the vehicle tank is containing the marked quantity or not. In this case we put the dipstick at the datum plate hold it firmly in the vertical position and deliver water from the finishing measure slowly till water comes up to the graduation mark. Wait for 5 minutes more and ensure that water level is up to the graduation mark. See the reading of the finishing measure and see if the tank is containing the marked volume of water within the permissible error allowed for the vehicle tank. This process is called as verification of the tank or more correctly one of the marks on the dipstick. While taking delivery from the vehicle tank, it should be ensured that bottom end of the dipstick is not filed off. For this purpose a permanent arrow mark is engraved on each side of the dipstick, such that the tip of the arrow just reaches the end face of the dipstick. A small amount of filing will become evident from the inspection of the arrow mark on the dipstick. In transit it is possible to change the capacity of the tank by tampering with the deadwood of the vehicle tank. That is why, only a non-removable deadwood is allowed.

13.8.5 Temperature Corrections

Normally temperature at which verification/calibration is carried is far from the reference temperature of the tank. We in India do not apply correction due to temperature, should it be necessary to do so, following formula may be used

$$V_{tr} = V_{tu}[1 + \alpha_s(t_s - t_r) - \alpha_u(t_u - t_r)]\rho_{st}/\rho_{ut}$$

Where

V_{tr} is volume of tank at reference temperature t_r

V_{tu} is the volume of water measured at the gage plant, having applied all necessary corrections to standard measures

α_s is coefficient of cubical expansion for the material of standard measure

α_u is coefficient of cubical expansion for the material of tank under calibration

t_u is the temperature of water in the tank under calibration

t_s is the temperature of water in the standard measure

t_r is the reference temperature for the vehicle tank and calibration tanks

ρ_{st} is the density of water in at the temperature t_s,

ρ_{ut} is the density of water at the temperature t_u.

13.9 INTERMEDIATE MEASURE

Quite often, in between the gage tanks of water calibration plant, and standard measures maintained by the laboratory, a measure or a set of measures is maintained, which are called intermediate measures. These measures are calibrated against the standard measures maintained by the laboratory and in turn are used to calibrate the gage tanks.

13.9.1 Construction and Shape

It has a cylindrical body mounted by a frustum of a cone and terminating in to a cylindrical measuring neck on the upper side. A typical one is indicated in Figure 13.4.

This type of measures we have already described in Chapter 12. It has a small neck on each side. It is provided with a draw off valve at the bottom and an overflow device at the top. The overflowing top neck is surrounded with outer transparent cylinder, which is to collect the water overflowed from the neck and to send it through a drainpipe if not to be measured or be collected in a measuring cylinder through the globe valve. Such measures should be inspected frequently for any accumulation of foreign matter like dirt and rust. If it cannot be inspected, because of its construction, it should be calibrated more often.

13.9.1.1 Preparing the Intermediate Measure

Before calibrating it, level it accurately, fill it with clean water and then drain it, allowing 2 minutes for this operation, close the valve.

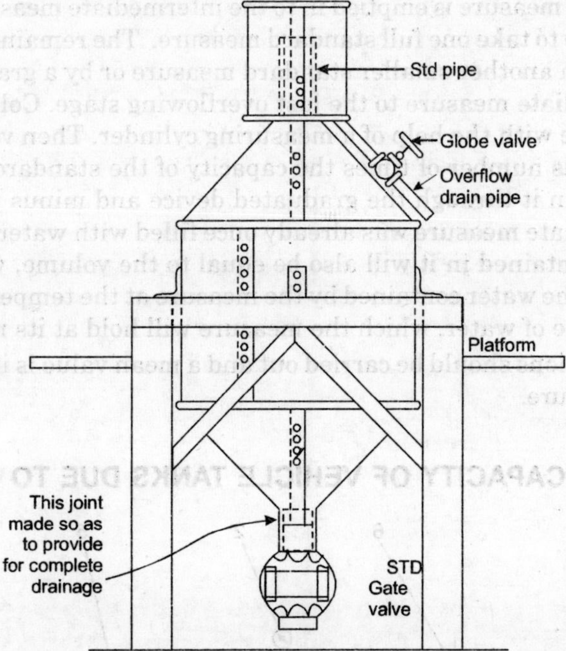

Figure 13.4 Intermediate measure

13.9.1.2 Filling of the Standard Measure

Fill the standard measure, which in this case is a content measure by using a soft flexible tube, which is inserted in to the measure in such a way that that while filling, the end remains below the surface of water. After the standard measure is filled, remove the rubber tube and make up for the volume of the tube by putting some more water to it, strike the measure gently with the palm of your hand so that any air bubble sticking to the wall of the measure comes out. Slide the glass disc ground side down, over the opening, observe if any air bubble is visible through the glass. If so slide back the glass plate a bit and with a small pipette add water until no bubble can be seen by closing the opening again. It may be noted that standard measure is of single capacity defined by the sliding (striking) glass and is in the form of a cylinder with no outlet valve as described in Chapter 2. This is the normal construction of a standard volume /capacity measure at the state level.

13.9.1.3 Emptying the Standard Measure in to the Intermediate Tank

Hold the glass firmly in place, fully covering the opening. With some filter paper or small sponge remove all the excess water from its outside including from the pouring lip if any. Ask the operator to lift the measure with glass in place in proper position for pouring. The operator should hold the glass firmly over the opening and as the measure is being tilted to pour, he should slid back the glass only slightly to have a small opening so that a stream of water starts pouring in to the intermediate tank. While the water from the measure is being poured, the standard should not rest on the rim of the intermediate measure, lest it dents the rim of the intermediate measure. To avoid its own denting, the measure should remain supported totally in the hands of the operator. Water should be poured very slowly to avoid dissolution of air and splashing. A drainage time of 30 seconds should be allowed.

After the standard measure is emptied in to the intermediate measure a sufficient number of times till it is not able to take one full standard measure. The remaining part of the measure is filled by filling it with another smaller standard measure or by a graduated device which is able to fill the intermediate measure to the just overflowing stage. Collect the overflow water and measure its volume with the help of a measuring cylinder. Then volume contained in the intermediate measure is number of times the capacity of the standard measure plus the net volume of water filled in it through the graduated device and minus the volume of overflow water. As the intermediate measure was already once filled with water and drained empty, so the volume of water contained in it will also be equal to the volume, which the measure can deliver. The volume of the water contained by the measure at the temperature of measurement is reduced to the volume of water, which the measure will hold at its reference temperature.

Three such repetitions should be carried out and a mean value is its capacity to deliver at the reference temperature.

13.10 INCREASE IN CAPACITY OF VEHICLE TANKS DUE TO PRESSURE

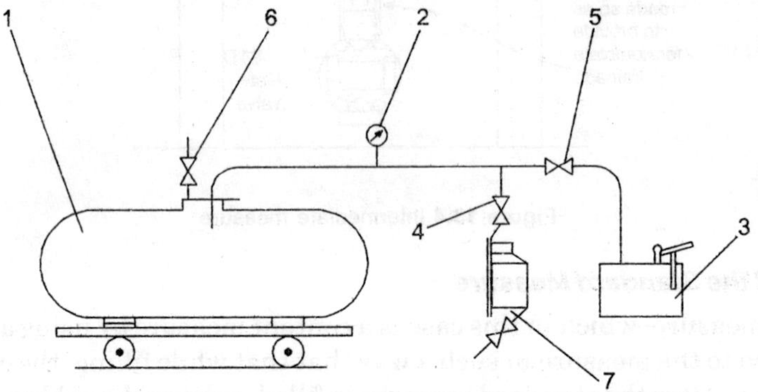

Figure 13.5 Expansion in capacity of tank under pressure

To measure the increase in capacity of the tank when subjected to pressure, fill the tank with water up to the brim of the dome, the orifice is then closed by means of cover plate. By means of the hydraulic pump 3, extra water is filled in the tank under pressure till the maximum service pressure $P1$ bar, above atmospheric pressure, is reached. The pressure inside the tank is indicated by pressure gauge 2. The filling valve 5 (Figure 13.5) is then closed. Ensure that there is no air pocket. In this situation the shell of the tank will be in maximum expanded position so will hold maximum water. With the reduction in pressure the tank will shrink so water will come out. Using the pressure control ball valve 4, reduce the pressure slowly; the water will start coming out. Receive this water in a standard measure. The standard measure is fitted with a scale, continuously indicating the volume of water received. Decrease the pressure to *P*2. The volume of water flown out from the tank and collected in the standard measure is noted down say it is V_1.

Decrease pressure *P*2 to new lower value of *P*3, collect the water from the tank in the measure and let addition in volume is V_2. Continue decreasing the pressure in steps till you arrive at the atmospheric pressure in the vehicle tank. The pressure is measured by the pressure gauge 2. The corresponding addition of water collected in the measure is noted down every time, thus giving a relationship between volume of water collected and pressure inside the tank.

The results are reported either in graphical form or in the tabulated form or both.

13.10.1 Example

Let a tank is subjected to pressure of 7 bars above atmospheric pressure and we wish to know the expansion in the tank with respect to excess pressure. By manipulating the ball valve 4, the pressure is reduced in steps of 1 bar and the volume of the water overflowing from the vehicle tank (measured with the help of standard measure) is recorded in the table as given below.

Table 13.2A

P reduced From-To bar	*Water collected in the measure V dm*^{3}	*Cumulative Volume dm*^{3}
7-6	10	150
6-5	10	140
5-4	15	130
4-3	15	115
3-2	20	100
2-1	30	80
1-0	50	50

From the above data, we can deduce expansion in the tank by adding successive volume of water received in the measure giving us the values given in Table 13.2B.

Table 13.2B

P bar above Atmosphere	*Expansion of the tank dm*^{3}
1	50
2	80
3	100
4	115
5	130
6	140
7	150

These results may be reported in the form of a graph as shown in Figure 13.6.

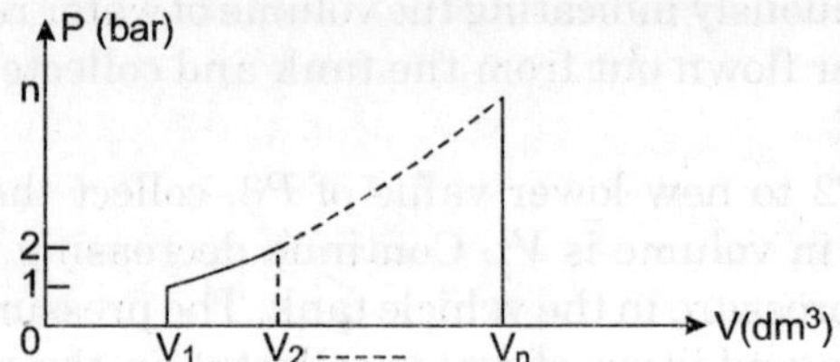

Figure 13.6 Increase in volume against press above atmospheric

The increase in capacity of the tank with excess pressure is shown in Figure 13.7.

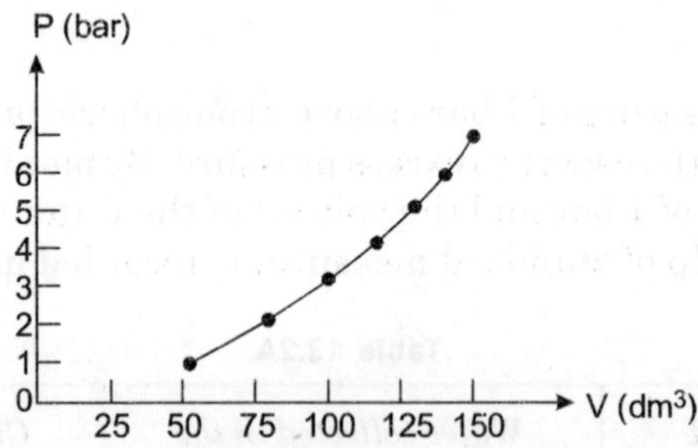

Figure 13.7 Increase in capacity versus excess of inside pressure

13.11 WATER-WEIGHING METHOD FOR VERIFICATION OF TANKS

Where water calibration plant (rig) as discussed above is not available, another method to calibrate/verify the vehicle tank is to weigh the water contained in the vehicle tank and divide by the density of water to get the volume of water contained in the tank. If necessary apply corrections for density of water at reference temperature, air buoyancy and coefficients of expansion of vehicle tank to indicate the volume contained at reference temperature. As the vehicle tank was rinsed with water and properly drained, it is assumed that it would deliver the same volume of water as it contained.

Following steps are taken to verify the vehicle tank:

Step 1: Before filling, inspect the tank for any foreign material, rust, or dirt and remove it if any.

Step 2: Fill the tank up to the top of its shell or on the top mark on its dipstick and empty it and allow it to drain off for 2 minutes or for a period specified in the relevant national standard specification.

Step 3: Weigh the vehicle with its tank empty. The vehicle should not carry with it any removable tools and fixtures. Let us call it as net weight.

Step 4: Weigh the tank, with water full up to the top of the tank shell, or to the point indicated by the dipstick. Last 500 dm^3 of water should be added slowly. Rate addition of last 500 dm^3 of water should not be greater than 20 dm^3 per minute. Leave the vehicle in this position for 15 minutes so that water is settled down. See the level of water if it has gone down fill more water to bring it to the desired level.

Step 5: Measure the temperature of water at the centre of the tank.

Step 6: Weigh the tank with it contents.

Step 7: Obtain the weight of the water by subtracting from it the net weight. Apply temperature correction due to difference in temperature of measurement and reference temperature. Further divide it by the density of water in kg/dm^3 when the weight of water is calculated in kg. This gives the capacity of the tank at reference temperature in dm^3.

If mass M of the water at t°C, then volume of water or capacity of the tank at t°C is V_t at t°C and is given by

$$M = V_t(\rho_t - \sigma),$$

Where ρ_t is the density of water at t°C and σ is the density of air.

Giving us $V_t = M/(\rho_t - \sigma) = M(1 + \sigma/\rho_t)/\rho_t$.

$$V_t = MV_1(1 + KV_1),$$

Where V_1 is the volume of 1 kg of water at t°C and $K = \sigma = 0.0012$ kg/m^3, a factor due to air buoyancy.

The values of density of water [6] and volume per kg in dm^3 are given in Table 13.3 for temperatures from 4°C to 48 °C.

Once V_t is known, the following relation gives V_{tref} at reference temperature

$$V_{tref} = V_t[1 + \alpha(t - t_{ref})]$$

Where α is the cubical expansion of the sheet metal used for the tank shell.

Table 13.3 Volume of Water at Different Temperatures

Temp °C	*Volume dm³/kg*	*Temp °C*	*Volume dm³/kg*
4	1.000025	5	1.000033
6	1.000057	7	1.000095
8	1.000149	9	1.000216
10	1.000297	11	1.000392
12	1.000499	13	1.000620
14	1.000753	15	1.000898
16	1.001055	17	1.001223
18	1.001403	19	1.001594
20	1.001796	21	1.002008
22	1.002231	23	1.002465
24	1.002708	25	1.002961
26	1.003224	27	1.003496
28	1.003778	29	1.004069
30	1.004369	31	1.004678
32	1.004996	33	1.005323
34	1.005659	35	1.006002
36	1.006354	37	1.006715
38	1.007083	39	1.007460
40	1.007844	41	1.008237
42	1.008636	43	1.009044
44	1.009458	45	1.009880
46	1.010310	47	1.010746
48	1.011189		

13.12 STRAPPING METHOD FOR CALIBRATION OF THE VEHICLE

Any vehicle tank may be taken as a horizontal tank as shown in Figure 13.8. So the method of strapping discussed in the Chapter 10 on horizontal tanks may also be employed in calibration and verification of vehicle tanks. Gage table may similarly be prepared to verify and calibrate the dipstick

Various points labelled with alphabets indicate the following:

Points E indicate the ends of the tank.

Point X indicates the end of head of the tank.

Points A and B indicate ends of the ringed plate inside the tank.

Points S indicate the intersection of spherical head with cylindrical head flange board.

E to X represents the overall projection of the head.

X to A and B to B represents the length of overlapped portion of the ring.

A to B represents length of under-lapped ring.

X to X represents total length of the main cylinder.

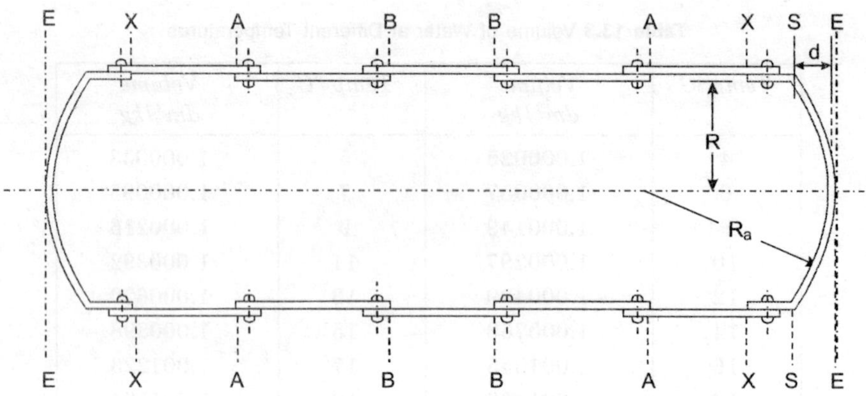

Figure 13.8 Un-mounted tank of a vehicle

S to X represents length or depth of head cylinder.

R is the internal radius of the cylindrical shell.

Ra is the radius of the spherical head.

13.13 SUSPENDED WATER

If water absorbed by petroleum liquids is in the suspension state and also there are solid impurities in the suspension state then corrections are also applied due these facts. If sufficient time is allowed, water settles down in the bottom and drains out via separate drain pipe, and then volume of this water is also measured and added to volume of liquid. Hence volume of water suspended in petroleum liquid is taken as part of the liquid.

If the liquid is paddled under pressure, without gaseous phase, the pressure is measured and the volume is corrected for (1) the compressibility of the liquid and (2) deformation of tank.

REFERENCES

[1] Gupta S.V. 2003. A treatise on standards of Weights and Measures, Commercial Law Publishers, New Delhi, India.

[2] API 2554. Measurement and Calibration of Tank Cars.

[3] OIML R 80: 1989. Road and Rail Tankers.

[4] BS 3441:1995. Tankers for Liquid Milk.

[5] OIML R-85 (1998), Automatic Level Gauges for Measuring the Level of Liquid in Fixed Storage Tanks.

[6] Gupta S.V. 2002, Density Measurements and Hydrometry, Institute of Physics, U.K., pp101.

BARGES AND SHIP TANKS

14.1 INTRODUCTION

Petroleum products are among most important commodity, which are transported through international borders. Most of the times, these are for custody transfer against payment or any other financial obligations. Transportation device may be a ship or a barge. Ships exclusively used for transporting crude oil are often called as tankers. So ships/tankers or barges do fall in the purview of legal metrology. Hence the capacity of these ships and volume of liquid products transferred by them should be measurable and verifiable. As the transactions involved are between two countries, all measurements should be mutually acceptable so should be carried out as per international standards.

Some international or accepted national standards used for measurement of some specific commodities (hydro-carbons) are given from 1 to 10 under the heading references.

A variety of hydrocarbon compounds are transported through ships. Some gases, like natural gas, which can be liquified under low temperature and high pressure, are also transported through ships.

Barges are used to carry smaller quantity, of the order of only few hundred cubic metres (few thousand barrels) in volume. They receive transportable items from the standing ship in international waters and are used to carry it to the shores of the interested countries. In other words barges are links in between international water to the Country water and shore.

When in a particular country, barges and lighters are used for the reception and delivery of part loads for measurements, the additional errors that may occur are taken care off. It is possible in such cases, to specify a minimum measuring height usually (500 mm) or a minimum "measurable volume".

14.1.1 Some Definitions

14.1.1.1 Ullage Height

The distance between the free surface of the liquid and the upper reference point measured along the axis of measurement is ullage height.

14.1.1.2 Axis of Measurement

The vertical line for a compartment along which all distances like ullage, innage or reference heights are measured from the axis of measurement.

14.1.1.3 Bulkhead

Watertight division or dividing wall in a ship is a bulkhead.

14.1.1.4 Fore

The fore side is the front side of the ship and is also known as bow side.

14.1.1.5 Aft

Back of the ship is the aft side. It is also known as stern side of the ship.

14.1.1.6 Camber

Normally for better drainage of water the deck has a slight slope away from the centreline bulkhead. The height of the point on the centreline bulkhead above the point where the normal, from the point on the shell, meets the central bulkhead is known as camber.

14.1.1.7 Inboard Height

Height of the centreline bulkhead is known as inboard height.

14.1.1.8 Outboard Height

Height of the tank on shell side is outboard height.

14.1.1.9 Corrected Heights

The corrected heights are those that have been extended on a determined slope to the point at which they are intended to be used, i.e. the centreline bulkhead and shell.

14.1.1.10 Port

All the items on the left of centreline bulkhead are on port side of the ship.

14.1.1.11 Starboard

All the items on the right of centreline bulkhead are on starboard side of the ship.

14.1.1.12 Closed Line Capacity of Pipeline

Closed line capacity of the pipeline is the outside volume of the pipeline. This volume is deducted from the capacity of the tank if pipeline is running through it.

14.1.1.13 Open Line Capacity of Pipeline

Open line capacity of the pipeline is the volume of liquid, which the pipeline contains when full.

14.2 BRIEF DESCRIPTION

Maritime authorities have detailed requirements for the construction and inspection after completion of a tanker. The sketches of a ship [11] in Figure 14.1 are given for information only.

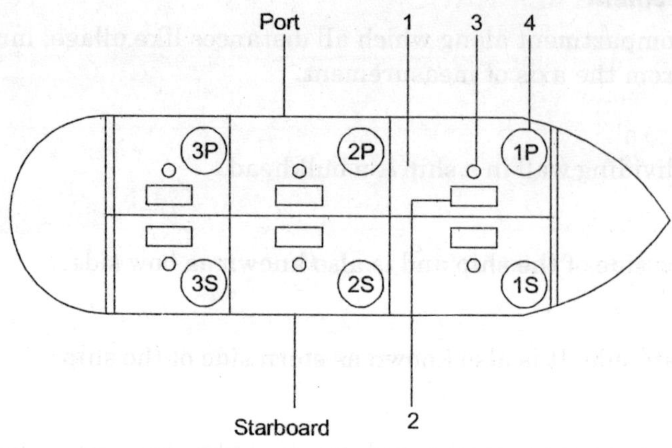

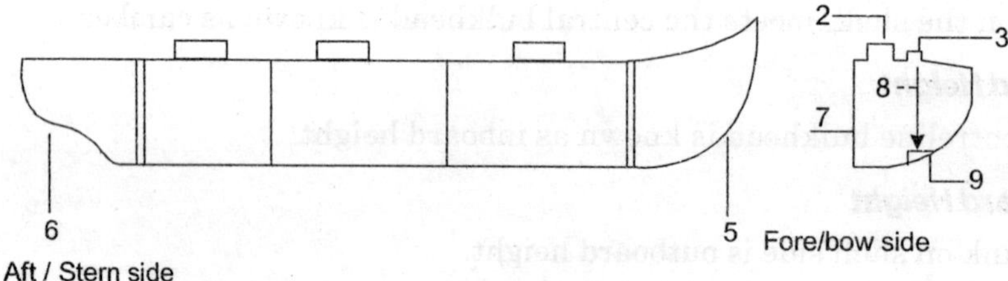

Figure 14.1

14.2.1 Sketch of a Tanker

In general, tanks are numbered from fore to aft (from front to back), with qualifications of "Port" (P) and "starboard" (S) or "centre" (C). In some countries a reverse order is adopted for certain categories of ships, in that case the fact is boldly mentioned in the relevant documents.

Here we may note that 1 indicates tank or a compartment, 2 is the inspection hatch for inspection of the interior of the tank. Starboard side and port side are shown in words at the end of arrows. The numeral 4 represents the cardinal number of the tank, the letter *P* or *S* following it gives the position of the tank relative to the central bulkhead (central dividing wall).

Lower sketch is the elevation of the ship, in which 5 indicates the front-fore or bow side, and 6 as the back-aft or stern side of the ship.

Side sketch-Vertical section of half a ship perpendicular to the elevation, gives a basic idea of a particular tank, showing the gauge hatch as 3, Inspection hatch as 2, longitudinal bulkhead (wall) is represented by 7. The numeral 8 indicates the vertical measurement axis and 9 the datum point.

Some tankers are provided with their own

- Pumping installations as well as
- Metering system.

14.3 MEASUREMENT AND CALIBRATION

Calibration and measurements of ships, tankers, barges and lighters are practically the same, except magnitudes of parameters to be measured may vary.

As in case of vehicle tanks and rail tankers, the methods of measurement and calibration in these cases also are

1. Strapping: Measuring of different dimensions at specified points and assuming the most appropriate geometrical shape and finally calculating capacity and finding volume height relationship.
2. Liquid calibration: In which liquid like water is used to find out the volume/ capacity of the tank and also volume height relationship.
3. Using the drawings and data supplied by the manufacturers. This method is used for in service ships, tankers or barges.

In fact both the methods, enumerated at 1 and 2, are used for complete measurement and calibration process of assigning capacity or establishing volume of liquid-height relationship.

14.4 STRAPPING METHOD

All measurements, in case of ship tanks, are inside, so this method may also be termed as dimensional method. Moreover there is nothing to strap around.

14.4.1 Equipment

We have discussed the equipment needed for strapping method in Chapters 8 and 9, here also same equipment at least by name is required, only difference is that here the length of the length measuring tape may be longer. Just to name, the equipment needed is:

Tapes for height and length measurement; Tapes to measure height will have dip weights; All tapes should be calibrated by a competent laboratory at specific temperature and when stretched with specified tension. For height measurements 20 metres tape may do, while for length, it may be of 50 m in length. Together you will need tape clamps, spring tension gauge or a dynamometer.

A 4 m measuring stick is also required for dip measurement. The stick is made of straight grains, hard wood and is graduated in mm. Measuring stick is fabricated in two sections each of 2 m, one section slide over the other so that they be extended to 4m.

Levelling device and line levels are often required.

In addition density hydrometer(s) along with a hydrometer jar and temperature measuring instruments and devices are also needed. The hydrometer must be calibrated as per national or International standard [12, 13]

14.4.2 Location of Measurements

Consider a transverse vertical plane at the centre of a tank. Height of the tank is to be measured at places (1) close to centreline bulkhead AB and closed to the shell CD. The two measured heights are indicated as KL and MN respectively in Figure 14.2. Horizontal distance between these two heights is also measured.

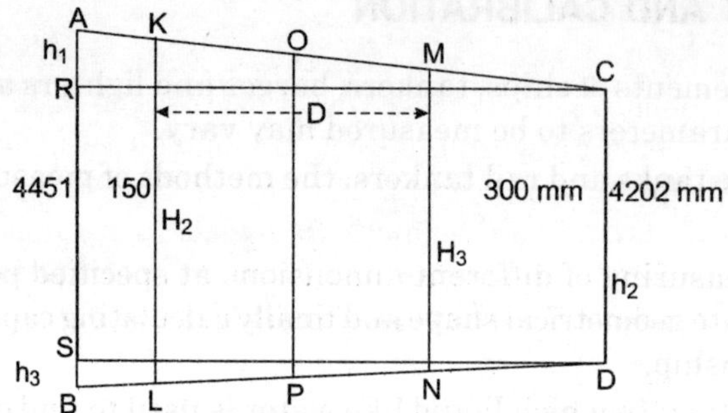

Figure 14.2 Central transverse vertical section of the tank

Height measurement

Three vertical transverse planes, one near the fore bulkhead, second midway and third near the aft bulkhead of the tank, are chosen. The midway transverse plane is shown in the Figure 14.2. Height H_2 at a known distance of say 150 mm from *AB*-centreline bulkhead and H_3 near the shell, 300 mm from *CD* are measured. These two heights are measured in each of the other two planes also.

From the measurements of height H_2 and H_3 and distance *D*, we calculate the total camber (camber of deck plus camber of bottom).

We know the distances of H_2 and H_3 from the centreline bulkhead so we can calculate the corrected inboard height i.e. the height of *AB*-centreline bulkhead. Similarly we know the distances of H_3 and H_2 from shell, so we calculate the corrected outboard height *CD* of the shell.

To measure camber of deck alone, elevation of the point *A* at the centreline bulkhead with respect to the point *C* on the shell is measured (see section 14.4.3.3). The elevation is a deck camber, which in fact is the height of the wedge of the deck. The deck camber subtracted from the total gives the camber of the bottom, i.e. height of the bottom wedge.

Length measurement

For length measurement of the tank, two horizontal planes are chosen one just above the bottom and second about 2 metres high above the bottom. In each plane, three lengths one near the centreline bulkhead, second along the middle and third near outboard bulkhead are measured. For details see section 14.4.3.2. Mean of all six measurements is taken as the length of the tank.

Width measurement

For width measurement, two planes as stated above are chosen, and measurement of the widths are taken near the fore bulkhead, middle of the tank and near the aft bulkhead. For details see section 14.4.3.2. Mean of all six measurements give the width of the tank.

14.4.3 Linear Measurement Procedure

All tanks must be clean, gas free and safe to enter. All linear measurements except that of deadwood are carried nearest to mm or in terms of 2 mm. The location and size of dead wood is measured, read and recorded within 1 mm.

14.4.3.1 Preliminary Measurements

Before entering the tank, measurements are to be taken on the deck to determine the camber. To measure it a straight line is stretched transversely across the deck, and the distances from the line to the deck at the centre line and at each side of the barge is taken. This camber is established at two different places on the deck. The forward and aft tanks on some barges may have a longitudinal shear so this should be measured with a straight line fore to aft. Measure the size and heights of the expansion hatches and locate them with respect to the central line bulkhead and fore and aft bulkheads. The total gage heights are measured and recorded.

14.4.3.2 Internal Measurements of Tank

Upon entering the barge tank, choose a tape path [14] that will permit an unobstructed measurement in the following locations:

I. Length measurements

1. Just above the bottom stiffeners
   - (*a*) Near the centreline bulkhead $L1$
   - (*b*) Middle of the tank $L2$ and
   - (*c*) Near the outboard bulkhead or shell $L3$
2. About two metres high above the bottom
   - (*a*) Near the centreline bulkhead $L4$
   - (*b*) Middle of the tank $L5$ and
   - (*c*) Near the outboard bulkhead or shell $L6$

II. Width measurement

3. Just above the bilge
   - (*d*) Near the forward bulkhead $W1$
   - (*e*) Middle of the tank $W2$ and
   - (*f*) Near the aft bulkhead $W3$
4. About 2 metres high above the tank bottom
   - (*a*) Near the forward bulkhead $W4$
   - (*b*) Middle of the tank $W5$ and
   - (*c*) Near the aft bulkhead $W6$

III. Height measurement

5. At the forward bulkhead
   - (*a*) Near the centreline bulkhead $H1$
   - (*b*) At measured distance from the shell (usually just inside the bilge) $H2$
6. At halfway back from the forward bulkhead
   - (*a*) Near the centreline bulkhead $H3$
   - (*b*) At measured distance from the shell (usually just inside the bilge) $H4$
7. Height at the aft bulkhead
   - (*a*) Near the centreline bulkhead $H5$
   - (*b*) At measured distance from the shell (usually just inside the bilge) $H6$

Let

L denotes the mean of $L1$, $L2$, $L3$, $L4$, $L5$ and $L6$ and

W denotes the mean of $W1$, $W2$, $W3$, $W4$, $W5$ and $W6$.

Mean of $H1$, and $H5$, the heights near the centreline bulkhead and mean of $H2$, and $H6$ give the total camber as the distance between the two heights is known.

14.4.3.3 Deck Camber

To measure the height of the deck wedge, heights from the centreline bulkhead and the shell are measured. Optical level is used for this purpose. Fix a horizontal line by levelling the optical level. Then measure the height along a vertical graduated scale placed at the centreline bulkhead and at the shell turn by turn. The difference in two heights gives the deck camber i.e. height h_1 of the deck wedge is AR in Figure 14.2. With the help of the total camber already measured and measurement of heights $H1$ and $H2$, the height of bottom wedge h_3 is calculated. SB represents h_3. It may be noted that outboard height h_2 is the height of the box RCDS.

14.4.3.4 Box Volume, Volume of Liquid Partially Filling It

The volume of the box then simply will be the product of the length, breadth and height h_2, giving us

$$V2 = L.W.h_2$$

As the horizontal section of the box will be a rectangle with constant sides, which in fact are length L and breadth W of the tank, so volume of liquid filling it partially up to the height z will be directly proportional to z-the height to which box is full.

14.4.3.5 Volume of Bottom Wedge and its Partial Volume

Wedge may be considered as a prism of height equal to length of the tank and base as right-angled triangle with one side equal to the width of the tank and the other side the height h_3 of the wedge. Please refer Figure 14.3, from the figure, we see that width W is given as

$$W = h_3.\tan(\alpha)$$

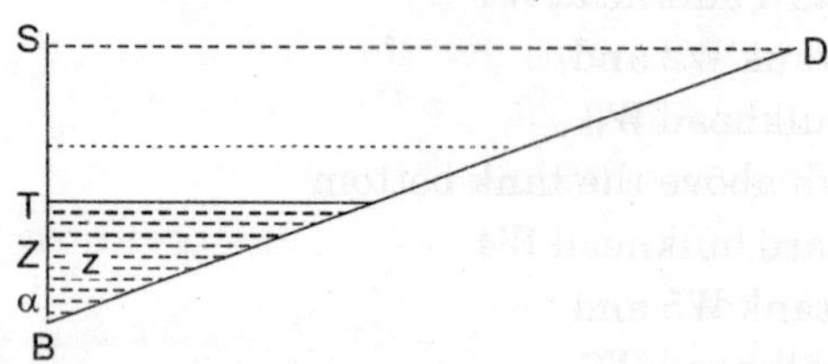

Figure 14.3 Vertical section of the lower edge

So volume V_1 of the wedge is given by

$$V_1 = W.L.h_3/2 = L.\, h_3^2.\tan(\alpha)/2 \qquad \text{...(1)}$$

Let the wedge is full up to the height z, then its horizontal section will be a rectangle with one side equal to length of the tank and the other side will be the base of the triangle with height z, which is variable. So the volume of liquid partially filling it up to the height z will not be a linear function of z.

Area of the partially filled wedge $z^2.\tan(\alpha)/2$,

Giving V volume of liquid partially filling the wedge as

$$V = L.\, z^2.\tan(\alpha)/2 \qquad \text{...(2)}$$

Combining (1) and (2) we get

$$V = (z/h_3)^2 V_1.$$

Or $$V/V_1 = k = (z/h_3)^2 \qquad \text{...(3)}$$

So for construction of the gage table in smaller steps, we can tabulate the values of k in terms of z/h_3 in steps of 0.001 or any other step commensurate to the required intervals. A similar method was adopted in case of horizontal storage tanks in Chapter 10.

14.4.3.6 Volume of Upper Wedge and Its Partial Volume

When the liquid is enough to fill the bottom wedge and rectangular box and partially fill the deck wedge. Then volume of liquid will be $V_1 + V_2 + KV_3$.

The value of K may be determined as follows:

Please refer Figure 14.4.

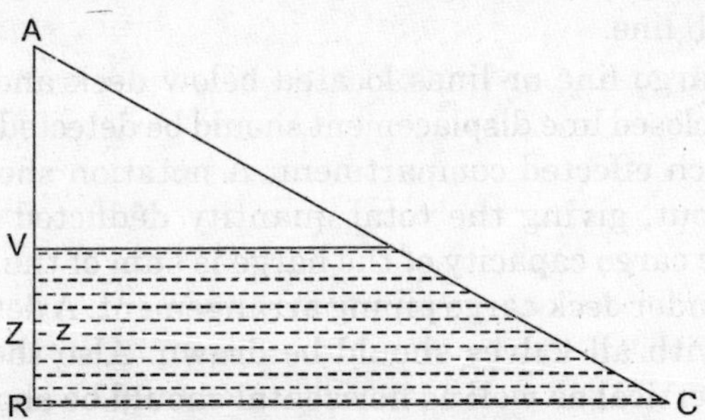

Figure 14.4 Vertical section of deck wedge

Let the liquid fills up to the height z, then

Volume/capacity of the deck wedge $V_3 = L.h_1^2 \tan(\beta)/2$

Partial volume $V = V_3 - Lx(h_1 - z)^2 \tan(\beta)/2$

$= V_3 - V_3 (h_1 - z)^2/h^2_1$

$= V_3[1 - (1 - z/h_1)^2]$ ...(4)

Giving the value K, such that $V = KV_3$

or $K = [1 - (1 - z/h_1)^2]$ ...(5)

These values of K can also be tabulated as a function of z/h_1 insteps of 0.001 or any other suitable step, which is required for given steps.

So partial volumes of liquid

For	$z < h_3$,	$V = kV_1$	...(6)
For	$z > h_3$ but $z < (h_3 + h_2)$	$V = V_1 + V_2 (z - h_3)/h_2$	...(7)
For	$z > h_2 + h_3$	$V = V_1 + V_2 + KV_3$	...(8)

K is given by $K = V_3[1 - \{1 - (z - h_3 - h_2)/h_1\}^2]$ ...(9)

Here z is the total height of the liquid levels from the lowest point of the tank i.e. the point B.

So we see that transverse section of a barge tank consists of:

1. Bottom wedge,
2. Box section
3. Deck wedge
4. Expansion hatch

The area of the vertical sections are given by

1. $W.h_3/2$ for bottom wedge

2. $W.h_2$
3. $W.h_1/2$

The volumes of tank portions bounded by the above sections are obtained by multiplying each area by L - the mean value of length.

14.4.3.7 Other Measurements

Thickness is measured of the strike plate, if present. Deadwood is measured giving the location from the bottom of the tank, if necessary, number its parts for identification. If drawings are available then deadwood can be calculated from that also. Measure the bilge radius if present. For this suspend a plumb line to point of tangency with the bottom and measure from the side of the barge tank to the plumb line.

For barges with main cargo line or lines located below deck and running through the compartments, the applicable closed line displacement should be detected at the proper elevation from the gross capacity of each effected compartment. A notation should be made that this deduction has been carried out, giving the total quantity deducted from each designated compartment. In this case the cargo capacity of the barge is sum of the net capacities of cargo compartments and the total under deck cargo piping arrangement. A detail sketch of the under deck pipeline arrangement with all valves should be drawn. Also the net capacities of the pipeline between the valves vertical as well as horizontal should be shown.

Barge tanks that are constructed in such a way that linear measurements are not practical then such barges should be calibrated by liquid calibration method.

14.4.3.8 Correction Due to Trim

Due to the nature of their use, barges sometimes decline towards the stern side. This is commonly known as trim by the stern, and the resultant change in the position of the liquid surface can be observed. Should the barges tanks are gagged in any location other than the centre with respect to fore and aft bulkheads, then these gages must necessarily be corrected to allow for rise or fall of the liquid surface. That is, the surface rises at the aft end with a trim by the stern. The correction is made to the indicated gage height and is calculated as follows:

$$X = TD/L$$

Where X = correction due to trim

T is total trim i.e. the difference in elevation between the fore and aft draft marks

D is distance from the centre of the tank to gage point

L is length between draft marks.

When the trim is by stern and gage is being taken at the centre of the tank, this correction is negative and is subtracted from the indicated gage height before reading the capacity tables. Correction for trim is not used if the liquid surface does not completely cover the tank bottom, or if the liquid surface is touching any portion of the tank.

14.4.4 Temperature Correction and Deadwood Distribution

14.4.4.1 Temperature Correction

The dimensions of all transverse sections of the tank are obtained from the measurement data. The dead rise is obtained by the measured deck camber from the difference in the corrected inboard and outboard tank heights. Linear measurements are carried with working tapes, which have been calibrated against some standard tape with 20°C as reference temperature. However the measurements, made by taking 20°C as basis, are reduced to 15.5°C, the reference temperature for volume of petroleum liquids.

14.4.4.2 Deadwood Distribution

All items or shapes in barge tanks, which displace liquid is to be listed. Then size of the deadwood is distributed along the height in small steps say of 2 mm, when measurements are taken in SI units or 1/8″ in FPS system.

Having obtained distributed deadwood, height-wise net volume is calculated after applying necessary corrections due to camber, trim and taking in to account of distributed volume of expansion hatch. Where the hatch is small and not considered as part of the tank, its capacity in terms of height taken in small steps is just indicated in the lower portion of the calibration certificate.

14.4.5 Format of calibration certificate

Name of the Barge: SVG star
Owner: India Shipping Limited
Location: Bombay port
Built by: Hindustan Shipyard
Built in: 2001
Measurement Date: 28th March 2003
Nominal capacity: 1908 m^3
Type of capacity table:
Innage: dm^3 per 2 mm
Ullage: dm^3 per 5 mm
Compartment No: 2P
Height of gage reference point above deck: 781 mm
Location of gage point
From aft bulkhead: 4877 mm
From centreline bulkhead: 2895

Note: There are six to 12 such tanks in a barge

Gage/calibration table: Gage table is generated from the data of the Table 14.4 given at the end of section 14.4.6.

14.4.6 Numerical Example

14.4.6.1 Measured Data

Total gage height: 520 cm
Height gage reference point above deck: 80 cm
Location of gage point: From aft bulkhead: 488 cm
From centre bulkhead: 290 cm
Size of expansion hatch: 76 cm
Above bottom stiffeners 180 cm above bottom

Length measurement

	cm	cm
Near centreline bulkhead	975.3	975.0
Centre of tank	974.8	975.2
Near outboard bulkhead	974.5	975.2
Average	**975.00 cm**	

Width measurement

Near forward bulkhead	579.2	579.8
Centre of tank	579.1	579.8
Near aft bulkhead	579.5	579.6
Average	**579.5**	

Height measurement

Height measurement	Near centreline bulkhead 15 cm from bulkhead	Near outboard bulkhead 30 cm from shell
	cm	cm
At forward end	444.9	421.0
Near aft bulkhead	445.1	422.0
Average	**445.0**	**421.5**
At centre	445.0	421.5

We see here that the section is a trapezium, one parallel side of which is outboard height at the shell and other parallel side is the inboard height at the centreline bulkhead.

Deck camber

FOR MEASUREMENT OF DECK HEIGHT (DECK CAMBER), THE DATA IS:

Height of horizontal line above the shell	30.5 cm
Elevation of line level at centreline bulkhead	10.5 cm
Subtracting the two heights gives height of deck wedge	20.0 cm

14.4.6.2 Calculations

Inboard and outboard heights

Horizontal distance between the two heights measurements is 533.4 cm.

Average length	975.00 cm
Average width	579.5 m
Average outboard height	421.5 cm
Average inboard height	444.5 cm

Horizontal distance between the two measured heights

579.5 – (15 + 30) = 534.5 cm

Slope of the deck and bottom wedge (444.5 – 421.5)/534.5 = 0.043030869

Outboard heights is taken at 30 cm from shell 421.5 cm

Less (30x0.04303) – 1.291 cm

Corrected outboard height (vertical side of the trapezium section of the tank is 420.209 cm

Inboard height is taken at 15 cm from centreline

In board height	444.5 cm
Plus (15 × 0.04303)	+ 0.645 cm
Corrected inboard height	445.145 cm

Vertical height (other vertical side of the trapezium section) is 4451 mm

This is also the height of the centreline bulkhead.

Total camber and dead-rise (445.145 – 420.209) = 24.936 cm

Calculation of Deck Camber

Height of horizontal transverse line at shell = 30.5 cm

Height of horizontal transverse line at centreline =10.5 cm

Deck camber = 20 cm

Height of the deck wedge

Height of the deck wedge = 20 cm = 200 mm

Dead rise

(24.936 – 20) = 4.936 cm

49.36 mm is the height of bottom wedge.

Dead volume

For height of the wedge from the foot of the gage height, we know that foot of the gauge height is 2895 mm from the centreline bulkhead. Hence height h_4 of the wedge from the foot of the gage height is given as

$$h_4 = 4.9 \times 2895/9750 = 1.455 \text{ mm} \approx 1.5 \text{ mm}$$

This means that while constructing the calibration table, the volume of liquid up to the height of 1.5 mm will not be taken into account. The value of such a volume is given by

$$29.82 \times (1.5)^2/10^2 = 0.67 \text{ dm}^3,$$

For constructing the gauge Table 14.4, The dead volume, if less than 1 dm^3 may be ignored but should be taken in to account if it is more than 1dm^3. In the example as it is too small to make any significant difference, in preparing the gauge Table 14.4, we have not taken it in to account.

Effect of bilge

The vertical outboard bulkhead meets the bottom not at right angles but forms a quadrant of the circle. This circular formation is called bilge. Due to this, there would be some reduction in volume of the tank to a height roughly up to the radius r of the bilge.

The reduction in volume is area between the two mutually perpendicular tangents at the points where quadrant of the circle meets the vertical and horizontal bulkheads. This area A is given by

$$A = r^2 - \pi r^2/4 = (1 - 0.785398)r^2 = 0.214602 \times r^2.$$

So reduction in volume is A times the Length of tank

In this particular case bilge radius is 300 mm and length L is 9750 mm

Hence Volume reduction = 188.31 dm^3.

Or 188.31/300 = 0.627 dm^3 per mm up to the height of 300 mm

This reduction in volume has also not been taken in to account in preparing the gauge table.

Volume of first 10 mm of bottom wedge is 28.92 dm^3.

Height of bottom wedge

Height of the bottom wedge = 4.9 cm = 49 mm

Box height of the tank

Box height of the tank is the corrected outboard height = 420.20 cm = 4202 mm

Volume distribution of bottom wedge

Height of wedge = 4.9 cm

Volume of bottom wedge = $L \times W \times$ height/2

(975.0 × 579.5 × 4.9/2) cm^3 = **1384.280 dm^3**

To spread volume over height, we see that volume is proportional to the square of height in increasing order.

So spread factor per 2 mm = 1384.280/(5 × 4.9 × 4.9) = 11.531

Table 14.1 Volume Distribution along the Height of the Bottom Wedge

Tank height	*Increments*	*Difference in squares*	*Volume per 2 mm in dm^3*	*Volume per cm in dm^3*
0 – 10 cm	5	1	11.531	57.655
10 – 20 cm	5	3	34.592	172.96
20 – 30 cm	5	5	57.654	288.27
30 – 40 cm	5	7	80.716	408.58
40 – 49	4.5	8.01 × 1.11	102.6144	461.763

Number of steps at the rate of every 2 mm in 49 mm = 24.5

Volume of Box section

Height of the tank is from 49 mm to 4251 mm (Number of steps are 20101)

Volume = 975.0 × 579.5 × 420.2 cm^3 = 237 700.7588 dm^3

Volume per 2 mm = 113.1347 dm^3

No of steps 2101

Volume distribution of deck wedge

Height of the wedge 4251 – 4451 mm

Height of the wedge 20.0 cm

No of steps 100

Volume = 975.0 × 579.5 × 20/2 cm^3 = 5650.125 dm^3.

Here also, we see that volume is proportional to the square of height in decreasing order.

Spread factor 5650.125/5 × 20 × 20 = 2.825

Table 14.2 Volume Distribution along the Height of the Deck Wedge

S.No.	*Tank height*	*Increments*	*Difference in squares*	*Volume per 2 mm in dm^3*	*Volume per cm in dm^3*
1	423.8 – 424.8	5	39	110.175	550.875
2	424.8 – 425.8	5	37	104.525	522.625
3	425.8 – 426.8	5	35	98.875	484.375
4	426.8 – 427.8	5	33	93.225	466.125
5	427.8 – 428.8	5	31	87.675	437.875
6	428.8 – 429.8	5	29	81.925	409.625
7	4298 – 4302	2	—	76.275	—
8	4302 – 4308	3	27	76.275	381.375
9	4308 – 4318	5	25	70.625	353.125
10	4318 – 4328	5	23	64.975	324.875
11	4328 – 4338	5	21	59.325	296.655
12	4338 – 4348	5	19	53.675	268.405
13	4348 – 4358	5	17	48.025	240.155
14	4358 – 4368	5	15	42.375	211.905
15	4368 – 4378	5	13	36.725	183.625
16	4378 – 4388	5	11	31.075	155.375
17	4388 – 4398	5	9	26.425	127.125
18	4398 – 4408	5	7	18.775	98.875
19	4408 – 4418	5	5	14.125	70.625
20	4418 – 4428	5	3	8.475	42.375
21	4428 – 4438	5	1	2.825	14.125

14.4.6.3 Deadwood

Data obtained from the drawings

Deadwood is either measured in steps of smaller height for example 2 mm or may be calculated from the data given by the manufacturers. An example of deadwood volumes distributed equally over height is given below:

Table 14.3A Deadwood Data obtained from the Drawings

S.No.	*Items*	*Height range mm*	*No. Steps*	*Volume dm^3*	*Deadwood volume /2 mm in dm^3*
1	Bottom stiffener	0-260 mm	130	432.8	3.332923
2	Miscellaneous	0 – 370 mm	185	414.6	2.233836
3	Heating coils pipes	25 – 140 mm	58	160.8	2.772414
4	Suction line	85 – 107 mm	11	58.5	5.327273
5	Stripping line	69 – 685	308	14.1	0.045779
6	Transverse bulkhead stiffener	65 – 4451	2193	50.9	0.023210
7	Transverse stiffener	203 – 4302	2050	23.7	0.011 560
8	Deck stiffener	3930 – 4451	262	142	0.541 985

14.4.6.4 Deadwood Distribution

To prepare deadwood distribution from the above data, write all the numerals of class interval in ascending order and make new set of intervals formed by consecutively increasing numerals as shown in Table 14.3B. Write the contribution of each deadwood from table above and write column wise. Add all the items in a row, which will give the deadwood for the corresponding interval.

Table 14.3B Deadwood Distribution

Interval	1	2	3	4	5	6	7	8	*Total*
0–25	3.333	2.234	—	—	—	—	—	—	5.567
25–69	3.333	2.234	2.772	—	—	—	—	—	8.339
69–85	3.333	2.234	2.772	0.0458	—	0.0232	—	—	8.408
85–107	3.333	2.234	2.772	0.0458	5.327	0.0232	—	—	13.735
107–140	3.333	2.234	2.772	0.0458	—	0.0232	—	—	8.408
140–203	3.333	2.234	—	0.0458	—	0.0232	—	—	5.636
230–260	3.333	2.234	—	0.0458	—	0.0232	0.0116	—	5.648
260–372	—	2.234	—	0.0458	—	0.0232	0.0116	—	2.315
372-685	—	—	—	0.0458	—	0.0232	0.0116	—	0.081
685–3930	—	—	—	—	—	0.023	0.0116	—	0.035
3930–4302	—	—	—	—	—	0.023	0.0116	0.542	0.577
4302–4451	—	—	—	—	—	0.023	—	0.542	0.565

14.4.6.5 Gauge / Calibration Table

By now, we have collected all the necessary data to prepare a gauge table. For brevity we have chosen an interval of 10 mm, except the interval has been changed at discontinuity in vertical section, like starting of box section or deck wedge. Last two columns (7 and 8) give the volume versus height relationship. From these two columns we may prepare a more detailed gauge table in smaller steps of say 2 mm.

Table 14.4 Calibration Table (Height versus Volume Relationship)

Height Range mm 1	*Steps* 2	*Deadwood /2mm* dm^3 3	*Volume/2 mm* dm^3 4	*Volume/2mm* dm^3 5	*Volume* dm^3 6	*Height mm* 7	*Volume* dm^3 8
—	—	—	—	—	—	0	0
0–10	5	5.567	11.531	5.964	29.82	10	29.82
10–20	5	5.567	34.592	29.025	145.125	20	174.945
20–25	2.5	5.567	57.654	52.087	130.2175	25	305.163
25–30	2.5	8.339	57.654	49.315	123.2875	30	428.45
30–40	5	8.339	80.716	72.377	361.885	40	790.335
40–49	4.5	8.339	102.614	94.275	424.2375	49	1214.57
49–69	10	8.339	113.135	104.796	1047.96	69	2262.53
69–85	8	8.408	113.135	104.727	837.816	85	3100.35
85–107	11	13.735	113.135	99.4	1093.4	107	4193.75
107–140	16.5	8.408	113.135	104.727	1727.9955	140	5921.74
140–200	30	5.636	113.135	107.5	3225	200	9146.4
200–203	1.5	5.636	113.135	107.5	161.25	203	9307.99
203–260	28.5	5.648	113.135	107.49	3063.44	260	12371.4
260–372	56	2.315	113.135	110.82	6205.92	372	18577.4
372–685	156.5	0.081	113.135	113.054	17692.951	685	36270.4
685–3930	1622.5	0.035	113.135	113.1	183504.75	3930	219775
3930–4251	160.5	0.577	113.135	112.558	18065.559	4251	237841
4251–4261	5	0.577	110.175	109.598	547.99	4261	238389
4261–4271	5	0.577	104.525	103.948	519.74	4271	238908
4271–4281	5	0.577	98.875	98.298	491.49	4281	239400
4281–4291	5	0.577	93.225	92.648	463.24	4291	239863
4291–4301	5	0.577	87.675	87.098	435.49	4301	240299
4301–4311	5	0.577	81.925	81.348	406.74	4311	240705
4311–4315	2	0.577	76.275	75.698	151.396	4315	240857
4315–4321	3	0.565	76.223	75.658	226.974	4321	241084
4321–4331	5	0.565	70.625	70.06	350.3	4331	241434
4331–4341	5	0.565	64.975	64.41	322.05	4341	241756
4341–4351	5	0.565	59.325	58.76	293.8	4351	242050
4351–4361	5	0.565	53.675	53.11	265.55	4361	242315
4361–4371	5	0.565	48.025	47.46	237.3	4371	242553
4371–4381	5	0.565	42.375	41.81	209.05	4381	242762
4381–4391	5	0.565	36.725	36.16	180.8	4391	242943
4391–4401	5	0.565	31.075	30.51	152.55	4401	243095
4401–4411	5	0.565	26.425	25.86	129. 33	4411	243224.
4411–4421	5	0.565	18.775	18.21	91.05	4421	243315
4421–4431	5	0.565	14.125	13.56	67.8	4431	243383
4431–4441	5	0.565	8.475	7.91	39.55	4441	243423
4441–4451	5	0.565	2.825	2.26	11.3	4451	243434
					243434		

14.5 LIQUID CALIBRATION METHOD

14.5.1 Shore Tanks and Meters

Another method of calibration of barge tanks is the liquid filling method [17]. The method is to transfer known volumes of liquid to the tank under calibration. Mostly water is used as the transfer liquid, however light oil or kerosene is also used. The volume of known liquid is transferred either with the help of standard measures or through calibrated meters [16].

The barge tanks are calibrated by transferring of liquid from a standard measure as many times as it is necessary to fill the tank. The standard measures, in this case, are called as shore tanks. The capacity of these measures depends upon those of barge tanks. These measures are pre-calibrated either by mobile tanks or are to be taken to mobile calibration rig similar to the one discussed in chapter 13 on Vehicle tanks.

Shore tanks must have smaller cross-sectional area so that change in volume per unit length of shore tanks is much smaller than the barge tanks to be calibrated against them. The shore tanks must be accompanied not only with calibration certificate but also a calibration table (Volume versus height relationship). In other words shore tanks must be fitted with gauge tube with proper scale so that partial volumes of water transferred can also be found out. Further incremental volume for the shore tank must be smaller than that of the tank under calibration.

For international trade it is necessary that the only mutually accepted laboratories calibrate all such standard measures. All calibration, in this case should be traceable to national and international standards with unbroken chain of measurements. For providing the mutually accepted laboratories in specific fields, both International Organisation of Legal Metrology (OIML), and International Bureau of Weights and Measures (BIPM) are making sustaining efforts. The laboratories are recognised in a specific area for a parameter after ensuring that they are following International standards made for this purpose.

Instead of shore tanks as standard measures, positive displacement meters of adequate accuracy are also used. The method adopted for calibration must conform to national or international standards like API 2555. The method has been discussed in brief in Chapter 7 on Storage tanks. Meters should be provided with meter proving tanks. Working meters should be frequently calibrated either through the meter prover tank or through a master meter.

It is also permissible to use the mixture of two methods, i.e. a shore tank of a fixed volume, so that multiple volume transfer method is used to fill the tank under calibration till less than one full tank is required for water level to reach the desire gauge height. In such situations the meter is profitably used to fill the rest of the tank.

Shore tanks or meters should be as close to the barge tank as possible. All pipe line from shore tank or meter should always remain full. It is preferable to use gravity flow for easier control of water.

Fill the barge tank to various locations shown in Figure 14.5. At each location, two measurements are taken one preliminary and the other final.

14.5.2 Filling Locations of the Tank

Numerals are representing various parts, locations and pipeline etc. Numerals 1 and 3 represent the top and bottom wedge. Numeral 4 indicates the juncture of side with bilge. 5 are the pipelines. Numeral 10 indicates the juncture of the side and deck. 7 and 8 indicate expansion and gauges hatches. 9 is the gauge axis. The numeral 11 represents the centreline bulkhead. Numeral 13 is first filling position below the zero of the gauge point. All other filling lines are marked with a numeral 2. The volume of liquid below first filling line below this line is not transferred and measured in any transaction.

14.5.3 Filling Procedure

The preliminary gauge height is taken when the water has been transferred into the barge tank to the approximate location. The final gauge height is taken, when all the tanks of the barge are filled up to the same level approximately. This is necessary to take into account of the expansion of the tank when adjoining tanks are empty. In actual use, the volume of liquid, the barge is supposed to be carrying is the sum of volumes of liquid contained in all the tanks. So sender will charge the receiver with that volume. However if each tank is filled and gauge table prepared, when the adjoining tanks are empty then each tank according to the calibration certificate will supposed to be carrying the amount, which will be less than the sum of actual volumes when adjoining tanks are full. Hence the receiver will get lesser amount than purported to be sent to him.

Temperature [15] is taken and recorded at each recorded gauge height both at the shore tank or the input side of the measuring meter as well as at the barge tank.

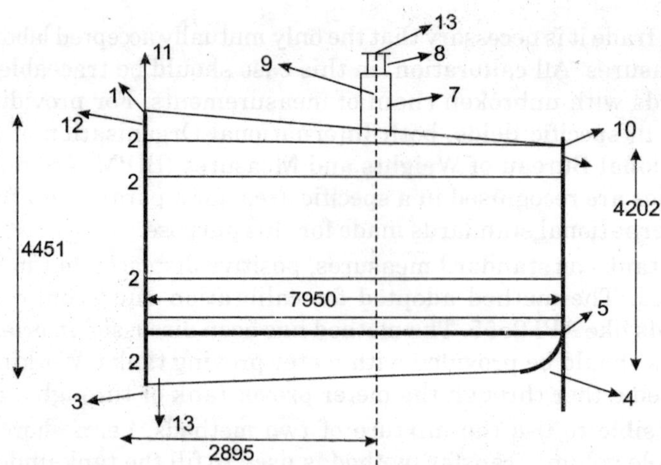

Figure 14.5 Locations of filling of liquid and gauge table for a barge tank
1 the deck wedge, 2 all thin lines are the filling locations, 3 bottom wedge,
13 first increment below the zero of the gauge point, 4 juncture of side with bilge,
5 is the pipeline, 6 liquid touches deck members, 10 juncture of side and deck, 7 is expansion hatch,
8 is gauge hatch, 9 is gauge axis, 11 is centreline bulkhead, 13 is the gauge reference point.

Necessary corrections are applied using the expressions given in chapter 5 in section on the volume transfer method. The exact position of the tank visa vice centreline and transverse bulkhead should be known and reported in the calibration certificate, for this reason, barge tanks are numbered as indicated in Figure 14.1.

14.5.4 Net and Total Capacities of the Barge

For barges with main cargo line or lines located below the deck and running through cargo apartments, the applicable closed line displacements are deducted. A notation is made on the calibration certificate or on gauge table indicating that this deduction has already been made. The amount of total volume due to cargo pipes running through the tanks should be indicated in the certificate both sum total of all deductions and deductions made for individual tanks. Therefore the total cargo capacity of the barge is the sum of all the cargo compartment net capacities, plus the total under deck piping capacity. It is better to make sketches of the under deck pipeline arrangement with all the valves shown, and also the actual calculated pipeline capacities, horizontal and vertical, between each valve should be reported.

14.6 CALCULATING FROM THE DETAILED DRAWINGS OF THE TANKS AND THE BARGE

For barges in active service and for which it is not safe to enter into the tanks, sometimes it is permitted to use the detailed drawing of the barge and tanks to construct gauge table for individual tanks.

The drawings must be such that one can calculate the deadwood and its distribution along the gauge height and also show the cargo pipelines if running through the tanks. Any changes if made after the original construction of the barge must be known and modified drawings should be made available. Then the data from such drawings may be used to prepare gauge tables. The procedure is the same as if all the data available has come from the field.

The total gauge height is measured and recorded for each tank.

This procedure of preparing capacity gauge tables is restricted to barges that are in service and not safe to enter.

REFERENCES

[1] DIS 6578. Static measurement and Calculation Procedure for Refrigerated Hydrocarbon Fluids.

[2] DP 7394. Conversion to Equivalent Liquid Volumes Natural Gas Liquids and Vapours.

[3] DP 8309. Measurement of Liquid Levels in Tanks Containing Liquefied Gasses in Bulks-Refrigerated Hydrocarbon Fluids- Electrical Capacitance Gauges.

[4] ISO 8311–Refrigerated Hydrocarbon Fluids Calibration of Membrane tanks and Independent Prismatic Tanks in Ships- Physical Measurement.

[5] DP 8310–Refrigerated Hydrocarbon Fluids- Measurement of Temperature in Tanks Containing Liquefied Gasses- Thermocouple and Resistance Thermometers.

[6] TC/28SC5 W12–148 DOC 44–Refrigerated Hydrocarbon Fluids-Measurement of Liquid Levels in the Tank Carrying Liquefied Gases in Bulk–Float Type Level Gauges.

[7] DP 9091/1–Refrigerated Hydrocarbon Fluids–Calibration of Spherical Tanks in Ships-Part 1–Stereo Photogrammetry.

[8] DP 9091/2. Refrigerated Hydrocarbon Fluids–Calibration of Spherical Tanks in Ships-Part 2–Triangulation Method.

[9] ISO/TR 8338. Crude Petroleum Oil–Transfer Accountability-Method for Estimation on Ships of Total Quantity Remaining on Board.

[10] DP 8697. Crude petroleum oil– Transfer Accountability-Method for Estimation on Ships of total on Board Quantity.

[11] OIML R–95, 1990. Ship's Tanks–General Requirements.

[12] Indian Standard Specifications for Density Hydrometers, IS: 3104, 1965.

[13] Alcoholometry and Alcohol Hydrometers, OIML Recommendations R-44, 1980.

[14] API 2553: Measurement and Calibration of Barges.

[15] API standard 2545: Measuring the Temperature of Petroleum Products.

[16] API 1101: Measurement of Petroleum Liquids Hydrocarbons by Positive Displacement Meters.

[17] API 2555: Liquid Calibration of Tanks.

INDEX

C

D

E

F

G

H

I

J

K

L

P

R

S

U

V

W